Seminar Nasional Fisika (SNF 2017)

Journal of Physics: Conference Series Volume 997

Surabaya, Indonesia
25 November 2017

Editors:

Dr. Budi Jatmiko
Dr. Madlazim
Dr. Munasir
Dr. Z.A. Imam Supardi
Nadi Suprapto
Utama Alan Deta

ISBN: 978-1-5108-6166-4
ISSN: 1742-6588

Printed from e-media with permission by:

Curran Associates, Inc.
57 Morehouse Lane
Red Hook, NY 12571

proceedings.com

Some format issues inherent in the e-media version may also appear in this print version.

Printed by Curran Associates, Inc. (2018)

For permission requests, please contact the Institute of Physics
at the address below.

Institute of Physics
Dirac House, Temple Back
Bristol BS1 6BE UK

Phone: 44 1 17 929 7481
Fax: 44 1 17 920 0979

techtracking@iop.org

Additional copies of this publication are available from:

Curran Associates, Inc.
57 Morehouse Lane
Red Hook, NY 12571 USA
Phone: 845-758-0400
Fax: 845-758-2633
Email: curran@proceedings.com
Web: www.proceedings.com

Table of contents

Volume 997

Seminar Nasional Fisika (SNF) 2017

25 November 2017, Surabaya, Indonesia

Accepted papers received: 19 March 2018
Published online: 12 April 2018

Preface

Papers

IOP Conf. Series: Journal of Physics: Conf. Series **997** (2018) 011001 doi:10.1088/1742-6596/997/1/011001

Preface
Seminar Nasional Fisika (SNF) 2017
Universitas Negeri Surabaya, Surabaya, Indonesia
November 25th , 2017

Nadi Suprapto, Abd. Kholiq, Utama Alan Deta and Nurita Apridiana Lestari

This conference, which is organized by the Department of Physics, Faculty of Mathematics and Natural Sciences, Universitas Negeri Surabaya, is held at Auditorium Prof. Slamet Dajono, FMIPA, UNESA.

The theme of this conference is "Menghilirkan Penelitian-Penelitian Fisika dan Pembelajarannya". Two experts as keynote speakers in physics domain will share their experiences.

1. Prof. Dr. Bayong Tjasyono H.K., DEA., (Professor in physical meteorology, Institut Teknologi Bandung)
2. Dr. Ing. Setyawan Purnomo Sakti, M.Eng., (Expert in material instrumentation, Department of Physics, Universitas Brawijaya Malang)

This Conference on Physics and Physics Education is attended by more than 350 participants. There are about 110 papers is presented orally covering wide-variety subjects of physics (material physics, geophysics, theoretical and computation physics, instrumentation physics, and applied physics) and physics education (physics learning, assessment, innovative learning, and media). Totally, there are more than 35 institutions from Sumatra, Java, Madura, Borneo, Celebes, Maluku, and Papua participated in this conference. We thank you all for your participation.

We thank the Organizing Committee, Reviewers, and Steering Committee for having been working hard. Finally, we would also like to thank the Rector of Universitas Negeri Surabaya, Dean of Faculty of Mathematics and Natural Sciences, and Bank Tabungan Negara (BTN) for their support for this conference.

We hope you will enjoy our togetherness. Thank you.

Surabaya, November 25th , 2017
Best Regards,

SNF 2017 Committee

IOP Conf. Series: Journal of Physics: Conf. Series **997** (2018) 011002 doi:10.1088/1742-6596/997/1/011002

Committee
Seminar Nasional Fisika (SNF) 2017
Universitas Negeri Sutabaya, Surabaya, Indonesia
November 25th, 2017

Person in charge	: Dekan FMIPA Unesa
	Prof. Dr. Madlazim, M.Si
	Prof. Dr. Budi Jatmiko, M.Pd
	Dr. ZA Imam Supardi, M.Si
Chairman	: Nadi Suprapto, Ph.D.
Vice Chairman	: Abdul Kholiq, S.Pd., M.T.
Secretary	: Utama Alan Deta, M.Pd., M.Si
Treasurer	: Nurita Apridiana Lestari, M.Pd.

Reviewer Team

1. Prof. Dr. Budi Jatmiko, M.Pd.
2. Prof. Dr. Madlazim, M.Si.
3. Dr. Wasis, M.Si.
4. Drs. Rudy Kustijono, M.S.
5. Dr. Munasir, M.Si.
6. Tjipto Prastowo, Ph.D.
7. Dra. Suliyanah, M.Si.
8. Dr. Frida U. Ermawati, M.Sc.

Editorial Team

1. Prof. Dr. Budi Jatmiko, M.Pd.
2. Prof. Dr. Madlazim, M.Si.
3. Dr. Munasir, M.Si.
4. Dr. Z.A. Imam Supardi, M.Si.
5. Nadi Suprapto, Ph.D.
6. Utama Alan Deta, S.Pd., M.Pd., M.Si.

Event Organizer

1. Woro Setyarsih, S.Pd., M.Si.
2. Dra. Titin Sunarti, M.Si.
3. Drs. Dwikoranto, M.Pd.
4. Drs. Supardiyono, M.Si.
5. Eko Hariyono, M.Pd.
6. Nugrahani P. Putri, M.Si.

Registration and Secretarial

1. Meta Yantidewi, M.Si.
2. Dr. Binar Kurnia Prahani, M.Pd.
3. Mukhayyarotin Niswati R.J, M.Pd.
4. Ilynda Noviati, S.Pd.

Food Services

1. Dra. Hermin Budiningarti, M.Pd.
2. Supardi, ST.

Publications and Documentation

1. Lydia Rochmawati, M.Si.
2. Abu Zainuddin, S,Pd.

Sponsorship

1. Diah Hari Kusumawati, M.Si.
2. Endah Rahmawati, M.Si.
3. Asnawi, M.Si

Logistics

1. Setyo Admoko, M.Pd.
2. Drs. Imam Sucahyo, M.Si.
3. Drs. Hainur Rasyid Achmadi, M.S.
4. Dzulkiflih, S.Si, M.T.
5. Afria Nur Hidayat
6. H. Chanaki, S.T.
7. Agus Supriono, S.T.

Security

1. Sanadji (Koordinator)
2. Petugas parkir FMIPA

Sponsor
Seminar Nasional Fisika (SNF) 2017
Universitas Negeri Surabaya, Surabaya, Indonesia
November 25th, 2017

The committee of SNF 2017 would like to thank the sponsors who have provided funding to this seminar. Sponsors from SNF 2017 are:

1. **Bank Tabungan Negara (Bank BTN)**

   Bank BTN

   Sahabat Keluarga Indonesia

   Website: https://www.btn.co.id/

2. **Universitas Negeri Surabaya**

   UNESA

   Universitas Negeri Surabaya

   Website: http://www.unesa.ac.id/

The committee of SNF 2017 also would like to thank the Editors and the Reviewers of **Jurnal Penelitian Fisika dan Aplikasinya (JPFA)** for supporting this seminar.

JPFA-UNESA

Jurnal Penelitian Fisika dan Aplikasinya (JPFA) – Journal of Physics Research and Its Application, publish every June and December, receives research articles with topics: (1) Physics Education, (2) Instrumentation Physics and Measurement, (3) Material Sciences, (4) Theoretical and Computational Physics, and (5) Earth Sciences. JPFA is indexed by Google Scholar, DOAJ Greentick and SINTA S3 (one grade below the Nationally Accredited Journal). Starting from Volume 8 Number 1, all articles will be written in English.

Website: https://journal.unesa.ac.id/index.php/jpfa

IOP Conf. Series: Journal of Physics: Conf. Series **997** (2018) 011004 doi:10.1088/1742-6596/997/1/011004

Photograph

Seminar Nasional Fisika (SNF) 2017

Universitas Negeri Surabaya, Surabaya, Indonesia

November 25th, 2017

Figure 1. Welcoming Address from Vice Rector of Academic Affairs Universitas Negeri Surabaya

Figure 2. Keynote Speaker and Moderator of Seminar Nasional Fisika (SNF) 2017

Figure 3. Prof. Dr. Bayong Tjasyono H.K., DEA (Keynote Speaker SNF 2017)

Figure 4. Dr.-Ing. Setyawan Purnomo Sakti, M.Eng. (Keynote Speaker SNF 2017)

Figure 5. Questions and Answers Session

Figure 6. Parallel Session

IOP Conf. Series: Journal of Physics: Conf. Series **997** (2018) 011004 doi:10.1088/1742-6596/997/1/011004

Figure 7. SNF 2017 Committee (1)

Figure 8. SNF 2017 Committee (2)

IOP Conf. Series: Journal of Physics: Conf. Series **997** (2018) 011004 doi:10.1088/1742-6596/997/1/011004

Figure 9. SNF 2017 Committee (3)

Peer review statement

All papers published in this volume of *Journal of Physics: Conference Series* have been peer reviewed through processes administered by the proceedings Editors. Reviews were conducted by expert referees to the professional and scientific standards expected of a proceedings journal published by IOP Publishing.

Determination of source parameters of the 2017 Mount Agung volcanic earthquake from moment-tensor inversion method using local broadband seismic waveforms

Madlazim[1,2], T Prastowo[1,2], Supardiyono[1,2] and T Hardy[3]

[1] Physics Department, Faculty of Mathematics and Natural Sciences, Universitas Negeri Surabaya, Kampus Unesa Ketintang, Surabaya 60231, Indonesia
[2] Center for Earth Science Studies, Faculty of Mathematics and Natural Sciences, Universitas Negeri Surabaya, Kampus Unesa Ketintang, Surabaya 60231, Indonesia
[3] Indonesian Agency for Geophysics, Climatology, and Meteorology, Jakarta 10720, Indonesia

E-mail: madlazim@unesa.ac.id

Abstract. Monitoring of volcanoes has been an important issue for many purposes, particularly hazard mitigation. With regard to this, the aims of the present work are to estimate and analyse source parameters of a volcanic earthquake driven by recent magmatic events of Mount Agung in Bali island that occurred on September 28, 2017. The broadband seismogram data consisting of 3 local component waveforms were recorded by the IA network of 5 seismic stations: SRBI, DNP, BYJI, JAGI, and TWSI (managed by BMKG). These land-based observatories covered a full 4-quadrant region surrounding the epicenter. The methods used in the present study were seismic moment-tensor inversions, where the data were all analyzed to extract the parameters, namely moment magnitude, type of a volcanic earthquake indicated by percentages of seismic components: compensated linear vector dipole (CLVD), isotropic (ISO), double-couple (DC), and source depth. The results are given in the forms of variance reduction of 65%, a magnitude of M_w 3.6, a CLVD of 40%, an ISO of 33%, a DC of 27% and a centroid-depth of 9.7 km. These suggest that the unusual earthquake was dominated by a vertical CLVD component, implying the dominance of uplift motion of magmatic fluid flow inside the volcano.

1. Introduction

Indonesia is a country geographically located in the circumference of the Pacific ring of fire that leads to the presence of a series of active volcanoes along the Sunda arc in the country, and thus potential to earth-related disaster due to volcanic eruption. Efforts for minimising impacts and reducing risks of such a catastrophic disaster have been implemented by the government and non-governmental bodies, but much has not yet been achieved. In this context, monitoring of volcano activities has therefore been important for the development of knowledge in the discipline and hazard mitigation study [1].

We provide here a volcanic earthquake driven by recent magmatic events of Mount Agung, Bali that occurred on September 28, 2017 as a case study to consider. When magmatic processes naturally occurring in the deep down a volcano generate vertical movement of viscous fluids that passes through crustal rocks up to the caldera leading to an eruption, many phenomena associated with earthquakes of volcanic origin are possible to observe, such as cracks on the ground, lava flows, massive transports of mass and volume of volcanic materials, and volumetric changes inside the volcano [2,3,4].

Careful analysis of earthquakes of volcanic origin allows us to examine the dynamics of a magma and its complex geophysical and geochemical processes inside a volcano [5]. This is important in part to predict eruption that is possible to occur and estimate the possible impacts of such a volcanic hazard on the environment. Although most of these earthquakes are relatively small in magnitude and hence recorded by nearby stations of local network, in some circumstances such earthquakes may induce events with large magnitudes possibly observed by regional and global networks of observatories [6]. In many examples of these events, it has been found that the source shows anomalous behaviours, such as unusual patterns of seismic radiation or frequency content [7]. In addition, previous work on this topic [8] found that earthquakes of volcanic origin with magnitudes ~ 5 dominated by a vertical CLVD component were empirically identified within 6 volcanic zones around US and that after a time period of magma activities, the corresponding volcanic activities became weakening in intensity.

In the present work, we investigate an unusual earthquake observed from Mount Agung, Bali that occurred on September 28, 2017. We hypothesize such an earthquake as a vertical CLVD component. Vertical CLVD events have deviatoric moment-tensors with nondouble-couple (NDC) components are evidence-based seismic phenomena and are dominated by uplift motion of source deformation due to lateral compressions and vertical dilatations [9]. This type of earthquake is rare in nature but possible to occur and frequently associated with an active volcano. Thus, this study is of importance aimed at determining the source parameters of the earthquake effectively driven by recent underground events from Mount Agung magmatic activities.

2. Methods

Mount Agung is one of the two largest volcanoes in Bali island, geographically located at 8.29° S and 115.40° E with its elevation reaches to a 3,031 m high. This volcano has been recently in debate by many whether it would erupt since a sequence of unusual magma-driven earthquakes beginning from September 13, 2017 to date. The one we used in this study for determination of the source parameters is that to occur on September 28, 2017.

2.1. Data collection

The data used in this study were obtained from broadband seismic waves recorded by the IA network of local seismic stations officially operated by the Indonesian Agency for Geophysics, Climatology, and Meteorology (BMKG) which provides 5 seismic stations: SRBI, DNP, BYJI, JAGI, and TWSI. These stations with their geographical positions relative to the source covering a full 4-quadrant region surrounding the epicenter are listed in table 1 and schematically illustrated in figure 1. All of the data used are freely available at http://202.90.198.100/webdc3/.

Table 1. Names of stations used in this study with their epicentral distances.

Network code	Seismic station	Latitude	Longitude	Epicentral distance (km)
IA	SRBI	8.08° S	115.20° E	31
	DNP	8.68° S	115.21° E	48
	BYJI	8.21° S	114.36° E	115
	JAGI	8.47° S	114.15° E	139
	TWSI	8.74° S	116.88° E	170

2.2. Data analysis

For data analyses in the current study, we first used a seismic wave velocity model for providing variations of both *P*- and *S*-wave speeds with depth D and their corresponding wave qualities Q_P and Q_S

for each body wave component, respectively, as well as the local density ρ_L. All of these quantities, seen in table 2, were used for a seismic data inversion method. The wave qualities denoted by Q_P and Q_S are non-dimensional numbers showing measures of how reliable the model is (given in the forms of variations of the body wave speeds and the densities with depth for seismic data inversion techniques). As also used by [6], a lower limit value of 500 for the *P*-wave speed distribution over the depth is a threshold number, above which corresponding *P*-waveforms from recorded seismograms are reliable to analyse owing to a minimum level of noises and for the same reason a value of 250 for the depth-distributed *S*-wave speed is a threshold number, above which corresponding *S*-waveforms are reliable.

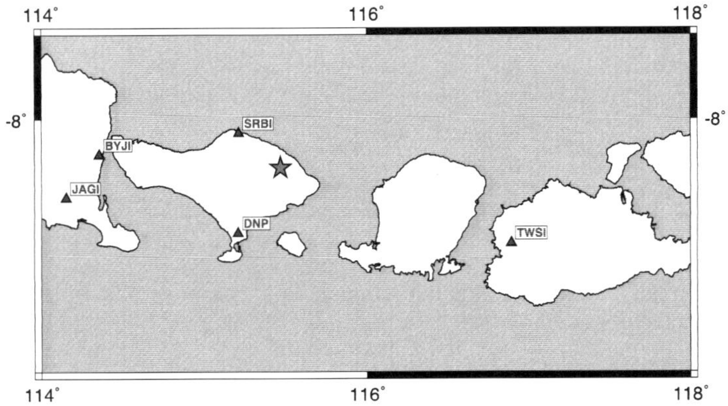

Figure 1. Map of the source (red star) and local stations (blue triangles) in this study.

Table 2. A seismic velocity model for use of seismic inversion method.

D (km)	v_P (km/s)	v_S (km/s)	Q_P	Q_S	ρ_L (g/cc)
0.0	4.8	3.1	500	250	2.4
13.0	5.6	3.2	500	250	2.6
45.0	6.1	3.5	500	250	2.8
700	8.0	4.6	1000	500	3.2

The analysis of broadband signals from recorded seismograms with a low-pass filter at frequencies ranging from 0.054 to 0.072 Hz obtained from a network of 5 local seismic stations for determination of the source parameters was performed using a software package of MTINV 3.05 (used in this study with permission from Dr. Ichinose). The use of observed seismic signals at low frequencies is reliable for estimating source parameters as these signals are insensitive to the effects of lateral wave velocity and density variations inside the Earth [8].

3. Results and discussions

We used the moment-tensor inversion method using the local broadband seismograms to determine the source parameters of unusual events nearby Mount Agung that occurred on September 28, 2017. The method utilized a simple model of a one-dimensional seismic wave velocity profile. The results of seismic data inversions are then provided in the forms of three consecutive figures 2, 3, and 4, where each figure is correlated one to another to provide a complete description of the events.

Figure 2 describes waveform fitting from full moment-tensor inversions of the 28 September 2017 volcanic earthquake at and nearby Mount Agung recorded by 5 seismic stations listed in Table 1. In each seismic station, the 3 local components observed and synthetic displacement seismograms are symbolized as T for a tangential component, R for a radial component, and Z for a vertical component.

In all stations, black-colored waveforms represent observed waveforms while red-colored waveforms in stations SRBI, DNP, and JAGI are for synthetic waveforms as well as the blue ones in stations BYJI and TWSI.

The results of the full seismic data inversions for all the broadband seismograms from the 5 stations can be seen on the top right of figure 2. The seismic inversion method yields a variance reduction of 65.2% from statistical analysis of the data fitting that corresponds to an earthquake of M_w 3.64 in size, a vertically oriented compensated linear vertical dipole (CLVD) component of 39.9%, an isotropic (ISO) component of 33.1%, a double-couple (DC) component of only 26.9%, and a centroid-depth of 9.7 km below the sea surface level. These features, to some extent, confirm the relative importance of vertical magmatic movement inside the volcano in the presence of highly viscous, magmatic fluid flow under the influence of uplift volcanic pressure. This geophysical process is complicated in nature as claimed by [5, 6] for the 2000 Miyakejima unusual swarms. As is the case discussed by [4], this study also finds that the relative dominance of uplift motion indicated by a major of a vertically oriented CLVD component, following a weak DC source mechanism, is accompanied by volumetric change inside the volcano. This finding is different from that reported by [10], in which unusual earthquakes that occurred in the Bardarbunga, Iceland in 1996 were to occur with no net volumetric component in the source mechanism.

Figure 3 shows a simple description of 3 interconnected graphics on the cross correlation between variance reduction and a double-couple (DC) source mechanism (given in per cent), and a centroid-depth (given in km). It is clear that the 28 September 2017 Mount Agung unusual events with a 65.2% variance reduction of magnitude 3.64 were visibly accompanied by many anomalous earthquakes with relatively large nondouble-couple (NDC) components of approximately 73%, in oppose to almost 27% double-couple source mechanism. Natural earthquake sources with a net volumetric component are commonly modeled using a DC source mechanism [5, 6] but it does not work for the case of the 2017 Mount Agung events. Although in a particular example, such as the 1996 Bardarbunga earthquake, volumetric component was absent [10], the present case clearly shows the relative importance of NDC components with a compensated linear vector dipole (CLVD) component is measured slightly greater, in percentage, than an isotropic (ISO) component. It suggests that the full moment-tensor inversions as well as the cross correlation between the corresponding source parameters in the present study reveal magmatic motion of viscous fluid flow in the vertical direction [7, 8] all the way up to the caldera before eruption to occur.

We have modeled broadband seismograms from the IA network of the 5 stations surrounding Mount Agung. The local broadband seismograms extracted from the occurrence of 28 September 2017 yield a vertically oriented CLVD component, a relatively dominant feature in the source mechanisms (as seen in figure 4) as a result of seismic moment-tensor inversions, the same method as utilized by [7, 8]. This suggests that vertically directed motion of magmatic fluid flow inside the volcano with a net volumetric change, as implied by [4, 5, 6], is a leading uplift force although it is not yet clear why the volcano remains active with no eruption. However, the full moment-tensor inversions give a better result in terms of a centroid-depth instead of depth determined by other methods of depth estimation, in particular when dealing with active volcanoes. Knowledge of the source process and its component dominance is directly derived from examining and analyzing differences in percentage of CLVD, ISO, and DC components (see figures 2 and 3).

It is interesting to make direct comparison of all aspects of focal and source mechanisms between this study and previous work [10]. The case considered here applied different broadband seismograms in comparison to those used in [10]. While a local network of seismic stations were provided by BMKG for use of the 2017 Mount Agung earthquake, regional broadband waveforms were used in [10] although the full moment-tensor inversions were applied in both. The results for these studies were nondouble-couple (NDC) source mechanisms. However, [10] found a volumetric change to be absent in the dominance of a vertically linear vector dipole (CLVD) component. In contrast, the present study has a clear (isotropic) net volumetric component that is comparable to the vertical CLVD component. The

IOP Conf. Series: Journal of Physics: Conf. Series **997** (2018) 012001 doi:10.1088/1742-6596/997/1/012001

fact that the vertical CLVD component (39.9%) is not much greater than the ISO component (33.1%) is likely to be the cause for Mount Agung being active with no eruption.

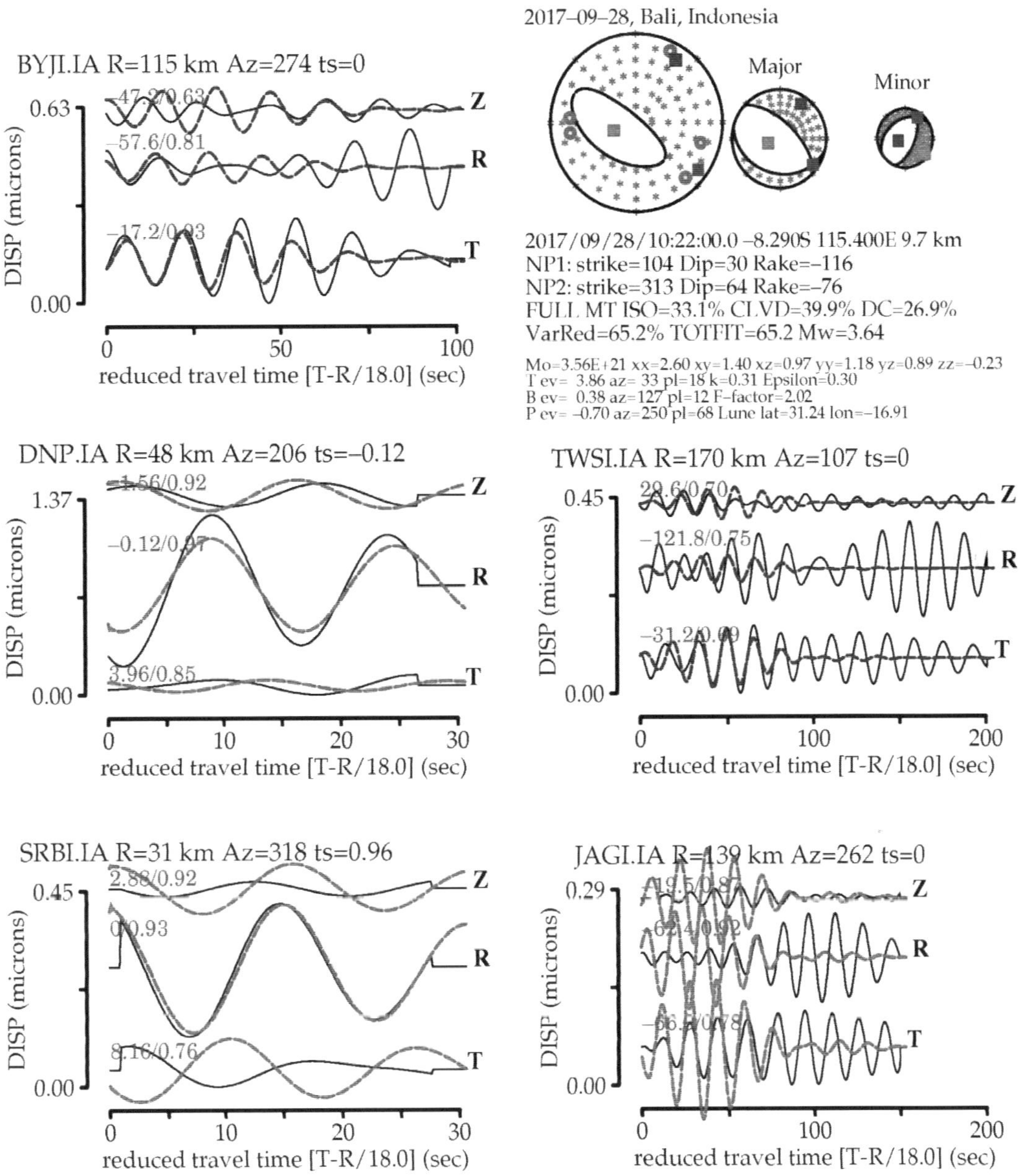

Figure 2. Broadband seismogram data with the 3 local Z, R, and T components showing fittings of observed and synthetic waveforms from the network of 5 local stations (SRBI, DNP, BYJI, JAGI, and TWSI) used in this study, where black colors denote observed waveforms and both red and blue colors represent synthetic waveforms. On the top right are the results of full seismic moment-tensor inversions, where all the computed source parameters, including source mechanisms are provided.

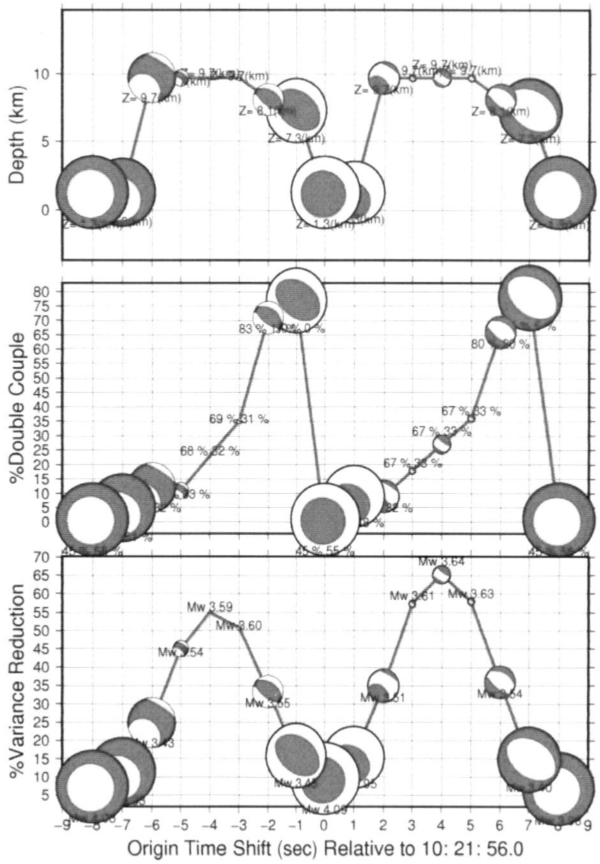

Figure 3. Three graphics in one, describing a simple cross correlation of variance reduction, double-couple component, and centroid-depth with respect to time taken for the full seismic moment-tensor data inversions. The variance reduction reaches the highest value at almost 65% corresponding to an unusual earthquake with a magnitude of M_w 3.64 (bottom). This relates to a 27% DC component (middle) and then a 9.7 km centroid-depth (top). The cross correlation shows that all the inter-correlated 'beach-balls' for the 2017 Mount Agung events are dominated by nondouble-couple source mechanisms.

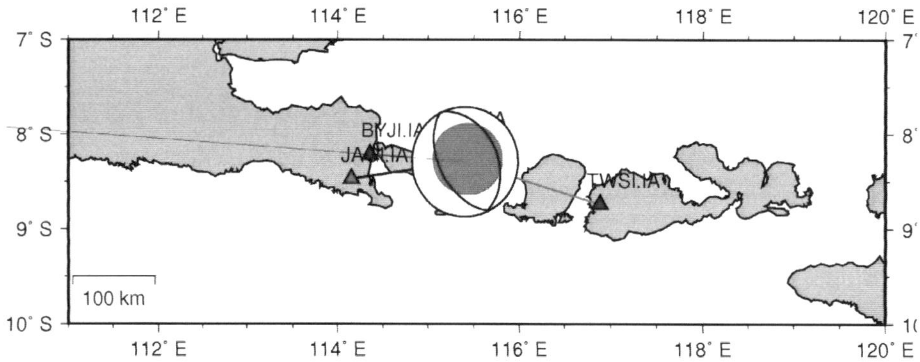

Figure 4. A simple map, showing a type of the Mount Agung unusual earthquakes (given in the form of a beach-ball dominated by NDC source mechanisms) that occurred on September 28, 2017 with all of the 5 local seismic stations (some stations are unseen due to an excessive size of the beach-ball) used for recording local broadband seismogram data are positioned located nearby the epicenter.

4. Conclusions

We have used seismic moment-tensor inversions for examining an unusual, magma-driven earthquake of Mount Agung, Bali that occurred on September 28, 2017 to determine the source parameters involved. The model of a one-dimensional velocity structure used in this study fits the local broadband data with statistical analysis of the 65.2% variance reduction that corresponds to the earthquake size of M_w 3.64, where the earthquake is found to be primarily characterized by the 39.9% vertical CLVD component as well as a less dominant part, the 33.1% isotropic (ISO) component and the relatively unimportant 26.9% DC component, and the hypocenter at a centroid-depth of 9.7 km, suggesting that such an earthquake is of a nondouble-couple source mechanism. The results indicate that the volcanic earthquake under consideration is dominated by uplift motion of magmatic fluid flow accompanied by a volumetric change inside the volcano. These two features are in common for volcanic earthquakes before eruption. Knowledge of which seismic component is dominant is necessary for decision making about volcano alert status. Because the vertical CLVD component (~40%) is not too large relative to the ISO component (~33%), we therefore speculate for this case that the volcano has not yet erupted although it remains active. However, as Indonesia is a country vulnerable to geophysical hazards, such as volcanic eruption, monitoring of volcanoes is of fundamental significance to prevent community from disastrous events. The methods developed are possibly evaluated for a wide range of applications to similar cases with different source mechanisms, for example, detection and characterization of nuclear explosions and natural tectonic earthquakes, as also discussed in [11, 12], for future work.

Acknowledgments
The authors firmly acknowledge supports for secondary data in the forms of seismograms available from the Indonesian Agency for Geophysics, Climatology, and Meteorology (BMKG), Jakarta for use in this work and sincerely thank Dr. Ichinose for his blessing permission to allow the authors to use a programming package of MTINV 3.05 for implementation of seismic data inversion method.

References

[1] Barclay E J, Renshaw C E, Taylor H A and Bilge A R 2011 *J. Geosci. Edu.* **59** 85-92
[2] Kumagai H 2009 Source quantification of volcano seismic signals *Encyclopedia of Complexity and Systems Science* ed R A Meyers (New York: Springer) 9899
[3] Zobin V M 2011 *Introduction to Volcanic Seismology* 2nd ed (Amsterdam: Elsevier) 474
[4] Kanamori H, Ekström G, Dziewonski A, Barker J S and Sipkin S A 1993 Japan *J. Geophys. Res.* **98** (B4) 6511
[5] Passarelli L, Rivalta E, Cesca S and Aoki Y 2015 *J. Geophys. Res. Solid Earth* **120** 4130
[6] Minson S E, Dreger D S, Bürgmann R, Kanamori H and Larson K M 2007 *J. Geophys. Res.* **112** B10308
[7] Shuler A and Ekström G 2009 *J. Volcanol. Geotherm. Res.* **181** 19
[8] Ichinose G A, Anderson J G, Smith K D and Zeng Y 2003 *Seismol. Soc. Am.* **93** 61
[9] Taira T, Smith R B and Chang W L 2010 *J. Geophys. Res.* **115** B02301
[10] Tkalcić H, Dreger D S, Foulger G R and Julian B R 2009 *Bull. Seismol. Soc. Am.* **99** 77
[11] Dreger D and Woods B 2002 *Tectonophys.* **356** 139
[12] Cho K-H 2014 D *Appl. Geophys.* **11** 429

The construction of partner potential from the general potential anharmonic in *D*-dimensional Schrodinger system

Suparmi[1], C Cari[1], K N Wea[1] and Wahyulianti[1]

[1]Physics Department, Graduate School, Universitas Sebelas Maret, Indonesia

E-mail: soeparmi@staff.uns.ac.id

Abstract. The Schrodinger equation is the fundamental equation in quantum physics. The characteristic of the particle in physics potential field can be explained by using the Schrodinger equation. In this study, the solution of 4 dimensional Schrodinger equation for the anharmonic potential and the anharmonic partner potential have done. The method that used to solve the Schrodinger equation was the ansatz wave method, while to construction the partner potential was the supersymmetric method. The construction of partner potential used to explain the experiment result that cannot be explained by the original potential. The eigenvalue for anharmonic potential and the anharmonic partner potential have the same characteristic. Every increase of quantum orbital number the eigenvalue getting smaller. This result corresponds to Bohrn's atomic theory that the eigenvalue is inversely proportional to the atomic shell. But the eigenvalue for the anharmonic partner potential higher than the eigenvalue for the anharmonic original potential.

1. Introduction

The Schrodinger equation is the equation that representation of electron or particle. The solution of Schrodinger equation is eigenfunction and eigenvalue a particle that used to describe the characteristic of a particle in-universe. The Schrodinger equation was developed into *D*-dimensional Schrodinger system to solve the complex system in quantum mechanic. So the Schrodinger equation can explain more about a system. There are many methods that used to solve the Schrodinger equation. For example the ansatz wave function method [1-4], the Nikiforov-Uvarov method[5-7], the the the asymptotic iteration method (AIM)[8-10], supersymmetric method [11-13] and other [14-15]. In this study, we used the wave function ansatz method and the supersymmetric method. The ansatz wave function method was used to solve the Schrodinger equation and the supersymmetric method was used to construction partner potential.

The partner potential was used to explain the experiment result that cannot be explained by the original potential. To construction the partner potential we using the ground state eigenfunction, the original potential and the supersymmetric operators. Some study about the construction of partner potential has been done. For example the construction partner potential from the Hylleraas potential [16], the construction partner potential from the Hulthen potential [17-18] and the construction partner potential from the Manning Rosen and Rosen-Morse potential [19]. In this study, we choose potential anharmonic as the original potential. The general form of anharmonic potential that we used is

$$V(r) = ar^2 + br + cr^{-1} + dr^{-2} \tag{1}$$

where a, b, c and d are potential parameters.

The first section of this paper we will review the supersymmetric method and the wave function ansatz method. The next section we will explain the Schrodinger equation in the D-dimensional system. We will present the solution of *D*-dimensional Schrodinger equation with anharmonic potential using the ansatz wave function method at three section. The next section we will present about construction partner potential from origin anharmonic potentials in *D*-dimensional Schrodinger system and then we will present the solution of *D*-dimensional Schrodinger equation with partner potential anharmonic.

2. Method

2.1. *The Supersymmetric method*

Definition of supersymmetric Hamiltonian H_{ss} partner for changing operator that commute with H_{ss} by Witten as follows [20]

$$H_{ss} = \begin{pmatrix} -\frac{d^2}{dx^2} + \frac{d\phi(x)}{dx} + \phi^2(x) & 0 \\ 0 & -\frac{d^2}{dx^2} - \frac{d\phi(x)}{dx} + \phi^2(x) \end{pmatrix} = \begin{pmatrix} H_+ & 0 \\ 0 & H_- \end{pmatrix} \tag{2}$$

The Hamiltonian partner $H_- = H_1$ $H_+ = H_2$ and partner potential $V_- = V_1$ and $V_+ = V_2$ are

$$V_-(x) = V_1 = \phi^2(x) - \phi'(x) \text{ and } V_+(x) = V_2 = \phi^2(x) + \phi'(x) \tag{3}$$

where ϕ is superpotential supersymmetric.

The relation between effective potential V_{ef} and first potential V_1 as follow

$$V_{ef}(x) = V_-(x;a_0) + E_0 = V_1(x;a_0) \tag{4}$$

with E_0 is ground state energy for effective potential. By setting the new supersymmetric operator are the lowering operator A and the rising operator $A^\dagger$ as

$$A^\dagger = -\frac{d}{dx} + \phi(x), \qquad A = \frac{d}{dx} + \phi(x) \tag{5}$$

so the Hamiltonian partner in Eq. (2) can be rewritten as

$$H_-(x) = H_1 = A^\dagger A, \qquad H_+(x) = H_2 - AA^\dagger \tag{6}$$

and also

$$A\psi_0^{(-)} = A\psi_0 = 0 \tag{7}$$

The new partner potential V_2 from Eq. (4) and Eq. (7) we get the relation [16]

$$V_2(x) = V_1 - 2\frac{d^2}{dx^2}\ln\psi_0 = V_{ef} - E_0 - 2\frac{d^2}{dx^2}\ln\psi_0 \tag{8}$$

This equation will be used to construct the partner potential from the original potential in Eq. (1).

2.2. *The Wave Function Ansatz*

The wave function ansatz was used to solve the radial part Schrodinger equation. The first step of ansatz wave function is we must determine the form of wave function ansatz that corresponds with potential that we used. From work [4] the wave function ansatz that suitable for Eq. (1) is

$$R_n(r) = h_n(r)\exp(P(r)) \tag{9}$$

with

$$h_n(r) = \begin{cases} 1 & \\ \prod_{j=1}^{n}\left(r - \sigma_j^n\right) & \end{cases} \text{when} \begin{array}{l} n = 0 \\ n = 1, 2, 3, \ldots \end{array} \tag{10}$$

The parameter $P(r)$ in Eq. (9) was determined by superpotential supersymmetric. So we have

$$P(r) = \frac{1}{2}Ar^2 + Br + C\ln r \tag{11}$$

The Eq. (9) with $P(r)$ in Eq. (11) is the wave function ansatz that we used to solve the Schrodinger equation with potential anharmonic.

3. Results and discussion

3.1. *The Schrodinger equation in D-dimensional system*

The general Schrodinger equation in D-dimensional system [21]

$$-\frac{\hbar^2}{2m}\nabla_D^2\Psi(r,\Omega) + V(r,\Omega)\Psi(r,\Omega) = E\Psi(r,\Omega) \tag{12}$$

with D is dimensional and the Laplacian operator as [22]

$$\nabla_D^2 = \frac{1}{r^{D-1}}\frac{\partial}{\partial r}(r^{D-1}\frac{\partial}{\partial r}) + \frac{1}{r^2}\sum_{j=1}^{D-2}\frac{1}{\sin^2\theta_{j+1}\sin^2\theta_{j+2}\cdots\sin^2\theta_{D-1}}\times$$
$$\left\{\frac{1}{\sin^{j-1}\theta_j}\left(\frac{\partial}{\partial\theta_j}\sin^{j-1}\theta_j\frac{\partial}{\partial\theta_j}\right)\right\} + \frac{1}{r^2}\left\{\frac{1}{\sin^{D-1}\theta_{D-1}}\left(\frac{\partial}{\partial\theta_{D-1}}\sin^{D-2}\theta_{D-1}\frac{\partial}{\partial\theta_{D-1}}\right)\right\} \tag{13}$$

To simplify calculation we set $\hbar = 1, 2m = 1$. By using variable separation method from Eq. (12) and (13) we get the D-dimensional Schrodinger equation radial part and polar part. The radial part of D-dimensional Schrodinger equation as follows

$$\frac{\partial^2 R(r)}{\partial r^2} - \left[\frac{\frac{3}{4} - \lambda_{D-1}}{r^2}\right]R(r) = -2(E - V(r))R(r) \tag{14}$$

where by Dong [21] λ_{D-1} is describe as

$$\lambda_{D-1} = l_{D-1}(l_{D-1} + 2) \tag{15}$$

with l_{D-1} is angular momentum.

3.2. *The Solution of D- dimensional Schrodinger equation with anharmonic potential*

First calculation, we solve the radial part of Schrodinger equation with anharmonic potential by using the wave function ansatz. The anharmonic potential in Eq. (1) substituting into the radial part of Schrodinger equation in Eq. (14) so we get

$$\frac{\partial^2 R(r)}{\partial r^2}-\left[2\left(E-ar^2-br-cr^{-1}-dr^{-2}\right)+\frac{\frac{3}{4}-\lambda_{D-1}}{r^2}\right]R(r)=0 \tag{16}$$

The Eq. (16) is radial part of Schrodinger equation with anharmonic potential. To solve Eq. (16), substituting the wave function ansatz in Eq. (19) into Eq. (16), so we have

$$P''(r)+\left(P'(r)\right)^2+\frac{h_n''(r)+2h_n'(r)P'(r)}{h_n(r)}=-2E+2ar^2+2br+2\frac{c}{r}+\frac{\left(2d+\frac{3}{4}-\lambda_{D-1}\right)}{r^2} \tag{17}$$

From Eq. (11), we obtained

$$P'(r)=Ar+B+\frac{C}{r},\qquad P''(r)=A-\frac{C}{r^2} \tag{18}$$

For $n=0$ from Eq. (10) $h_1=1$, and by substituting Eq.(18) into Eq. (17), we have

$$A^2r^2+2ABr+\frac{2BC}{r}+\frac{C^2-C}{r^2}+A+2AC+B^2=-2E+2ar^2+2br+2\frac{c}{r}+\frac{\left(2d+\frac{3}{4}-\lambda_{D-1}\right)}{r^2} \tag{19}$$

And then, by using the separation variable method, from Eq. (19) we obtained the equations as follow

$$A^2=2a \tag{20a}$$
$$AB=b \tag{20b}$$
$$BC=c \tag{20c}$$
$$C^2-C=2d+\frac{3}{4}-\lambda_{D-1} \tag{20d}$$
$$A+2AC+B^2=-2E \tag{20e}$$

From Eq. (20) the parameters A, B, C can be determined, so we can get the eigenvalue and the eigenfunction for $n=0$. From Eq. (20) we have

$$A=-\sqrt{2a}\,,\; B=-\frac{b}{\sqrt{2a}}\;\text{ and }\; C=\frac{1}{2}\left(1+\sqrt{4+8d-4\lambda_{D-1}}\right) \tag{21}$$

Then, by using Eq. (21) and Eq. (20e) we obtained the eigenvalue as follow

$$E_0=\frac{\sqrt{2a}}{2}\left(2+\sqrt{4+8d-4\lambda_{D-1}}\right)-\frac{b^2}{4a} \tag{22}$$

And the eigenfunction as

$$R_0(r)=N_0 r^{\frac{1}{2}\left(1+\sqrt{4+8d-4\lambda_{D-1}}\right)}\exp\left[-\sqrt{2a}r-\frac{b}{\sqrt{2a}}r\right] \tag{23}$$

For $n=1$ from Eq. (10) we have $h_1=\left(r-\sigma_1^{(1)}\right)$, by substituting Eq. (18) into Eq. (17), we get

$$A-\frac{C}{r^2}+A^2r^2+B^2+\frac{C^2}{r^2}+2ABr+2AC+\frac{2BC}{r}+\frac{2\left(Ar+B+\frac{C}{r}\right)}{r-\sigma_1^{(1)}}=$$
$$-2E+2\left(ar^2+br+cr^{-1}+dr^{-2}\right)+\frac{\left(\frac{3}{4}-\lambda_{D-1}\right)}{r^2} \tag{24}$$

By following the steps that we used in our calculation when $n=0$ we have the parameters A,B,C as

$$A=-\sqrt{2a}\,,\; B=-\frac{b}{\sqrt{2a}}\; C=\frac{1}{2}\left(1+\sqrt{4+8d-4\lambda_{D-1}}\right) \tag{25}$$

and

$$A\left[1+2\left(C+1\right)\right]+B^2=-2E \tag{26}$$

And then, by substituting Eq. (25) into Eq. (26), we obtain the eigenvalue for $n=1$ as

$$E_1=\frac{\sqrt{2a}}{2}\left(4+\sqrt{4+8d-4\lambda_{D-1}}\right)-\frac{b^2}{4a} \tag{27}$$

and the eigenfunction as follows

$$R_1\left(r\right)=\left\{N_1\left(r-\sigma_1^{(1)}\right)r^{\frac{1}{2}\left(1+\sqrt{4+8d-4\lambda_{D-1}}\right)}\exp\left[-\frac{1}{2}\sqrt{2a}r-\frac{b}{2\sqrt{2a}}r\right]\right\} \tag{28}$$

By using the same steps, we can determine the eigenvalue and the eigenfunction for $n=2$ and more. So we can generalisation the eigenvalue and the eigenfunction for n as follow

$$E_n=\frac{\sqrt{2a}}{2}\left(2+2n+\sqrt{4+8d-4\lambda_{D-1}}\right)-\frac{b^2}{4a} \tag{29}$$

and

$$R_n\left(r\right)=\left\{N_{nl}\prod_{i=1}^{n}\left(r-\sigma_i^{(n)}\right)r^{\frac{1}{2}\left(1+\sqrt{4+8d-4\lambda_{D-1}}\right)}\exp\left[-\frac{1}{2}\sqrt{2a}r^2-\frac{b}{\sqrt{2a}}r\right]\right\} \tag{30}$$

with $n=0,1,2,3,\cdots$, N_{nl} is normalization constant.

3.3. The construction of partner potential from origin potential anharmonic in Schrodinger 4 dimensional system

From Eq. (16) the effective potential for *D*-dimensional Schrodinger equation with anharmonic potential is

$$V_{ef}=2ar^2+2br+2\frac{c}{r}+\frac{\left(2d+\frac{3}{4}-\lambda_{D-1}\right)}{r^2} \tag{31}$$

To construction the partner potential we use Eq. (8). So we obtained the new partner potential as follow

$$V_2\left(r\right)=2ar^2+2br+2\frac{c}{r}+\frac{\left(2d+\frac{3}{4}-\lambda_{D-1}\right)+2C}{r^2}-E_0-2A \tag{32}$$

where E_0 in Eq. (22), A in Eq (21) dan C in Eq (21).

The Eq. (31) is partner potential anharmonic that will be used as the new potential. The new partner potential solved in the D-dimensional Schrodinger system by using the wave function ansatz method. Substituting Eq. (31) into Eq. (14) we have the Schrodinger equation with partner potential anharmonic in the D-dimensional system as following

$$\frac{\partial^2 R(r)}{\partial r^2}-\left(\frac{\frac{3}{4}-\lambda'_{D-1}}{r^2}\right)R(r)=(-2E-2E_0-4A)R(r)+4\left[ar^2+br+\frac{c}{r}+\frac{\left(d+C+\frac{1}{2}\left(\frac{3}{4}-\lambda_{D-1}\right)\right)}{r^2}\right]R(r) \quad (33)$$

By using supersymmetric approximation we determine the wave function ansatz that corresponds with partner potential anharmonic as

$$R_n(r)=h_n(r)\exp(P(r)) \quad (34)$$

with

$$P(r)=\frac{1}{2}A'r^2+B'r+C'\ln r \quad (35)$$

And then, substituting the wave function ansatz in Eq. (33) into Eq. (32) and by using the same steps that we were used to solving the Schrodinger equation with anharmonic potential, we have the solution of Schrodinger equation with the new partner potential anharmonic as follow

$$E_n=\left[\frac{\sqrt{4a}}{2}\left(2+2n+\sqrt{4\left(4d+4C+\frac{9}{4}-\lambda_{D-1}-\lambda'_{D-1}\right)}\right)-E_0-2A-\frac{b^2}{a}\right] \quad (36)$$

and

$$R_n(r)=\left\{N_{nl}\prod_{i=1}^{n}\left(r-\sigma_j^{(n)}\right)r^{\frac{1}{2}\left(1+\sqrt{4\left(4d+4C+\frac{9}{4}-\lambda_{D-1}-\lambda'_{D-1}\right)}\right)}\exp\left[-\frac{1}{2}\sqrt{4a}r^2-2\frac{b}{\sqrt{4a}}r\right]\right\} \quad (37)$$

with $n=0,1,2,3,\cdots$, E_0 in Eq (22), A and C in Eq. (21).

To see the characteristic eigenvalue of the anharmonic potential and the new partner potential anharmonic, we calculate the numeric eigenvalue by using MATLAB program. We present the numeric eigenvalue in Table 1.

Table 1. The eigenvalue of anharmonic potential and the anharmonic partner potential with the value of parameters potential are $a=1/32$, $b=1$, $c=4$, $d=0$ and the angular momentum $l=0$.

n	**Eigenvalue anharmonic potential *(1/fm)***	**Eigenvalue anharmonic partner potential *(1/fm)***
0	15.750	30.2819
1	15.500	30.0769
2	15.250	29.8750
3	15.000	29.6772
4	14.750	29.4846

In Table 1 we can see that the eigenvalue of the original anharmonic potential and the partner potentials anharmonic have the same characteristic. Every increase of quantum orbital number the eigenvalue getting smaller. This result corresponds to Bohrn's atomic theory that the eigenvalue is inversely proportional to the atomic shell. But the eigenvalue for the anharmonic partner potential higher than the eigenvalue for the anharmonic original potential. This phenomenon happens because the parameters of the origin potential in Eq (1) and the parameters of the partner potential in Eq (31) are different.

4. Conclusion

We have constructed the partner potential from the original anharmonic potential in Schrodinger *D*-dimensional system using the supersymmetric method. The partner potentials that we obtained is solvable. By using the wave function ansatz method we were obtained the eigenvalue and the eigenfunction of Schrodinger equation with anharmonic potential and the anharmonic partner potential. The eigenvalue of the original anharmonic potential and the partner potentials anharmonic have the same characteristic. Every increase of quantum orbital number the eigenvalue getting smaller. This result corresponds to Bohrn's atomic theory that the eigenvalue is inversely proportional to the atomic shell. But the eigenvalue for the anharmonic partner potential higher than the eigenvalue for the anharmonic original potential.

Acknowledgments

This work partly supported by PUT-MRG of LPPM Sebelas Maret University.

References

[1] Child M S, Dong S H and Wang X G 2000 *J. Phys. A: Math Gen.* **33**
[2] Wea K N, Suparmi, Cari and Anwar F 2017 *J. Phys: Conf. Series* **795** 012023
[3] Wahyulianti, Suparmi, Cari and Anwar F 2017 *J. Phys: Conf. Series* **795** 012022
[4] Wea K N, Suparmi, Cari and Wahyulianti 2017 *AIP Conf. Proc.* **1868** 060005
[5] Nikiforov A F and Uvarov V B 1998 *Special Functions of Mathematical Physics* (Birkhauser: Boston)
[6] Cari and Suparmi 2012 *IOSR J. App. Phys.* **2** 13
[7] Cari and Suparmi 2013 *J. Phys: Conf. Series* **423** 012031
[8] Pratiwi B N, Suparmi, Cari and Anwar 2016 *J. Phys: Conf. Series* **776** 012090
[9] Barakat T 2005 *J. Phys Lett. A* **344** 411
[10] Ikot A N 2012 *J. Chinese Phys. Lett.* **29** 090302
[11] Mitftahul M, Suparmi and Cari 2016 *J. Phys: Conf Series* **776** 012079
[12] Suparmi 2011 *Mekanika kuantum II* (Surakarta: Sebelas Maret University) 57
[13] Siregar, Suparmi, Cari and Yuliani H 2013 *Int. J. Engineering and Science* **2** 14
[14] Bayrak O, Bostoosum I and Cifti H 2006 *Int. J. Quantum Chemistry* **107**
[15] Ikhdair S M and Sever R 2008 *J. Mol. Struct.* **855** 12
[16] Suparmi, Cari, Pratiwi B N and Nugraha D A 2017 *J. Phys: Conf. Series* **820** 012023
[17] Bhoi J and Laha U 2016 *Brazilian J. Phys.* **49** 129
[18] Wahyulianti, Suparmi, Cari, and Wea K N 2017 *AIP Conf. Proc.* **1868** 060011
[19] Wea K N, Suparmi, Cari and Wahyulianti 2017 *AIP Conf. Proc.* **909** 012028
[20] Ikot A N, Awoga O A and Antia D A 2013 *Chin. Phys. B* **22** 020304
[21] Witten E 1981 *Nucl. Phys. B* **188** 513
[22] Dong S H 2011 *Wave Equation in Higher Dimensions* (New York: Springer) p 42

IOP Conf. Series: Journal of Physics: Conf. Series **997** (2018) 012003 doi:10.1088/1742-6596/997/1/012003

Variation sweep rate cyclic voltammetry on the capacitance electrode activated carbon/PVDF with polymer electrolyte

L Rohmawati[1], W Setyarsih[1] and T Nurjannah[1]

[1]Physics Department, Universitas Negeri Surabaya, Ketintang Surabaya

E-mail: lydiarohmawati@unesa.ac.id

Abstract. Sweep rate of the process voltammetry cyclic characterization is very influential towards the electrode capacitance value, especially on activated carbon electrodes/PVDF. A simple method of this research by use a mixing for electrode activated carbon /10 wt. % PVDF and the separator is made of a polymer electrolyte (PVA/H_3PO_4) by a sol gel method. The prototype supercapacitor is made in the form of a sandwich with a separator placed between two electrodes. Electrodes and separators are arranged in layers at a pressure of 1500 psi, then heated at 50°C for 10 minutes. Next done cyclic voltammetry in a potential range of -1 V to 1 V with a sweep rate of 5 mV/s, 10 mV/s, 20 mV/s, 25 mV/s and 50 mV/s. This results of curves voltammogram is reversible, the most wide curve on the sweep rate of 5 mV/s and most narrow curve on a sweep rate of 50 mV/s. Supercapacitor capacitance values obtained by 86 F/g, 43 F/g, 21 F/g, 16 F/g, and 8 F/g.

1. Introduction

Supercapacitor is one of the best energy storage devices, besides being economical and environmentally friendly. Supercapacitor is also large storage electric energy devices by large density, huge loading capacity, fast and durable charging and discharging processes when compared to other energy storage devices such as generally capacitors [1]. Supercapacitor constructions are similar to generally capacitors consisting of pair electrodes filled by electrolyte and separated of dielectric material serves as a separator [1]. In his research [2], supercapacitor devices consist of electrodes, separators, electrolytes and current collector. The supercapacitor devices consists of two electrodes, wherein one electrode is connected to a positive electric current source so as to be called a positive electrode, while the other electrode is connected to a negative pole and an electrolyte solution. The electrolyte solution used in this supercapacitors are H_3PO_4 and Na_2SO_4 [3], [4]. Usually the materials used in the supercapacitor electrode are graphene, carbon nanotube, carbon aerogel, porous carbon (carbon-active), and mineral-carbon composites. Conditions for electrodes, such as low electronic resistance, good electrolyte accessibility, good chemical and mechanical stability, and high capacitance values [5].

Electrodes wich made from active carbon nanopore have advantages, including relatively cheap, easy to obtain, polarized, stable to acidic and alkaline compounds, regularly arranged natural pores and large of surface area about 400-2200 m^2/g [6]. Nanopore activated carbon can be made from natural materials, one of them from coconut shells, with carbon content of coconut shell fixture that is 66.79% [7]. In the study [8] about the manufacture of active carbon electrodes from coconut shells, obtained carbon surface area 414 m^2/g and belongs to the mesopore category which has a capacitance of 10 F/g. When compared with *Kluwak* shells (*Pangium edule*), which has higher carbon content than coconut shell that is 77.1% but has a much smaller capacitance value ie 0.192 F/g [9]. However, the shortage of electrodes made from activated carbon is the value of relatively small capacitance and poor mechanical stability. To meet the drawback the active carbon electrode needs to be added with the polymer binder. One of the polymers used is PVDF (polyvinylidenedifluoride).

PVDF is a pure thermoplastic fluropolymer and is not very reactive. PVDF has advantages such as lightness, flexibility, high resistance to strong impact, has an easier melting process because of its relatively low melting point and low density compared to other fluropolymers. PVDF is known as a binder polymer which requires volatile organic solvents, for example N-Methyl-2-Pyrrolidone (NMP). Based on [10] in his research on the manufacture of electrodes made of 90 wt. % activated carbon/ 10 wt. % PVDF obtained the value of capacitance of 325 F/g. According to [11], preparation of electrodes 30 mg activated carbon / 1 wt. % PVDF which tested voltammetry cyclic with variations sweep rate (5 mV/s, 10 mV/s, 20 mV/s, 25 mV/s dan 50 mV/s) obtained the highest capacitances value at sweep rate 5 mV/s amount 47 F/g. At low sweep rate, electrons will diffuse evenly into the electrode surface until mesopore, resulting in an increase in the width of the current curve which indicates a large capacitance value, and vice versa at high sweep rate, the electrons can diffuse rapidly but only up to the surface of the electrode, the electron just enter the macropore [12]. The electrode pores of unelectable electrons produce narrow curves that indicate low capacitance values.

In the supercapacitor, the main components to consider in addition to the electrode are the electrolyte and the separator. Separators are placed between two electrodes to prevent the direct transfer of electrons between two electrodes. Separators should be thin enough and have a porous layer. This is to ensure the smooth flow of ions in the electrolyte to diffuse into the pores of both electrodes [13]. One of the polymers that can be used as an electrolyte polymer is a polymer PVA (polyvinyl alcohol) with phosphoric acid electrolytes (H_3PO_4). PVA is a polymer that has advantages that are cheap, non-toxic, and chemically stable. In terms of electrical properties, PVA is an insulator material with high resistivity of 3.1–3.8 x 107 Ωcm [14]. PVA dissolved in polar solvents such as water, dimethyl sulfoxide, acetamide and dimethylformamide.

Bases on the reference [4] about the supercapasitor electrode of coconut shell activated carbon with the electrolyte polymer separator (PVA/H_3PO_4), ratio of carbon and activator solution (1:1, 1:2, and 1:3) capacitance obtained of 15 F/g, 18 F/g and 26 F/g. Research on supercapacitor with commercial activated carbon electrodes/PVDF and electrolyte polymer separators (PVA/H_3PO_4) obtained capacitance value of charging and discharging value of capacitance respectively 34.7 F/g and 38.39 F/g [3]. According to these studies, it is necessary to conduct research on the manufacture of active carbon electrodes/PVDF with electrolyte polymer whose results were tested with cyclic voltammetry of sweep rate variations between 5 mV/s to 50 mV/s. The results of this study expected that the electrode have an optimal capacitance value.

2. Materials and Methods

The materials used in this study: gading coconut shells, natrium oxide (NaOH 0,5M), aquades, aquades, hydrochloric acid (HCl 1M), polyvinylidene flouride (PVDF), N-Methyl-2-pyrrolidone (NMP), polyvinyl alcohol (PVA), and ortho-phosphorid acid (H_3PO_4). There are several processes for making activated carbon and composites.

2.1. Preparation of Activated Carbon

Preparation of activated carbon from coconut shells consists of 3 processes, they are dehydration, carbonation and activation. Preparation of activated carbon in the carbonation process by simple heating methods and activation process with chemical activation. In the dehydration process, the coconut shell is dried under sunlight for ± 7 days. The carbonization process uses simple heating method by burning a coconut shell in a furnace at a temperature 400°C in 2 hours. Then crushed carbon powder, after that sieved until the carbon powder get away from the filter (200 mesh). Activation process used the activator NaOH 0.5 M, carbon powder soaked in NaOH 0.5 M for 24 hours, ratio of carbon:NaOH = 1:3, then heated at temperature 800°C in 5 hours. After that the carbon is washed by HCl 1M and aquades repeatedly up to pH 7 remove the alkaline salt compound, dried at temperature 110°C in 5 hours to remove water vapor that is trapped inside the pores [8].

2.2. Preparation of Activated Carbon Electrodes/PVDF

Preparation of activated carbon electrodes supercapacitor/PVDF used 90 wt. % activated carbon of coconut shells and 10 wt. % polyvinylidene flouride (PVDF). PVDF dissolved by N-Methyl-2-pyrrolidone (NMP) at temperature 40°C for 30 minutes, after that mix 90 wt. % activated carbon on PVDF solution by mixing methods [10], stirring by magnetic stirrer in 30 minutes to homogeneous composite. Activated carbon electrodes/PVDF characterized by SEM to know the morphological of composite.

2.3. Preparation of Electrolytic Polymer Separator

Preparation of electrolyte separator used sol-gel methods, the first dissolves 2 grams PVA to aquades and stirring by magnetic stirrer at temperature 70°C. After homogeneous solution, added H_3PO_4 by 50% concentration which is then printed on the petri plate. The next dried at room temperature for 14 days [3]. The electrolytic polymer membrane is characterized by Brunauer Emmet Teller (BET) to determine the porosity of the membrane.

2.4. Preparation of Prototype Supercapacitor

Prototype supercapacitor made in sandwich shape. Between the two activated carbon electrodes/PVDF (in pellets shape) added electrolyte membrane of PVA (Figure 1). Further, two electrodes and the electrolyte membrane are simultaneously compressed at the pressure1500 psi, after that heating at temperature 50°C in 30 minutes. The function of this process to produces good contact between activated carbon electrodes/PVDF and membrane. Furthermore, supercapacitor with weight of 0.6 gram is tested with cyclic voltammetry in the potential range -1,0 V to 1,0 V by sweep rate 5 mV/s, 10 mV/s, 20 mV/s, 25 mV/s and 50 mV/s. The electrolyte solution used in this study is ortho-phosphorid acid (H_3PO_4) 85% solution.

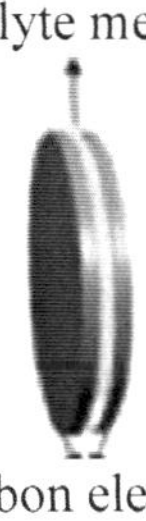

Figure 1. Illustrations of supercapacitor devices [1].

2.5. Calculating The Electrode Capacitance Value

Data of cyclic voltammetry test for all samples on sweep rate variation are 5 mV/s, 10 mV/s, 20 mV/s, 25 mV/s and 50 mV/s can be processed by Matlab software. Furthermore, to calculate of specific capacitance value supercapasitor using equations (1) and (2) [1], ie:

$$C_{sel}(F) = \frac{\int i dV}{\Delta V \times V_s} \quad (1)$$

$$C_s(F/g) = \frac{2C_{sel}}{m} \quad (2)$$

Information:

Csel = cell capacitance (F)
i = discharge current (A)

Vs	= sweep rate (V/s)
ΔV	= range of potensial (V)
Cs	= specific capacitance (F/g)
m	= mass (g)

3. Result and Discussion

3.1. The Result of BET

In this research the sample tested is electrolyte polymer separator sample (PVA/H_3PO_4) by using characterization tools BET Quantrachrome NovaWin.

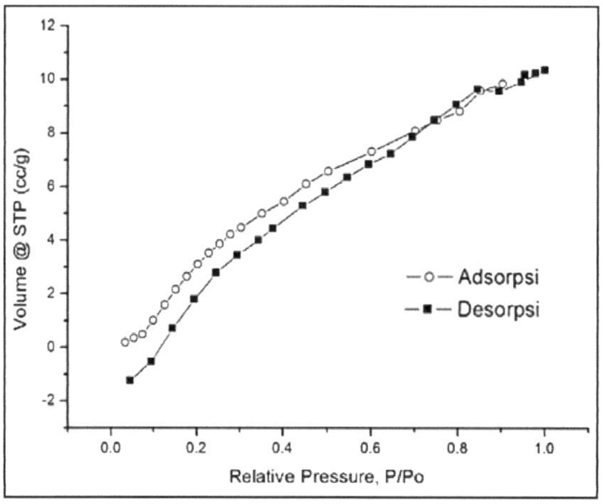

Figure 2. Isothermal graphics of electrolyte polymer separator.

Figure 2 showed isothermal graphics of electrolyte polymer separator (PVA/H_3PO_4) which shows the relationship graph between relative pressure at temperature 77.35 K and volume of gas adsorbate ie nitrogen gas (N_2). The isotherm graph shows that the electrolytic separator adsorption-desorption is a type IV of adsorption-desorption which forms a hysteresis loop over a relative pressure range is 0.75–0.85 P/P_o or Langmuir isotherm for the IUPAC classification which is a type IV of adsorption-desorption occurring in mesopore materials [15]. The open shape of the end curve shows that the volume of gas absorbed and released when the adsorption-desorption not the same. This indicates that not all nitrogen gas is removed during the adsorption process, which means that nitrogen gas is retained within the pore. When analogous to charge storage, the charge stored in the supercapacitor is proportional to the amount of gas volume difference absorbed and released during the adsorption-desorption process.

To find the size pore of the electrolyte polymer separator by looking at the largest pore size distribution in the data BJH (Barret-Joyner-Hallenda) adsorption and desorption. Pore distribution (Figure 3) as mathematic written $\frac{dV_p}{dr_p}$ or $\frac{dA_p}{dr_p}$ as the function r_p, where V_p = volume of pore, r_p = radius of pore, A_p = surface area or wall of pore.

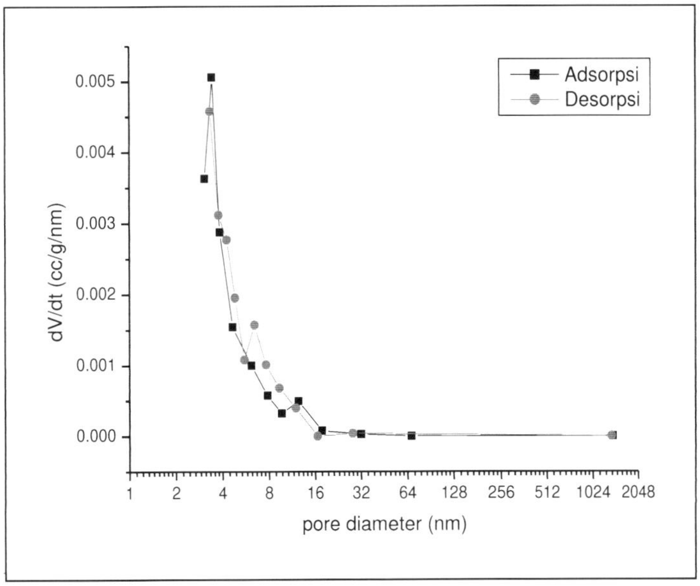

Figure 3. Pore Size Distribution of Electrolytic Polymer Separator.

Figure 3 The largest adsorption pore distribution is 0.005059 cc/g/nm at the pore diameter ie 3.423 nm, while the largest desorption pore distribution is 0.004579 cc/g/nm at the pore diameter is 3.308 nm. So obtained pore size in the range 3.3–3.4 nm, so included in the category mesopore (2 nm – 50 nm) [16]. Separators with mesopore structures can perform rapid charge transfers of electrolyte ions to diffuse into the pores of the electrodes in the process of charging and discharging the supercapacitor [15].

3.2. The Result of Voltammetry Test

Voltammetry cyclic is a voltammetric technique in which the current is measured during sweeping potential from initial potential to final potential and back to initial potential. The material tested by voltammetry cyclic in this study is supercapacitor composed of activated carbon electrodes/PVDF with electrolyte polymer separator (PVA/H_3PO_4).

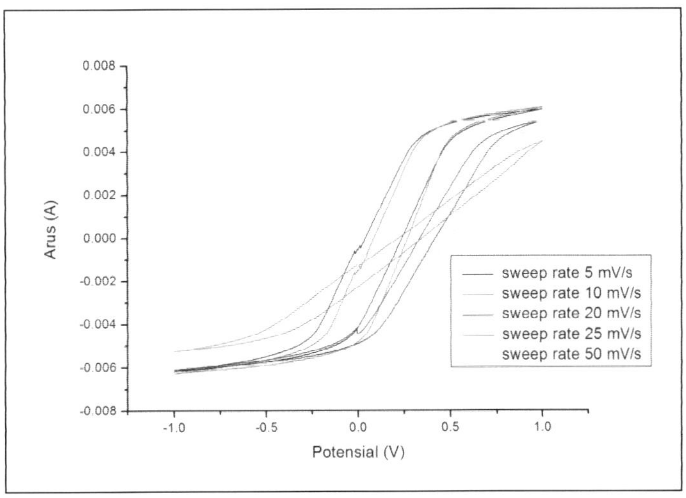

Figure 4. The voltammogram curve of supercapacitors electrode by sweep rate variation.

Figure 4 showed voltammogram curve of supercapacitors electrode by electrolyte polimer. The figure show that the resulting voltammogram curve is reversible. The curve at the sweep rate of 5 mV/s and 10 mV/s closer to the ideal curve of the capacitor means that the current density is almost constant (at the currents ± 5mA) along by the given of difference potential, thus allowing large stored loads.

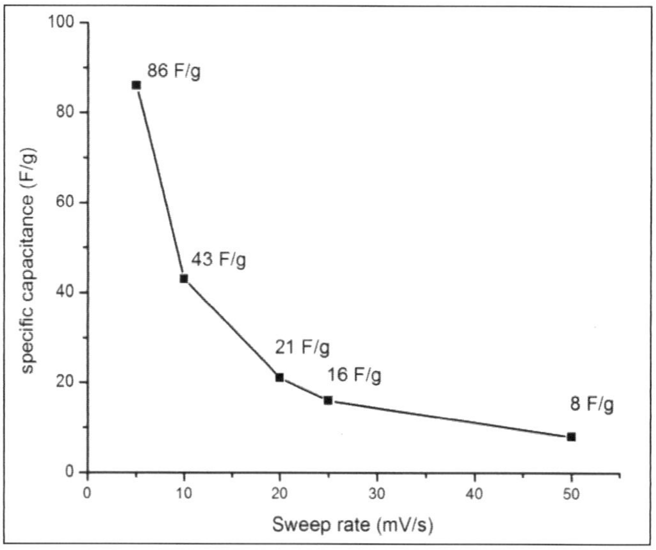

Figure 5. Supercapacitor Capacitance Value.

Figure 5 shown the capacitance value of supercapacitor on sweep rate 5 mV/s, 10 mV/s, 20 mV/s, 25mV/s, and 50 mV/s each is 86 F/g, 43 F/g, 21 F/g, 16 F/g and 8 F/g. The largest capacitance value is obtained at sweep rate 50 mV/s is 8 F/g. Capacitances value in this study better than the results of research [3] and [11]. However, the resulting capacitance value is smaller than [10] which uses one electrode. In the voltammetry cyclic test of one electrode, the ions from the electrolyte will go directly into the pores of the electrode leading to rapid charging and discharging process while in the voltammetry cyclic test of two electrodes, there is separator which acts as an insulator to prevent the direct transfer of electrons between the two electrodes, which leads to fast charging process and long evacuation process. Separators should be sufficiently thin and have high porosity so as to increase the drainage of ions in the electrolyte to diffuse into the pores of the electrode [13]. Sweep rate of voltammetry cyclic affect the shape of the voltammogram curve and the capacitance value of the supercapacitor. At low sweep rates, electrons have long time to diffuse evenly into the electrode surface and the charge transfer (stored and given charge) gets larger so that the voltammogram curve forms wider and higher capacitance values, while the higher the sweep rate curve shape the voltammogram is narrower and the capacitance value is smaller. This happens because at a high sweep rate, the electrons can diffuse rapidly but the charge transfer on the electrode surface is limited.

3.3. The Result of SEM

Based on SEM- micrographs (Figure 6) by magnification 7500 × can be seen that activated carbon electrodes/PVDF porous, but there is still lump indicating that the sample has not been mixed evenly. Some PVDF particles are well distributed into the pores of activated carbon, but others just stick to the surface. This can affect the diffusion process of ions or electrons. It affects the diffusion process of ions or electrons.

Figure 6. SEM- micrographs of activated carbon/PVDF.

4. Conclusion

The capacitance value of the supercapacitor by the voltammetry cyclic characterization sweep rate 5 mV/s, 10 mV/s, 20 mV/s, 25 mV/s and 50 mV/s obtained the largest capacitances value in sweep rate 5 mV/s ie 86 F/g, while the smallest capacitance value on sweep rate 50 mV/s is 8 F/g. The smaller of sweep rate used, the capacitance value of electrodes supercapacitor activated carbon/PVDF is greater.

References

[1] Rosi M, Ekaputra M, Iskandar F, Abdullah M and Khairurrijal 2012 *Pros. Semin. Nas. Mater.* (Bandung: ITB) p 42-45
[2] Fitriana VN 2014 *Superkapasitor Berbasis Nanokomposit TiO_2/C* (Malang: Universitas Negeri Malang)
[3] Hasyim M, Sa'adu L and Dasuki K 2012 *Inter. J. Sustain. Energy Environ. Res.* **1** 1
[4] Rosi M, Iskandar F, Abdullah M and Khairurrijal 2013 *Pros. Semin. Nas. Mater.* (Bandung: ITB) 74-77
[5] Drobny D, Tychyna S, Maletin Y, Stryzhakova N and Zelinsky S 2013 *Proc. Int. Conf.* vol 2 (Korsakova: Sumy State University) p 04NEA17(1)-04NEA17(2)
[6] Rosi M, Abdullah M and Khairurrijal 2009 *J. Nanosains & Nanoteknologi* **1** 26
[7] Purwanto D 2011 *J. Penelit. Has. Hutan* **29** 57
[8] Wati GA, Rohmawati L and Putri NP 2015 *J. Inov. Fisika Indonesia* **4** 6
[9] Habibah MD, Rohmawati L and Setyarsih W 2016 *J. Inov. Fisika Indonesia* **5** 19
[10] Ruiz V, Blanco C, Granda M, Menendez M and Santamaria R 2007 *Influence of Electrode Preparation on The Electrochemical Behaviour of Carbon-Based Supercapacitors* (Spain: Instituto Nacional del Carbon)
[11] Natalia M, Sudhakar Y and Selvakumar M 2013 *Indian. J. Chem. Technol.* **20** 392
[12] Taer E, Zulkifli, Sugianto, Syech R and Taslim 2015 *Pros. Semin. Nas. Fisika* (Jember: Universitas Jember) p 105
[13] Taer E, Sugianto, Sumantre MA, Taslim R, Iwantono and Dahlan D 2013 *Pros. Semin. Nas. Fisika* (Padang: Universitas Andalas) p 82
[14] Jihar SM 2008 *Study Film Polyvinyl Alcohol (PVA) Dimodifikasi Dengan Acrylamide (AAM) Sebagai Material Sensitif Terhadap Kelembaban* (Jakarta: Universitas Indonesia)
[15] Hu Chijuan 2008 *Fluid Coke Derived Activated Carbon As Electrode Material For Electrochemical Double Layer Capacitor* (Toronto: University of Toronto)
[16] Jisha MR, Hwang Y, Shin J, Nahm K, Kumar T, Karthikeyan K, Dhanikaivelu N, Kalpana K, Renganathan N and Stephan A 2009 *J. Mater. Chem. Phys.* **115** 33
[17] Nurdiansah H and Susanti D 2013 *J. Teknik ITS* **2** 13

Atmospheric study of the impact of Borneo vortex and Madden-Julian oscillation over Western Indonesian maritime area

R M Saragih[1], R Fajarianti[1] and P A Winarso[1]

[1]Department of Meteorology, Indonesian State College of Meteorology, Climatology and Geophysics, Pondok Aren, South Tangerang, Indonesia

E-mail: ruth.mandasari@stmkg.ac.id

Abstract. During the Asian winter Monsoon (November-March), the Indonesia Maritime Continent is an area of deep convection. In that period, there is a synoptic scale disturbance over Northwest of Borneo Island called Borneo vortex. In addition to the impact of Asian Winter Monsoon, Madden-Julian Oscillation (MJO) also have an impact on deep convection during an active period. This study aims to study the impact of interaction Borneo vortex and MJO (during MJO active period in phase 3, 4 and 5) and rainfall condition over the western part of Indonesia Maritime Continent using compositing technique in the period of November-March 2015/2016. The parameters used to identify the incidence of Borneo vortex, MJO, and its interaction is vertical velocity. When MJO is active, Borneo vortex occurs most often in phase 5 and at least in phase 3. However, Borneo vortex occurs most often when the MJO is inactive. The interaction between Borneo vortex and MJO seems may affect not so much rainfall occurrence in the western part of IMC.

1. Introduction

Indonesia is referred to as the Indonesian Maritime Continent (IMC) because Indonesia is an archipelago state located in the tropics. IMC located between two continent namely Asia and Australia, and two ocean namely Pacific and Indian. The impact of the geographical location is experiencing an Indonesia maritime continent seasonal wind blow from the continent of Asia (the Asian Monsoon) and the continent of Australia (the Australia Monsoon) which changed its direction periodically [1]. The air circulation in the influence by the motion of the Sun, all at a time when the position of the Sun in the northern hemisphere then high pressure will be in South hemisphere and Australia as well as the Monsoon occurs otherwise at the time the position of the Sun in the South (also known as with the boreal winter) is going to happen to the movement of air masses from Asia towards an Indonesia maritime continent which is called Monsoon Asia. The characteristics of air masses originating from the continent of Asia is both cold and dry. At the time of the air, masses interact with IMC's warm causes deep convection in that area.

During the period of the Asian Monsoon, the synoptic scale weather disruption, there is either a closed circulation of the cyclone in the western part of the island of Borneo known as the Borneo vortex[4]. The phenomenon of the Borneo vortex associated with the incidence of heavy rain around the region of Sumatera, Borneo, and Java [5]. The strength and position of the Borneo vortex are sensitive to the presence of cold surge. The convection area around of the Borneo vortex will be strengthened when cold surge strengthens.

The IMC region also experienced the phenomenon in the form of a moving oscillation of the Indian Ocean towards the East in the area of Equator with a period of 30-60 days known as the Madden Julian Oscillation (MJO). According to Holton [8], the phenomenon of convection, thus reviving MJO effect on the increase in rainfall. The movement of the MJO is divided into 8 phases starting from the Indian Ocean to the Western Pacific and is in the region of Indonesian maritime continent on phase 4 and 5.

The phenomenon of the MJO and the Borneo vortex cause the IMC's area become an active area of deep convection, but there have been no studies specifically observed the existence of this phenomenon at the same time. The purpose of this research is to see the influence of MJO activity against Borneo vortex and its impact on rainfall in the western part of the Indonesia maritime continent. In this journal will be discussed regarding the influence of activity of the Borneo vortex without the MJO and the interaction of the Borneo vortex with the MJO (in phase 3, 4, and 5).

2. Data and method

The data used in this study is an ECMWF (European Center for Medium-Range Weather Forecast) ERA-Interim reanalysis Interim data at 00.00 Universal Time Conversion (UTC) last November 2015 to March 2016 in the 925 hPa in the atmospheric layer with spatial resolution 2.5^0 x 2.5^0. The parameters used are vertical velocity (vertical motion). Data of Madden Julian Oscillation (MJO) used are sourced from Bureau of Meteorological with data period is November 2015 until March 2016. Satellite data from TRMM (Tropical rain shower Measuring Mission) is used to view the distribution of precipitation.

The region became research area is in the West part of IMC (95^0- 120^0 E 10^0 S 10^0 N) which includes the islands of Sumatera, Borneo, and Java (Figure 1).

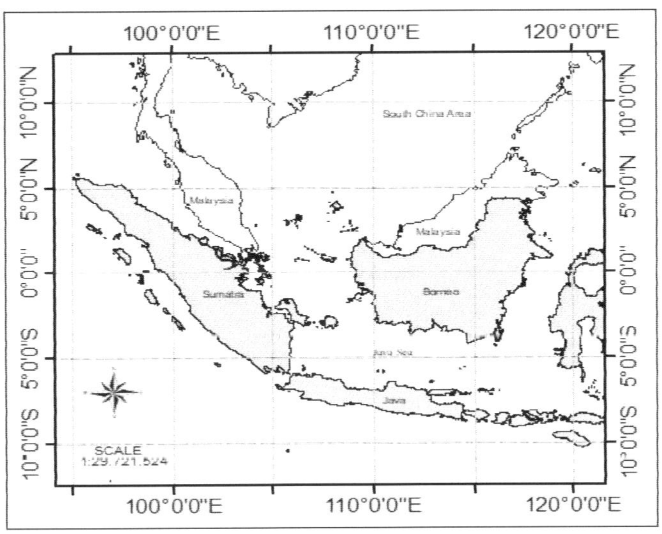

Figure 1. Western Indonesian maritime continent on 95^0- 120^0 E 10^0 S 10^0 N, includes Sumatera island, Borneo island and Java island.

Borneo vortex is identified with the way saw a counter-clockwise circulation in the 925 hPa layer in the area of 2.5^0 S-7.5^0 N and 102.5^0-117.5^0 E and at least one wind exceeds 2 ms^{-1} [3]. The phenomenon is observed when the MJO is in the Indonesia maritime continent, namely phases 4 and 5 (the maritime continent), as well as on phases 3 expansion when the MJO is starting to get to the West coast of Sumatera.

IOP Conf. Series: Journal of Physics: Conf. Series **997** (2018) 012004 doi:10.1088/1742-6596/997/1/012004

After identifying the incidence of Borneo vortex and the MJO, then performed in which the incidence of the Borneo vortex without the MJO and the interaction of the Borneo vortex with the MJO on phases 3, 4 and 5 and see its impact on rainfall in the western part of the IMC.

3. Result and Discussion

3.1 Identification of Borneo vortex

Formation of Borneo vortex happens due to the vorticity which is reproduced by wind shear and reinforced by the northeast monsoon wind convergence with the topography of the island of Borneo [11]. Figure 2 shows one Borneo vortex that is observed. During the period from November until March 2015/2016, there were 19 events of Borneo vortex, which mostly appears in November and March. Meanwhile, the Borneo vortex has the longest period in December (4 days) same with Chang [4].

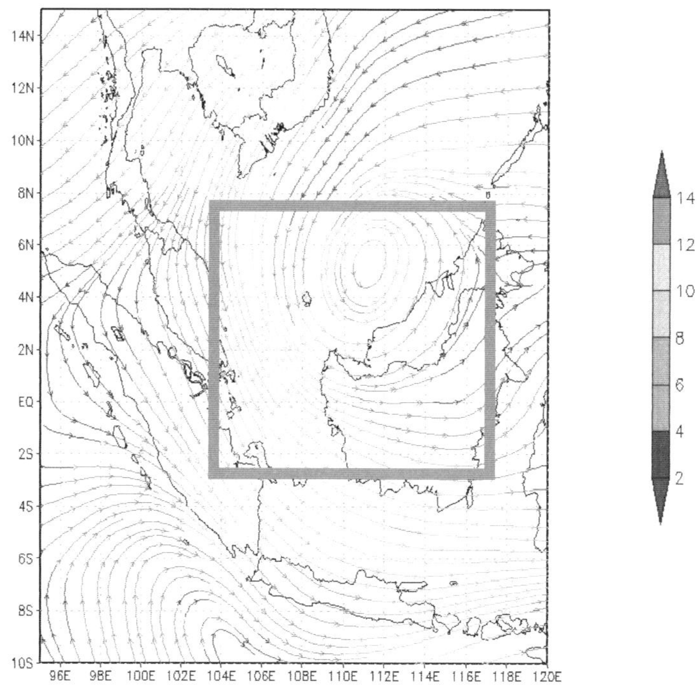

Figure 2. Winds at 925 hPa and identification area of Borneo vortex (green box).

3.2 Identification of Madden-Julian Oscillation

MJO can be distinguished by 2 conditions, strong and weak/ inactive. Through the phase diagram of the MJO (figure 3), active MJO conditions identified by the position of the line outside the centre (circle), the farther from the centre of the circle the more strong intensity of the MJO. When MJO is active in phase 4, it signifies the MJO appear in the western IMC and in phase 5 appear in the eastern IMC.

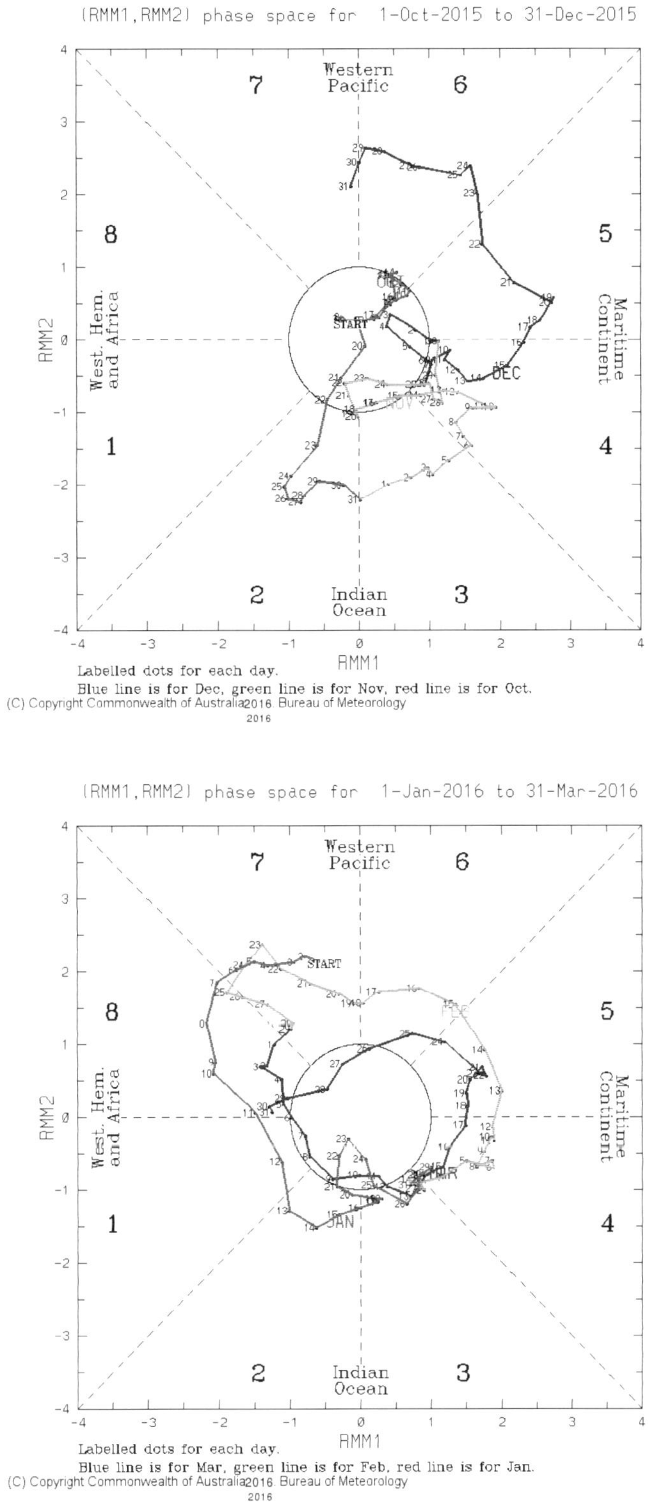

Figure 3. MJO diagram phase period October - December 2015 (above) and January - March 2016 (below).

From figure 3 we can be observed the appearance of strong MJO on phase 3, 4 and 5. During November – December 2015, MJO phenomenon appeared for 33 days and longest occurred in November at phase 4. Meanwhile, in the period January to March 2016, MJO phenomena appear for 35 days with the longest period (9 days appear in phase 4 in February).

3.3 The Impact study of MJO and Borneo vortex

After analyzing the genesis of Borneo vortex and MJO, then the data displayed into a graph (figure 4)

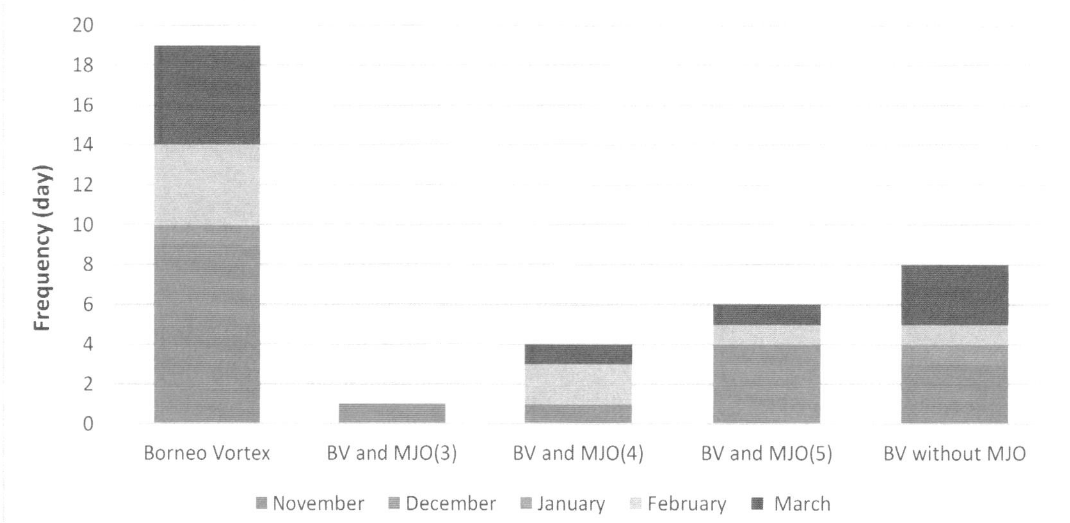

Figure 4. Frequency of occurrences Borneo vortex (total), Borneo vortex during MJO phases 3, Borneo vortex during MJO phases 4, Borneo vortex during MJO phases 5 and Borneo vortex without MJO.

Based on figure 4, the Borneo vortex phenomenon most occur when the inactive MJO on all phase (phase 3, 4 and 5) of 42.10% of the total incidence of the Borneo vortex. The frequency of occurrence of the phenomenon of Borneo vortex at most at a time when strong MJO phase 5 as 6 events (31.57%) and there is a period of Borneo vortex is most long 4 events in December. In otherwise, when active MJO is in phase 3, there is one case of Borneo vortex (5.26%) and in phase 4 there are 4 cases (21.05%). According to Chang [4] the impact of the existence of the MJO is weaken the cold surge that influence to the formation of Borneo vortex so that the frequency of occurrence of Borneo vortex at the moment the MJO is in phase 3 and 4 (Western of IMC) is smaller than when the MJO is strong in phase 5.

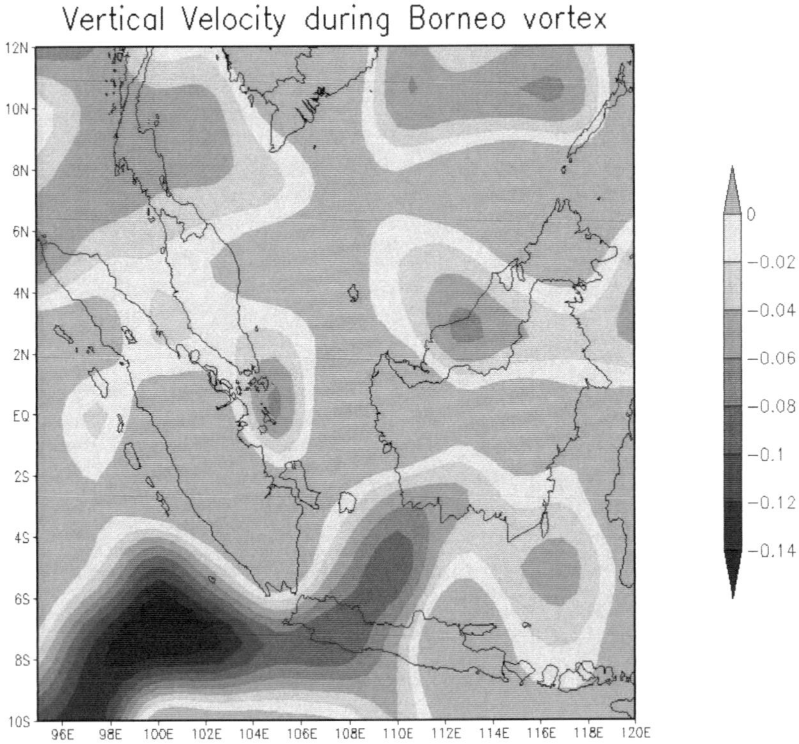

Figure 5. Vertical velocity (ms^{-1}) during Borneo vortex without active MJO.

By knowing the value of vertical velocity parameters, the movement of air masses surrounding the region of the vortex can be known. The cyclonic flow on Borneo vortex system will make the air mass moves upward. The movement of air masses over and commensurate convection causes the convection that occurs. In figure 5, the region of Peninsula Malaysia, Karimata Strait to the South of the island of Borneo, there is a current of rising air masses that are strong enough (marked with colours increasingly blue). The smaller the value of vertical speed indicates the movement of air masses upward. The movement of air masses rise will be higher in the centre of the vortex.

When the Borneo vortex appeared in the period of active MJO, especially in phase 3 and 4, the value of vertical velocity does not indicate the existence of a mass movement of a strong upward air as in genesis Borneo vortex only. This indicates the intensity of the vortex that is not too strong. Unlike the MJO on phase 5, there is a rising air mass movement in the environment near the centre of the vortex (the South China Sea northern Borneo).

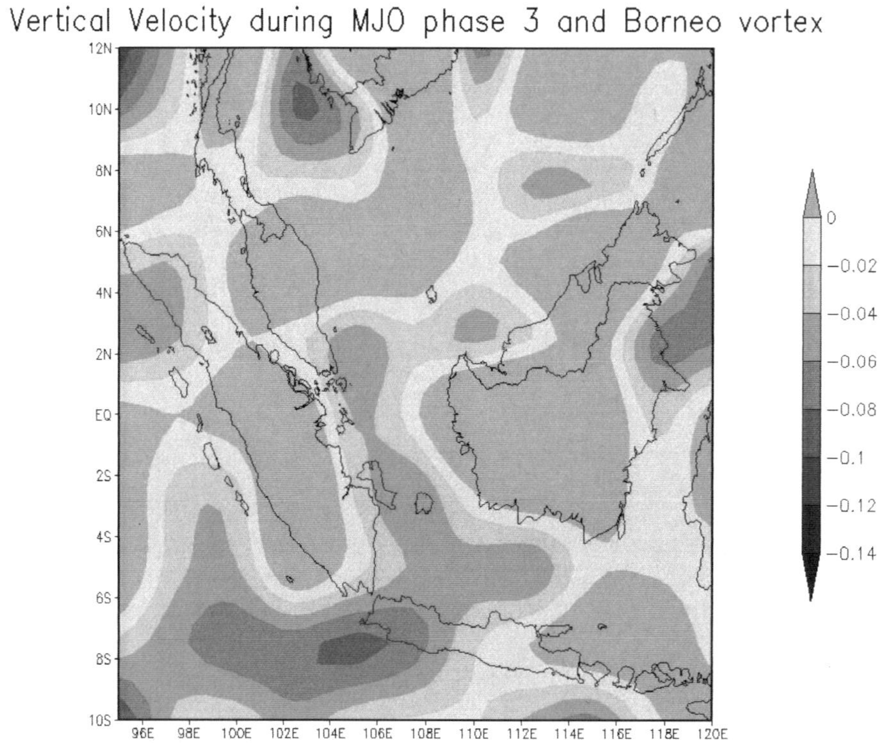
Vertical Velocity during MJO phase 3 and Borneo vortex
12N
10N
8N
6N
4N
2N
EQ
2S
4S
6S
8S
10S
96E 98E 100E 102E 104E 106E 108E 110E 112E 114E 116E 118E 120E
0
-0.02
-0.04
-0.06
-0.08
-0.1
-0.12
-0.14

(a)

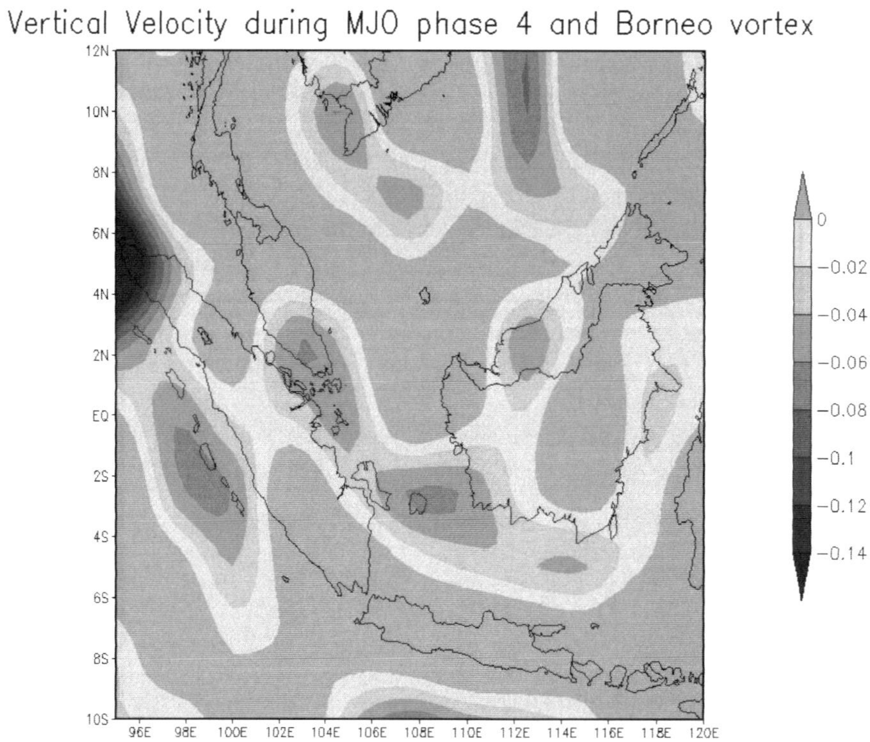
Vertical Velocity during MJO phase 4 and Borneo vortex
12N
10N
8N
6N
4N
2N
EQ
2S
4S
6S
8S
10S
96E 98E 100E 102E 104E 106E 108E 110E 112E 114E 116E 118E 120E
0
-0.02
-0.04
-0.06
-0.08
-0.1
-0.12
-0.14

(b)

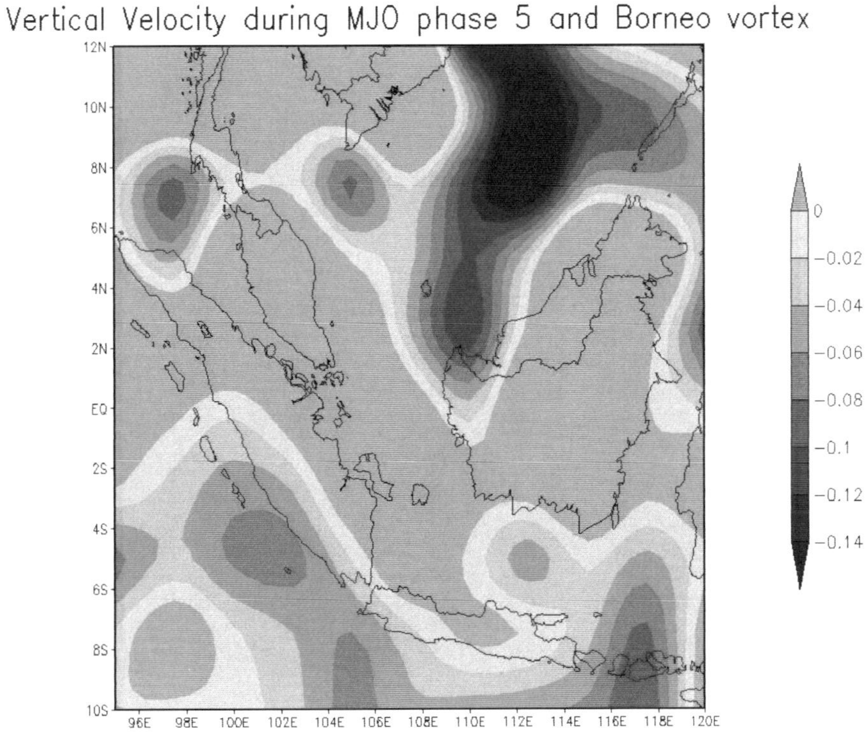

(c)

Figure 6. Vertical velocity (ms^{-1}) of Borneo vortex during strong MJO on phases 3 (a), phases 4(b) and phases 5 (c).

3.4 Rainfall study using TRMM

Based on the rainfall data from TRMM satellite during the occurrence of Borneo vortex without MJO (figure 7) can be known that rainfall at the time of occurrence of Borneo vortex without MJO recorded slightly to heavy rain near the vortex occurrence.

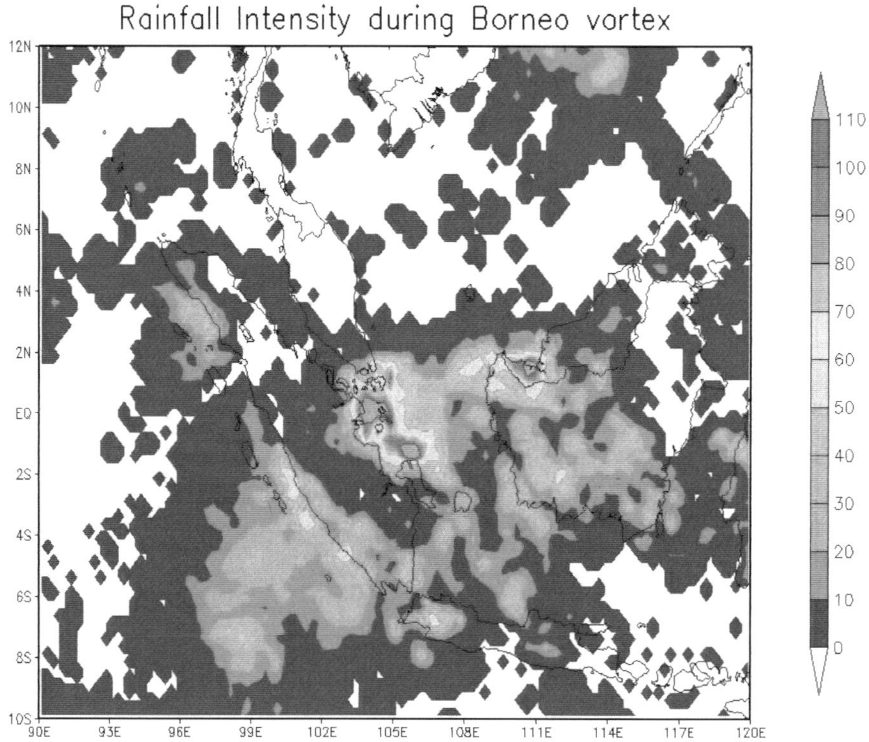

Figure 7. Rainfall intensity from TRMM's data (mm) during Borneo vortex without MJO.

Generally, rainfall is recorded with the intensity of the slight to moderate rain (10 - 40 mm) which includes the western part of Indonesia (Western and southern parts of Kalimantan, Java Sea, the southern part of the Sumatra Island and the western part of Java Island). Recorded data of heavy rain (50 - 100 mm) around the waters of Riau Islands and Bangka Belitung. From figure 8, there was heavy rain in the Riau Islands and northern Borneo island to the amount of precipitation more than 110 mm.

IOP Conf. Series: Journal of Physics: Conf. Series **997** (2018) 012004 doi:10.1088/1742-6596/997/1/012004

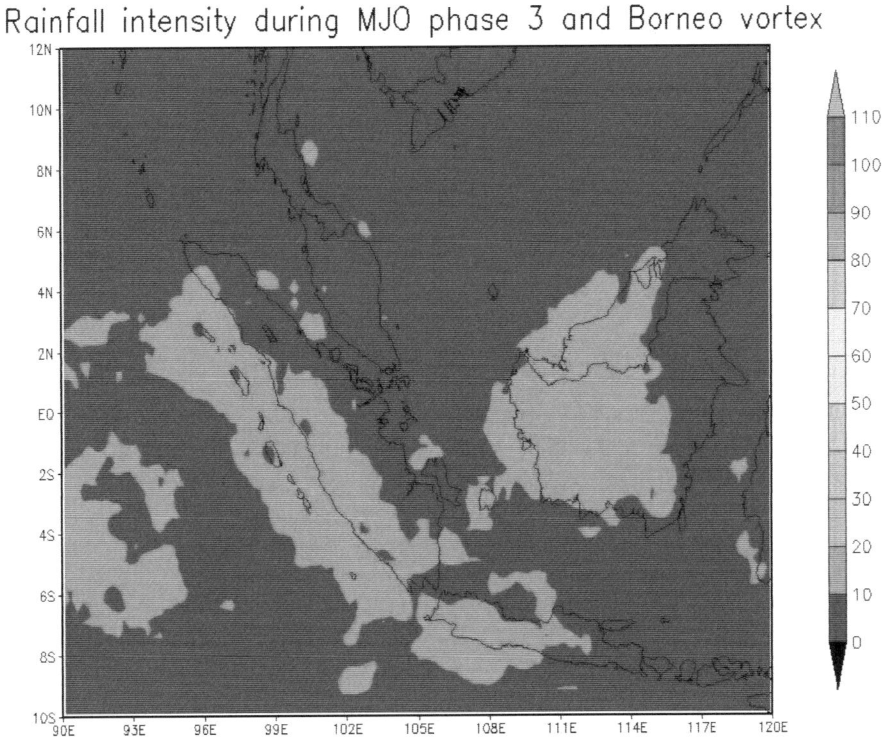

(a)

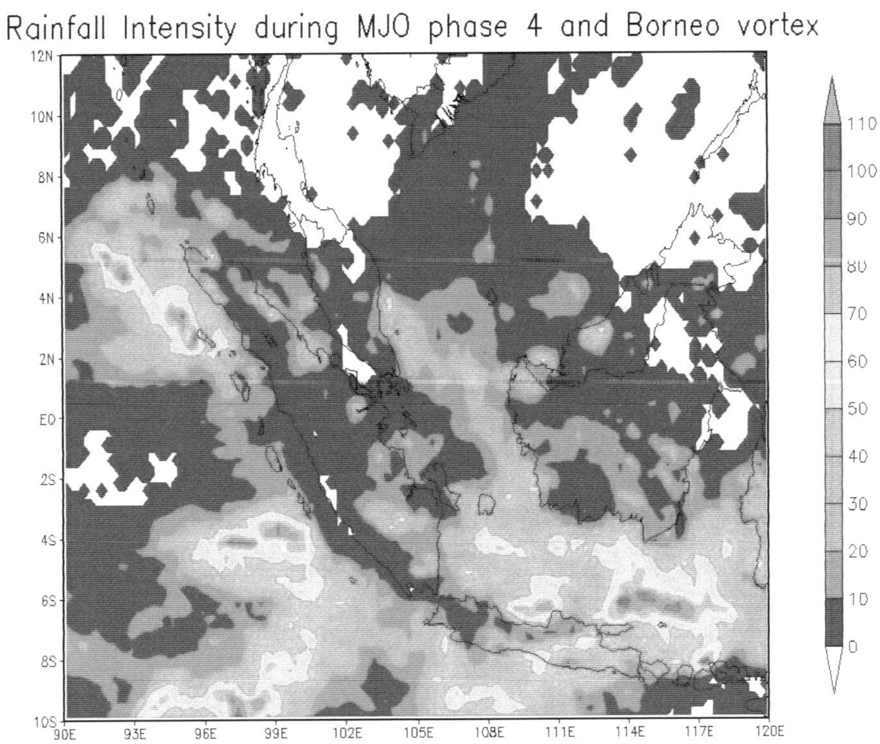

(b)

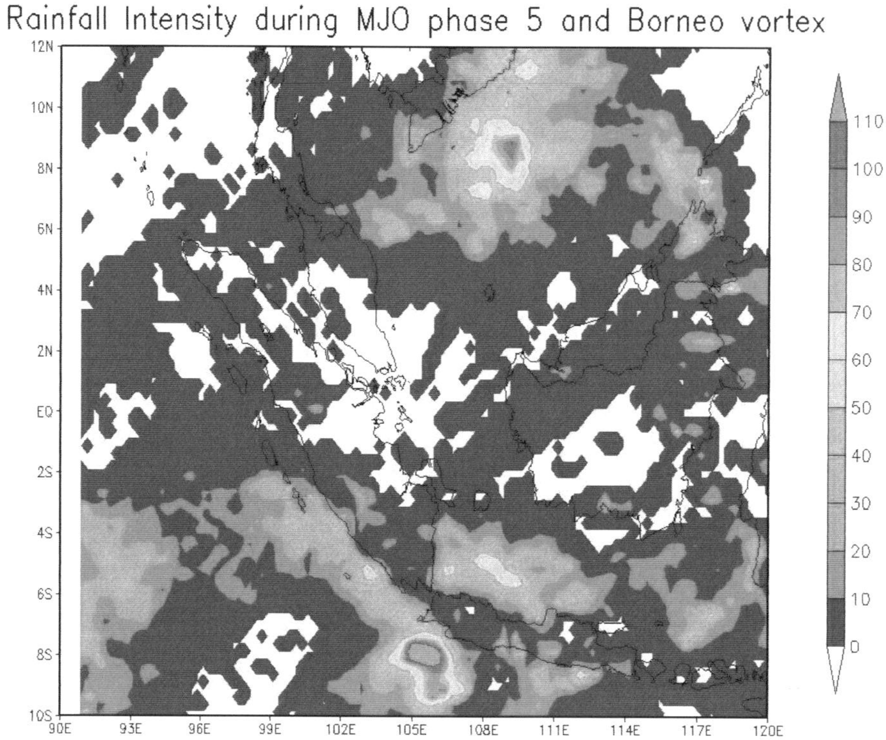

(c)

Figure 8. Rainfall data from TRMM's data (mm) of Borneo vortex during strong MJO in phase 3(a), phase 4 (b) and phase 5 (c).

Rainfall is recorded when the occurrence of Borneo vortex without MJO different with the amount of precipitation at the time of occurrence of Borneo vortex which appears simultaneously at the time of active MJO on phase 3. From figure 8 (a) can be known that rain occurred with slight rain (10 – 20 mm) with a wider area coverage of rain (West and South of the island of Borneo as a whole, Western Malaysia, Sumatra and the western part of Java island that extends to the Middle section). While the rainfall data recorded when the occurrence of Borneo vortex appeared concurrently with the active MJO on phase 4, rainfall increased to heavy intensity (10-90 mm) with a wider area of coverage again. However, the high intensity of rainfall does not occur in the western part of Borneo, but it occurs at the Java Sea and the western part of the island of Sumatra. While the coverage area in Sumatera's trench narrows with the slight rain. The air mass from southwest of Sumatera met the air mass from northeastern Sumatera. The effect is high rainfall intensity over the southern area of IMC. At the time of occurrence of Borneo vortex appeared concurrently with the active MJO on phase 5, heavy rain recorded by region with the center of the vortex in South China Sea and precipitation recorded over 110 mm, while in the West area of IMC, the rainfall intensity vary from 20-50 mm, which located in the central of the Java Sea.

4. Conclusion

From this research study can be summarized as follows the highest frequency occurrence of Borneo vortex was during inactive MJO phases 3, 4 or 5. MJO may decrease the frequency of Borneo vortex especially during MJO phases 3 and 4. When MJO at phases 5, frequency Borneo Vortex may increasing. Meanwhile, the interaction between MJO and Borneo vortex may enhance the indication of vertical motion as part of the convective process. Based on this study, tell us that the interaction between MJO (in phases 3 and 5) and Borneo vortex may affect the rainfall intensity over western IMC but not so much. The interaction between MJO phases 4 and Borneo vortex may increase rainfall intensity over western IMC.

Acknowledgement
In this study, we thank all those who have given us advice, motivation, and support, especially for our parents. We also thank Meteorological Climatological and Geophysical Agency of Indonesia (BMKG) and Indonesian State College of Meteorology, Climatology and Geophysics (STMKG).

References

[1] Agita V W 2016 *Kajian Dampak Interaksi Monsoon Asia, Cold Surge, dan Mjo Terhadap Dinamika Atmosfer dan Curah Hujan di Benua* Maritim *Indonesia* (Jakarta: Sekolah Tinggi Meteorologi Klimatologi dan Geofisika)
[2] Tangang F T, Juneng L, Salimun E, Vinayachandran P N, Seng Y K, Reason C J C, Behera S K, and Yasunari T 2008 *J. Geophys. Res. Lett.* **35** LI4507
[3] Anip M H M 2012 *Interannual and Interdecadal Variabillity of the Borneo Vortex during Boreal Winter Monsoon* (Missouri: University of Missouri)
[4] Chang C P, Harr P A and Chen H J 2005 *J. Am. Meteorol. Society* **133** 489
[5] Paulus A W and Shanas S P 2017 *J. Clim. Weather Forecast.* **5** 1000189
[6] Shanas S P 2016 *Pengaruh Cold Surge dan Borneo Vortex di Benua Maritime Bagian Barat*, (Jakarta: Sekolah Tinggi Meteorologi Klimatologi dan Geofisika)
[7] Mia V 2017 *Kajian Dampak Madden-Julian Oscillation (MJO) Terhadap Curah Hujan di Beberapa Wilayah Sumatera (Studi Kasus tahun 2016)* (Jakarta: Sekolah Tinggi Meteorologi Klimatologi dan Geofisika)
[8] Holton J R 2004 *An Introduction to Dynamic Meteorology* 4th ed (USA: Elsevier Publication)
[9] Ooi S H, Samah A A and Braesicke P 2011 *J. Geophys. Rev.* **116** D21116
[10] Zakir A, Sulistya W and Khotimah M K 2010 *Perspektif Operasional Cuaca Tropis* (Jakarta: Pusat Penelitian dan Pengembangan BMKG)
[11] Andarini D F 2012 *Analisis Cold Surge and Borneo Vortex Menggunakan Vortisitas Potensial* (Bandung: Institut Teknologi Bandung)

IOP Conf. Series: Journal of Physics: Conf. Series **997** (2018) 012005 doi:10.1088/1742-6596/997/1/012005

Dilatometric shrinkage study on magnesium titanate-based ceramic systems

F U Ermawati[1] and S Suasmoro[2]

[1]Physics Department, Faculty of Mathematics and Natural Sciences, Universitas Negeri Surabaya (UNESA), Surabaya 60231, Indonesia
[2]Physics Department, Faculty of Mathematics and Sciences, Institut Teknologi Sepuluh Nopember (ITS), Surabaya 60111, Indonesia

E-mail: frida.ermawati@unesa.ac.id

Abstract. The development of dielectric materials for applications in the microwave frequencies has been increasing with rapid progress in mobile and satellite communication systems. Magnesium titanate ($MgTiO_3$)-based ceramics have been the favourite candidates for such applications due to their excellent dielectric characteristics, *i.e.* a moderate level of dielectric constant together with a high-quality factor and high-temperature stability. These outstanding performance, however, can only be achieved when the ceramics are highly dense. The work reported in this paper discussed the study on the dilatometric shrinkage behaviour of pure and zinc-doped magnesium titanate ($Mg_{1-x}Zn_xTiO_3$ for x = 0–0.5) ceramic systems after the systems following the heating passage up to 1300 °C. The results were discussed based on the phase formation data recorded from powder X-ray diffraction (XRD). An additional 2 wt. % V_2O_5 to the MZT0.2 system has increased the shrinkage of the system, and hence the relative density. The V_2O_5 addition also prevented the grain growth and did not alter the structure. From 100 Hz to 20 MHz, the dielectric permittivity is constant; which varies from (15.4 – 17.0) ± 0.1 % throughout the samples, these values are therefore frequency independent.

1. Introduction

Magnesium titanate ($MgTiO_3$) has been recognised to have a relative permittivity and other dielectric properties that are suitable for applications at high frequencies. Research on this dielectric ceramic material has been started since the last two decades, and the dielectric-performance reports on this ceramic in various microwave-frequencies have been available after that [1-4]. Many efforts have been taken to improve the characteristic of the ceramic [5-8]. One of the aspects to be considered in fabricating a good ceramic is the density. Ceramic with high relative density is advantageous as the measured permittivity represents the material permittivity without the inclusion of the vacuum permittivity.

The work reported in this paper is intended to study the dilatometric shrinkage on the magnesium titanate-based dielectric ceramics ($Mg_{1-x}Zn_xTiO_3$ for x = 0 – 0.5). The aim was to see the dielectric performance of the ceramics. This paper also presents the discussion on the similar shrinkage from another $Mg_{0.8}Zn_{0.2}TiO_3$ system after the inclusion of V_2O_5. A fraction of V_2O_5 was added into the system as an effort to reduce the sintering temperature of the system. The resulting density, the structure, the microstructure and the relative permittivity of the developed ceramics were discussed.

2. Experiment

Six different compositions of $Mg_{1-x}Zn_xTiO_3$ powders for $x = 0 - 0.5$ (MZT0 – MZT0.5) were synthesised following the liquid mixing method [9-10]. The starting materials and the solvent were Mg, Zn and Ti metal powders (Merck) and hydrochloric acid (HCl). For each composition, each metal powder was weighed carefully according to the $Mg_{1-x}Zn_xTiO_3$ stoichiometry and dissolved individually in HCl using a magnetic stirring hotplate. The resulting liquids were homogeneously mixed and dried. The dried powder was crushed to produce homogeneous-grain-size powder. The zinc-doped MZT0.1 – MZT0.5 powders were subsequently calcinated at 550 °C for 2 hours at the heating rate of 10 °C/min using a Carbolite RHF-1400 furnace, while the zinc-free MZT0 powder was calcined at 700 °C for 2 hours.

Dilatometric shrinkage behaviour of MZT0 – MZT0.5 ceramics was investigated by preparing the green bodies of the respective calcined powders with the dimensions of 5×5×20 mm^3 using a rectangular uniaxial die and pressing at 100 MPa in the atmospheric environment using a hydraulic hand press. The green bodies were heated from room temperature to 1300 °C using an L76 Platinum Series Dilatometer (Linseis) at the heating rate of 10 °C/min. The shrinkage behaviour of another MZT0.2 system, *i.e.* the MZT0.2 system with the inclusion of 2 weight (wt.) % V_2O_5, was also examined. The addition of such a liquid agent into the system was intended to see the possibility of reducing the sintering temperature of the ceramic due to the formation of the liquid phase of V_2O_5 at the melting point which is below the sintering temperature. For that purpose, both MZT0.2 and V_2O_5 powders were weighed at the prescribed wt. % ratio and were mixed using zirconia balls for 2 hours in a planetary ball mill under ethanol environment. The ratio between the powder mass, the zirconia balls and the ethanol were 1: 5: 5/3. The mixed slurry was subsequently separated from the ethanol using a rotary evaporator operated at 20 rotations per minute for 1 hour and 90 °C. The powder product was stored in the oven for overnight to ensure that all ethanol evaporated. The dried powder was crushed manually using a mortar and pestle for gaining the homogeneous grain size powder. The resulting powder was then ready for the dilatometric shrinkage study.

For dielectric property characterisation, the green bodies of all the samples were prepared using a uniaxial die in the form of a disc and the hydraulic hand press. The powders were pressed at the same pressure (100 MPa) and sintered using a Carbolite RHF-1400 furnace at the temperature suggested by the dilatometric shrinkage curves. Sintering was performed in the atmospheric environment and at the same heating rate (10 °C/min). The bulk density of the ceramics was measured following the Archimedes technique. The structural analysis was performed using a Philips X'Pert Multi-Purposes Diffractometer System with CuK_α radiation. The microstructural data was recorded using a scanning electron microscope (SEM) CARL ZEISS EVO MA 10. The dielectric property in the form of real part of the relative permittivity (ε'_r) was measured at room temperature from 1 Hz to 20 MHz using an SI 1260 Impedance/Gain-Phase Analyzer (Solartron Analytical).

3. Results and Discussion

3.1. Dilatometric shrinkage of green bodies

The following subsubsections provide the dilatometric shrinkage data measured from all the samples (MZT0 – MZT0.5 and MZT0.2 + 2 wt. % V_2O_5) recorded using the L76 Platinum Series Dilatometer (Linseis) and the analyses.

3.1.1. MZT0 – MZT0.5. Figure 1 shows the intended shrinkage curves ($\Delta L/Lo$) as a function of temperature recorded from MZT0 – MZT0.5 samples. As observed, all the curves experience contraction as indicated by dropping of the curves, starting at 800 - 850 °C and this lasts up to 1300

°C. The total shrinkage varies from 12.8 – 14.0 %. At the temperature around 1100 °C, the shrinkage rate slows down as signified by the presence of a dip on the curves. This particular concave-shaped curve is seen in the samples containing zinc (MZT0.1 - MZT0.5), but not for the one without zinc (MZT0).

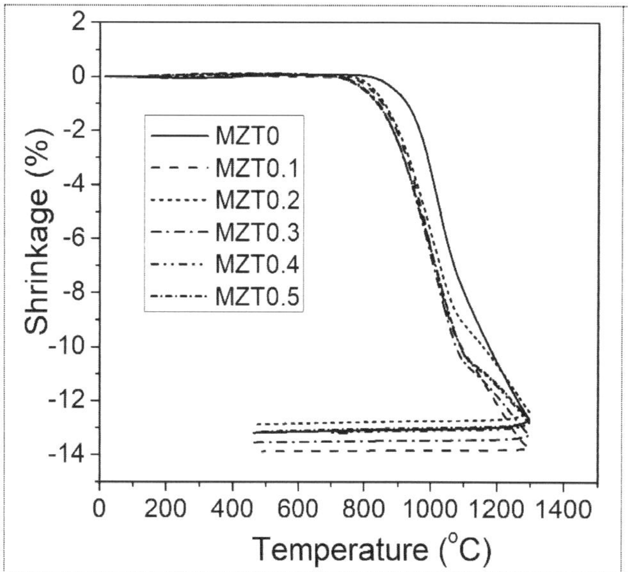

Figure 1. The dilatometric shrinkage curves of MZT0 – MZT0.5 green bodies measured from room temperature to 1300 °C with the step size of 10 °C/min using the L76 Platinum Series Dilatometer.

To understand the dip co-presence in Figure 1, the MZT0.2 (without V_2O_5) was sintered at two different temperatures, each at 900 °C (*i.e.* before the dip position in Figure 1) and at 1300 °C (after the dip) for 8 hours. Figure 2 depicts the XRD patterns of the intended MZT0.2 after sintering at 900 and 1300 °C.

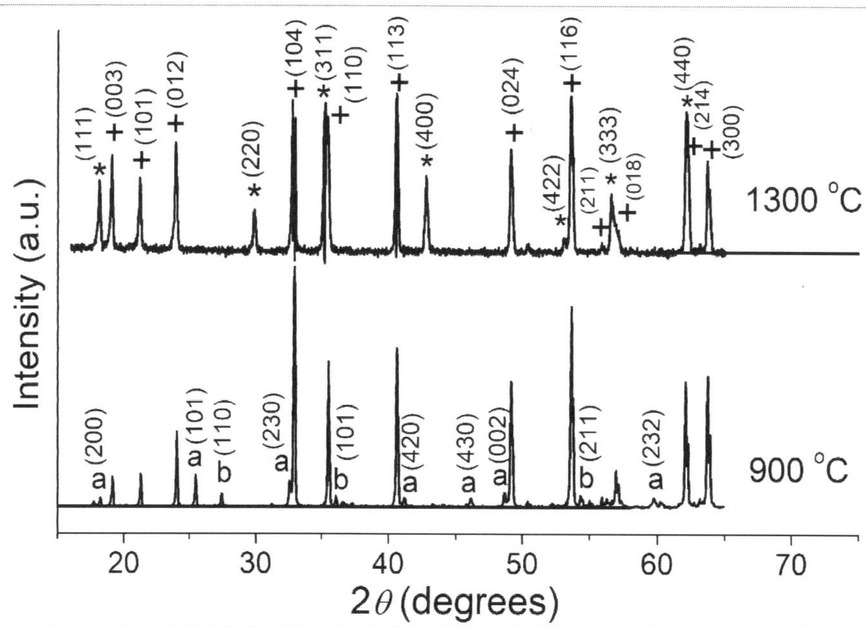

Figure 2. The XRD patterns of MZT0.2 ceramics sintered at 900 and 1300 °C. + =

$Mg_{0.8}Zn_{0.2}TiO_3$, * = Mg_2TiO_4, a = $MgTi_2O_5$, b = TiO_2

In Figure 2, the 900 °C pattern consists of the main $Mg_{0.8}Zn_{0.2}TiO_3$ phase (the peaks labelled '+') accompanied by $MgTi_2O_5$ ('a') and TiO_2 rutile ('b') as the minor phases. The identified $MgTi_2O_5$ phase coincides with the powder diffraction file (PDF) No. 35-0792, while the TiO_2 phase agrees with the PDF No. 21-1276. The detected $Mg_{0.8}Zn_{0.2}TiO_3$ phase matches the $MgTiO_3$ (PDF No. 06-0494). The X-ray diffractometer also identified the $Mg_{0.8}Zn_{0.2}TiO_3$ phase on the 1300 °C pattern. The accompanying secondary phase on this particular pattern, however, is different, *i.e.* Mg_2TiO_4 (the peaks labelled '*'). This crystalline phase matches the PDF No.25-1157. The fact is that, after sintering at 1300 °C, the X-ray diffractometer only detected the minor Mg_2TiO_4 phase, while the $MgTi_2O_5$ and TiO_2 phases were not. This fact gives a strong indication that during the sintering process, a reaction between TiO_2 and an amorphous 'phase' containing magnesium oxide took place to form the secondary Mg_2TiO_4 phase, possibly via the process of $2MgO + TiO_2 \rightarrow Mg_2TiO_4$. This side reaction is considered to be detrimental because this has slowed the shrinkage rate, and the loss is indicated as the concave-shaped curve at 1100 °C in Figure 1. Based on this result, all the samples were subsequently sintered at 1400 °C for 8 hours using the Carbolite RHF-1400 furnace at the heating rate of 10 °C/min.

3.1.2 MZT0.2 with V_2O_5 addition. Figure 3 shows the shrinkage curve of MZT0.2 + 2 wt. % V_2O_5 sample, together with that of the MZT0.2 without V_2O_5 addition (in Figure 1). As observed, both curves started to shrink at the similar temperature, 800 °C. Above that temperature, the curve with V_2O_5 experiences a total shrinkage of up to 14.4 %. This percentage is higher than that of the curve without V_2O_5 (12.8 %). During heating, V_2O_5 with low melting point (656 - 690 °C) formed a liquid phase. The presence of such liquid caused the capillary force to attract fluid into the pores. This attraction gave rise the grains to rearrange the positions to be more stable. In the areas where the capillary pressure is high, *i.e.*, in the places where the particles are close together, the atoms were likely to enter the solution and resided in the areas with a lower chemical potential. The smaller particles moved toward the solid-solution structure and stayed on the larger particles causing compaction. This compacting process spread throughout the system. Due to the presence of V_2O_5, the shrinkage process occurred more quickly and the total percentage is, therefore, higher as depicted in Figure 3. Based on that, the MZT0.2 sample with V_2O_5 addition was subsequently sintered at 1100 °C for 4 hours. The subsection 3.2 provides the density, the structure, the microstructure and the dielectric characteristic data on this particular ceramic, together with that of the other ceramics without V_2O_5.

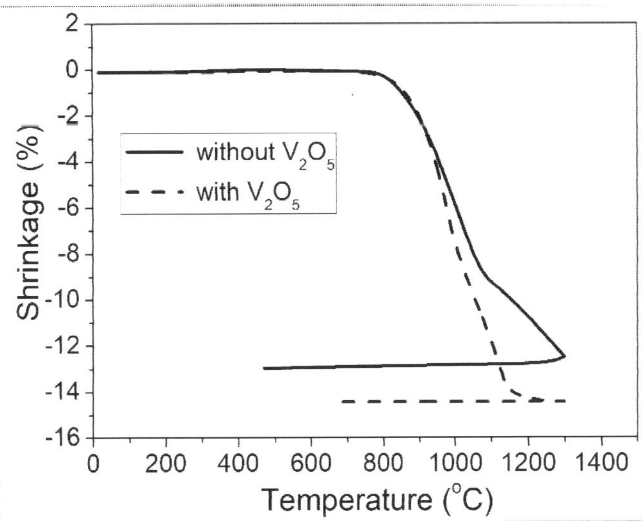

Figure 3. The dilatometric shrinkage curves of MZT0.2 both with and without 2 wt. % V_2O_5 addition recorded from room temperature to 1300 °C at the heating rate of 10 °C/min using the L76 Platinum Series Dilatometer.

3.2. Ceramic density, structure, microstructure and dielectric characteristic

Figure 4 depicts the relative density of MZT0 – MZT0.5 ceramics after sintering at 1400 °C for 8 hours and of MZT0.2 with V_2O_5 after sintering at 1100 °C for 4 hours. As shown, the relative density of the MZT0 – MZT0.5 ceramics, in general, increases with the zinc content, *i.e.* from 90.0 % of the theoretical density in MZT0 to 92.4 % in MZT0.5. The zinc content has increased the relative density of the systems. In Figure 4, the percentage of the MZT0.2 with V_2O_5 addition is 94.5 % which is considerably higher than that of the same ceramic without V_2O_5 (89.9 %). The V_2O_5 content in MZT0.2 has played a role in increasing the density.

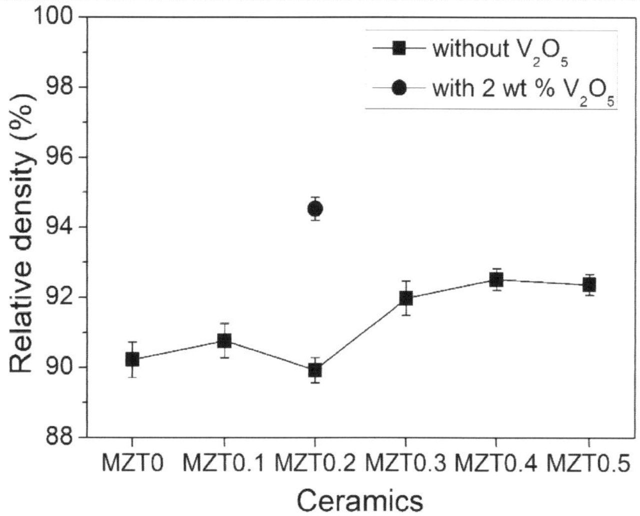

Figure 4. The relative density of MZT0 – MZT0.5 ceramics (without V_2O_5) after sintering at 1400 °C for 8 hours and of MZT0.2 with V_2O_5 after sintering at 1100 °C for 4 hours.

The content of V_2O_5 was not found to alter the structure of the MZT0.2 system, as depicted in Figure 5. As observed, the two patterns consist of the same phases, *i.e.* $MgTiO_3$ (the peaks with *hkl*) and Mg_2TiO_4 (the peaks labelled '*'). There is no phase containing zinc or vanadium was recorded on the patterns. This disappearance suggested that the zinc ions entered the magnesium sites because the radius of magnesium and zinc ions are similar, and the resulting phase is $Mg_{0.8}Zn_{0.2}TiO_3$. The vanadium ions, as mentioned before, has turned to be the liquid phase at its melting point and might remain as an amorphous phase at higher temperatures.

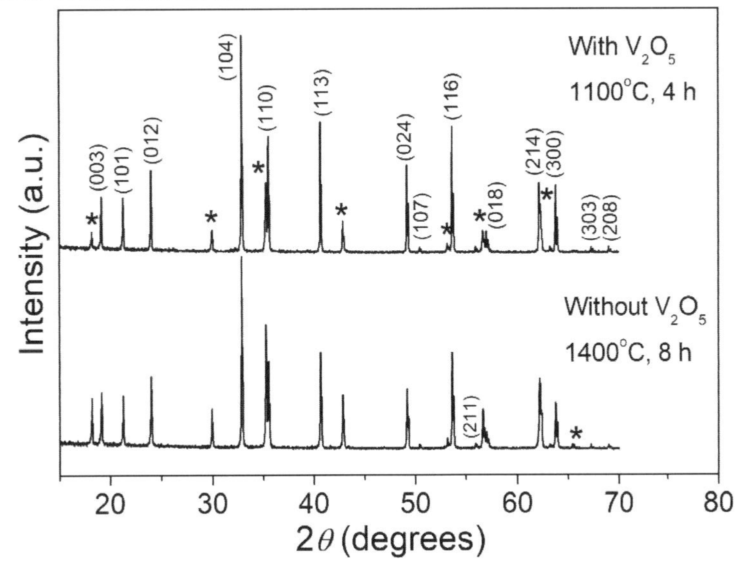

Figure 5. The XRD patterns of MZT0.2 ceramics with and without V_2O_5 addition after sintering at 1100 °C for 4 hours and at 1400 °C for 8 hours, respectively. The peaks with (*hkl*) = $MgTiO_3$, * = Mg_2TiO_4.

Figure 6 compares the XRD patterns recorded from all the samples (MZT0 – MZT0.5 and MZT0.2 with V_2O_5). As depicted, the spectra are composed of the diffraction peaks derived from the two phases, *i.e.* $Mg_{1-x}Zn_xTiO_3$ (the peaks with *hkl*) and Mg_2TiO_4 ('*'). This result confirmed that variation of the zinc content (*x*-values) in the $Mg_{1-x}Zn_xTiO_3$ without V_2O_5 and the inclusion of 2 wt. % V_2O_5 in the $Mg_{0.8}Zn_{0.2}TiO_3$ does not change the structure of $Mg_{1-x}Zn_xTiO_3$ ceramics. Instead, the co-presence of Zn and V_2O_5 improved the density of the systems.

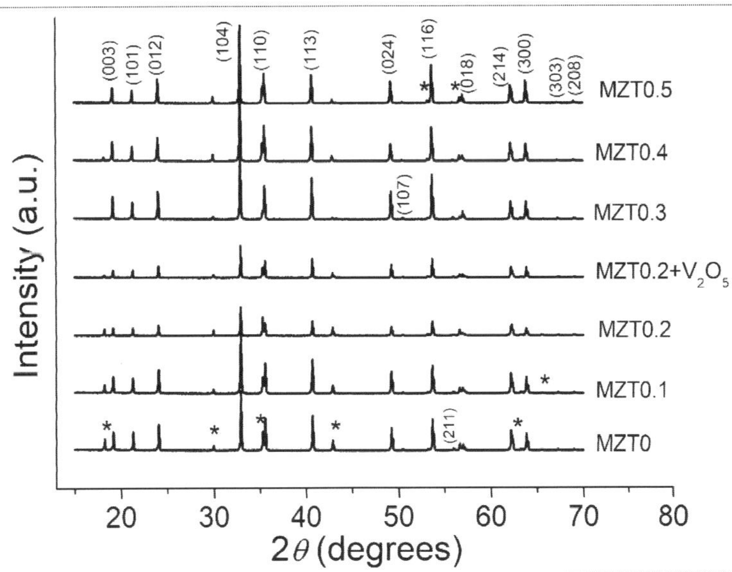

Figure 6. The XRD patterns of MZT0-MZT0.5 ceramics sintered at 1400 °C for 8 ours and of MZT0.2 with 2 wt.% V_2O_5 sintered at 1100 °C for 4 hours. The peaks with (*hkl*) = $MgTiO_3$, * = Mg_2TiO_4.

Figure 7 provides the microstructure of MZT0.2 ceramic with and without V_2O_5 addition recorded from the SEM CARL ZEISS EVO MA 10. As observed, the morphology and the grain size of the two samples are different. The grains in the sample containing V_2O_5 (Figure 7a) elongate in one direction with the thickness of about 1 μm. The tiny grains are also pictured. In the sample without V_2O_5 (Figure 7b), however, the grains had grown, and the size is almost uniform, *i.e.* 3-5 μm. The grain boundaries are visualised. The grain growth is due to the high sintering temperature (at 1400 °C). In addition, to reducing the sintering temperature, the co-presence of V_2O_5 in the MZT0.2 system was confirmed to prevent the grain growth.

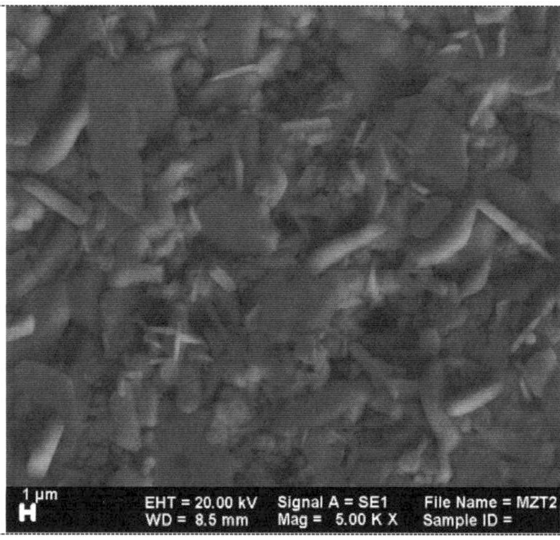

Figure 7a. The SEM image of MZT0.2 with 2 wt. % V_2O_5 after sintering at 1100 °C for 4 hours.

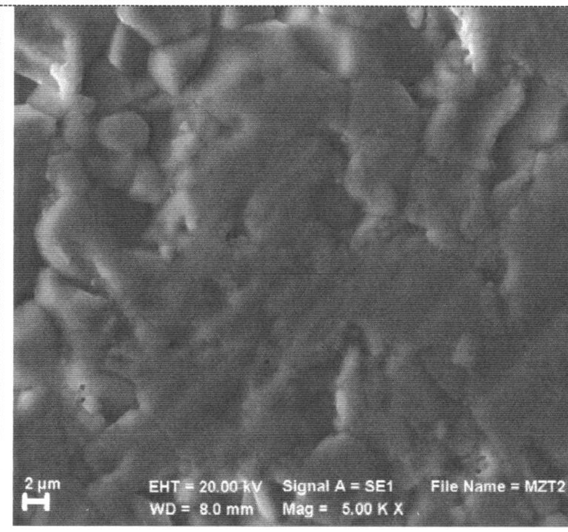

Figure 7b. The SEM image of MZT0.2 after sintering at 1400 °C for 8 hours.

Figure 8 shows the characteristic of real part of permittivity (ε'_r) of MZT0 – MZT0.2 and MZT0.2 with V_2O_5 addition measured from 1 Hz to 20 MHz using the SI 1260 Impedance/Gain-Phase Analyzer. All samples display the similar character, *i.e.* at 1 Hz, the ε'_r values reach the maximum before declining exponentially and constant afterwards. The maximum ε'_r at 1 Hz varies from 72 ± 0.1 % in MZT0.1 to 320 ± 0.1 % in MZT0.2 with V_2O_5 addition. From 100 Hz to 20 MHz, the ε'_r values are constant (15.4 to 17.0) ± 0.1 % throughout the samples. These findings are showing that the dielectric characteristics of the developed ceramics are similar to each other.

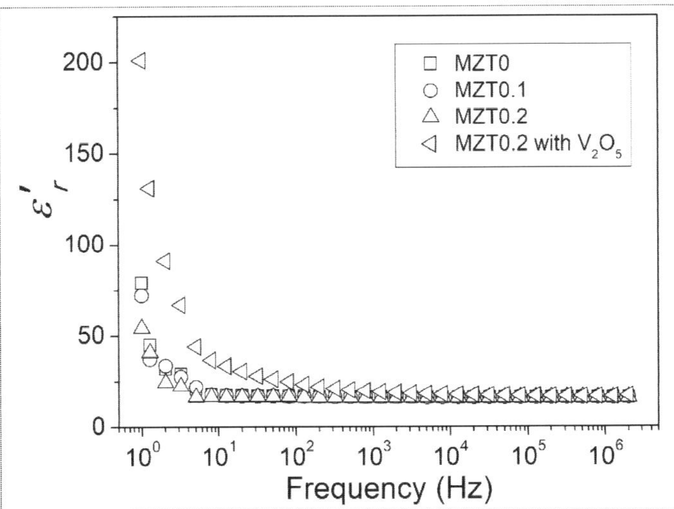

Figure 8. The characteristic of real part of the permittivity of MZT0 – MZT0.2 ceramics and of MZT0.2 with V_2O_5 addition measured from 1 Hz to 20 MHz using the SI 1260 Impedance/Gain-Phase Analyzer. The estimated measurement error is 0.1 %.

4. Conclusion

The dilatometric shrinkage study on the $Mg_{1-x}Zn_xTiO_3$ dielectrics has been carried out. The use of V_2O_5 is very significant in increasing the density of the ceramics and preventing the grain growth. This particular liquid agent does not alter the structure of the ceramics as well as the dielectric characteristic of the ceramic systems.

References

[1] Ferreira V M and Baptista J L 1994 *Mater. Res. Bull.* **29** 10
[2] Abothu I R, Rao A V P and Komarneni S 1999 *Mater. Lett.* **38** 3
[3] Suresh M K, Thomas J K, Sreemoolanadhan H, George C N, John A, Solomon A, Wariar P R S and Koshy J 2010 *Mater. Res. Bull.* **45** 7
[4] Sreedhar K and Pavaskar N R 2002 *Mater. Lett.* **53** 452-455.
[5] Deng Y F, Tang S D, Lao L Q and Zhan S Z 2010 *Inorg. Chim. Acta.* **363(4)** 827-829.
[6] Bernard J, Belnou F, Houivet D and Haussonne JM 2008 *J. Mater. Process. Tech.* **199** 1-3
[7] Xue W, Yu H and Xu G 2013 *J. Mater. Sci.-Mater. El.* **24** 4
[8] Tang B, Li H, Fan P, Yu S and Zhang S 2014 *J. Mater. Sci. Mater. El.* **25** 6
[9] Ermawati F U, Suasmoro S and Pratapa S 2015 *Adv. Mater. Res.* **1112** 47-52
[10] Ermawati F U, Pratapa S, Suasmoro S, Hübert T and Banach U 2016 *J. Mater. Sci. Mater. El.* **27** 6637-6645.

Acquisition of electrical signals using commercial electronic components for detection system of Lead ion in distilled water

Pujiyanto[1], M Yasin[1] and F Rusydi[1]

[1] Department of Physics, Faculty of Science and Technology, Airlangga University

E-mail: pujiyanto@fst.unair.ac.id

Abstract. Development of lead ion detection systems is expected to have an advantage in terms of simplicity of the device and easy for concentration analysis of a lead ion with very high performance. One important part of lead ion detection systems are electrical signal acquisition parts. The electrical signal acquisition part uses the main electronic components: non inverting op-amplifier, instrumentation amplifier, multiplier circuit and logarithmic amplifier. Here will be shown the performance of lead ion detection systems when the existing electrical signal processors use commercial electronic components. The results that can be drawn from this experimental were the lead ion sensor that has been developed can be used to detect lead ions with a sensitivity of 10.48 mV/ppm with the linearity 97.11% and had a measurement range of 0.1 ppm to 80 ppm.

1. Introduction

Several measurement methods of the chemical concentration have been carried out using optical fiber sensors [1,2.3]. The optical method developed to measure the concentration of the material using a displacement phenomenon between the two ends of optical fibers. The phenomenon of displacement between the two ends of optical fibers has been successfully developed as a displacement sensor that can measure object displacement to micron-scale resolutions [4,5,6]. Furthermore, the displacement sensor may be used as an attractive approach to the measurement of non-contact hydroxyapatite thickness of the material [7,8,9]. The displacement sensor has been shown to be used as a good approach to measuring transparent plate thickness based on transmission and reflection phenomena [10]. Fiber optic sensors using tapered fiber optics can be used to detect potassium in distilled water [11]. Fiber optic displacement sensors using bundled probe can be used for medal detection [12].

The overall development of the method of measuring the concentration of a chemical by using fiber optic sensor that has been done still has not produced a direct reading of its concentration value. This work was attempted to develop optical fiber spectrophotometry technique that can measure the concentration of lead ions with direct concentration reading. To allow for a direct reading of the value of concentration, a signal acquisition system is required using commercial electronic components that available on the market. Conceptually the signal acquisition process can be shown through the block diagram as Figure 1. Each block is described by a transfer function (TF) which expresses the relationship between output and input. If the input is expressed with V_i and the output is expressed by V_k, the statement $TF = V_k / V_i$. After going through the entire of the acquisition process, it is expected output voltage V_o is proportional to concentration (c).

The sensor probe consist of a light source, directional coupler and silicon photo detector. A violet 405 nm light from Light Emitting Diode (LED) is launched into directional coupler and directed to a region where the light interact with lead ion liquid. This interaction results in a modulation of optical intensity and modulated light is collected by the another optical fiber to be transmitted to the optical detector. By an optical detector, the received light intensity was converted into electrical voltage signal and then forwarded to acquisition system. The interaction of light with lead ions is an absorption phenomenon. Therefore, based on lambert-beer law, V_k / V_i is proportional to e^{-c}. At non-inverting operational amplifier, V_k / V_i is a gain of amplifier so that the output V_2 is proportional to e^{-c}. In amplifier intrumentation, one of the input voltages is given a certain reference voltage (V_r) which has greater value than V_2 so that V_k / V_i is proportional to (V_r-V_2). The output voltage V_3 of the amplifier intrumentation is proportional to e^c. The output voltage V_3 after multiplied by the number k (selected k = 100) and passed the amplifier logarithmic will be obtained output voltage V_o is proportional to *c*. This is because V_k / V_i for logarithmic amplifiers proportional to *ln* (V_4). Since the output voltage of V_o proportional to *c*, it is possible to read the results of direct ion concentration detection directly.

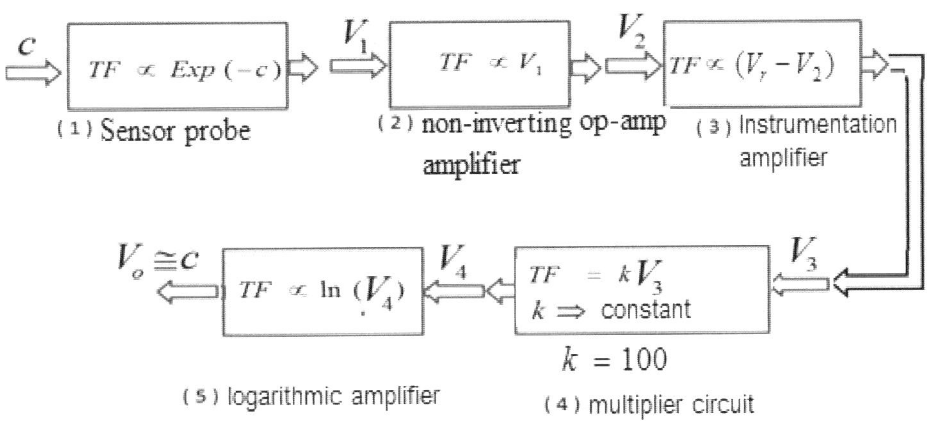

Figure 1. Block diagram of the electrical signal acquisition process.

2. Methods

This work was attempted to develop optical fiber spectrophotometry technique that can measure the concentration of lead ions with direct concentration reading. The main focus of this work was electric signal acquisition systems design using commercial electronic components and of optical probes. Probe sensor using asymmetric directional coupler *1 x 4* tree.

Procedures of this work focused on the design of electrical signal acquisition system devices using commercial electronic components for lead ion detection systems in distilled water were carried out through the following stages: 1) the design and manufacture of electric signal acquisition systems using commercial electronic components 2) the design and manufacture of optical probes 3) design of experiment setup for detection of concentrations of lead ions in distilled water, 4) sample preparation for testing, 5) data collection and data analysis. Conceptually the electrical signal processing in acquisition system was designed following the block diagram in Figure 1. The non-inverting amplifier output voltage (V_2), the output voltage of the multiplier circuit (V_4) and the output voltage of the logarithmic amplifier (V_o) were measured as the function of the lead ion concentration.

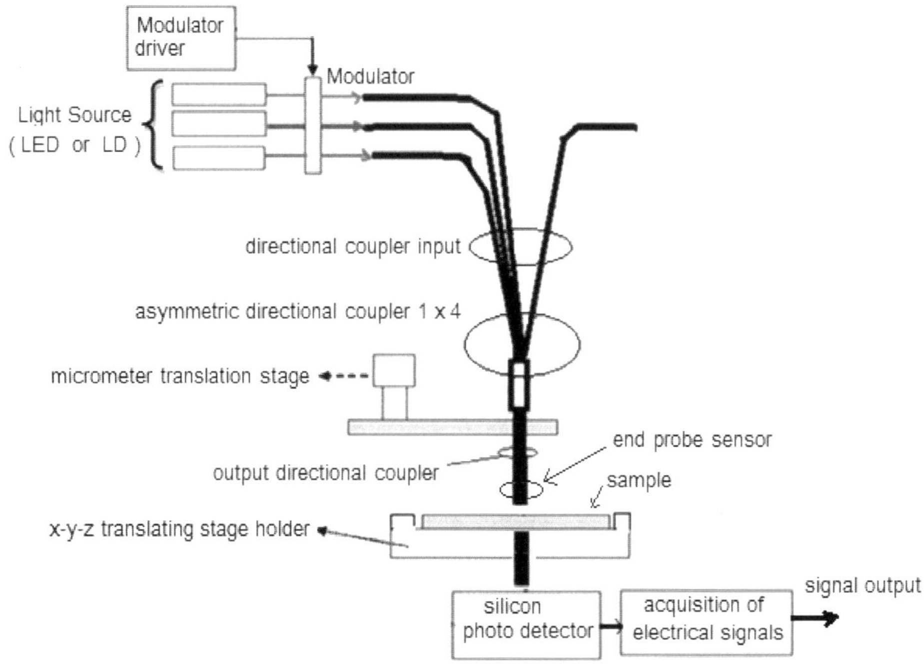

Figure 2. Experimental setup for concentration detection lead ions in water.

The design of the main optical system was the design of the probe sensor detecting the concentration in distilled water. Probe sensor concentration of several heavy metal ions were made from optical fiber displacement system using asymmetric directional coupler *1 x 4* tree. The directional coupler output was enabled to transmit light from the light source to the sample. The sensor probe integrated with the cuvette where the sample placed.

The experimental setup for the detection of concentrations of heavy metal ions in the water were shown in Figure 2. The main components of the experimental setup were light sources (violet LED) and concentration sensor probes.

In Figure 2, the light from a modulated LED light source (405 nm) through an electronic modulator with a certain frequency were fed into directional coupler input and transmitted to the sample through directional coupler output. The light passing through the sample guided through fiber optics to be transmitted to the optical detector. By an optical detector, the received light intensity was converted into electrical voltage signal. This voltage signal was amplified by a non-inverting OP-amplifier and then forwarded to the instrumentation and logaritmic amplifier. Output voltage non-inverting amplifier, output voltage of multiplier circuit and output voltage of logarithmic amplifier will be determined for the value as a function of the sample concentration.

The experiment was begun with the preparation of samples of lead ions. The lead sample of nitrate $Pb(NO_3)_2$, dissolved in water. Samples of $Pb(NO_3)_2$ were fed ethylene tetra acetate acid reagent (EDTA = $C_{10}H_{16}N_2O_8$). Each sample was made variation of concentration from 0.1 ppm to 80 ppm.

The data taken were the output voltage value of each part of the electrical signal acquisition system as its input function. The main data taken and to be analysed is the final output voltage value as a function of the concentration of lead ions in water.

Theoretically, the final output voltage as a function of concentration is linear. The result of the analysis gave the same data with a linear function.

3. Results and Discussion

Data obtained from this research were non-inverting amplifier output voltage, output line voltage multiplier and output voltage of logarithmic amplifier as a function of concentration of lead solution in water. The output voltage data as a function of lead ion concentration for Op-Amp amplifier, Multiplier Circuit and logarithmic amplifier presented in Table-1 and regression for each data presented in Figure 3. In Figure 3, it showed that the relationship between the output voltage Op-Amp amplifier (V_2) to the lead ion concentration can be approximated by function which decays exponentially with the equation $V_2 = 369.89\ e^{-0.0003\ c}$ with the coefficient of determination value 96.59 %. The relationship between the output voltage Multiplier Circuit (V_4) to the lead ion concentration can be approximated by function exponensial with the equation $V_4 = 157.13\ e^{0.0105\ c}$ with the value of coefficient of determination value 94.32%. The coefficient of determination value stated the level of data conformity with a particular function model. The form of voltage function V_2 and V_4 corresponded to the expected form of transfer function in the Op-amp amplifier and the multiplier circuit.

Table 1. Output voltage as a function of lead ion concentration (c) for Op-Amp amplifier, multiplier circuit and logarithmic amplifier.

(c) Concentrations of lead ions (ppm)	(V_2) Op-amplifier output voltage (x10 mV)	(V_4) Multiplier Circuit output voltage (x 0.01 mV)	(V_o) Logarithmic amplifier output voltage (x 0.1 mV)
0.1	371.0	136.2	309.0
1	369.4	164.5	497.7
10	368.7	178.7	580.8
20	367.3	206.9	727.4
40	365.2	253.3	929.3
60	363.1	306.8	1121.0
80	362.0	336.1	1212.0

In Figure 3, it showed that the relationship between the output voltage to the lead ion concentration can be approximated by linearity function with the value $R^2 = 0.9431$ (regression coefficient $R = 0.9711$) with the equation $V_o = (1.0483\ c + 45.205)$. In this case, V_o presented as the magnitude of the output voltage of the logarithmic amplifier, whereas c presented as the concentration of lead ions. Based on the value of R, it could be concluded that there were linearity correlation between the output voltages with lead ions. From the results of linear regression test can be known the sensors performance including sensitivity, linearity and measurement range. The performance of the lead ion sensor can be expressed in Table 2.

Table 2. Performance of lead ion sensor.

Parameter	Parameter Value for sample:
Sensitivity	10.48 mV/ppm
Measurement range	0.1 – 80 ppm
Linearity	> 97 %

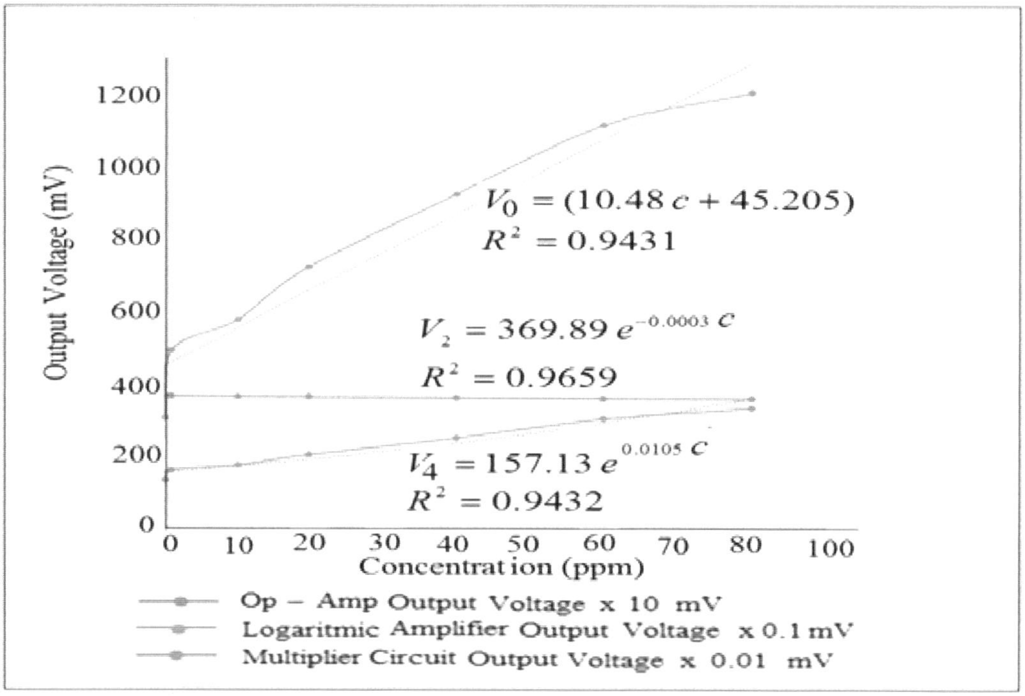

Figure 3. Voltage output of Op-Amp, Instrumentation and Logarithmic Amplifier as a function of Lead ion concentration.

4. Conclusion

Development of lead ion detection systems for concentration analysis of a lead ion is presented and demonstrated. The output voltage of the detector shows a linear relationship to the change of lead concentration in distilled water. The 1 x 4 tree asymmetric directional coupler sensor probes were used and it enables to detect the concentration of several heavy metal ions simultaneously. The results that can be drawn from this experimental were the lead ion sensor that has been developed can be used to detect lead ions with a sensitivity of 10.48 mV/ppm with the linearity 97.11% and had a measurement range of 0.1 ppm to 80 ppm.

Acknowledgment

This research was supported by grant of PTUPT from Ministry of Research, Technology, and Higher Education of The Republic Indonesia.

References

[1] Rahman H A, Harun S W, Yasin M and Ahmad H 2011 IEEE *J. Sel. Top. Quantum Electron.* **18** 5
[2] Yasin M, Harun S W and Ahmad H 2011 *J. Optoelectronics Adv. Mater.* **13** 8
[3] Yasin M, Harun S W, Yang H Z and Ahmad H 2010 *Optoelectron. Adv. Mater. Rapid Com.* **4** 8
[4] Yasin M, Harun S W, Abdul-Rashid H A, Kusminarto, Karyano and Ahmad H 2008 *Las. Phys. Lett.* **5** 55
[5] Yasin M, Harun S W, Pujiyanto, Ghani Z A and Ahmad H 2010 *Laser Phys.* **20** 10
[6] Ahmad H, Yasin M, Thambiratnam K, Harun S W 2012 *Sensor Review* **32** 3
[7] Chen Y, Zhang X, Zhang P and Liu C 2009 *Proc. of the 2009 IEEE Int. Conf. Infor. Autom.*

951-956
[8] Zhang X, Bai W and Chen Y 2007 *Proc. of the 2007 IEEE Int. Conf. Mechatron. Autom.* 3015
[9] Kim D H and Ilev I K 2010 *Electronics Letters* **46** 24
[10] Abdullah M, Bidin N, Yasin M, Pujiyanto, Harun S W and Shaharin M S 2014 *J. Optoelectron. Adv. Mater.* **16** 5
[11] Yasin M, Pujiyanto, Tanjung M and Apsari R 2015 *SPIE Proc. Photonics Opt. Apl.* **9444** 94440F
[12] Yasin M, Samian, Supadi, Pujiyanto and Yhuwana Y G Y 2016 *AIP Conf. Proc.* **1718(1)** 100003.

IOP Conf. Series: Journal of Physics: Conf. Series **997** (2018) 012007 doi:10.1088/1742-6596/997/1/012007

Local site effect of soil slope based on microtremor measurement in Samigaluh, Kulon Progo Yogyakarta

U N Prabowo[1], A F Amalia[1] and F E Wiranata[2]

[1] Department of Physics Education, Universitas Sarjanawiyata Tamansiswa, Kusumanegara No.157, Yogyakarta 55167, Indonesia
[2] Postgraduated of Physics Department, Universitas Gadjah Mada, Sekip Utara, Bulaksumur, Yogyakarta 55281, Indonesia

E-mail: urip.nurwijayanto@ustjogja.ac.id

Abstract. This paper investigated soil slope-local site effect of earthquake inducing landslide by using microtremor Horizontal to Vertical Spectral Ratio (HVSR) method. Microtremor measurements of 15 sites which were recorded for 45 minutes at each site were carried out in Ngargosari village, Samigaluh, Kulon Progo-Indonesia. Microtremor analysis using HVSR method was performed using Geopsy software. HVSR method resulted in predominant frequency values that ranges between 2,77 to 13,82 Hz and amplification factors varied from 0,46 to 5,70. The predominant frequency is associated with the depth of bedrock and the amplification factor reflects the geological condition of soil (sedimentary layer). The soil vulnerability index (Kg) varied from 0,08 to 5,77 and the higher value (Kg>3,4) in the south of the research area was identified as the weak zone of earthquake inducing landslide.

1. Introduction

The local site effect is a response to local geological conditions (the dynamic characteristic of surface sediment layers and rocks beneath) if there was an earthquake [1]. It can lead amplification of the ground motion during the earthquake [2]. The previous research has suggested that the local site effect became one of the determinants of the destruction level of a region due to an earthquake [3][4]. In addition to cause destruction to buildings, earthquakes which occurred on the slopes can also cause the landslide [5][6]. Earthquakes can induce soil/rocks movement which changes the slope condition susceptible to move into critical condition then it finally moves [7].

This research reviewed the slope local site effects in Ngargosari village, Samigaluh, Kulonprogo. It is a part of D.I Yogyakarta located in the southern of Java Island and close to encounter between the tectonic plate of Eurasia and Hindia-Australia which had high seismic activities. In the peak ground acceleration (PGA) of the bedrock map [8], the research area had PGA value between 0,2-0,3g.

The assesment of disaster mitigation (earthquake, landslide, tsunami etc) and city development plan can use geophysical and geological survey to achive the subsurface information of the area [9]. In this research, we used microtremor measurement as geophysical survey to determined the local site effects. The microtremor measurements was analyzed by using Horizontal to Vertical Spectral Ratio (HVSR) method [10] in order to obtain predominant frequencies, amplification and seismic vulnerability index value which described the slope local site effect and the potential of landslides. HVSR method is efficient method to determine the contact between the stable and unstable layer (bedrock) which become the slip field of a landslide [11].

2. Geological Setting of Research Area

The research took place in Ngargosari village, Samigalih, Kulon Progo district, Yogyakarta (Figure 1). It was downhill area in the east of Nglinggo village and part of Menoreh Mountains. Based on the geological map of Yogyakarta [12], the research area is a part of Kebobutak formation intruded by andesite rocks. Kebobutak formation is composed by andesitic breccia, tuff, lapilli, agglomerate, and the insertion of andesitic lava flows.

At the outcrops that were located around the measurement, it was seen that the surface layers were the soil weathering effects with the insertion of weathering andesitic breccia (Figure. 2 and Figure. 3).

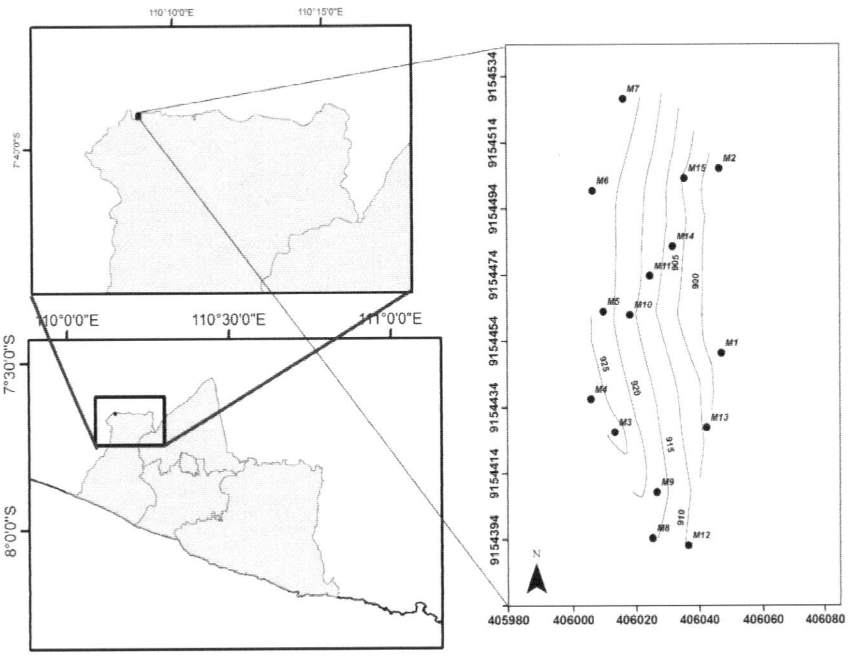

Figure 1. The location of research area.

Figure 2. The outcrops located around research..

Figure 3. Weathering stones.

3. Microtremor Measurements

The micrometer measurement of 3 components (vertical, North-South, and West-East) took place at the slopes of the hill about 15 points with 45 minutes of each point and 100 Hz for sampling rate. The measurements were executed in a quiet condition by avoiding the noise of human activities and other disturbances.

The results of micrometer maesurement were processed based on the HVSR method using Geopsy software by following the criteria of Sesame Project [13]. The data of each measured component was filtered by using a bandpass filter at the frequency of 1-20 Hz and hereafter a windowing process was done with window length about 10 s. The data of each window was transformed to frequency domain (spectrum) using Fast Fourier Transform method and the curve of transformation was smoothed by a smoothing process using Konno-Omachi method [14] with a constant value b=20. The spectrum results of each window in the vertical/V and horizontal components (North-South/NS and West-East/WE) were compared to the equation 1 [15].

$$HVSR = \frac{\sqrt{US + BT}}{V} \tag{1}$$

The result of the HVSR curve used was the mean value of all windows that described the predominant frequency and amplification of surface sediment layers.

4. Results and Discussion

4.1. *Predominant Frequency (f_0)*

Figure. 4 shows the predominant frequency value of the research area ranged between 2,77 – 13,82 Hz. The predominant frequency value associated with the topographic conditions [16] and the thickness of sediment layers (H) could be estimated based on the equation 2 [17].

$$H = \frac{V}{4f_0} \tag{2}$$

where V is the velocity of the secondary wave in sedimentary layer/soil. So, the lower predominant frequency values were associated with the deeper bedrock layers.

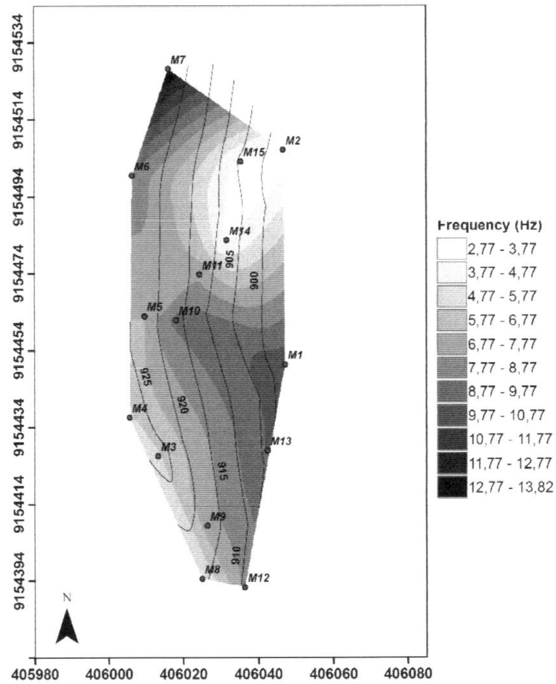

Figure 4. Predominant frequency values of research area.

4.2. *Amplification*

In Figure 5, the value of research area amplification ranged between 0,46 - 5,70. Figure 6 shows that there was no correlation between predominant frequency values to amplification so the values of amplification were not affected by the sediment thickness but they were affected by other factors. The factors that influence amplification values according to some previous researches include the impedance contrast between sediment and bedrock layers [15], the type of sedimentary layers such as shear strain, damping ratio, density[18], the water saturation of sedimentary layers [19], class of soil types [20], and quality factor of wave propagation [21].

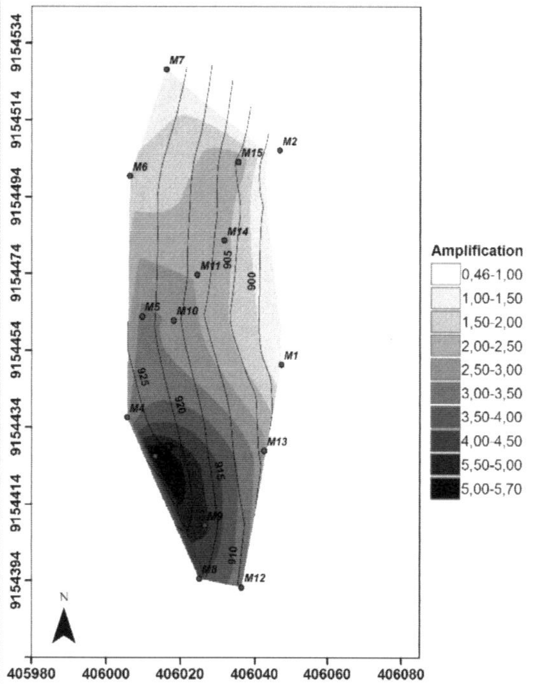

Figure 5. Amplification values of research area.

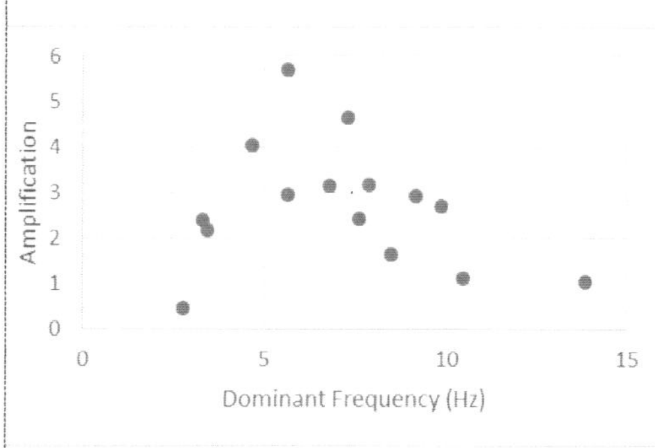

Figure 6. Graph of amplification values vs predominant frequency.

4.3. *Seismic Vulnerability Index (Kg)*

Seismic Vulnerability Index (*Kg*) of an area describes the level of soil damage due to local site effect [22]. Furthermore, the value of *Kg* also corresponds with the level of building destruction when the earthquake came through [23]. In Figure 7, *Kg* value of the research area was about 0,08-5,77.

The seismic vulnerability index could be correlated with the ground shear strain (γ) values describing the condition of soil dynamics when the earthquake came through [24]. The sediment layers would be inelastic and experiencing the surface deformation in the form of soil cracks which declined at γ = 1000 x 10^{-6}. The sediment layers were experiencing the massive deformation of landslides and liquefactions if γ = 10.000 x 10^{-6} [3]. The calculation of the ground shear strain used equation 3 [22].

$$\gamma = e\frac{A^2}{f_0}\frac{\alpha}{\pi^2 v_b} \tag{3}$$

With the assumption that PGA of bedrock (α) was maximum with the value of 0,3 g [8], shear wave velocity in bedrock layer (vb) = 600 m/s and the efficency of soil dynamic force (e) was 60% [22]. So, the value of γ = 1000 x 10^{-6} was equal to *Kg* > 3,4 and γ = 10.000 x 10^{-6} was equal to *Kg*>34.

The measurement point with *Kg*>3,4 was in the south of the research area. It was at M3 and M8 points. So, it could be deformed to be soil cracks and became weak zone of the landslide induced by the earthquake.

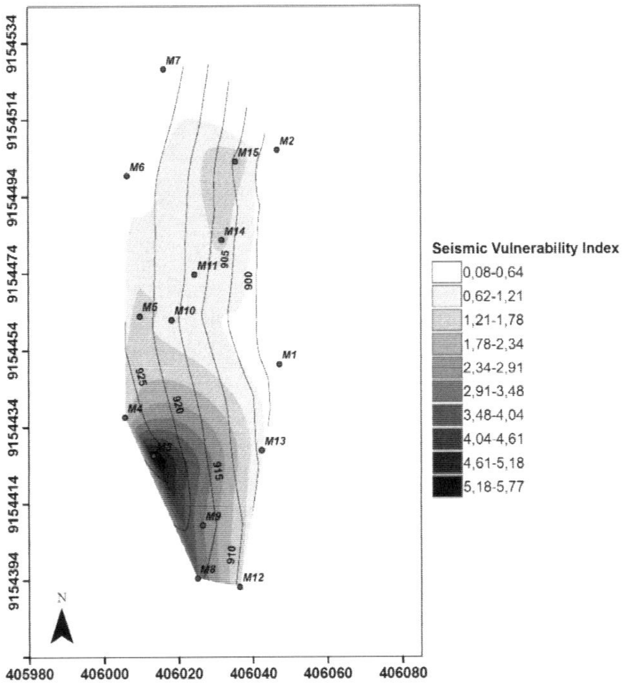

Figure 7. The seismic vulnerability index values of research area.

5. Microtremor Measurements

The predominant frequencies in the research area were varied between 2,77 – 13,82 Hz which correspond to the thickness of sedimentary layers. The value of amplification was between 0,46 – 5,70. It did not depend on the thickness of sediment but it was more affected by the geological condition of sedimentary layers and bedrock. The research area with the value of *Kg*>3,4 was a weak zone of the landslide induced by an earthquake.

Acknowledgements

Acknowledgements are given to DPRM DIKTI who has funded this research through PDP (Penelitian Dosen Pemula) scheme 2017

References

[1] Ngadmanto D, Susilanto P, Nurdiyanto B, Pakpahan S and Masturyono 2013 *J. Metereologi Geofisika* **14** 109
[2] Warnana D, Soemitro R and Utama W 2011 *Int. J. Basic Appl. Sci.* **11** 73
[3] Nakamura Y, Sato T and Nishinaga M 2000 *Proc. the Sixth Int. Con. on Seismic Zonation EERI*, (California: Palm Springs)
[4] Daryono, Sutikno, Sartohadi J, Dulbahri and Broptopuspito K 2009 Efek tapak lokal (local site effect) di Graben Bantul berdasarkan pengukuran mikrotremor *Int. Conf. Earth Science And Technology* (Yogyakarta) pp. E041-E046
[5] Karnawati D 2008 Earthquake induced landslide at Sengir, Prambanan district Yogyakarta *The Yogyakarta Earthquake of May 27, 2006* ed D Karnawati, S Pramumijoyo, R Anderson and S Husein (Yogyakarta: Star Publishing Company) part 1 pp 9.1-9.11
[6] Saputra A, Sartohadi J, Hadmoko D S and Gomez C 2015 *Forum Geografi* **29** 99
[7] Karnawati D 2007 *Dinamika Teknik Sipil* **7** 179

[8] Asrurifak M, Irsyam M, Budiono B, Triyoso W and Hendriyawan 2010 *Civil Engineering Dimension* **12** 52
[9] Bahri S and Madlazim 2012 *J. Penelit. Fis. Apl.* **2** 23
[10] Nakamura Y 1989 *Quarterly report of Railway Technical Research Institute* **30** 25
[11] Meric O, Garambois S, Malet J P, Cadet H, Gueguen P and Jongmans 2007 *Bul. Soc. Geol.* **178** 137
[12] Rahardjo W, Sukandarrumidi and Rosidi H 1995 *Peta geologi lembar Yogyakarta* (Bandung: Pusat Penelitian dan Pengembangan Geologi, Badan Geologi Indonesia)
[13] Sesame 2004 *Guidelines for the implementation of the H/V spectral ratio technique on ambient vibrations measurements, processing and interpretation* (SESAME Europan Research Project : European Commision Research General Directorate)
[14] Konno K and Ohmachi T 1998 Ground motion characteristics estimated from spectral ratio between horizontal to vertical components of microtremor *Bul. Seismological America* **88** 228
[15] Nakamura Y 2000 Clear identification of fundamental idea of Nakamura's technique and its applications *Proc XII World Conf. Earthquake Eng.* pp 1-8
[16] Soemitro R A A, Warnana D D, Utama W and Asmaranto R 2011 *J. Basic App. Scientific Res.* **1** 412
[17] Seht I and Wohlenberg J 1999 Microtremor used to map thickness of soil *Bul. Seismological Soc. America* **89** 250
[18] Wang S and Hao H 2002 *Soil Dynamics and Earthquake Eng.* **22** 551
[19] Jun Y 2006 Frequency-Dependent amplification of unsaturated surface soil layer *J. Geotech Geoenvironmental Eng.* **132** 526
[20] Partono W, Irsyam M, Prabandiyan R W S and Maarif S 2013 *Media Komunikasi Teknik Sipil* **19** 125
[21] Prabowo UN, Marjiyono and Sismanto 2016 *Science Tech* **2** 112
[22] Nakamura Y 1997 Seismic vulnerability indices for ground and structures using microtremor *World Congress on Railway Research* (Florence, Italy) pp 1-7
[23] Sungkono and Santosa B 2011 *J. Neutrino* **4** 1
[24] Ishihara K 1982 Evaluatian of soil properties for use in earthquake response analysis *Proc. Int. Symp. On Numerical Model in Geomech* (Zurich) pp 273-259

IOP Conf. Series: Journal of Physics: Conf. Series **997** (2018) 012008 doi:10.1088/1742-6596/997/1/012008

The effectiveness of collaborative problem based physics learning (CPBPL) model to improve student's self-confidence on physics learning

B K Prahani[1], N Suprapto[1], Suliyanah[1], N A Lestari[1], M N R Jauhariyah[1], S Admoko[1] and S Wahyuni[2]

[1]Department of Physics, Faculty of Mathematics and Natural Science, Universitas Negeri Surabaya, Surabaya 60231, Indonesia
[2]Department of Physics Education, Faculty of Training and Teacher Education Jember University, Jember 68121, Indonesia

E-mail: binarprahani@gmail.com

Abstract. In the previous research, Collaborative Problem Based Physic Learning (CPBPL) model has been developed to improve student's science process skills, collaborative problem solving, and self-confidence on physics learning. This research is aimed to analyze the effectiveness of CPBPL model towards the improvement of student's self-confidence on physics learning. This research implemented quasi experimental design on 140 senior high school students who were divided into 4 groups. Data collection was conducted through questionnaire, observation, and interview. Self-confidence measurement was conducted through Self-Confidence Evaluation Sheet (SCES). The data was analyzed using Wilcoxon test, n-gain, and Kruskal Wallis test. Result shows that: (1) There is a significant score improvement on student's self-confidence on physics learning (α=5%), (2) n-gain value student's self-confidence on physics learning is high, and (3) n-gain average student's self-confidence on physics learning was consistent throughout all groups. It can be concluded that CPBPL model is effective to improve student's self-confidence on physics learning.

1. Introduction

Self-confidence is one of the most important aspects of student learning and motivation [6,7]. Self-confidence refers to the expectations and beliefs for success in one's life [6, 8] and influences the success of student learning processes and outcomes [9-13]. Self-confidence that learners have the belief that there is a chance to succeed and make them motivated [13]. Self-confidence is explicitly contained in Permendikbud No. 20 of 2016 on the competency standards of secondary education graduates that students are required to have self-confidence in interacting effectively with the surrounding social and natural environment in placing itself as a reflection of the nation in the association of the world [14]. It shows the learning of physics in senior high school as a medium to increase self-confidence.

The expectation of the importance of self-confidence is contrary to the reality in Indonesian education. The results of preliminary studies show that self-confidence in senior high school students in Indonesia is generally low in the process of physics solving problems [15, 16]. Supported by TIMSS study results, Indonesia has the lowest level for student's self-confidence in subject physics, chemistry, biology, and earth science [17]. The results of previous research indicate a serious problem related to the self-confidence of senior high school students. Therefore the need to increase student's self-confidence in physics solving problems, especially in senior high school physics level.

In the previous research, Collaborative Problem Based Physic Learning (CPBPL) has been developed to improve student's self-confidence on senior high school [16]. The CPBPL model has been specially designed to increase the self-confidence of senior high school students in physics problems solving. CPBPL model has 5 phases, namely: (1) motivation and problem orientation; (2)

collaborative problem solving activities; (3) presents; (4) non-routine problem solving; and (5) evaluation. The previous research developed a device of learning physics as an operational form of CPBPL model developed. CPBPL model learning tool is a novelty based on the need for current and future. This is in accordance with the novelty of [18] that "there is a need for the intervention and its design is based on state-of-the-art knowledge." Before developing the learning tool, analysis, and literature review to obtain features as an improvement of needs, difficulties and weaknesses in the field [19]. The learning tools developed have key features in CPBPL model, namely (1) integrating teacher books and warp student books, (2) basic and integrated science process skills, (3) student cantered learning, (4) collaborative problem solving skills and self-confidence is emphasized in physics learning. The CPBPL model and learning tools developed by researchers have fill the needs and demands of the curriculum in Indonesia and 21st Century skills. Therefore, it is hoped that the implementation of the CPBPL model that have been developed are expected to contribute to improving student's quality in Indonesia.

A model is said to be of high quality when it meets the criteria of validity, practicality, and effectiveness [20]. In previous research has been developed CPBPL model that proved valid to increase self-confidence of high school students in physics learning [21]. The purpose of this research is to analyze the effectiveness of CPBPL model to increase the self-confidence confidence of senior high school students in physics learning. The focus of the problem in this study included: (1) whether there was a significant increase (statistically) of student's self-confidence before and after the CPBPL model was applied, (2) how much level of confidence increased before and after applied CPBPL model there is an average difference in the level of student's self-confidence after learning with CPBPL model in all four groups.

2. Methodology of Research

2.1. General Background of Research

This research was conducted at Senior High School State 19 Surabaya and Senior High School State 1 Tegaldlimo, Indonesia from October to December 2016. The scope of this research is on X Class senior high school students taking physics learning in academic year 2016/2017. Basic physics materials used in physics learning in class include (1) Heat: Effect of heat on changes in temperature of objects; the influence of heat on the change of being, and Black's principle. (2) Elasticity material: (a) material elasticity, (b) Hooke's law, and (c) spring series. This research is emphasized on the analysis of the fulfilment of the effectiveness of CPBPL model by analysing the improvement of student's self-confidence before and after following the process of problem solving on physics learning with CPBPL model. The effectiveness of the CPBPL model was determined based on a significant increase in scores (statistically) between pre-test and post-test of student's self-confidence, as well as the mean of n-gain determined by criteria: low, medium and high.

2.2 Sample of Research

The sample in this study were 140 students of Senior High School State 19 Surabaya and Senior High School State 1 Tegaldlimo, Indonesia; which is in the four groups, namely: group-1 (students of class X1 Senior High School State 19 Surabaya), group-2 (students class X2 Senior High School State 19 Surabaya), group-3 (students of class X1 Senior High School State 1 Tegaldlimo) 4 (students of class X2 Senior High School State 1 Tegaldlimo). Each group consists of 30 students who take physics learning in academic year 2016/2017.

2.3 Instrument and Procedures

Self-Confidence Evaluation Sheet (SCES) is an instrument used to measure student's self-confidence in the process of physics problems solving adapted from Problem Solving Confidence Questionnaire [22]. Self-Confidence Evaluation Sheet (SCES) includes self-confidence indicators in this research, namely motivation, ability, and perseverance [23]. Basic Physics materials used in physics learning include: (1) Heat: Effect of heat on changes in temperature of objects; the influence of heat on the change of object, and Black principle. (2) Elasticity material: a) Material elasticity, b) Hooke's Law, and c) spring series. This research uses one group pretest-posttest design that is O1 X O2 [24]. The learning process begins by giving pre-test (O1). Each student is required to complete the Self-Confidence Evaluation Sheet (SCES). After the pre-test, the teacher applies CPBPL model and learning tool in each group (X). Learning tools are specifically designed to be integrated with self-confidence indicators consisting of: motivation, ability, and perseverance at each learning phase. The process of physics learning ends with post-test (O2). Every student is required to post-test of self-confidence by filling Self-Confidence Evaluation Sheet (SCES).

2.4 Data Analysis

Student's self-confidence is analysed based on the assessments obtained by students before and after learning using the CPBPL model. The pre-test, post-test, and n-gain data of student's self-confidence were further analysed using inferential statistical tests with the help of SPSS and reinforced by qualitative descriptive analysis. The score level for self-confidence is based on indicators of motivation, ability, and perseverance [22, 23]. The criteria for achieving the score for self-confidence are as follows. $0 \leq x \leq .2$: Very Low; $.2 < x \leq .4$: Low; $.4 < x \leq .6$: Moderate; $.6 < x \leq .8$: Height; $.8 < x \leq 1$: Very High. The effectiveness of the CPBPL model is determined by: (1) there is a statistically significant increase in the level of significance, α = 5%, (2) the level of increase determined by the normalized gain value (n-gain) at least moderate, and (3) the increase in the four groups statistically did not differ significantly. The n-gain value is determined by the equation: n-gain = (score post-test - score pre-test) / (maximum score - pre-test score) [25]. According to the following criteria: (1) if n-gain $\geq .7$ (high), (2) if $.3 <$ n-gain $< .7$ (moderate), and (3) if n-gain $\leq .3$ (low).

3. Result of Research

The learning outcomes of all groups related to the self-confidence are presented in Figures 1 and Table 1. Black bars represent the mean of pre-test, shape bar scores represent the mean post-test scores, and vertical bar scores represent the n-gain scores. Figure 1 shows the average post-test scores of self-confidence in the Physics learning for all groups is greater than the pre-test score. The average pre-test, post-test, and n-gain scores associated with self-confidence indicators for all groups are presented in detail in Table 1. Figure 1 shows the average n-gain value of self-confidence for group-1, group-2, group-3, and group-4 are respectively .72; .74; .71; and .71. The average n-gain value of self-confidence for all groups is in the high category.

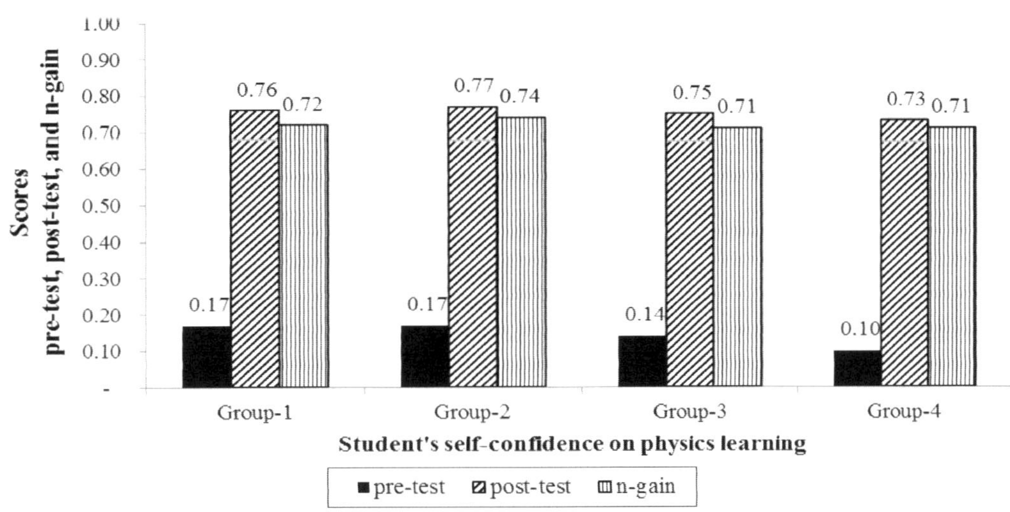

Figure 1. The average pre-test, post-test, and n-gain scores of self-confidence in all groups.

IOP Conf. Series: Journal of Physics: Conf. Series **997** (2018) 012008 doi:10.1088/1742-6596/997/1/012008

Table 1. The average score of pre-test, post-test and n-gain of self-confidence in all groups.

School	Groups	Scores	Self-confidence Indicator		
			Motivation	Ability	Perseverance
Senior High School State 19 Surabaya	1	Pre-test	.13	.14	.25
		Post-test	.70	.76	.83
		n-gain	**.66**	**.72**	**.77**
	2	Pre-test	.12	.13	.25
		Post-test	.77	.75	.80
		n-gain	**.74**	**.71**	**.73**
Senior High School State 1 Tegaldlimo	3	Pre-test	.13	.12	.17
		Post-test	.75	.71	.80
		n-gain	**.71**	**.67**	**.76**
	4	Pre-test	.11	.11	.09
		Post-test	.73	.70	.77
		n-gain	**.70**	**.66**	**.74**

Table 1 shows that the self-confidence test scores of each indicator include: motivation, ability, perseverance are low for all groups and the post-test score of self-confidence in each indicator is high for all groups. The n-gain of self-confidence score of each indicator includes: motivation, ability, and perseverance in all groups are moderate/high.

Table 2. Wilcoxon test Result of self-confidence for all groups.

School	Group	Data	N	Z	p
Senior High School State 19 Surabaya	1	*Pretest-Posttest*	35	-4.11	< .01
	2	*Pretest-Posttest*	35	-4.11	< .01
Senior High School State 1 Tegaldlimo	3	*Pretest-Posttest*	35	-4.11	< .01
	4	*Pretest-Posttest*	35	-4.11	< .01

p < .05 (2-tailed)

Table 2 shows the improving in pre-test and post-test self-confidence tested using Wilcoxon test. The Z score gives a value of -4.11; -4.11; -4.11; -4.11 for group-1; group 2; group 3; group 4. Each score is considered significant, because $p < .05$. Because Z the result of the calculation is negative so it shows that there is an increase of high school students after applied learning with CPPBL model for all groups. Furthermore, the consistency of the impact of the implementation of the CPPBL model on improving self-confidence was analyzed using the Kruskal-Wallis test shown in Table 3.

Table 3. Kruskal-Wallis test Result of self-confidence for all groups.

	N	Chi square	df	p
Group 1, group 2, group 3, group 4	140	2.17	3	.54

Table 3 shows the significance value $p = .54 > .05$. This clearly indicates that there is no difference in the increase the self confidence after using the CPPBL model for all groups. This indicates a consistent increase in self confidence in all groups after using the CPPBL model.

4. Discussions

A more complete explanation concerning the effectiveness of the CPPBL model for enhancing student's self-confidence is described as follows. The effectiveness of the CPPBL model on improving student's self-confidence in physics learning can be seen from: (1) an increase in pre-test and post-test scores, (2) n-gain self-confidence value, and (3) (consistent), as shown in Figure 1, Table 1, and Table 2. Before the CPPBL model is applied; students less master the self-confidence, average score of students is under the standard score (minimum score .20 in score range 0-1), that is the average score of self-confidence for group-1, group-2, group-3, and group-4 respectively are .17; .17; .14; and .14. All this time, students are not used to motivation, ability, and perseverance. This is supported by [15, 16] that the self-confidence of high school students in Indonesia is generally still low in the process of physics problems solving. Supported with TIMSS study result, Indonesia has lowest level for student's self-confidence in subject physics, chemistry, biology, and earth science [17].

Conversely, after the CPPBL model was applied to the Physics learning, the mastery of student's self-confidence rises above average and becomes high; the average score of scientific self-confidence for groups-1, group-2, group-3, and group-4 respectively to .76; .77; .75; .73 (well beyond the minimum score of .70 in the 0-1 score range). The increase in self-confidence is allegedly influenced by the role of the student's self-confidence as the scenario in phase of the CPPBL. Each phase emphasizes self confidence in collaborative physics problem solving activities. Self-confidence refers to the expectations and beliefs for success in one's life [6, 7] and influences the success of student learning processes and outcomes [9-12]. Supported with other research results by [26] that problem-solving group has the highest learning outcomes compared to individual and competition learning.

The increase of self-confidence score in all groups was significant and did not differ (consistent) at the 5% significance level with n-gain of .72 for group-1; .74 for group-2; .71 for group-3; and .71 for group-4. This means that there is an increase in student's self-confidence after the application of physics learning with CPPBL model. Increased self-confidence is supported by the availability of self-confidence learning needs as suggested in phase of the CPPBL model. Students must work collaborative and are interdependent positive to optimize all their knowledge in designing and implementing problem-solving strategies. It supported with theory self-regulated learning [27]. Self-confidence is explicitly contained in Permendikbud Article 20 of 2016 on the competency standards of primary and secondary education graduates that students are required to have self confidence in interacting effectively with the surrounding social and natural environment in placing itself as a reflection of the nation in the association of the world [14]. Based on the above description, the implementation of CPPBL model is able to develop the role of student's self confidence in collaborative problem based physics learning.

The level of self-confidence improvement for all groups due to the application of learning with the CPPBL model is significant at 5% real level and consistent in the high category. It is suspected because a teacher device that has been valid supports it. In addition, the increase in self-confidence is strongly influenced by the success of phase of CPPBL model. Teachers should be able to increase student's self-confidence by providing many varied and challenging experiences that can improve learning success. The importance of collaborative investigation according to the John Dewey flow of problem solving [28] that the class should be a laboratory for solving real-life problems [29-30]. It is based on cognitive constructivist theory by Piaget (1954, 1963), any student of any age actively involved in the process of obtaining information and constructing their own knowledge [28]. Students should be actively involved in individual problem solving before contributing in collaborative problem solving. Based on the above description, the implementation of CPPBL model is able to develop the role of student's self confidence in physics learning.

5. Conclusions
Conclusion of this research shows that: (1) There is a significant score improvement on student's self-confidence on physics learning (α=5%), (2) n-gain value student's self-confidence on physics learning is high, and (3) n-gain average student's self-confidence on physics learning was consistent throughout all groups. It can be concluded that CPBPL model is effective to improve student's self-confidence on

physics learning. The implication of this research is that CPPBL model can be used as an alternative to overcome the low of student's self-confidence on physics learning. To strengthen the result of this research, it is necessary to do further research in various education levels and countries.

References

[1] Shubert W C and Meredith C D 2015 *Phys. Rev. ST Phys. Educ. Res.* **11** 020138
[2] Teodorescu E R, Bennhold C, Feldman G and Medsker L 2013 *Eur. J. Phys. Educ.* **5** 15
[3] Teodorescu E R, Bennhold C, Feldman G and Medsker L 2014 *Phys. Rev. ST Phys. Educ. Res.* **9** 010103
[4] Sawtelle A, Brewe E, Goertzen M R and Kramer H L 2012 *Phys. Rev. ST Phys. Educ. Res.* **8** 020111
[5] Mason A and Singh C 2010 *Phys. Rev. ST Phys. Educ. Res.* **6** 020124
[6] Dowd E J, Araujo I and Mazur E 2015 *Phys. Rev. ST Phys. Educ. Res.* **11** 010107
[7] Keller M J 2010 *Motivational Design for Learning and Performance the ARCS Model Approach* (USA: Springer)
[8] Ackerman R and Zalmanov H 2012 *Psychol. Bull. Rev.* **19** 1187
[9] Roediger H L and Karpicke J D 2006 *Psychol. Scie.* **17** 249
[10] Kornell N, Hays M J and Bjork R A 2009 *J. Exp. Psychol.: Learn. Memory Cog.* **35** 989
[11] Richland L E, Kornell N and Kao L S 2009 *J. Exp. Psychol.: Applied* **15** 243
[12] Chasteen V S, Wilcox B, Caballero D M, Perkins K K, Pollock J S and Carl 2015 *Phys. Rev. Spec. Topics. Phys. Educ. Res.* **11** 02011
[13] Cheng C Y and Yeh T H 2009 *British J. Educ. Tech.* **40** 597
[14] Kemdikbud 2016 *Salinan Lampiran Peraturan Menteri Pendidikan dan Kebudayaan Nomor 20 Tahun 2016 Tentang Standar Kompetensi Lulusan Pendidikan Dasar dan Menengah* (Jakarta: Kemdikbud)
[15] Prahani B K, Yuanita L dan Soegimin W W 2015 *J. Pembelajaran Pendidik. Sains* **4** 503
[16] Prahani B K 2017 *Model Collaborative Physics Problem Based Learning (CPPPBL) untuk Meningkatkan Keterampilan Pemecahan Masalah Kolaboratif, Keterampilan Proses Sains, dan Kepercayaan Diri Siswa SMA* (Surabaya: Unesa)
[17] Martin M O, Mullis I V, Foy P and Stanco G M 2012 *TIMSS 2011: International Science Report* (Boston: TIMSS and PIRLS International Study)
[18] Nieveen N, McKenney S and van Akker 2007 *Educational Design Research* (New York: Routledge)
[19] Prahani B K, Yuanita L dan Nur M 2016 Analisis dan Solusi Penyempurnaan Buku Fisika Kelas X SMA/MA Kurikulum 2013 untuk Mendukung Implementasi Kurikulum Nasional *Pros. Semin. Nas. Universitas Negeri Surabaya* **4th** (Surabaya: Unesa) 197
[20] Plomp T and Nieveen N 2013 *Educational Design Research* (Enschede: SLO)
[21] Prahani B K, Nur M dan Yuanita L 2016 Validitas Model *Self Confidence Collaborative Problem Solving Pros. Semin. Nas. Universitas Negeri Surabaya* **4th** (Surabaya: Unesa) 192
[22] Gok T 2012 *Lat. Am. J. Phys. Educ.* **6** 21
[23] Mafakheri S, Malkhalifeh R M, Shahvarani A and Behzadi H M 2013 *Math. Educ. Trends. Res.* **5** 1
[24] Fraenkel J R, Wallen N E and Hyun H H 2012 *How to Design and Evaluate Research in Education* (New York: McGraw-Hill)
[25] Hake R R 1998 *Am. J. Phys.* **66** 64
[26] Kirschner F, Paas F, Kirschner A P and Janssen J 2011 *Learn. Instr.* **21** 587
[27] Moreno R 2010 *Educational Psychology* (New York: John Wiley & Sons Inc)
[28] Arends R I 2012 *Learning to Teach* (New York: Mc. Graw-Hill Companies)
[29] Budiningarti H, Suprapto N and Admoko S 2015 *J. Penelit. Fis. Apl.* **5** 56
[30] Azizah R, Yuliati L and Latifah E 2015 *J. Penelit. Fis. Apl.* **5** 44

Synthesis of polyaniline/cellulose composite as humidity sensor

N P Putri[1], D H Kusumawati[1], N Widiyanti[1] and Munasir[1]

[1]Department of Physics, Faculty of Mathematics and Natural Science, Universitas Negeri Surabaya, Surabaya 60231, Indonesia

E-mail: nugrahaniprimary@unesa.ac.id

Abstract. Water hyacinth is one weed plant that has cellulose content of 60% on the stem and is a good absorbent. in this study cellulose extraction from hyacinth has been done through several stages. Polyaniline/cellulose composite (PANi/cellulose) is prepared by an in-situ chemical method using cupric sulphate as an initiator. The representative PANi/cellulose samples are characterized by Fourier Transform Infrared (FTIR). On comparing it appears that spectra PANi/cellulose contains vibrational bands due to both PANi and cellulose. This may indicate the formation of PANi/cellulose composite. From the resistance measurement results, it can be seen that with the addition of cellulose to PANi can improve the sensitivity of the polyaniline based moisture sensor

1. Introduction

Humidity can affect the chemical, physical and biological process that occurs in everyday life. For example, food and medicines stored in a container require a certain level of moisture in order not to break down quickly. Human comfort is also influenced by humidity somewhere. Humans will feel less comfortable in places with high humidity because the high humidity will trigger the breeding of microorganisms that carry disease germs. However, low humidity (less than 20%) will also cause drying of membrane mucous membranes, therefore the human body needs an environment with sufficient moisture levels. Commercially available humidity sensors are fabricated by conventional sensing materials like ceramics [1] and electrolytic metal oxides. Organic materials due to light weight, flexibility, low cost and large surface area have developed the great interest for their use in humidity sensors. The humidity sensors based on organic sensing materials are classified into capacitive, resistive, optic [2], mechanical, and thermoelectric type of sensors depending on basic sensing principle [3].

Polyaniline is one of the most widely applied conductive polymers due to its highly regulated, stable and easily produced. The use of polyaniline as a humidity sensor has been carried out by previous researchers. L. Qianqian et.al which synthesize polyaniline nanofiber to be used as a sensor for pavement. From this research, it is found that nanofiber polyaniline can be applied to good humidity sensor, and its sensitivity increases with the addition of PEO which is hydrophilic [4]. Milind V.K, et al also uses polyaniline as a humidity sensor by ink-printed technique to obtain a flexible humidity sensor [5]. Pablo Cavallo et.al report the mechanism of resistivity changes on humidity sensors made by polyaniline [6].

Cellulose is one of the biopolymers found in plants, such as water hyacinth, which has cellulose content of 60% on its stem [7]. One of the most common characteristics of cellulose is hydrophilicity which is an important property to develop humidity sensors due to better adsorption sites for H_2O molecules [8]. The present study reports the grafting of cellulose onto polyaniline chain to improve its adsorption. The electrical resistance of PANi-cellulose film on PCB substrate was monitored under controlled humid environment to explore its use for the humidity sensor.

2. Experimental Procedures

2.1. Extraction of cellulose

Cellulose extraction from water hyacinth consists of several stages, the first stage cleaning and drying stem of water hyacinth. A dried water hyacinth then cut, blended and sieved to obtain a homogeneous fibre size. The second stage is the dewaxing process using soxhlet apparatus. The result of the dewaxing process then soaked with $NaClO_2$ and allowed to stand for 3 hours in a water bath at 80°C. This process repeated up to 2 times then washed with distilled water. The next step is a process of hemicellulose removal, where the precipitate reacted with NaOH and soaked in the water bath for 4 hours. Thereafter, the precipitate was immersed in the $NaClO_2$ solution for 3 hours. To obtain pure cellulose fibre, the precipitate was immersed in NaOH for 4 hours at 60°C. Pure cellulose fibres soaking results then filtered and washed and then dried in the sunlight for 2 days.

2.2. Polymerization of Polyaniline (PANi)

Polyaniline was obtained by the chemical oxidative polymerization process. The oxidative polymerization reaction of aniline occurred by adding oxidant monomers like ammonium persulfate (APS) to aniline dissolved in an acid solution. For the aniline aqueous solution, 1.82 ml aniline was dissolved in 50 ml of 1 M HCl doping acid. For the oxidant solution, APS (5.71 gr) was dissolved in 50 ml of 1 M HCl doping acid. The oxidant solution was quickly poured into the aniline solution at room temperature followed by immediate magnetic stirring. The stirring was stopped after 2 h, and the solution left undisturbed for 24 h. The green precipitates were filtrate with filler paper and wash with distilled water. The product was finally dried out for FTIR characterization.

2.3. Synthesis of PANi/cellulose composite

Aniline (1.82 ml) and cellulose (0.3; 0.4; 0.5 g) were dissolved in 50 ml of 1 M HCl solution. For the oxidant solution, APS (5.71 gr) was dissolved in 50 ml of 1 M HCl doping acid. The oxidant solution was then quickly poured into the aniline solution at room temperature followed by immediate magnetic stirring. In resultant solution, 5 ml 1 M $CuSO_4$ was slowly added with continuous stirring for 1 h. The resultant solution was then left for 3 – 4 h and a dark-green precipitate was obtained. The precipitate was filtered then washed with distilled water and dried for characterization.

2.4. Characterization

FTIR spectra were recorded with Shimadzu FTIR spectrometer. Change in resistance of PANi/cellulose was measured with CD800a Sunwa digital multimeter. In order to evaluate the humidity-sensing behaviour of PANi/cellulose, a film was coated on PCB surface by spin coating technique after making 3% solution of PANi/cellulose in m-cressol.

3. Result

3.1. Spectra study

FTIR spectra of PANi, cellulose and PANi/cellulose composite are shown in Figure 1. Polymerization of PANi has been successfully performed in emeraldine salts (ES) form, which has quinoid and benzenoid groups which are characteristic of polyaniline (Table 1). Cellulose extracted from water hyacinth has the -O- bond which is one of the characteristic attributes of cellulose, so it can be concluded that water hyacinth is one of the cellulose sources found in nature. On comparing it appears that spectra PANi/cellulose contains vibrational bands due to both PANi and cellulose. Further, the peak appeared in PANi spectra at 1640 cm^{-1} shifted at 1637 cm^{-1} in the composite. Otherwise, IR peak of cellulose appearing in spectra at 1317 cm^{-1} and 1157 cm^{-1} shifted at 1315 cm^{-1} and 1155 cm^{-1} in PANi/cellulose composite. These may indicate the formation of PANi/cellulose composite [8].

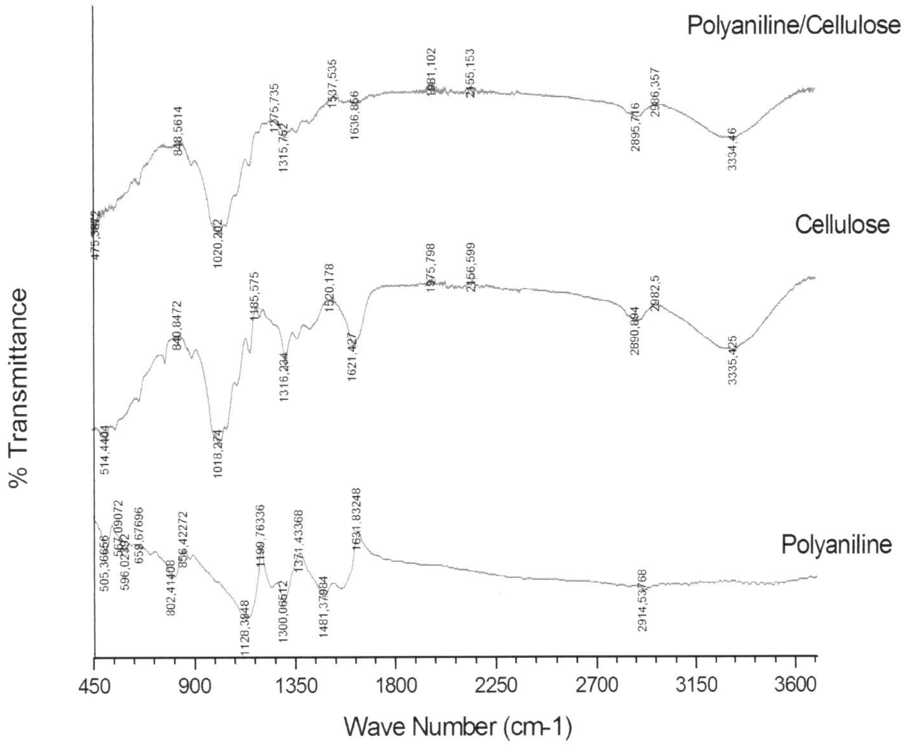

Figure 1. IR-spectra for PANi, cellulose and PANi/cellulose.

Table 1. Bond type for PANi, cellulose and PANi/cellulose.

No	Wave number (cm^{-1})				Bond type
	Data sheet	PANi	Cellulose	PANi/Cellulose	
1	1650-1560	1640.44	1621,52	1636,91	C=C stretch (Q)
2	1500-1400	1463.00	-	-	C=N stretch (Q),
3	1400-1300	-	1316,60	1315	–O- bonding
4	1335-1250	1298.00	-	-	C-H bending (Q)
5	1250-1020	1224.55	1157,28	1155	C-N stretch,
		1124.18	-	1020,21	C-C stretch,
					C-H bending (B)
6	850-550	593.00	-	551,93	C-H bending

3.2. Humidity sensing behaviour

The measurement of humidity sensing behaviour is done in a confined space with a saturated salt solution [9]. The variation in resistance of PANi, cellulose and PANi/cellulose with an increase in humidity is shown in Figure 2. The higher the relative humidity percentage, the lower the resistance value obtained. This indicates there is an interaction between the PANi/cellulose film with the moisture content in the air. With increasing moisture in the air, more water molecules are absorbed to the film surface (swelling effect). The water molecules bonded in the film will be ionized into H + and OH- ions. This will decreases the resistivity of the film [8]. In this study, the electrical resistance of the film decrease can be caused by the pores, initially filled with dry air, are progressively filled with water which has a higher dielectric constant than air, although there is no supportive morphological data.

From Figure 2 it can be seen that there is a decrease in PANi resistivity, from 1320 Ω (at 64% RH) to 100 Ω (at 80% RH) and has a good linearity. Cellulose is an insulator material that has much greater resistance when compared with polyaniline (4520 Ω at 64% RH). A decrease in resistance can be caused by an increase in water molecules binding to the -O- group on cellulose. With the addition of varying cellulose, it can be seen that the more cellulose is added, the lower the resistivity of the PANi/cellulose composite (when measured at 64% RH). With increasing humidity, the decrease in resistance is no longer linear but forms a polynomial curve. From humidity measurements of 64% to 69%, a decrease in resistance is very far when compared to humidity measurements at other points. Decreasing resistance at humidity measurements from 74% to 80% RH is lower than other points, since saturation has occurred, where the water molecules attached to the sample have reached the maximum. Grafted cellulose that has strong hygroscopic properties will improve the hydrophilic properties of PANi. The more cellulose that is grafted onto the PANi the more matrix as the absorbent of water vapour molecules. This may cause decreased resistivity of PANi/cellulose.

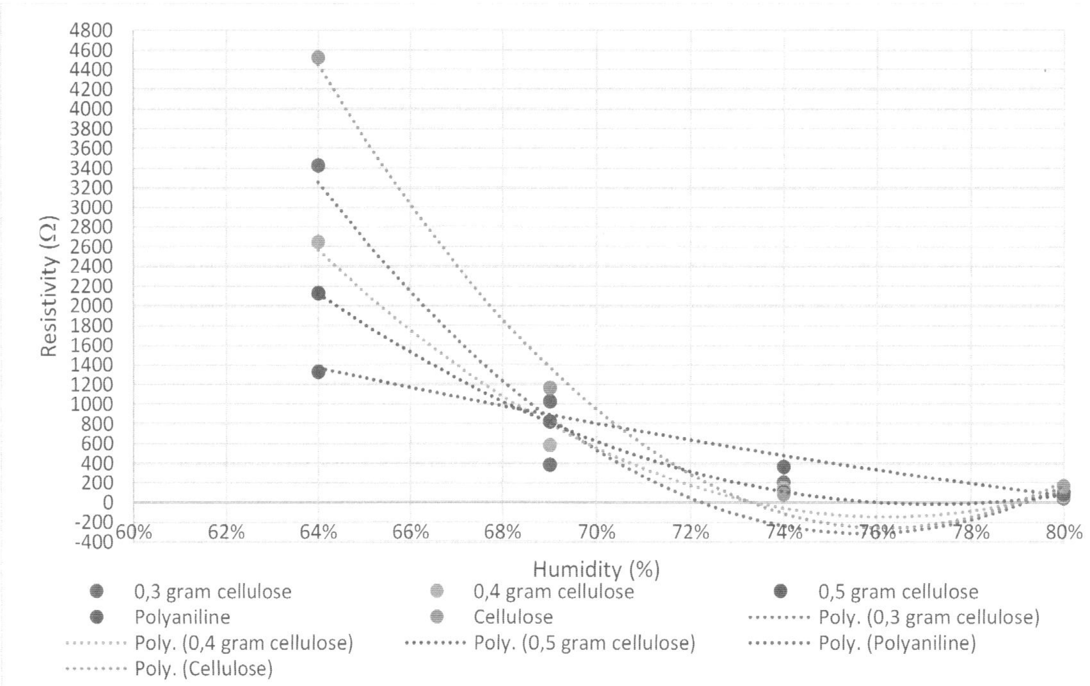

Figure 2. Change in resistance of PANi, cellulose and PANi/cellulose composite.

4. Conclusion

From this research can be concluded that cellulose has been successfully grafted onto PANi as seen in IR spectra. From a measurement of humidity sensing behaviour can be seen that PANi/cellulose composite can be used as humidity sensor material.

IOP Conf. Series: Journal of Physics: Conf. Series **997** (2018) 012009 doi:10.1088/1742-6596/997/1/012009

References

[1] Parthibavarman M, Hariharam V and Sekar C 2011 *J. Mater. Sci. Eng. C* **31** 840
[2] Maddu A, Modjahidin K, Sardy S and Zain H 2006 *J. Makara Teknol.* **10** 45
[3] Cahni M T S, Karimof K S, Khalid F A and Moiz S A 2013 *J. Solid State Sci.* **18** 78
[4] Lin Q, Li Y and Yang M 2012 *J. Sensors Actuators, B Chem.* **161** 967
[5] Kulkarni M V, Apte S K, Naik S D, Ambekar J D and Kale B B 2013 *J. Sensors Actuators, B Chem.* **178** 140
[6] Cavallo P, Acevedo D F, Fuertes M C, Soler-Illia G J A A and Barbero C A 2015 *J. Sensors Actuators, B Chem.* **210** 574
[7] Putera R D H 2012 *Ekstraksi Serat Selulosa Dari Tanaman Eceng Gondok (Eichornia Crassipes) Dengan Variasi Pelarut* (Depok: Universitas Indonesia)
[8] Shukla S K 2012 *J. Eng. Mater. Sci.* **19** 417
[9] Mandela R and Muttaqin A 2016 *J. Fis. Unand.* **5** 2

IOP Conf. Series: Journal of Physics: Conf. Series **997** (2018) 012010 doi:10.1088/1742-6596/997/1/012010

Analysis of mineral content of iron sand deposit in Bontokanang Village and Tanjung Bayang Beach, South Sulawesi, Indonesia

V A Tiwow[1], M Arsyad[1], P Palloan[1] and M J Rampe[2]

[1]Department of Physics, Universitas Negeri Makassar, Kampus UNM Parangtambung, Makassar 90224, Indonesia
[2]Department of Chemistry, Universitas Negeri Manado, Kampus UNIMA, Tondano 95618, Indonesia

E-mail: vistatiwow@unm.ac.id

Abstract. Iron sand found in coastal areas in Sulawesi Selatan has not been optimally utilized. In this study, to analysis the mineral content of iron sand deposit, extraction, and characterization process were conducted. Iron sand was obtained from Bontokanang Village Takalar district and Tanjung Bayang Beach Makassar. Chemical composition analysis was performed of on samples by means of Tescan Vega3SB Scanning Electron Microscopy (SEM) coupled with Bruker Quantax System Energy Dispersive Spectroscopy (EDS). Analysis of the crystalline level and mineral composition were performed on the sample by using Rigaku MiniFlex II X-Ray Diffraction (XRD). The SEM-EDS analysis showed that iron sand in Bontokanang Village and Tanjung Bayang Beach was dominated by iron oxide (FeO) with concentration of 66,70 wt.% and 79,56 wt.%. The XRD analysis showed that iron sand from Bontokanang Village contains magnetite, chromium (VI) oxide, and aluminum oxide with concentrations of 59 wt.%, 30 wt.%, And 11 wt.%. While the iron sand from Tanjung Bayang Beach contains mineral iron silicon oxide and magnetite with concentrations of 77 wt.% And 23 wt.%. Iron sand deposit from Bontokanang Village and Tanjung Bayang Beach contain magnetite minerals that can be further explored and extracted as commercial minerals.

1. Introduction

Iron is an element which is naturally associated with other elements such as nickel, cobalt, manganese, chromium, molybdenum, and titanium [1]. Iron could be found in nature either in the form of iron ore [2–4] or iron sand [5]. Iron sand is abundant resources in Indonesia and could be found in beach, river and volcanic mountains.

The potential of iron in South Sulawesi is quite large. Iron ore potential can be found in District Bontocani. Its constituent minerals such as pyrite, hematite, magnetite, and also are found in minor amounts of molybdenum minerals [6,7]. While the potential of iron sand in South Sulawesi was located in the River Jeneberang, coastal of Bontokanang Village, Beach Tanjung Bayang, coastal of Jeneponto Regency and others that have not been optimally utilized.

Figure 1. Iron sand deposit in (a) Bontokanang Village Takalar District dan (b) Tanjung Bayang Beach Makassar.

Iron sand contains magnetic minerals such as magnetite (Fe_3O_4), hematite (α-Fe_2O_3), maghemite (γ-Fe_2O_3), ilmenite ($FeTiO_3$), and titanomagnetite [8–10]. However, iron sand was not obtained the pure form in nature. There are some impurities such as Si, Ca, Mg, Al, etc. The magnetic minerals could be extracted or purified using magnetic separation as well as chemical separation [11,12]. Magnetic minerals can be used as the basic material for the production of toner [13], pigment [14], permanent magnet, and others metal industries.

Research on iron sand mineral content using X-Ray Fluorescence (XRF) and X-Ray Diffraction (XRD) method has been done on iron sand from Tapunggaya, Southeast Sulawesi dominated by Fe_2O_3 and followed by TiO_2, MgO, SiO_2 [15]. Analysis of mineral content iron sand by XRD method in Central Java from eight locations namely Srandil, Munggangharjo, Congot, Bayuran, Bandungharjo, Benteng Portugis, Pundenrejo, and Gunung Wungkal obtained magnetite and in minor composition hematite and ilmenit [16]. The similar study was also carried out at coastal areas at Syiah Kuala Beach, Banda Aceh. The iron sand was manually extracted with magnetic rod and characterized by XRD and XRF methods. The results showed that magnetic phase was dominant and the minor phase included TiO_2 and SiO_2 [17].

In this paper, analysis of the mineral content of iron sand deposit in Bontakanang village Takalar district and Tanjung Bayang beach Makassar Sulawesi Selatan will be reported. Scanning Electron Microscopy (SEM) method coupled with Energy Dispersive Spectroscopy (EDS) method was used for chemical composition analysis and XRD method for analysis of crystalline level and mineral composition.

2. Experimental

2.1. Materials

Iron sand was obtained in coastal areas of South Sulawesi, Indonesia, namely in Bontokanang Village Takalar District and Tanjung Bayang Beach Makassar, in the depth of 5 cm from the precipitate surface of iron sand representatively.
Some text.

2.2. Sample preparation

Iron sand was cleaned of dirt and washed with aquadest. Then dried at room temperature. Subsequently, the sample was weighed of 5 g, extracted manually using magnet rod [18–21] 40 times, and crushed using mortar for 6 hours.

2.3. Characterization using Scanning Electron Microscopy-Energy Dispersive Spectroscopy (SEM-EDS) method

The iron sand samples were characterized using Tescan Vega3SB SEM to investigate the morphology of material surfaces and particle distribution. This microscope works on the voltage of 20 kV using Back Scattered Electron (BSE) detector. This detector is sensitive to differences in chemical composition on the sample surface. Elemental testing with EDS was performed on a sample with Bruker Quantax System, where EDS Detector was equipped with the Esprit 1.9 software [22].

2.4. Characterization using X-Ray Diffraction (XRD)
Iron sand samples were characterized using X-Ray Diffraction (XRD) Rigaku MiniFlex II type. XRD measurements were done by powder method [22]. This XRD machine works on measurement conditions voltage of 30 kV, current of 15 mA, scan width of 0.02°, scan speed/duration time of 4°/min., and the scan range of 10°-70°. The qualitative analysis was done by search and match technique using PDXL2 software coupled with ICDD card (International Center for Diffraction Data) ver. 2011. Quantitative analysis of weight percent (wt.%) of crystalline phase was tested by RIR (Reference Intensity Ratio) method [6]. Furthermore, XRD measurements are compared with elemental measurement results using EDS.

3. Results and Discussion
Iron sand in Bontokanang Village, Takalar District and Tanjung Bayang Beach Makassar was obtained representatively, prepared sample and characterized by using SEM-EDS and XRD. Figure 1 shows that SEM image of iron sand samples from Bontokanang Village and Tanjung Bayang Beach were characterized using SEM Tescan Vega3SB type with 1000x magnification of the original size. Figure 1 shows that the surface of iron sand samples from Bontokanang Village and Tanjung Bayang Beach is gray with different levels caused by differences in the composition of the constituent elements. While the distribution of grains of sand was not uniform size caused by crushing techniques.

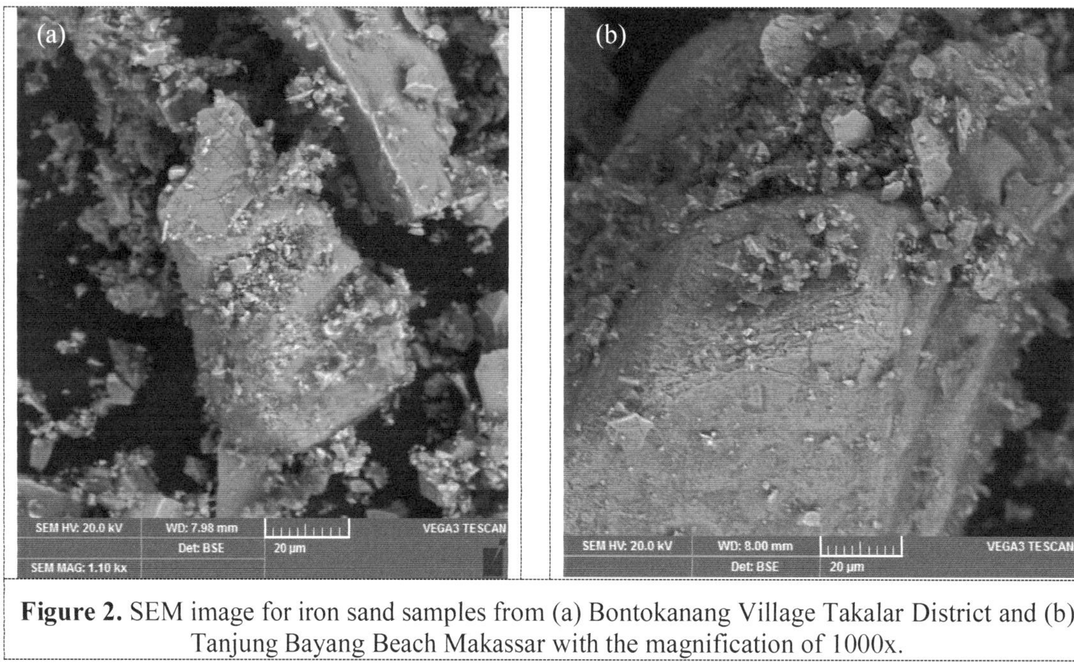

Figure 2. SEM image for iron sand samples from (a) Bontokanang Village Takalar District and (b) Tanjung Bayang Beach Makassar with the magnification of 1000x.

The chemical composition analysis obtained from EDS analysis of iron sand samples from Bontokanang Village Takalar District and Tanjung Bayang Beach Makassar is shown in Table 1. Based on EDS analysis, iron sand deposit from Bontokanang Village dominated by iron oxide (FeO) with a concentration of 66.70 wt.%, as well as iron sand from Tanjung Bayang Beach dominated by iron oxide (FeO) with concentration of 79.56 wt.%. Other compounds with minor compositions are titanium

IOP Conf. Series: Journal of Physics: Conf. Series **997** (2018) 012010 doi:10.1088/1742-6596/997/1/012010

(TiO_2), silicon (SiO_2), aluminum (Al_2O_3), magnesium (MgO), sodium (Na_2O), calcium (CaO), phosphorus (P_2O_5), and potassium (K_2O).

Table 1. The chemical composition of iron sand samples produced from two different places above.

	Mass percentage (wt%)	
Oxide	Bontokanang Village	Tanjung Bayang Beach
Sodium (Na_2O)	1.61	0.75
Magnesium (MgO)	3.60	0.92
Aluminium (Al_2O_3)	5.27	3.80
Silicon (SiO_2)	6.01	5.68
Phosphorus (P_2O_5)	0.85	-
Potassium (K_2O)	0.19	0.08
Calcium (CaO)	1.36	0.67
Titanium (TiO_2)	14.42	8.53
Iron (FeO)	66.70	79.56

The iron content is large enough caused by the extraction stage using magnet rod. The element of iron is attracted by the magnet rod so that the iron sand looks black [23–25]. This shows that iron sand has magnetic properties [26], namely ferromagnetic that can be pulled strongly by the magnet. Elements in minor compositions were obtained because it was also attracted to the iron when extracted. This shows that TiO_2, Al_2O_3, MgO, Na_2O, K_2O has paramagnetic properties. While SiO_2, CaO, P_2O_5 has diamagnetic properties. Table 1 also shows that these elements were derived from weathering of rocks at Lompobattang Volcano that is no longer active and carried by river currents to form alluvial deposits in the Jeneberang River area. At the lower reaches of the Jeneberang River, alluvial deposits were carried by the ocean currents and deposition occurs on the southern coast of Bantaeng District up to the west coast of Makassar.

XRD analysis results using XRD tool Rigaku MiniFlex II type operating on voltage of 30 kV and current of 15 mA. Measurements were made in the range of diffraction angle (2θ) of 10°-70°. Figure 2 shows that the diffraction pattern of iron sand samples from Bontokanang Village. In the samples were found three minerals with different compositions, namely magnetite (Fe_3O_4), chromium (VI) oxide (Cr_2O_3) and aluminum oxide (Al_2O_3). Figure 3 shows that iron silicon oxide ($Fe_{5.36}Si_{0.64}O_8$) and magnetite (Fe_3O_4) were found in iron sand samples from Tanjung Bayang Beach Makassar. The results showed that in both research sites magnetite minerals were obtained.

Mineral composition of iron sand samples based on XRD analysis with RIR method were shown in table 2. The magnetite mineral composition on the iron sand samples from Tanjung Bayang Beach is small. This is caused by more iron elements to form iron silicon oxide than magnetite. This result is in accordance with the SEM-EDS results obtained that the main element in the sample is iron. In addition, no mineral was obtained where the constituent elements are titanium. This is indicated by the crushing step carried out for 6 hours causing the minerals to be destroyed. Magnetite minerals (Fe_3O_4) are found in both research sites that can be further explored and extracted to obtain high purity magnetite, hence it can be used as commercial minerals.

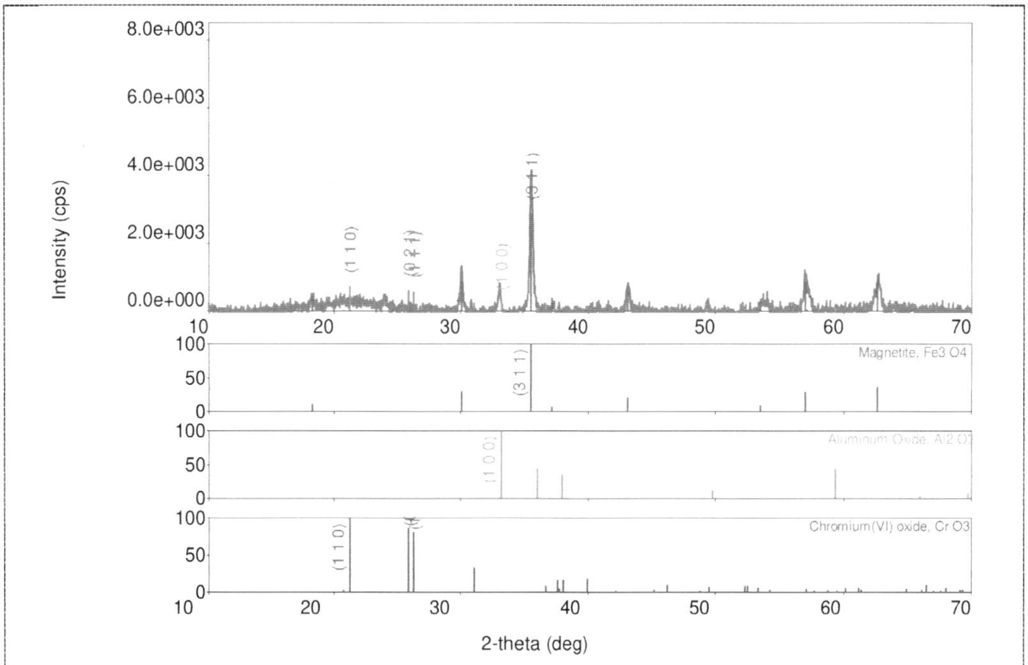

Figure 3. The diffraction pattern of iron sand from Bontokanang Village, Takalar District.

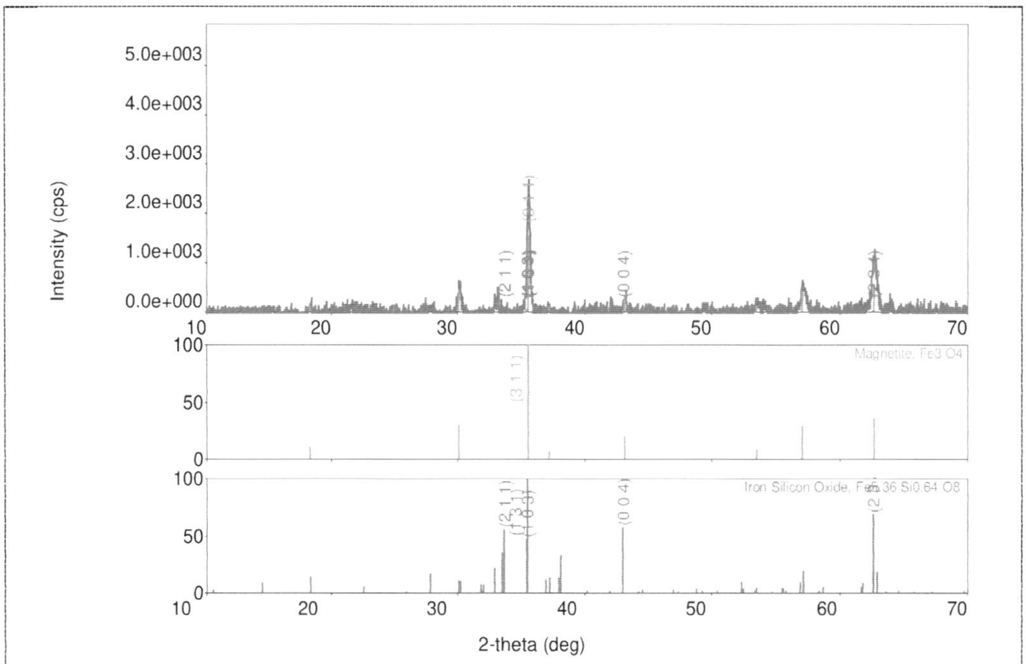

Figure 4. The diffraction pattern of iron sand from Tanjung Bayang Beach Makassar.

Table 2. Mineral composition of iron sand samples from Bontokanang Village and Tanjung Bayang Beach based on XRD analysis.

	Mass percentage (wt%)	
Mineral	Bonto Kanang Village	Tanjung Bayang Beach
Magnetite (Fe_3O_4)	59	23
Iron silicon oxide ($Fe_{5.36}Si_{0.64}O_8$)	-	77
Aluminum oxide (Al_2O_3)	30	-
Chromium (IV) oxide (Cr_2O_3)	11	-

4. Conclusion

Analysis result of the mineral content of iron sand deposit in Bontokanang Village Takalar District and Tanjung Bayang Beach Makassar showed that iron oxide (FeO) was the major element with the concentration of 66,70 wt.% and 79,56 wt.%, respectively The magnetite of 59 wt.%, Chromium (IV) oxide of 30 wt.%, and aluminum oxide of 11 wt.% were obtained in iron sand samples of Bontokanang Village, while iron silicon oxide of 77 wt.% and magnetite of 23 wt.% were obtained in iron sand samples from Tanjung Bayang Beach. From both locations, the magnetite mineral was found to be potential as an industrial raw material. Therefore, further research is required to purify it.

Acknowledgements

We would like to thank Muhammad Rizal Fahlepy for helping the preparation iron sand samples. Acknowledgments also were given to the operator of Microstructural Laboratory of Department of Physics, Universitas Negeri Makassar, Nurfadillah, S.Si., Syamsidar, S.Si., and Kharisma Noer Afifah, S.Si. in assisting data analysis.

References

[1] Kramadibrata S 2013 *Procedia Earth Planet. Sci.* **6** 6–7
[2] Muwanguzi A J B, Karasev A V, Byaruhanga J K and Jonsson P G 2012 *ISRN Mater. Sci.* **2012** 1–9
[3] Rudmin M, Reva I, Gunko A, Mazurov A and Abramova R 2015 *IOP Conf. Ser. Earth Environ. Sci.* **27** 12026
[4] Kiptarus J J, Muumbo A M, Makokha A B and Kimutai S K 2015 *Int. J. Min. Eng. Miner. Process.* **4** 8–17
[5] Vasudevan G 2016 *Proc. the 3rd Int. Conf. Civil, Biol. Environ. Eng.* 10–13
[6] Subaer, Nurhayati, Nurhasmi and Nurfadillah 2014 *Indones. J. Appl. Phys.* **4** 134–141
[7] Haris A, Amin B D, Yusuf A M and Nurhasmi 2014 *J. Sains dan Pendidik. Fis.* **10** 263–268
[8] Findorak R, Froehlichova M and Legemza J 2014 *Metalurgija* **53** 9–12
[9] Nugraha P A, Sari S P, Hidayati W N, Dewi C R and Kusuma D Y 2016 *Proc. The 2016 Conf. on Fundamental and Applied Science for Advanced Technology (Yogyakarta)* **1746** (New York: AIP Conf. Proc.) p 20028
[10] Bassez M-P 2017 *Procedia Earth Planet. Sci.* **17** 492–5
[11] Sunaryono, Taufiq A, Mashuri, Pratapa S, Zainuri M, Triwikantoro and Darminto 2015 *Mater.*

Sci. Forum **827** 229–34
[12] Setiadi E A, Sebayang P, Ginting M, Sari A Y, Kurniawan C, Saragih C S and Simamora P 2016 *J. Phys. Conf. Ser.* **776** 12020
[13] Ataeefard M, Ghasemi E and Ebadi M 2014 *Sci. World J.* **2014** 706367
[14] Mufti N, Atma T, Fuad A and Sutadji E 2014 *Proc. the 3rd Int. Conf. on Theoretical and Applied Physics (Malang)* **1617** (New York: AIP Conf. Proc.) p 165–169
[15] Nurdin M, Watoni A H and Abdillah N 2016 *Int. J. ChemTech Res.* **9** 483–91
[16] Yulianto A, Bijaksana S, Loeksmanto W, Sekaran J R and Pati G 2003 *Indones. J. Phys.* **14** 1–4
[17] Jalil Z, Rahwanto A, Mustanir, Akhyar and Handoko E 2017 *Proc. the 2nd Int. Symp. on Current Progress in Mathematics and Sciences (Depok)* **1862** (New York: AIP Conf. Proc.) 30023-1 030023–4
[18] McCubbin D, Leonard K S, Young A K, Maher B A and Bennett S 2004 *J. Environ. Radioact.* **77** 111–131
[19] Yulianto A, Sulhadi, Azis A L I and Dayati E 2013 *Malaysian J. Fundam. Appl.* **9** 211–215
[20] Rahmawati R, Melati A, Taufiq A, Sunaryono, Diantoro M, Yuliarto B, Suyatman S, Nugraha N and Kurniadi D 2017 *IOP Conf. Ser. Mater. Sci. Eng.* **202** 12013
[21] Rahmawati R, Permana M G, Harison B, Nugraha, Yuliarto B, Suyatman and Kurniadi D 2017 *Procedia Eng.* **170** 55–59
[22] Subaer 2015 *Pengantar Fisika Geopolimer* (Jakarta: Direktorat Jenderal Pendidikan Tinggi)
[23] Ige A and Rehren T 2003 *Iams* **23** 15–20
[24] Fouda M F R, Amin R S, Saleh H I, Labib A A and Mousa H A 2010 *Aust. J. Basic Appl. Sci.* **4** 4540–4553
[25] Chaussard E and Kerosky S 2016 *Remote Sens.* **8** 100
[26] Sehah and Raharjo S A 2017 *J. Penelit. Fis. Apl.* **7** 79–88.

Exploring physics concepts among novice teachers through CMAP tools

N Suprapto[1], Suliyanah[1], B K Prahani[1], M N R Jauhariyah[1] and S Admoko[1]

[1]Department of Physics, Faculty of Mathematics and Natural Science, Universitas Negeri Surabaya, Surabaya 60231, Indonesia

E-mail: nadisuprapto@unesa.ac.id

Abstract. Concept maps are graphical tools for organising, elaborating and representing knowledge. Through Cmap tools software, it can be explored the understanding and the hierarchical structuring of physics concepts among novice teachers. The software helps physics teachers indicated a physics context, focus questions, parking lots, cross-links, branching, hierarchy, and propositions. By using an exploratory quantitative study, a total 13-concept maps with different physics topics created by novice physics teachers were analysed. The main differences of scoring between lecturer and peer-teachers' scoring were also illustrated. The study offered some implications, especially for physics educators to determine the hierarchical structure of the physics concepts, to construct a physics focus question, and to see how a concept in one domain of knowledge represented on the map is related to a concept in another domain shown on the map.

1. Introduction

Some previous researchers indicated that students' understanding of physics concepts still low (e.g. Newtonian mechanics [1-4], electric circuit [5], etc.). Indeed, the problem about conceptions in physics happens not only for students, but also it occurs for prospective teachers and inservice teachers. A concept map is one of the solutions for facilitating them to arrange the hierarchical structure of the physics concept [6]. Especially for physics teachers, the use of concept map in practice helps students in understanding physics concepts and improving the quality of the end-products like graphical structures [7]. On the other hand, novice teachers need to learn and to understand the most important part on constructing concept maps since their role as beginner teachers. In other words, they should be given opportunities to reflect on the processes and to examine the qualities of concept maps [6]. Creating a concept map can be facilitated through manually via paper and pencil or utilised a software.

CmapTools software was employed in this study [6]. The software was initiated by "the Institute for Human and Machine Cognition" that proposing the contribution of information and technology in learning by focusing on concept mapping [6][8]. Generally, working with cmap tools deals with some terms: focus question, context, cross-links, parking lots, and propositions. Every concept map is characterised by a focus question that covers the context of map. Context is shorter expression of focus question. It supports lecturer or students to illustrate the hierarchical structure of the map. Meanwhile, cross-link represents the relations among concepts in different aspects [6]. It guides users (lecturer and students) grasp on how a concept in one aspect of knowledge designated to another domain. The list of concepts can be figured out via parking lots. A parking lot contains some concepts due to the users will arrange and fit in these concepts into the concept map target (final concept map).

If there is not good connection among concepts then these concepts may keep remain on the original parking lots [6]. In other words, a parking lot means the list of concepts before it moves to its position in map.

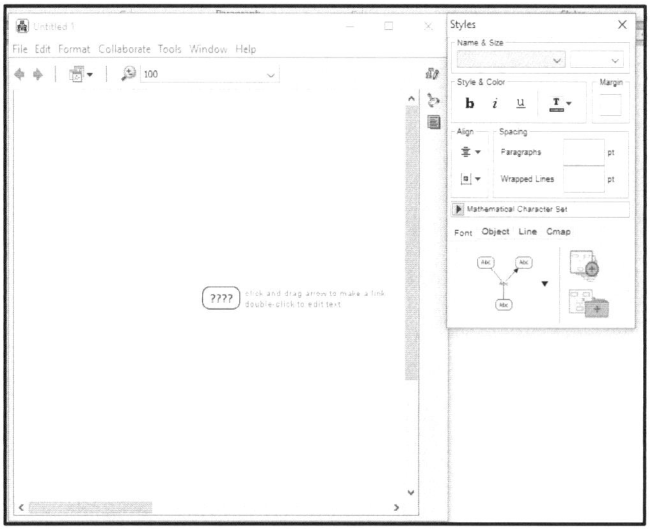

Figure 1. The display of CMAP tools.

The propositions are used to link one concept to another concepts or some different concepts. This feature measures the differentiation level of fundamental concepts. This process has ensued with respect to the levels of the concept meaning quality [9]. Cross-links or grouping of concepts are the ways of concepts can be interconnected together. This feature is "a further way of indicating interrelationships among concepts" [9]. There are three types of cross-links: "(i) point grouping: a number of single concepts eminate from one concept; (ii) open grouping: three or more concepts are linked in a single chain; and (iii) closed grouping: concepts from a closed system" [9]. Indeed, this feature can be seen as a portion of comprehensive understanding of meanings.

This article explores the understanding and the hierarchical structuring about physics concepts among novice teachers. The software helps physics teachers indicated a physics context, focus questions, parking lots, cross-links, and propositions. A total 13-concept maps with different physics topics created by novice physics teachers were analysed. Finally, the main differences of scoring between lecturer and peer-teachers' scoring were also illustrated according to the rubric which defined previously.

2. Method

The participant of this study was 13-novice teachers who participated in the teacher professional development program through PPG-SM3T, Universitas Negeri Surabaya (State University of Surabaya), Indonesia. The process of collecting data is summarised in Figure 2.

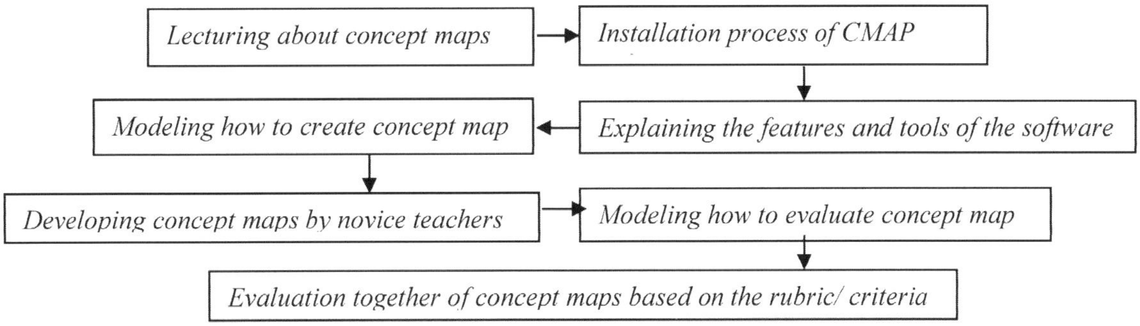

Figure 2. The procedure of conducting research.

Table 1. Evaluation criteria and procedure of scoring of a concept map (modification from [9]).

Criteria of Evaluation	Code	Definition	Procedure of Scoring
Focus question	A	A good focus question is used to overview the whole context for a concept map	Give 5 points for a good focus question, 3 points for focus question in the moderate level, no score for concept map without focus question.
Context	B	A context is aimed to examine the hierarchical structure of the map.	Give 5 points for a good context, 3 points for context in the moderate level, no score for concept map with non appropriate context.
Parking lots or concept recognition	C	Parking lots contain list of concepts, which are objects, events, or situations that are indicated by labels or symbols.	"Give 1 point for each appropriate concept. Count all concepts which are connected to other concepts through propositions" [9].
Cross-links or Grouping	D	Cross-links are the methods on how concepts can be associated together.	Point grouping = 1 point Open grouping = 2 points Closed grouping = 3 points
Hierarchy	E	The hierarchical structure of map indicates "the correction level of concepts, whether the main concepts are at the top position and the specific concepts are at the lower end position" [9].	"Give 4 points for each concept correctly assigned to a level, 2 points for each concept on level removed from an assigned level, and 0 point or no score for concepts that are two levels removed" [9].
Branching	F	Branching denotes "the level differentiation among concepts and the extent the more specific concepts are associated with more general concepts" [9].	"Give 1 point for each branching which has at least two declaration lines" [9].
Proposition	G	Proposition indicates "the association between concepts, which is represented by connecting words or phrases within link-words" [9].	"Give 1 point for each word or phrase as simple propositions, ½ point for repeated use of simple propositions, 2 points for scientific propositions, and 1 point for repeated propositions" [9].

It was noted that there was no an 'ideal' map nor one 'correct' map [9]. Therefore, the evaluation criteria or rubrics for each concept map attempts to overcome the quality and the problem of idiosyncratic and a large number of concept should be established. The scoring rubric yields data with a interactive structure [11]. Additionally, there are some difficulties in examining concept maps which comprise a huge number of concepts. Therefore, a smoothly criterion and guidance for scoring are needed, as well as the following features [9]: (a) the scoring rubric encompasses criteria that are easier and more quickly to evaluate a concept map and (b) the scoring rubric has a scoring scheme that distinguishes between different stages of concept development. Table 1 depicts the evaluation criteria or rubrics in evaluating concept maps.

3. Result and Discussion

Novice teachers need to resonate the operational definitions and skills associated with concept mapping. Additionally, they also need to integrate the epistemology behind concept mapping into physics courses. The practice of concept mapping which integrated through the design of lesson plans and in microteaching are possible boulevards for reflecting on the task of constructing physics' concept maps [7]. Figure 3 illustrates the concept map created by a group of novice teacher with assisted of Cmap tools with topics of thermodynamic process. Meanwhile, concept maps scores with different physics topics created by novice physics teachers are shown in Table 2.

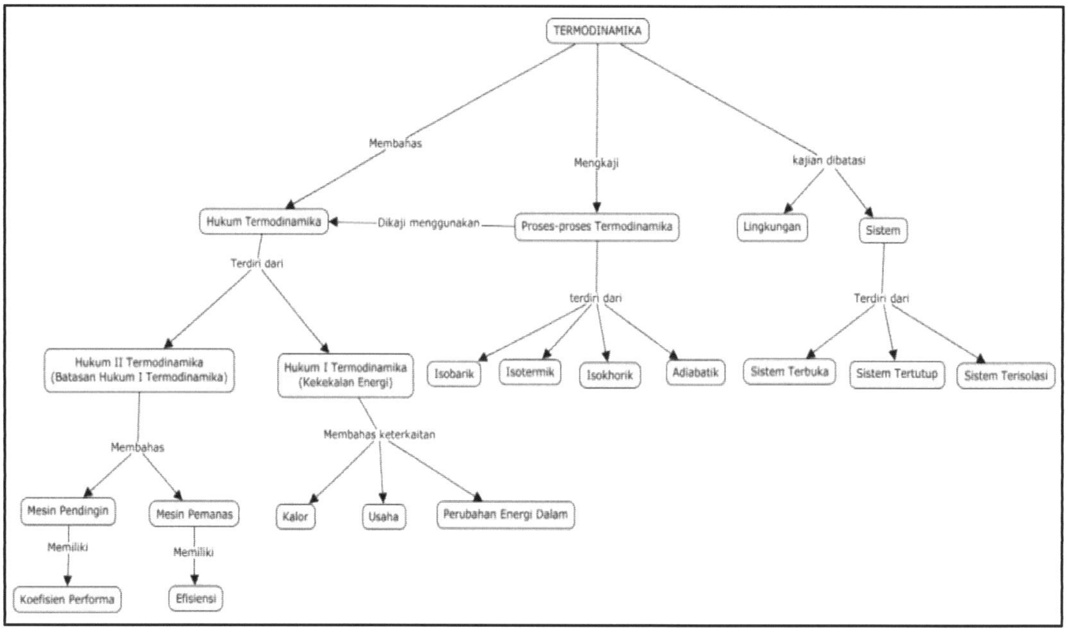

Figure 3. An example of concept map about thermodynamic process.

According to Table 2, the average scores in creating concept maps varied in different topics. The top three among the concepts were heat, thermodynamic process, and vector with average score 87, 83, and 78.5, respectively. Meanwhile, novice teachers performed in a moderate level of these topics, such as rectilinear motion, Newton's law, trajectory, circular motion, dynamical fluid, statical fluid, and measurement. In contrast, they had problems regarding the concepts of the theory of gas kinetic, elasticity, and rigid body with average score 36.5, 41.5, and 50, respectively. Novice teachers have not much idea in developing concept map of these topics. Therefore, instructor or lecturer need more pay attention to these three concepts. However, CMAP tools offer "the use of concept map as an evaluation tool and a learning" [10]. Indeed, it's role as learning instruction overlapped with it's role as assessment [11-12]. It has great potential as a classroom technique for illuminating the conceptual understanding [12] and identify misunderstanding of students [15]. Thus, a concept map is as reflectors of conceptual understanding. As an implication, a master concept map should be created by the instructor or lecturer that could be used for tutoring and testing guidance. Then, participants or novice teachers can be encouraged to generate their own maps, and to check their work with the instructor [10].

Table 2. Concept maps scores with different physics topics created by novice physics teachers.

Topics	Sources of scores				Avg
	Peer teacher (1) "A;B;C;D;E;F;G"	Peer teacher (2) "A;B;C;D;E;F;G"	Peer teacher (3) "A;B;C;D;E;F;G"	Lecturer /Instructor "A;B;C;D;E;F;G"	
Measurement	"5; 3; 19; 8; 2; 2; 12"; Total= 51	"5; 5; 19; 12; 2; 2; 12"; Total=57	"5; 3; 19; 10; 2; 3; 11"; Total=53	"5; 3; 18; 8; 2; 3; 10"; Total=49	52.5
Vector	"0; 5; 29; 12; 4; 4; 24"; Total= 78	"0; 5; 30; 15; 2; 5; 25"; Total=82	"0; 5; 31; 10; 4; 5; 23"; Total=78	"0; 5; 31; 12; 2; 4; 22; Total=76	78.5
Rectilinear motion	"0; 5; 25; 14; 4; 6; 24"; Total=78	"0; 5; 25; 15; 4; 5; 20"; Total=74	"0; 5; 25; 18; 4; 6; 26"; Total=84	"0; 5; 25; 15; 4; 5; 20"; Total=74	77.5
Trajectory	"5; 5; 21; 14; 4; 6; 15"; Total=70	"5; 5; 21; 15; 4; 4; 12"; Total=66	"5; 5; 21; 18; 4; 3; 15"; Total=71	"5; 5; 21; 15; 4; 4; 12"; Total=66	68.25
Circular motion	"0; 5; 34; 3; 4; 5; 18"; Total=69	"0; 5; 34; 3; 4; 5; 14"; Total=65	" 0; 5; 34; 3; 2; 4; 16"; Total=64	" 0; 5; 34; 3; 4; 4; 16"; Total=66	66
Newton laws	" 0; 5; 31; 8; 4; 4; 28"; Total=80	" 0; 5; 31; 8; 4; 4; 26"; Total=78	"0; 5; 31; 6; 4; 5; 27"; Total=78	"0; 5; 30; 7; 4; 3; 25"; Total=74	77.5
Elasticity	"0; 5; 11; 4; 4; 3; 14"; Total=41	"0; 3; 11; 6; 4; 2 ; 18"; Total=44	"0; 5; 11; 6; 2; 3; 16"; Total=43	"0; 3; 11; 4; 2; 2; 16"; Total=38	41.5
Rigid body	"0; 5; 12; 6; 2; 3; 20"; Total=48	"0; 5; 12; 9; 4; 3; 20"; Total=53	"0; 3; 12; 9; 4; 3; 22"; Total=53	"0; 3; 12; 6; 2; 3; 20"; Total=46	50
Statical fluid	"0; 2; 14; 12; 4; 3; 24"; Total=59	"0; 5; 14; 12; 4; 3; 24"; Total=62	"0; 5; 14; 12; 2; 3; 20"; Total=56	"0; 3; 14; 12; 2; 3; 20"; Total=54	57.75
Dynamical fluid	"0; 3; 21; 10; 2; 3; 22"; Total=61	"0; 5; 20; 10; 4; 3; 20"; Total=62	"0; 5; 20; 10; 4; 3; 22"; Total=64	"0; 5; 20; 10; 2; 3; 20"; Total=60	61.75
Heat	"5; 5; 29; 9; 4; 4; 34"; Total=90	"5; 5; 30; 9; 4; 4; 30"; Total=87	"5; 5; 30; 9; 4; 4; 28"; Total=85	"5; 5; 29; 9; 4; 4; 30"; Total=86	87
Theory of gas kinetic	"0; 3; 15; 4; 2; 2; 10"; Total=36	"0; 3; 15; 4; 2; 2; 12"; Total=38	"0; 3; 15; 6; 2; 2; 10"; Total=38	"0; 3; 15; 4; 2; 2; 8"; Total=34	36.5
Processes of thermodynamic	" 0; 5; 21; 15; 4; 5; 34"; Total=84	" 0; 5; 21; 15; 4; 5; 34"; Total=84	" 0; 5; 21; 15; 4; 5; 32"; Total=82	" 0; 5; 21; 15; 4; 5; 32"; Total=82	83

Note: "The codes of (A, B, C, D, E, F, and G)" are based on Table 1.

Some problems regarding the process of developing concept map via Cmap tools included cross-links and propositions [15]. Some novice teachers were unskilled in the ways how concepts can be linked together. They had some problems regarding point grouping, open grouping, and closed grouping. Turning to propositions, the main problem was a level of proposition, either simple or compound. In fact, "the level complexity of the proposition criterion" is distinguished into two levels: (i) simple, and (ii) compound. "A simple proposition is well-defined as a proposition containing only one subject-object clause" [11][15]. Meanwhile, examples of simple propositions are 'isothermic belongs to process of thermodynamic', 'thermodynamic deals with pressure', and 'Carnot machine is part of ideal machine'. Then, "a compound proposition is well-defined as a proposition containing one or more dependent clauses" [11][15]. The following are some examples of compound propositions: 'thermodynamic explores processes of isobaric, isochoric, isothermic, adiabatic and verifies the law of thermodynamics. Being expert for teachers in emerging their concept maps influence the conception of physics learning [13], self-efficacy for physics learning [13], self-efficacy for physics teaching, and pedagogical content knowledge [14]. Additionally, since concept map is built from focus question in initial stage then it needs to integrate with some innovative models which is questioning as the basic feature, such as socratic model [16] and LBQ model [17]. Finally, the concept mapping exercise itself is something that we would endorse to (novice) teachers.

4. Conclusion

Through Cmap Tools software, it can be explored the understanding and the hierarchical structuring about physics concepts among novice teachers. It appeared to have some essential effects on students learning. The software facilitated physics teachers indicating a physics context, focus questions, parking lots, cross-links, hierarchy, branching, and propositions. Through this research, it can be noted that novice teachers performed at moderate level in creating concept maps, such as heat, thermodynamic process, and vector. In contrast, they have problems regarding the concepts of elasticity, rigid body, and theory of gas kinetic. The study offered some significances, especially for physics educators to determine the hierarchical structure of the physics concepts, to construct a physics focus question, and to comprehend how a concept in one aspect of knowledge signified to another aspect presented on the map. It seems concept mapping has great potential to increase the conceptual change of physics concept. Thus, concept map is as indicators of conceptual understanding on physics.

Acknowledgments

This paper is part of the teacher professional development program through PPG-SM3T, *Universitas Negeri Surabaya* (State University of Surabaya), Indonesia. Therefore, the authors would like to thank all novice teachers who took part in the study.

The image of a concept map shown in Figure 3 was produced using Cmap software, available from http://cmap.ihmc.us/.

References

[1] McDermott L C 1984 *Phys. Today* **37** 24
[2] Yeo S and Zadnik M 2000 *Aust. Sci. Teach J.* **46** 9
[3] Bayraktar S 2009 *Int. J. Sci. & Math. Educ.* **7** 273
[4] Suprapto N, Syahrul D A, Agustihana S, Pertiwi C A and Ku C-H 2016 *Chemistry* **25** 718
[5] Engelhardt P V and Beichner R J 2004 *Am. J. Phys.* **72** 98
[6] Cañas A J, Hill G, Carff R, Suri N, Lott J and Eskridge T 2004 CmapTools: A knowledge modeling and sharing environment. In A J Cañas, J D Novak and F M González (Eds.), *Concept maps: Theory, methodology, technology*. Proc. of the first Int. Conf. on concept mapping (Vol. I, pp. 125-133) (Pamplona, Spain: Universidad Pública de Navarra)
[7] Subramaniam K and Harrell P E 2015 *Educ. Res.* **57** 217
[8] Cañas A J, Hill G, Lott J, and Suri N 2003 *Permissions and access control in CmapTools* (Pensacola, FL: Institute for Human and Machine Cognition)
[9] Cronin P J, Dekkers J and Dunn J G 1982 *Res. Sci. Educ.* **12** 17
[10] Elhelou M-W A 1997 *Educ. Res.* **39** 311
[11] Stoddart T, Abrams R, Gasper E and Canaday D 2000 *Int. J. Sci. Educ.* **22** 1221
[12] Edwards J and Fraser K 1983 *Res. Sci. Educ.* **13** 19
[13] Suprapto N, Chang T-S and Ku C-H 2017 *J. Baltic Sci. Educ.* **16** 7
[14] Suprapto N and Ku C-H 2017 *Tiga dekade riset tentang PCK dalam Pendidikan Sains (Fisika)*

dan potensi pengembangan ke depannya [Three decades of research on PCK in Science (Physics) and it's prospects in the future]. Paper presented at Semin. Nas. Fisika dan Aplikasinya (Malang, Indonesia: Universitas Negeri Malang)
[15] Gourlay H 2017 *Phys. Educ.* **52** 014001
[16] Suprapto N and Dwikoranto 2011 *J. Penelit. Fis. Apl.* **1** 14
[17] Suprapto N, Suliyanah and Admoko S 2013 *J. Penelit. Fis. Apl.* **3** 1

IOP Conf. Series: Journal of Physics: Conf. Series **997** (2018) 012012 doi:10.1088/1742-6596/997/1/012012

Qualitative analysis of ionospheric disorders in Solok earthquake (March 6, 2007) viewed from anomalous critical frequency of layer F (f0F2) and genesis spread F

D Pujiastuti[1], S Daniati[1], E Taufiqurrahman[1], B Mustafa[2] and Ednofri[3]

[1] Physics Laboratory of Earth, Department of Physics, Andalas University
[2] Department of Civil Engineering, Andalas University
[3] National Aerospace and Space Agency SPD Kototabang

E-mail : dwi_pujiastuti@yahoo.com

Abstract. A qualitative analysis has been conducted by comparing the critical frequency anomalies of layer F (f0F2) and Spread F events to see the correlation with seismic activity before the Solok earthquake (March 6, 2007) in West Sumatra. The ionospherics data used was taken using the FMCW ionosonde at LAPAN SPD Kototabang, Palupuah, West Sumatra. The process of ionogramme scaling is done first to get the daily value of f0F2. The value of f0F2 is then compared with its monthly median to see the daily variations that appear. Anomalies of f0F2 and Spread F events were observed from February 20, 2007 to March 6, 2007. The presence of f0F2 anomalies was the negative deviation and the presence of Spread F before earthquake events were recommended as Solok earthquake precursors as they occurred when geomagneticsics and solar activities were normal.

1. Introduction

Along with the development of seismology, theories for earthquake prediction are increasingly emerging. Some methods used to predict earthquakes are seismogeologic, statistic analysis of seismicity, correlation analysis and precursor methods. Such theories are generally obtained by conducting case studies of major earthquakes that have occurred. One study of earthquake prediction ever done is to observe the earth's atmosphere.

Ionosphere is a region of Earth's upper atmosphere, from about 50 km to 1,000 km altitude from the earth's surface, and contains charged particles [1]. The morphology of the ionosphere layer is divided into layers D, E, E-Sporadic (Es), F_1 and F_2 [2]. Layer D has the properties to absorb high-frequency HF (High Frequency) radio and VHF (Very High Frequency) waves, while others reflect radio wave energy, thus the readable layers of the ionosonde are layers E, Es, F_1 and F_2, while layer D not [3].

Deformation in the earthquake preparation zone will cause radon deposited in the rock to break out into the air and decay. The radon decay energy ionizes atoms in the atmosphere and causes changes in ionic composition of ions [4-6]. Earthquake prediction can be done on all four layers because there are electromagnetic waves as the effect of plate interaction in the ionosphere layer

More research is done on the ionosphere F layer because this layer appears during the day and night. Based on a case study conducted by T. Xu [7], a decrease in the critical frequency (foF_2) in the ionosphere F layer from its monthly median 3 days before the Wenchuan earthquake, China on May 12, 2008. At the time of disturbance occurs geomagneticsic activity was normal. So also with solar activity. Therefore, the disturbance of the F layer is likely to be caused by an earthquake that will occur. In addition to the critical frequency change of foF_2 (up or down of the monthly median) of the ionosphere F layer, there is also a disturbance of Spread F. Spread F is the distribution of the radio wave traces reflected by ionosphere layer F. This trace spread occurs due to instability that occurs below the layer height F ionosphere [8]. According to Muslim [9], Spread F disruption rate is proportional to disturbance by solar activity. In the event of Spread F the ionosphere the critical frequency of layer F can not be observed anymore. Spread F observation can be done without long time, because it can be observed directly on an ionogramme of ionosonde recording without having to separate ionogramme parameters.

In this research will be seen the influence of the Solok earthquake (March 6, 2007) against the ionosphere F layer condition. Specific observations were made by comparing the critical frequency anomaly characteristics of layer F (foF_2) and the occurrence of Spread F prior to the earthquake that was detached from the effects of magnetics storm and solar storm.

2. Research Methods

The data used is the critical frequency data of layer F (foF_2) recorded on FMCW (Frequency Modular Continous Wave) ionosonde located in LAPAN Kototabang, Agam Regency, West Sumatera at Koto Tabang from February 20, 2007 to March 6, 2007. As data comparators were used magnetic storm data and solar activity data. The data of Dst Index is the number of units of magnetic activity every hour in 24 hours for 30 or 31 days according to the number of dates of the month. The solar activity data is the X-Ray Solar Flare data downloaded from the Space Weather Prediction Center (SWPC). The X-Ray Solar Flare data form is a graph that states the solar activity for each class for 3 days.

Scaling of the ionogramme to obtain the critical frequency value of layer F (foF_2) is carried out according to the method suggested by Jiyo [10]. In the ionogramme, foF_2 is the rightmost peak and is the highest peak in the ionogramme curve. To change the pixel value to frequency, the formula used is $f = 2 + (P_x - 80) \times 2/30$, with P_x being the pixel value on the x-axis.

After obtaining the value foF2 for 29 days every 15 minutes, then carried out statistical operations to get the median foF_2. Median used as a benchmark in viewing foF_2 variations that occur in every time. The foF_2 deviation (δfoF_2) is obtained from the median difference with the observed foF_2, and any deviations exceeding the standard deviation value ($+\sigma$) are marked as an anomaly.

To get the result of Spread F is done by observing all ionogramme data from February 20, 2007 until March 6, 2007 and record the time of Spread F. The results of this study separated spread F data from ionogramme data that is not disturbed. Furthermore, the anomalous results (foF_2) and Spread F obtained are compared with Geomagneticsic activity data and solar storm activity over that time span.

3. Results and Discussion

Observation of ionospheric data recorded by FMCW ionosonde, observation of geomagneticsic activity and observation of solar activity (sunspot number parameter) done before Solok earthquake starting from February 20, 2007 until March 6, 2007

3.1. Radius zone Preparation of the earthquake Solok March 6, 2007

Radius of the Solok earthquake preparation zone calculated by the Dobrovolsky equation [5] is 564.94 km. The earthquake preparation zone radius includes also Kototabang area where ionosonde equipment is located. Thus, it is certain that the ionosphere in the Kototabang region is interfering with seismic activity in the earthquake preparation zones.

IOP Conf. Series: Journal of Physics: Conf. Series **997** (2018) 012012 doi:10.1088/1742-6596/997/1/012012

3.2. Dst Index around the time of the Solok 2007 earthquake

The plot of the Dst index around the time of the Solok 2007 earthquake is shown in Figure 1. Although it looks fluctuating, the drop in the Dst index around the maximum Solok earthquake is only about 50 nT. The existence of disturbance in geomagneticsic activity can be known by the existence of fluctuation of the amplitude of geomagneticsic activity which decreased drastically enough to reach hundreds of nT [11]. So it can be said during 14 days before and after the Solok earthquake 2007 did not happened magnetic storm.

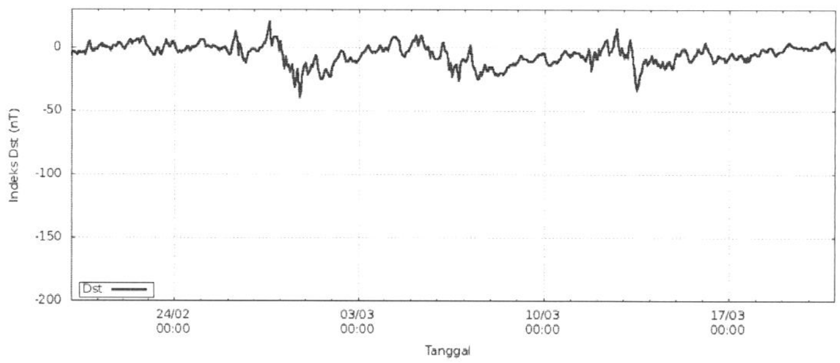

Figure 1. Dst index plot for February 20 - March 20, 2007, where there is no earth geomagneticsics field disturbance.

3.3 Critical Frequency (foF_2) before the 2007 Solok earthquake

Figure 2a shows plot foF_2, median, and plot deviation (δfoF_2) for the March 6th Solok earthquake, each for 7-14 days before the earthquake (Figure 2a) and 0 - 7 days before the earthquake (Figure 2b). Standard deviation used σ = 0; 84 MHz. Based on these limits, we select a significant decrease (exceeding the normal limits) during the day, ie on 20, 22, 26, 27 February, 05 and 06 March 2011.

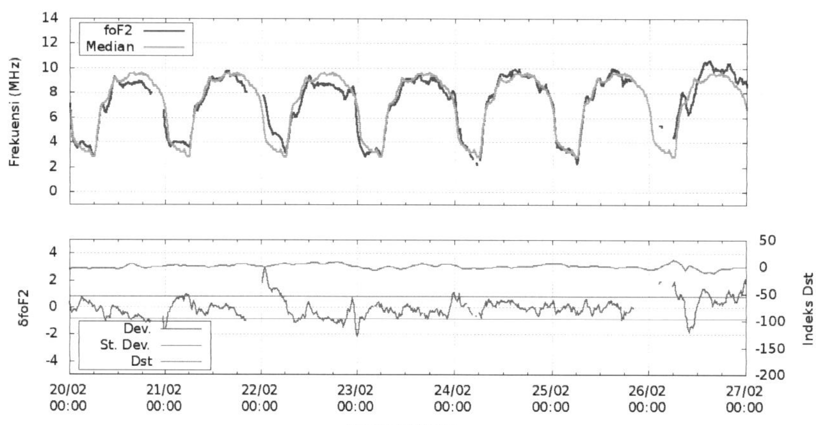

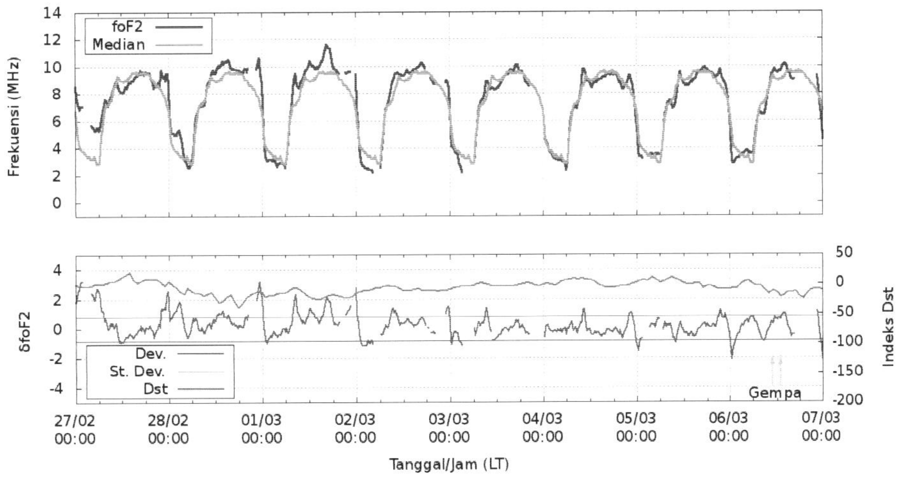

Figure 2. Plot foF_2 and its deviation for a) 14 - 7 days before the Solok earthquake and b) 7 - 0 days before the Solok earthquake.

The daily variation for February 20 is shown in Figure 3. In the early hours of February 20 foF_2 is quite stable with values that are not much different from the monthly median. Then, starting at 08.00 am foF_2 down and for the next 3 hours the value of foF_2 is below the median.

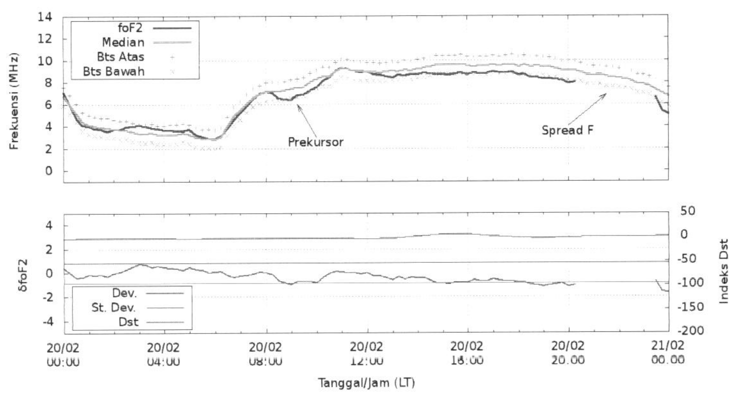

Figure 3. Plot foF_2 and deviation for February 20, 2007.

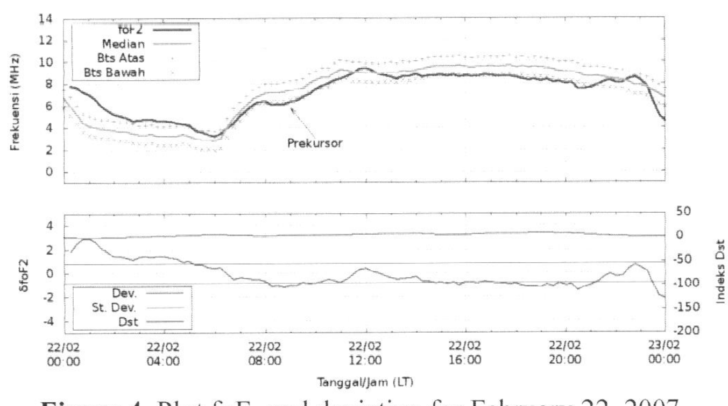

Figure 4. Plot foF_2 and deviation for February 22, 2007.

Next, the daily variations for February 22 are shown in Figure 4. Since the early days the foF_2 values look high. The critical frequency foF_2 has returned to its normal limit at 5:00 am, but continues to fall until it reaches a negative deviation. For February 26, 2007 or 8 days before the earthquake (Figure 5), foF_2 decrease occurs after the previous foF_2 exceeds the upper limit. Visible foF_2 has negative deviation starting at 09.00 WIB.

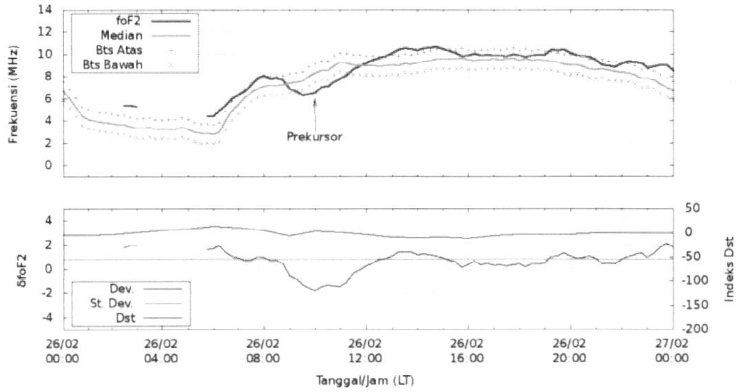

Figure 5. Plot foF_2 and deviation for February 26, 2007.

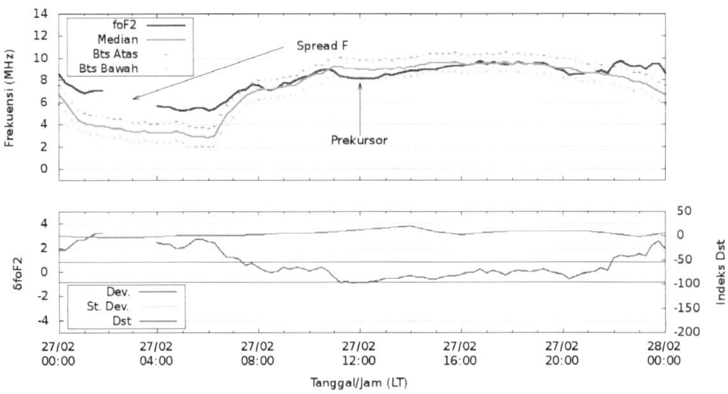

Figure 6. Plot foF_2 and deviation for February 27, 2007.

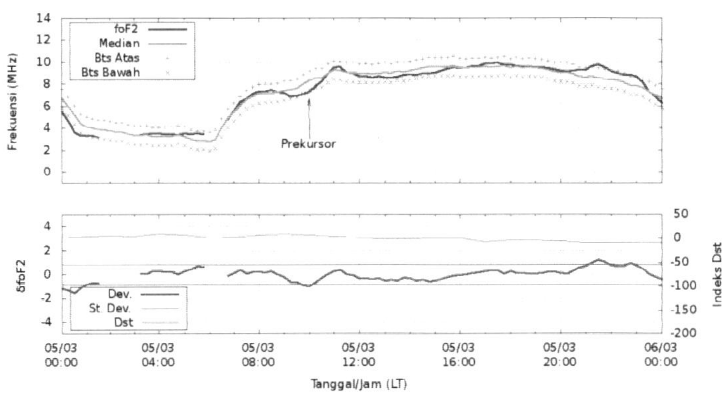

Figure 7. Plot foF_2 and deviation for March 5, 2007.

For February 27 (7 days before the earthquake), plot foF_2 and its deviation are shown in Figure 6. It appears that foF_2 is also high since the night, allegedly because there is Spread F. Around 11.00 WIB decreases, and at 16.00 WIB foF_2 is back in parallel with median. For one day before the earthquake,

March 5, 2007, plots foF_2 and foF_2 are shown in Figure 7. The deviation pattern for March 5 looks the same as on 20 February.

Last for the 6th of March, 2007 which is the day of the earthquake, the plot is shown with Figure 8. The first earthquake occurred at 1049 WIB, while foF_2 began to look down at 06.00 WIB, 5 hours before the earthquake.

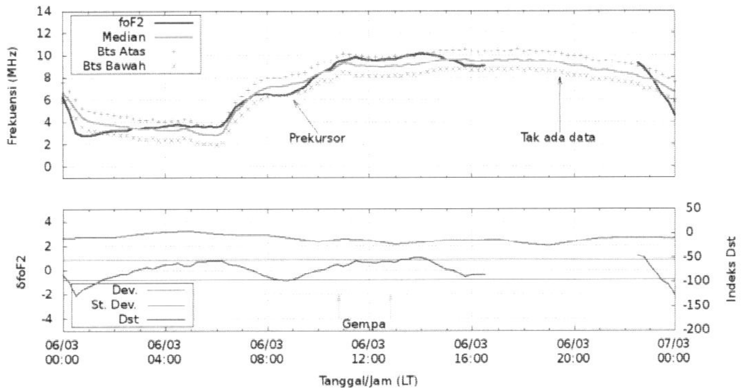

Figure 8. Plot foF_2 and deviation for March 6, 2007.

3.4 Spread F before the Solok 2007 earthquake

Spread F observations were performed starting from February 20, 2007. Figure 9 shows the critical frequency fluctuations of the uninterrupted layer of the ionosphere F (foF_2) and the critical frequency of the maximum ionosphere F (foF_2) layer occurring during the day, and decreasing in the early morning. This shows the activity of the ionosphere F coating under normal conditions. This ionospherics activity cycle will recur each day.

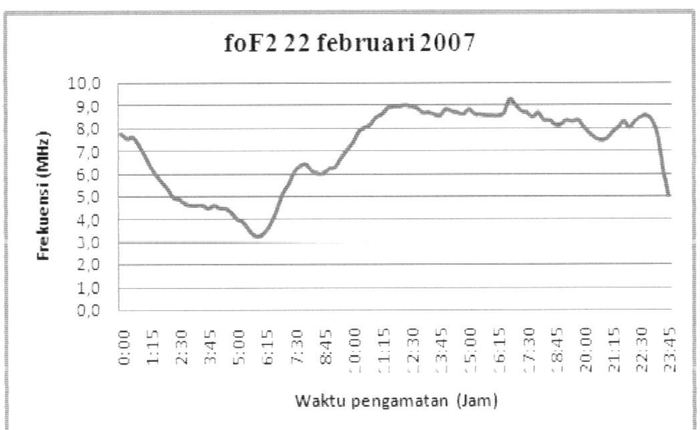

Figure 9. Critical Frequency of Ionosphere February 22, 2007.

On February 23, 2007, the ionosphere condition showed a low frequency at 00:00 to 06:15, which is small from 4 MHz and increased to 10 MHz during the day. This condition is still normal because the critical frequency pattern can be observed at any time as in Figure 10.

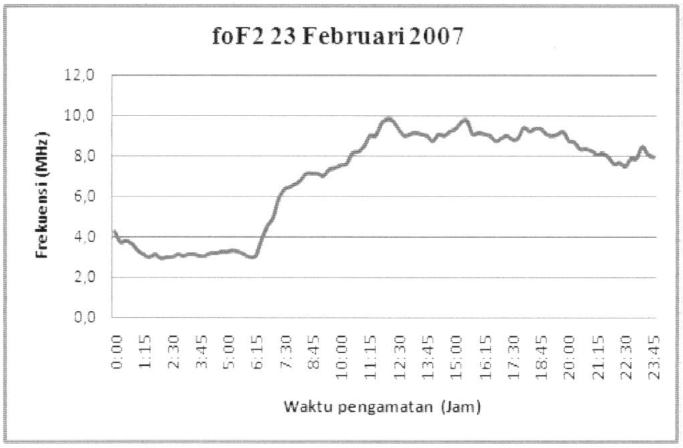

Figure 10. Ionosphere Critical Frequency February 23, 2007.

On February 24, 2007, the condition of the ionosphere F layer which did not undergo the ionosphere F layer trace spread (Spread F) but there was an E layer (E-Sporadic) disturbance at 5:00 am and the critical frequency of the ionosphere F layer (foF$_2$) was at outside the observed frequency limit so that it can not be observed on the ionogramme. At 07:00, the critical frequency of the ionosphere F layer begins to increase with increasing heat emitted by the sun, as shown in Figure 11.

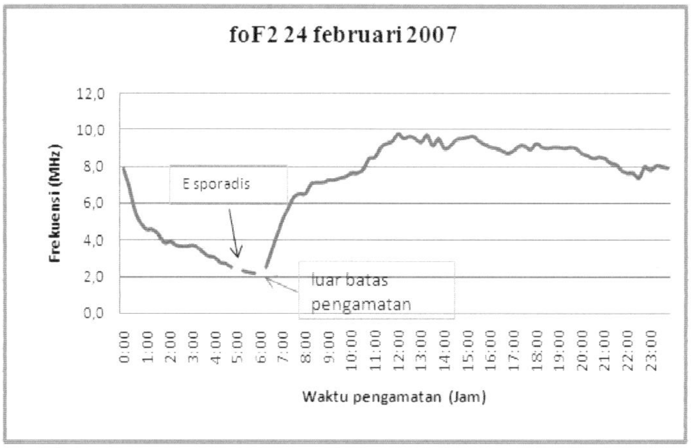

Figure 11. Critical Frequency of the ionosphere February 24, 2007.

Observations for the first period of ionospheric data from February 20, 2007 to February 26, 2007 showed that the ionosphere F layer is normal. As of 22, 23, February 24, 2007 there was no interruption. The observation of ionospherics data for this second period was conducted from February 27, 2007 to March 6, 2007. On March 2, 2007 the critical frequency of the ionosphere F layer was interrupted on04:30 until 05:30 because the critical frequency of the ionosphere F layer is beyond the observed frequency limit . This shows that the critical frequency of low ionosphere F coating is small from 2MHz. At 05:45 to 06:00 am accompanied by E-Sporadic (E_S) occurrence on layer E causes the reflected wave of ionosonde can not reach layer F. Then at 17: 45-19: 00 appears E-Sporadic again and continues with the incident Spread F at 19: 15-20: 00. After 45 minutes, Spread F occurs again from 20: 45-22: 45 as shown in Figure 12.

The ionosphere conditions on March 3, 2007 showed a decrease in frequency from 00:00 to 02:30 due to decreased solar intensity. From 03:00 to 06:15 and at 20:00, there is a disturbance in the E layer that is E-Sporadic. At 20:45 until 23:45, there is the appearance of Spread F as shown in Figure 13.

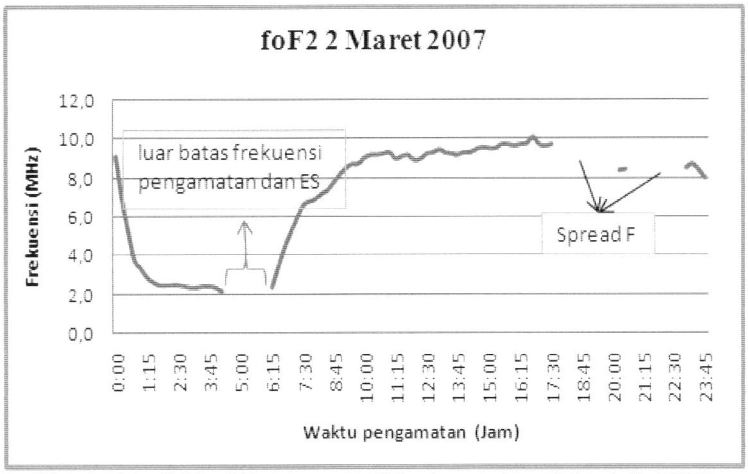

Figure 12. Critical Frequency of the ionosphere March 2, 2007.

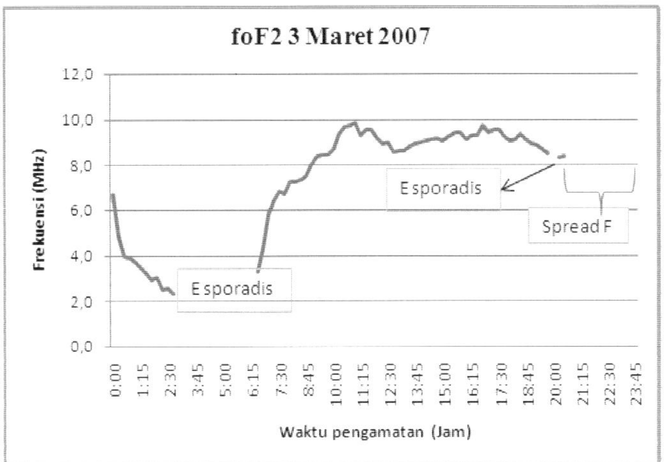

Figure 13. The Critical Frequency of the ionosphere March 3, 2007.

On March 5, 2007 Spread F took place at the end of the night at 02:00 to 03:00 as shown in Figure 14. Observations for the second period were from February 27, 2007 to March 6, 2007, ie, a week before the earthquake showed an emergence Spread F at ionosphere F layer. The appearance of Spread F is not influenced by geomagnetics activity, because from observation of geomagnetics data there is no interference at the geomagnetics so it does not affect layers of ionosphere F.

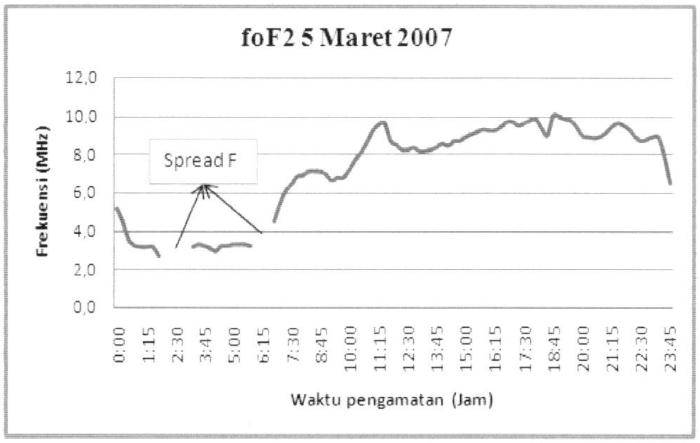

Figure 14. The Critical Frequency of the ionosphere March 5, 2007.

3.5 Sunspot Activity Chart based on Sunspot Numbers Around Time of Solok Earthquake

Figure 15 shows the solar activity based on the sunspot number. Sun activity decreased from February 19, 2007 to February 21, 2007 from sunspot number 20 to 10 and increased again on February 22, 2007, this condition repeated until February 27, 2007. On March 1, 2007 to March 3, 2007, the number the sunspot was 7, then increased on March 5, 2007 with the number of sunspot 17 and decreased until the sunspot became zero on March 7, 2007 until March 20, 2007. This number of sunspots is the number of black spots on the sun that can cause an explosion at the surface of the sun. The sunspot number has a positive correlation with Spread F. In the long run from 1992 to 1995 the annual solar activity has a strong enough relationship to Equatorial Spread F or ESF [9]. If the number of sunspots large then the frequency of Spread F will be large as well.

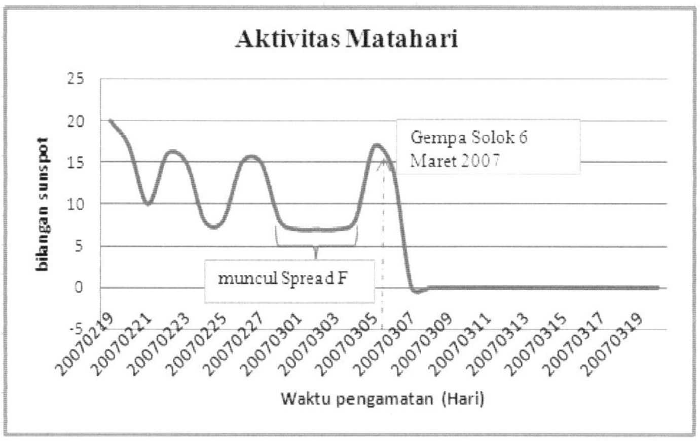

Figure 15. Sun activity with sunspot number parameter.

3.6 Comparison of critical Frequency (foF_2) and Spread F before The Solok earthquake 2007 as precursor of The Solok Earthquake

In this research, there is a disturbance of critical frequency in the F (foF_2) layer and Spread F before the earthquake of the Solok March 6, 2007. For Solok earthquake (2007), precursors appeared on 20, 22, 26, 27 Februari, 5 and 6 March 2007. The emergence the foF_2 anomaly identified as the precursor in the earthquake Solok dominant occurs during the day when there is a negative deviation of the foF_2 value whereas Spread F is the spread of radio waves in the F-region due to ionospherics irregularity in the area appear at night and in the Solok earthquake predominant about 5 days before the earthquake. On observation foF_2 indicated the possibility of Spread F on February 20 and 27 due to foF_2

fluctuation. This result is different from the result obtained on direct observation of fluctuation of layers F frequency.

From the above results, we can see that there is a time difference between precursors between precursor analysis with critical frequency foF_2 and Spread F. The time of the emergence of faster precursors in critical frequency analysis occurs because the observations are performed more thoroughly by using statistical analysis to find the anomaly of foF_2 deviation and separation parameter parameter of ionosphere layer F_2. Meanwhile, for Spread F analysis done by direct analysis of the critical frequency fluctuation of the ionosphere F layer where at the time of the occurrence of Spread F, the critical frequency of layer F can not be observed anymore. Spread F observation can be done without long time, because it can be observed directly on the ionogramme of ionosonde recording without having to separate ionogramme parameters.

From the comparison of two ways above analysis shows that the analysis of critical frequency fluctuations with the separation of layer F parameters ie on layer F_2 and the application of statistical methods resulted in the detection of the possibility of precursor earthquake earlier than observation of direct layer F-fluctuation fluctuations.

4. Conclusion

From the research that has been done this, some conclusions that can be taken that is: (1) Observation of ionospheric anomalies using critical frequency foF_2 appeared on 20, 22, 26, 27 February, 5 and 6 March 2007, (2) The occurrence of Spread F with observations of F layer frequency fluctuations is directly seen on 2, 3, and 5 March 2007, (3) The timing of faster precursors in critical frequency analysis occurs because the observations are performed more thoroughly by using statistical analysis to find the anomaly of the deviation value foF_2 and the parameter parameter of ionosphere, F_2 layer, is performed. While for analysis of Spread F is done by direct analysis of the critical frequency fluctuations of the ionosphere F layer without separating the layer F parameters where at the time of the occurrence of the critical frequency Spread F layer can not be observed again, (4) The existence of the foF_2 critical anomaly and the occurrence of Spread F are predicted as earthquake precursors because they occur when geomagneticsic and sun activities are normal.

Acknowledgment
We would like to thank the National Aeronautics and Space Agency (LAPAN) for allowing the use of FMCW ionosonde data from LAPAN SPD Kototabang.

References

[1] Kaloka S, Jiyo, Suhartini S, Perwitasari S, Mardiani A S and Dear V 2010 *Lapisan Ionosfer, Prediksi Frekuensi dan Teknis Komunikasi Radio (Jiyo, Ed)* (Bandung: Pusat Pemanafaatan Sains Antariksa LAPAN)
[2] Hunsucker R D and Hargreaves J K 2003 *The High-Latitude Ionosphere an its Effects on Radio Propagation* (United Kingdom: Cambridge University Press)
[3] Jiyo 2014 *Majalah Sains dan Teknologi Dirgantara* **9** 1
[4] Pulinets S A, Legen'ka A D, Gaivoronskaya T V and Depeuv V K 2003 *J. Atmos. Sol. Terestrial Phys.* **65**
[5] Pulinets S A and Boyarchuk K 2004 *Ionospheric Precursors of Earthquakes* (Berlin: Springer-Verlag)
[6] Pulinets S A 2004 *Ionospheric Precursors of Earthquake: Recent Advances Theory and Practical Applications* **15** 3
[7] T Xu, Y Hu, J Wu, Z Wu, Y Suo and J Feng 2010 *Giant disturbance in the ionospheric F2 region prior to the M 8.0 Wenchuan earthquake on 12 May* 2008 (China: Xidian University)
[8] Ednofri S S 2006 *Spread F di Atas Kototabang* (LAPAN: Kototabang)

[9] Muslim B 2001 *Sains Filsafat Sains Teknol.* **3** 12
[10] Jiyo 2008 *Berita Dirgantara* **2** 9
[11] Ruhimat M 2001 *Warta LAPAN* **3** 15

The effectiveness of CPI model to improve positive attitude toward science (PATS) for pre-service physics teacher

T Sunarti[1], Wasis[1], Madlazim[1], Suyidno[2] and B K Prahani[1]

[1]Department of Physics, Faculty of Mathematics and Natural Science, Universitas Negeri Surabaya, Indonesia
[2]Department of Physics Education, Faculty of Mathematics and Natural Science, Universitas Lambung Mangkurat, Indonesia

E-mail: titinsunarti@unesa.ac.id

Abstract. In the previous research, learning material based Construction, Production, and Implementation (CPI) model has been developed to improve scientific literacy and positive attitude toward science for pre-service physics teacher. CPI model has 4 phases, included: 1) Motivation; 2) Construction (Cycle I); 3) Production (Cycle II); and 4) Evaluation. This research is aimed to analyze the effectiveness of CPI model towards the improvement Positive Attitude toward Science (PATS) for pre-service physics teacher. This research used one group pre-test and post-test design on 160 pre-service physics teacher divided into 4 groups at Lambung Mangkurat University and Surabaya State University (Indonesia), academic year 2016/2017. Data collection was conducted through questioner, observation, and interview. Positive attitude toward science for pre-service physics teacher measurement were conducted through Positive Attitude toward Science Evaluation Sheet (PATSES). The data analysis technique was done by using Wilcoxon test and n-gain. The results showed that there was a significant increase in positive attitude toward science for pre-service physics teacher at α = 5%, with n-gain average of high category. Thus, the CPI model is effective for improving positive attitude toward science for pre-service physics teacher.

1. Introduction

Learning physics offers a way of thinking about a world that has real-life applications and rational explanations of the world in order to be easily understood, to support open, objective and fair thinking, and to rely on empirical evidence [1]. Physics education can prepare the competence of physics teacher graduate to master science literacy and have a positive attitude toward science [2]. Attitudes toward science determine the success rate of pre-service physics teacher in studying physics [3]. Pre-service physics teacher can give more attention to the learning process that lecturers do and more scientific participation in scientific investigation like a scientist in searching and finding science.

[4] Describes how to improve positive attitudes toward science in physics learning, namely: (1) increasing motivation or interest in learning science, motivation is an important construct in physics learning. Lecturers can increase their motivation in physics learning by making pre-service physics teacher passionate in physics subject discussion, happy to solve physics problems, enjoy reading physics lessons, enjoy acquiring new knowledge in physics, feel physics study influence career prospects in the future, there is a desire of career in physics or physics projects, (2) increasing confidence in physics learning, pre-service physics teacher' beliefs about their own ability can make

them more willing to support the success of the learning process and to alert the quality of their performance. The lecturer always assures the pre-service physics teacher that being able to study the ever-expanding physics can provide the best answers to test scientific questions, easily understand new ideas in physics, be sure to plan and carry out experiments well, and be sure to draw conclusions based on experimental data or reference studies, (3) support of scientific inquiry, the pre-service physics teacher feel the direct experience of conducting scientific investigations like a scientist in searching and finding physics material. Lecturers encourage the role of pre-service physics teacher in supporting imitative investigations to understand the environment, solving physics problems in life, developing science and technological innovation, and being responsible for resources and the environment. Pre-service physics teacher conditioned to assess the environmental situation and demonstrate willingness to act to preserve natural resources, as well as demonstrate a sense of personal responsibility for maintaining a sustainable environment.

In fact, the positive attitude toward science of pre-service physics teacher is still far from expectations. The results of the preliminary study of [5] found that most pre-service physics teacher of Unesa physics education program is less able to show a positive attitude toward science. Pre-service physics teacher found it difficult to learn physics, are less interested and less confident in learning physics, less participant and less convinced in group investigation activities, less daring to make decisions in using physics materials to solve real life problems, and have less responsibility for environmental problems. The results of previous research indicate a serious problem related to the positive attitude toward science of pre-service physics teacher.

In the previous research, Construction, Production, and Implementation (CPI) model has been developed to improve scientific literacy and positive attitude toward science for pre-service physics teacher. The CPI model has been specially designed to increase scientific literacy and positive attitude toward science for pre-service physics teacher. CPI model has 4 phases, namely: 1) Motivation; 2) Construction (Cycle I); 3) Production (Cycle II); and 4) Evaluation. The previous research developed a device of learning physics as an operational form of CPI model developed. The CPI model and learning tools developed by researchers have filled the needs of 21st Century skills. Therefore, it is hoped that the implementation of the CPI model that have been developed quality of pre-service physics teacher.

A model is said to be of high quality when fulfill the criteria of validity, practicality, and effectiveness [6]. Indicator of positive attitude toward science for pre-service physics teacher is motivation in science, belief in learning science, support in scientific investigation, and responsibility for resources and environment [4]. Fundamental physics materials used in physics learning. The purpose of this research is to analyze the effectiveness of the CPI model to increase the positive attitude toward science for pre-service physics teacher. The focus of the problem in this study included: (1) whether there was a significant increase (statistically) of positive attitude toward science for pre-service physics teacher before and after the CPI model was applied, (2) how much level of positive attitude toward science for pre-service physics teacher increased before and after applied CPI model.

2. Methodology of Research

2.1. General Background of Research

This research was conducted at Lambung Mangkurat University and University of Surabaya State (Indonesia). The scope of this research is on pre-service physics teacher in academic year 2016/2017. This research is emphasized on the analysis of the fulfillment of the effectiveness of CPI model by analyzing the improvement of positive attitude toward science for pre-service physics teacher before and after following the CPI model. The effectiveness of the CPI model was determined based on a significant increase in scores (statistically) between pre-test and post-test of positive attitude toward science for pre-service physics teacher, as well as the mean of n-gain determined by criteria: low, medium and high.

2.2 Sample of Research

The samples in this study were 160 pre-service physics teacher of Lambung Mangkurat and University of Surabaya State, Indonesia; which is in the four groups, namely: group-1 (class VIA Lambung Mangkurat University), group-2 (class VIB Lambung Mangkurat University), group-3 (class VIA Surabaya State University), and group-4 (class VIB Surabaya State University). Each group consists of 40 pre-service physics teacher in academic year 2016/2017.

2.3 Instrument and Procedures

Positive Attitude toward Science Evaluation Sheet (PATSES) is an instrument used to measure Positive Attitude toward Science [4]. PATSES includes motivation in science, belief in science learning, support in scientific inquiry, and responsibility for resources and environment [4]. Fundamental physics materials used in physics learning. This research uses one group pretest-posttest design, which is O1 X O2 [7]. The learning process begins by giving pre-test (O1). Each pre-service physics teacher is required to complete the PATSES. After the pre-test, the lecturer applies the CPI model and learning tool in each group (X). Learning science is consisting of: motivation in science, belief in learning science, support in scientific investigation, and responsibility for resources and environment at each learning phase. The process of physics learning ends with post-test (O2). Every pre-service physics teacher is required to post-test of positive attitude toward science by filling PATSES.

2.4 Data Analysis

Positive attitude toward science pre-service physics teacher is analyzed based on the assessments obtained by pre-service physics teacher before and after learning using the CPI model. The pre-test, post-test, and n-gain data of positive attitude toward science of pre-service physics teacher were further analyzed using inferential statistical tests with the help of SPSS and reinforced by qualitative descriptive analysis. The score level for positive attitude toward science pre-service physics teacher is based on indicators of motivation in science, belief in science learning, support in scientific inquiry, and responsibility for resources and environment [4]. The n-gain value is determined by the equation: n-gain = (score post-test - score pre-test) / (maximum score - pre-test score) [8]. According to the following criteria: (1) if n-gain $\geq$ .7 (high), (2) if .3 $<$n-gain $<$ .7 (moderate), and (3) if n-gain $\leq$ .3 (low).

3. Result of Research

The learning outcomes of all groups related to the positive attitude toward science of pre-service physics teacher are presented in Figures 1 and Table 1. Vertical bar represent the mean of pre-test, shape bar scores represent the mean post-test scores, and black bars scores represent the n-gain scores. Figure 1 shows the average post test scores of positive attitude toward science of pre-service physics teacher for all groups is greater than the pre-test score. The average pre-test, post-test, and n-gain scores associated with positive attitude toward science of pre-service physics teacher indicators for all groups are presented in detail in Table 1. Figure 1 shows the average n-gain value of positive attitude toward science for group-1, group-2, group-3, and group-4 are respectively .75; .73; .73; and .72. The average n-gain value of positive attitude toward science of pre-service physics teacher for all groups is in the high category.

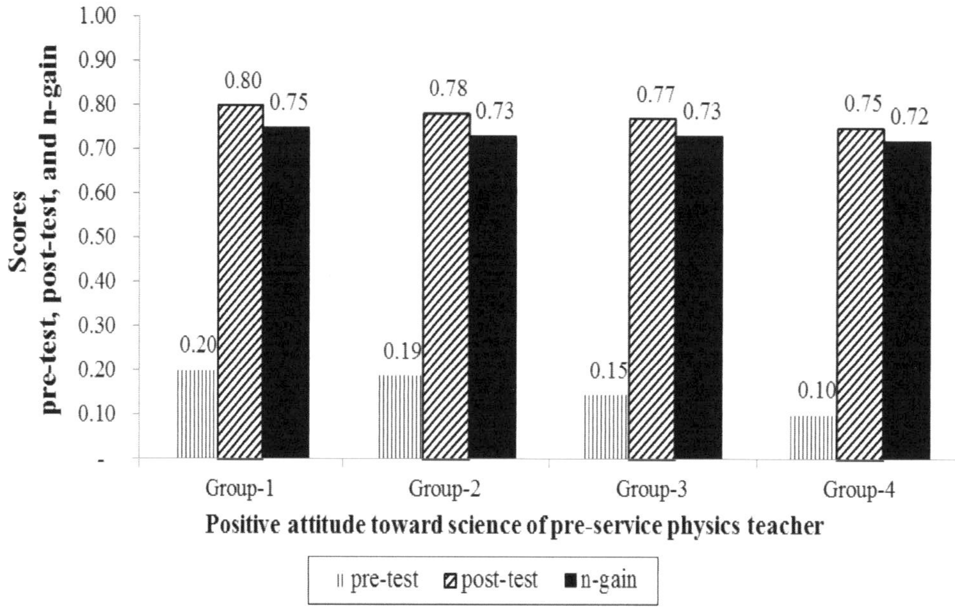

Figure 1. The average pre-test, post-test, and n-gain scores of positive attitude toward science of pre-service physics teacher in all groups.

Table 1. The average score of pre-test, post-test and n-gain of positive attitude toward science of pre-service physics teacher in all groups.

School	Groups	Scores	Positive Attitude Toward Science Indicator			
			Motivation in science	Belief in science learning	Support in scientific inquiry	Responsibility for resources and environment
Lambung Mangkurat University	1	Pre-test	.14	.15	.26	.25
		Post-test	.75	.77	.84	.83
		n-gain	**.71**	**.73**	**.78**	**.77**
	2	Pre-test	.12	.13	.25	.25
		Post-test	.77	.75	.80	.80
		n-gain	**.74**	**.71**	**.73**	**.73**
Surabaya State University	3	Pre-test	.13	.12	.17	.16
		Post-test	.75	.71	.80	.81
		n-gain	**.71**	**.67**	**.76**	**.77**
	4	Pre-test	.10	.11	.09	.10
		Post-test	.74	.70	.77	.78
		n-gain	**.71**	**.66**	**.75**	**.76**

Table 1 shows that the positive attitude toward science test scores of each indicator include: motivation in science, belief in science learning, support in scientific inquiry, and responsibility for resources and environment are low for all groups and the post-test score of positive attitude toward science in each indicator is high for all groups. The n-gain of positive attitude toward science score of each indicator includes: motivation in science, belief in science learning, support in scientific inquiry, and responsibility for resources and environment in all groups are moderate or high.

Table 2. Wilcoxon test result of positive attitude toward science for all groups.

University	Group	Data	N	Z	p
Lambung Mangkurat University	1	*Pretest-Posttest*	40	-4.10	< .01
	2	*Pretest-Posttest*	40	-4.11	< .01
Surabaya State University	3	*Pretest-Posttest*	40	-4.11	< .01
	4	*Pretest-Posttest*	40	-4.10	< .01

p < .05 (2-tailed)

Table 2 shows the improving in pre-test and post-test positive attitude toward science tested using Wilcoxon test. The Z score gives a value of -4.10; -4.11; -4.11; -4.10 for group-1; group 2; group 3; group 4. Each score is considered significant, because p < .05. Because Z the result of the calculation is negative so it shows that there is an increase of positive attitude toward science of pre-service physics teacher after applied learning with CPI model for all groups.

4. Discussions

The effectiveness of the CPI model on improving positive attitude toward science of pre-service physics teacher can be seen from (1) an increase in pre-test and post-test scores and (2) n-gain positive attitude toward science value, as shown in Figure 1, Table 1, and Table 2. Before the CPI model is applied; pre-service physics teacher less mastery the positive attitude toward science, average score of pre-service physics teacher is under the standard score (minimum score .20 in score range 0-1), that is the average score of positive attitude toward science for group-1, group-2, group-3, and group-4 respectively are .20; .19; .15; and .10. All this time, pre-service physics teacher are not used to motivation in science, belief in science learning, support in scientific inquiry, and responsibility for resources and environment. This is supported by [5] found that most pre-service physics teacher of Unesa physics education program is less able to show a positive attitude toward science. Pre-service physics teacher find it difficult to learn physics, are less interested and less confident in learning physics, less participant and less convinced in group investigation activities, less daring to make decisions in using physics materials to solve real life problems, and have less responsibility for environmental problems. The results of previous research indicate a serious problem related to the positive attitude toward science pre-service physics teacher.

After the CPI model was applied to the pre-service physics teacher, the mastery of positive attitude toward science rises above average and becomes high; the average score of positive attitude toward science for groups-1, group-2, group-3, and group-4 respectively to .80; .78; .77; .72 (well beyond the minimum score of .75 in the 0-1 score range). The increase in positive attitude toward science is allegedly influenced by design of scenario in phase of the CPI model. Each phase emphasizes positive attitude toward science in scientific activities. Lecturers are able to cultivate pre-service physics teacher 'positive attitude in constructing their own science literacy (phase 1 of CPI model), improving positive attitude toward pre-service physics teacher' science through scientific investigation, problem solving, and involvement in decision-making process related to science issues and solutions (phase 2 of CPI model) space to establish positive attitudes toward their own science by assigning responsibility for drafting the scientific literacy lesson plan well (phase 3 of CPI model), and involving them in the evaluation and reflection of well-executed outcomes and learning processes (phase 4 of CPI model). Reinforced interview data and observation that pre-service physics teacher feel lecturers can inspire pre-service physics teacher to grow and apply a positive attitude to science well. Pre-service physics teacher find it easy to apply self-motivation in science, feel confident in learning science, active in scientific inquiry, and responsible for resources and environment. Application of CPI model is able to facilitate positive attitude toward science not only accept, but also apply it to support pre-service physics teacher's success in constructing science literacy and designing science literacy learning well. Supported with other research results that Physics education can prepare the competence of physics teacher graduate to master science literacy and have positive attitude toward science [2].

Attitudes toward science determine the success rate of pre-service physics teacher in studying physics [3]. The increase of positive attitude toward science of pre-service physics teacher score in all groups was significant at the 5% significance level with n-gain of .75 for group-1; .73 for group-2; .73 for group-3; and .72 for group-4. This means that there is an increase in positive attitude toward science of pre-service physics teacher after the implemented CPI model. Increased positive attitude toward science is supported by the availability of positive attitude toward science of pre-service physics teacher learning needs as suggested in phase of the CPI model. It is supported by metacognition theory that capable pre-service physics teacher are aware of how they learn and take steps in maximizing their efforts to achieve better results [9]. Another support of Bandura's modeling theory is that the cognitive, affective, and behavioral changes that result from observing others' behaviors and explanations [10]. Self-regulated learning theory that pre-service physics teacher learn when engaged in the process of setting personal goals, combined with motivation, thought processes, strategies, and behaviors leads to the achievement of goals [9]. Theory of distributed cognition learning that pre-service physics teacher share ideas with others to improve their understanding, being encouraged to clarify and organize their own ideas, elaborate what they know, find weaknesses in reasoning, and enjoy alternative views that are just as valid as which they have known by the term [10]. Reflection as deepening comprehension including interrelation between the essential concepts of lecture materials can foster awareness of pre-service physics teacher attitude, provided that the lecturer exemplifies [11]. Application of CPI model is able to facilitate positive attitude toward science not only accept, but also apply it to support student's success in constructing science literacy and designing science literacy learning well [12-14]. Based on the above description, the implementation of CPI model is able to develop the role of positive attitude toward science of pre-service physics teacher.

5. Conclusions

This research shows that there was a significant increase in positive attitude toward science for pre-service physics teacher at α = 5%, with n-gain average of high category. Thus, the CPI model is effective for improving positive attitude toward science for pre-service physics teacher. It can be concluded that CPI model is effective to improve positive attitude toward science for pre-service physics teacher. This positive attitude toward science is clearly by design in every phase of the CPI model. The implication of this research is that CPI model can be used as an alternative to overcome the low of positive attitude toward science for pre-service physics teacher. To improve the result of this research, it is necessary to do further generalizations in various education levels and countries.

6. Acknowledgements

The author's gratitude goes to the Ministry of Research, Technology and Higher Education who has funded the Research (SK Number: 0001.82 / UN38.11-P/LT/2017 dated 06 December 2016). Likewise, the author's gratitude goes to the Surabaya State University and Lambung Mangkurat University that has provided research opportunities.

References

[1] Newton D P 2008 *A Practical Guide to Teaching Science in the Secondary School* (USA: Routledge)
[2] NRC 2011 *Inquiri and the National Science Education Standards. A Guide for Teaching and Learning* (Washington: National Academy Press)
[3] Lederman N G, Lederman J S and Antink L 2013 *Int. J. Educ. Mat. Sci. Technol.* **1** 138
[4] Thomson S, Hillman K and Bortoli L D *Programme for International Student Assessment: A Teacher's Guide to PISA Scientific Literacy* (Victoris: Acer Press)
[5] Sunarti T 2015 *Pemahaman Literasi Sains Mahasiswa Calon Guru Fisika Universitas Negeri Surabaya Pros. Semin. Fisika dan Pembelajarannya Pascasarjana Universitas Negeri Malang* (Malang: UM)

[6] Plomp T and Nieveen N 2013 *Educational Design Research* (Enschede: SLO)
[7] Fraenkel J R, Wallen N E and Hyun H H 2012 *How to Design and Evaluate Research in Education* (New York: McGraw-Hill)
[8] Hake R R 1998 *Am. J. Phys.* **66** 64
[9] Eggen P D and Kauchak D P 2012 *Educational Psychology: Windows on Clasrooms* (New Jersey: Pearson)
[10] Moreno R 2010 *Educational Psychology* (New York: John Wiley & Sons Inc)
[11] Rusli A 2013 *Simposium Nasional Inovasi dan Pembelajaran Sains* (Bandung: UPI)
[12] Moutinho S, Torres J, Fernandes I and Vasconcelos C 2014 *Proc. Soc. Behav. Sci.* **191** 1871
[13] Akenngi H and Sirin A 2013 *A. J. Educ. Res. Rev.* **8** 1882
[14] Arifin L and Sunarti T 2017 *J. Penelit. Fis. Apl.* **7** 68

IOP Conf. Series: Journal of Physics: Conf. Series **997** (2018) 012014 doi:10.1088/1742-6596/997/1/012014

The D-dimensional non-relativistic particle in the Scarf Trigonometry plus Non-Central Rosen-Morse Potentials

U A Deta[1], N A Lestari[1], M Yantidewi[1], A Suparmi[2] and C Cari[2]

[1]Department of Physics, Universitas Negeri Surabaya, Indonesia
[2]Department of Physics, Graduate Program, Sebelas Maret University, Indonesia

E-mail: utamadeta@unesa.ac.id and utamaalan@yahoo.co.id

Abstract. The D-Dimensional Non-Relativistic Particle Properties in the Scarf Trigonometry plus Non-Central Rosen-Morse Potentials was investigated using an analytical method. The bound state energy is given approximately in the closed form. The approximate wave function for arbitrary l-state in D-dimensions are expressed in the form of generalised Jacobi Polynomials. The energy spectra of the particle are increased when the dimensions are higher. The relationship between the orbital number in each dimension is recursive. The special case in 3 dimensions is given to the ground state.

1. Introduction

The solution of Schrodinger equations for some potentials is very essential. It provides the information of the quantum system. Some researchers investigated the approximate solution of the shape invariant potentials. This potential has been investigated using several techniques. For example, Coulomb potential using Laplace transformation [1], Morse potential using series expansion method [2], Modified Poschl-Teller, Hulten, Scarf hyperbolic Manning-Rosen plus Scarf potentials using NU method [3-7], q-deformed Rosen-Morse Potential using Supersymmetry Quantum Mechanics [8]. The solution of shape-invariant potentials in D-dimensions only in radial part [1-6]. Deta [9] try to solve the angular part with D-dimensional Poschl-Teller potential. The other D-Dimensional shape invariant potentials has not been solved yet.

The exact solution of Schrodinger Equation can be solved exactly only for $l = 0$ in 1 dimension. For the other value of l and D ($l \neq 0$ and $D > 1$), the Schrodinger equation might be solved approximately with appropriate approximation scheme. One of the suitable approximation schemes is proposed by Greene and Aldrich [10,11].

Recently, Deta [9] solved the combination between radial plus angular potential in D-dimensions. He solved Scarf hyperbolic plus non-central Poschl-Teller potential. Suparmi [12] solved four dimensional extended hyperbolic Scarf I plus three dimensional separable trigonometric noncentral potentials. However, other combinations of radial and angular potential in D-dimension have not been studied yet. One of the combinations that still has not been studied is Scarf Trigonometry plus Non-Central Rosen-Morse potentials.

In this research, we will try to solve the Schrodinger equation in D-Dimensions for Scarf Trigonometry plus Non-Central Rosen-Morse potential using Nikivorof-Uvarov method. The N-U method developed by Nikiforov-Uvarov [13]. This method based on solving a second order linear differential equation. The equation is reduced to a generalized equation of hypergeometric type differential equation by a suitable change of variable.

The Scarf potential describes the periodically arranged particles like a crystal [5,6,14,15]. The application of this potential is a crystal model in the solid state physics [5,6]. On the other hand, the Rosen-Morse potential describes the quark-gluon dynamics of Quantum Chromodynamics [16]. The solution of this potential remains in the radial part [16-18]. The combination of these potential in D-Dimensions with the centrifugal term is separable potential. Hence, this system can be solved by the separation of variable method. The solution of radial part of Schrodinger equation is one solution. Nevertheless, the solution of angular part of Schrodinger equation is D-1 solutions.

2. The Nikiforov-Uvarov Method

The Schrodinger equation in D-dimensions of any shape invariant potential can be reduced into hypergeometric-type of a differential equation using a suitable variable transformation [19-21]. The hypergeometric-type of differential equation in Nikiforov-Uvarov method is:

$$\frac{\partial^2\Psi(s)}{\partial s^2}+\frac{\bar{\tau}(s)}{\sigma(s)}\frac{\partial\Psi(s)}{\partial s}+\frac{\bar{\sigma}(s)}{\sigma^2}\Psi(s)=0 \tag{1}$$

where $\sigma(s)$ and $\bar{\sigma}(s)$ are polynomials, mostly in the second order, and $\bar{\tau}(s)$ is the first order polynomial. Equation (1) can be solved using separation of variable, that is:

$$\Psi=\phi(s)y(s) \tag{2}$$

By inserting equation (2) into equation (1), we get the hypergeometric type equation, that is:

$$\sigma\frac{\partial^2 y}{\partial s^2}+\tau\frac{\partial y}{\partial s}+\lambda y=0 \tag{3}$$

Where ϕ (s) is a logarithmic derivative. The solution of ϕ (s) is obtained from condition:

$$\frac{\phi'}{\phi}=\frac{\pi}{\sigma} \tag{4}$$

while the function π (s) and the parameter λ are defined as:

$$\pi=\left(\frac{\sigma'-\bar{\tau}}{2}\right)\pm\sqrt{\left(\frac{\sigma'-\bar{\tau}}{2}\right)^2-\bar{\sigma}+k\sigma} \tag{5}$$

$$\lambda=k+\pi' \tag{6}$$

The value of k in equation (5) can be found from the condition that the expression under the square root of equation (5) must be a square of a polynomial. This polynomial is mostly first-degree polynomial, therefore the discriminate of the quadratic expression is zero. A new eigenvalue of equation (3) is:

$$\lambda=\lambda_n=-n\pi'-\frac{n(n-1)}{2}\sigma''; \; n=0,1,2,\ldots \tag{7}$$

where

$$\tau=\bar{\tau}+2\pi \tag{8}$$

The new bound state energy obtained using the equation (6) and (7). To generate the bound state energy and the corresponding eigenfunction, the condition that $\tau' < 0$ is required. The solution of the second part of the wave function, y_n (s), is connected to Rodrigues relation [20]:

$$y_n(z)=\frac{C_n}{\rho(z)}\frac{d^n}{dz^z}\left\{\sigma^n(z)\rho(z)\right\} \tag{9}$$

where C_n is normalization constant, and the weight function $\rho(s)$ must satisfies the condition:

$$\frac{\partial(\sigma\rho)}{\partial s}=\tau(s)\rho(s) \tag{10}$$

The wave function of the system obtained from equation (4), (9) and (10).

3. Scarf Trigonometry plus Non-Central Rosen-Morse Potentials

The Scarf Trigonometry potential can be expressed as:

$$V = \frac{\hbar^2}{2m}\left\{\frac{b^2 + a(a-1)}{\sin^2 r} - \frac{2b\left(a - \frac{1}{2}\right)\cos r}{\sin^2 r}\right\} \tag{11}$$

The Non Central Rosen-Morse potential can be expressed as:

$$V = \frac{\hbar^2}{2mr^2}\left\{\frac{\upsilon(\upsilon+1)}{\sin^2 \Omega_D} - 2q\cot\Omega_D\right\} \tag{12}$$

The Schrödinger Equation in D-Dimensions for Scarf Hyperbolic plus Non-Central Pocshl-Teller Potential can be expressed as:

$$-\frac{\hbar^2}{2m}\nabla_D^2\Psi(r,\Omega) + \left[\begin{array}{l}\frac{\hbar^2}{2m}\left\{\frac{b^2 + a(a-1)}{\sin^2 r} - \frac{2b\left(a - \frac{1}{2}\right)\cos r}{\sin^2 r}\right\} \\ + \frac{\hbar^2}{2mr^2}\left\{\frac{\upsilon(\upsilon+1)}{\sin^2 \Omega_D} - 2q\cot\Omega_D\right\}\end{array}\right]\Psi(r,\Omega) = E\Psi(r,\Omega) \tag{13}$$

where

$$\nabla_D^2 = \frac{1}{r^{D-1}}\frac{\partial}{\partial r}\left(r^{D-1}\frac{\partial}{\partial r}\right) - \frac{\Lambda_D^2(\Omega_D)}{r^2} \tag{14}$$

with $\Lambda_D^2(\Omega_D)$ is hypersperical harmonics, that is, D-Dimensional angular momentum operator. For $2 \le k \le D-1$, we have:

$$\Lambda_D^2(\Omega_D) = L_k^2 = \sum_{a<b=2}^{k-1} L_{ab}^2 = -\frac{1}{\sin^{k-1}\theta_k}\frac{\partial}{\partial\theta_k}\left(\sin^{k-1}\theta_k\frac{\partial}{\partial\theta_k}\right) + \frac{L_{k-1}^2}{\sin^2\theta_k} \tag{15}$$

and for $k = 1$, we have:

$$L_1^2 = -\frac{\partial^2}{\partial\theta_1^2} \tag{16}$$

Subtitute (14), (15), (16), using variable substitution, and separation variable, that is,

$$\varepsilon^2 = -\frac{2m}{\hbar^2}E \tag{17}$$

$$\Psi(r,\Omega_D = \theta_1,\theta_2,...\theta_{D-1}) = \Psi(r)\Phi(\theta_1 = \varphi)H(\theta_2,...\theta_{D-1}) \tag{18}$$

we have the radial, polar, and azimuth parts of Schrodinger equation are:

$$\frac{1}{r^{D-1}}\frac{\partial}{\partial r}\left(r^{D-1}\frac{\partial}{\partial r}\right)\Psi(r) - \left\{\frac{b^2+a(a-1)}{\sin^2 r} - \frac{2b\left(a-\frac{1}{2}\right)\cos r}{\sin^2 r} + \varepsilon^2\right\}\Psi(r) = \frac{\lambda^*}{r^2}\Psi(r) \tag{19}$$

$$\frac{1}{\sin^{k-1}\theta_k}\frac{\partial}{\partial\theta_k}\left(\sin^{k-1}\theta_k\frac{\partial}{\partial\theta_k}H\right) - \frac{l_{k-1}\left(l_{k-1}+k-2\right)}{\sin^2\theta_k}H + l_k\left(l_k+k-1\right)H - \left\{\frac{\upsilon(\upsilon+1)}{\sin^2\Omega_D} - 2q\cot\Omega_D\right\}H = 0 \tag{20}$$

for $k = 2, 3, 4, \ldots, D-1$.

$$-\frac{1}{\Phi}\frac{d^2}{d\theta_1^2}\Phi = L_k^2 \tag{21}$$

3.1. Solution of azimuth part of Schrodinger equation in D-dimensions

The azimuth part of Schrodinger equation in D-Dimensions (equation 20) is the ordinary differential equation. The unnormalized solution of azimuth part of Schrodinger equation in D-Dimensions (21) is:

$$\Phi = Ae^{im\varphi} \tag{22}$$

where A is normalization constant.

3.2. Solution of radial part of Schrodinger equation in D-dimensions

The radial part of Schrodinger equation in D-Dimensions (Eq. 19) is hypergeometric differential equation. Equation (19) can be solved using some approximation and variable substitution, that is,

$$\Psi(r) = r^{-\frac{D-1}{2}}R(r) \tag{23}$$

$$r \approx \sin r \tag{24}$$

$$\cos r = s \tag{25}$$

Substitute equation (23), (24), and (25) to equation (19), we have:

$$\frac{d^2}{ds^2}R(s) - \frac{s}{\left(1-s^2\right)}\frac{d}{ds}R(s) - \left\{\frac{b^2+a(a-1)-2b\left(a-\frac{1}{2}\right)s + \varepsilon^2\left(1-s^2\right)+\lambda^* + \frac{(D-1)(D-3)}{4}}{\left(1-s^2\right)^2}\right\}R(r) = 0 \tag{26}$$

By comparing equation (1), (26), and using eigenvalue of Nikivorof-Uvarov method, we obtain the energy spectra of Scarf Trigonometry plus Non-Central Rosen-Morse potentials in D-Dimensions is:

$$E = \frac{\hbar^2}{2m}\left(p+n+\tfrac{1}{2}\right)^2 \tag{27}$$

where

$$p = \sqrt{\frac{\left\{b^2+\left(a-\tfrac{1}{2}\right)^2+\lambda^*+\frac{(D-1)(D-3)}{4}\right\} - \sqrt{\left[\left\{b+\left(a-\tfrac{1}{2}\right)\right\}^2+\lambda^*+\frac{(D-1)(D-3)}{4}\right]\left[\left\{b-\left(a-\tfrac{1}{2}\right)\right\}^2+\lambda^*+\frac{(D-1)(D-3)}{4}\right]}}{2}} \tag{28}$$

and A_{D-1} is centrifugal term which depends on the eigenvalue of polar part of Schrodinger equation in D-Dimensions.

By using eigenfunction of NU method, we obtain:

$$R_n = B_n\left(1-s^2\right)^{\frac{1}{2}p+\frac{1}{4}}\left(1+s\right)^{\frac{b\left(a-\frac{1}{2}\right)}{2p}} \left(1-s\right)^{-\frac{b\left(a-\frac{1}{2}\right)}{2p}} P_n^{(\alpha,\beta)}(s) \tag{29}$$

where $P_n(\alpha,\beta)$ is Jacobi Polynomial, that is,

$$P_n^{(\alpha,\beta)}(s) = \frac{(-1)^n}{2^n n!}(1-s)^{-\alpha}(1+s)^{-\beta} \frac{d^n}{dz^n}\left\{(1-s)^{\alpha}(1+s)^{\beta}\left(1-s^2\right)^n\right\} \tag{30}$$

and B_n is normalization constant.

Finally, from equation (23), (25), and (29), we have the radial wave function of Schrodinger equation for Scarf Hyperbolic plus Non-Central Pocshl-Teller Potentialin D-Dimensions is:

$$\Psi(r) = B_n r^{-\frac{D-1}{2}} B_n(\sin r)^{p_r+\frac{1}{2}}(1+\cos r)^{\frac{b\left(a-\frac{1}{2}\right)}{2p}} (1-\cos r)^{-\frac{b\left(a-\frac{1}{2}\right)}{2p}} P_n^{(\alpha,\beta)}(\cos r) \tag{31}$$

The effect of extra dimensions on bound state energy and radial wave function can be visualized as

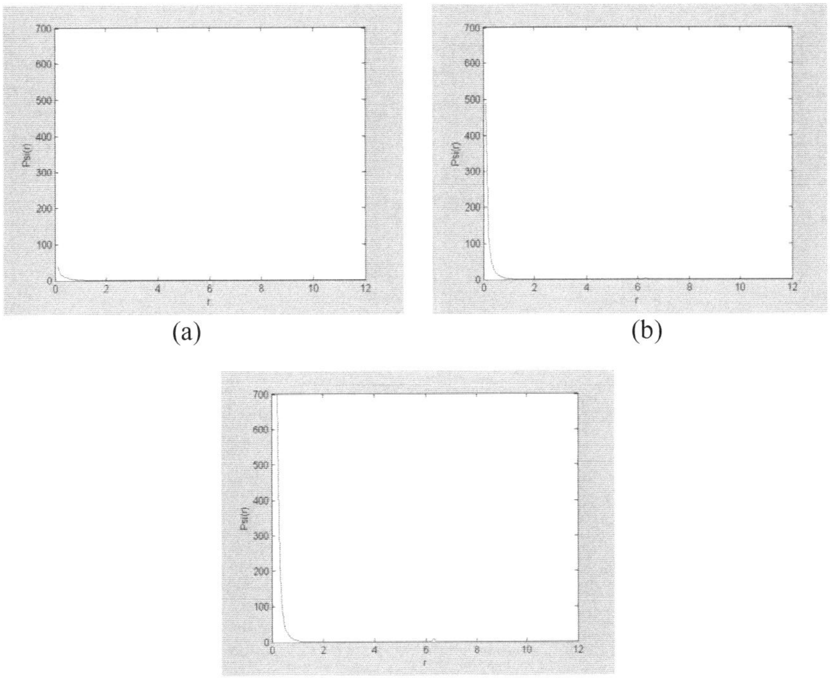

Figure 1. Radial Wave Function on Ground State for Scarf Trigonometry Potential (a = 2 and b = 2) in (a) 3D, (b) 4D, and (c) 5D.

From Figure 1, we have the effect of extra dimensions in radial wave function is increase the amplitude of radial wave function. It was found to agree with previous work [6]. Based on superstring theory [22], the number of dimensions in the universe restricted to 10-spatial dimensions and 1-time dimension. If the amount of spatial dimension more than 10, the universe unstable and collapse. Thus, the maximum value of D is limited to 10 spatial dimensions.

3.3. Solution of polar part of Schrodinger equation in D-dimensions

According to the value of k, there are two equations at polar part of Schrodinger equation in D-Dimensions. First, for $k = 2, 3, 4, \ldots, D-1$, we solve equation (20) using variable substitution, that is,

$$A_k = l_k\left(l_k + k - 1\right) \quad (32a)$$

$$A_{k-1} = l_{k-1}\left(l_{k-1} + k - 2\right) \quad (32b)$$

$$H = \frac{Q}{\left(\sin\theta_k\right)^{\frac{k-1}{2}}} \quad (33a)$$

$$s = \cot\theta_k \quad (33b)$$

Substitute equation (32a), (32b), (33a), (33b) to (21a), we have

$$\frac{d^2Q}{ds^2}+\frac{2s}{\left(1+s^2\right)}\frac{dQ}{ds}-\left\{\frac{\left[\frac{k-1}{2}\frac{k-3}{2}+A_{k-1}+\upsilon(\upsilon-1)\right]\left(1+s^2\right)-2qs-\left(\frac{k-1}{2}\right)^2-A_k}{\left(1+s^2\right)^2}\right\}Q=0 \tag{34}$$

By comparing equation (1), (34), and using eigenvalue of Nikivorof-Uvarov method, we obtain the eigenvalue of Scarf Hyperbolic plus Non-Central Pocshl-Teller potential in D-Dimensions:

$$A_k=\left(\sqrt{\frac{k-1}{2}\frac{k-3}{2}+\frac{1}{4}+A_{k-1}+\upsilon(\upsilon-1)}+n+\frac{1}{2}\right)^2-\frac{q^2}{\left(\sqrt{\frac{k-1}{2}\frac{k-3}{2}+\frac{1}{4}+A_{k-1}+\upsilon(\upsilon-1)}+n+\frac{1}{2}\right)^2}-\left(\frac{k-1}{2}\right)^2 \tag{35}$$

By using eigenfunction of NU method, we obtain:

$$H\left(\cot\theta_k\right)=C_n\frac{\left(1+\cot^2\theta_k\right)^{\frac{1}{2}p}}{\left(\sin\theta_k\right)^{\frac{k-1}{2}}}e^{-\frac{q}{p}\arctan\left(\cot\theta_k\right)}\frac{d^n}{d\cot\theta_k{}^n}\left\{\left(1+\cot^2\theta_k\right)^{n-p}e^{\frac{2q}{p}\arctan\left(\cot\theta_k\right)}\right\} \tag{36}$$

with

$$p=\sqrt{\frac{\left[\left(\frac{k-1}{2}\right)^2+A_k\right]+\sqrt{\left[\left(\frac{k-1}{2}\right)^2+A_k\right]^2+4q^2}}{2}} \tag{37}$$

Note that the value of $k = 2, 3, 4, \ldots, D-1$, depend on the number of dimensions.

4. Special cases for the ground and first excited state in 3 dimensions

Special cases for 3 Dimensional problems for Scarf Hyperbolic plus Non-Central Pocshl-Teller Potential, that is:

$$\Psi\left(r,\Omega_3\right)=\Psi(r)Y\left(\Omega_3\right) \tag{38}$$

with solution of radial part of 3 dimensional Schrodinger equation:

$$\Psi\left(r\right)=B_{n_r}r^{-1}B_{n_r}\left(\sin r\right)^{\frac{1}{2}p_r+\frac{1}{4}}\left(1+\cos r\right)^{\frac{b\left(a-\frac{1}{2}\right)}{2p_r}}\left(1-\cos r\right)^{-\frac{b\left(a-\frac{1}{2}\right)}{2p_r}}P_{n_r}^{(\alpha,\beta)}\left(\cos r\right) \tag{39}$$

$$P_{n_r}^{(\alpha_r,\beta_r)}(\cos r)=\frac{(-1)^{n_r}}{2^n n_r!}(1-\cos r)^{-\alpha_r}(1+\cos r)^{-\beta_r}\frac{d^{n_r}}{d(\cos r)^{n_r}}\left\{(1-\cos r)^{\alpha_r}(1+\cos r)^{\beta_r}\left(\sin^2 r\right)^{n_r}\right\} \tag{40}$$

$$\alpha_r=-p_r+\frac{b\left(a-\frac{1}{2}\right)}{p_r} \tag{41a}$$

$$\beta_r=-p_r-\frac{b\left(a-\frac{1}{2}\right)}{p_r} \tag{41b}$$

$$p_r=\sqrt{\frac{\left\{b^2+\left(a-\frac{1}{2}\right)^2+A_2\right\}-\sqrt{\left[\left\{b+\left(a-\frac{1}{2}\right)\right\}^2+A_2\right]\left[\left\{b-\left(a-\frac{1}{2}\right)\right\}^2+A_2\right]}}{2}} \tag{42}$$

From equation (39) to (42), we can make a visualization of radial wave function using Matlab, that is:

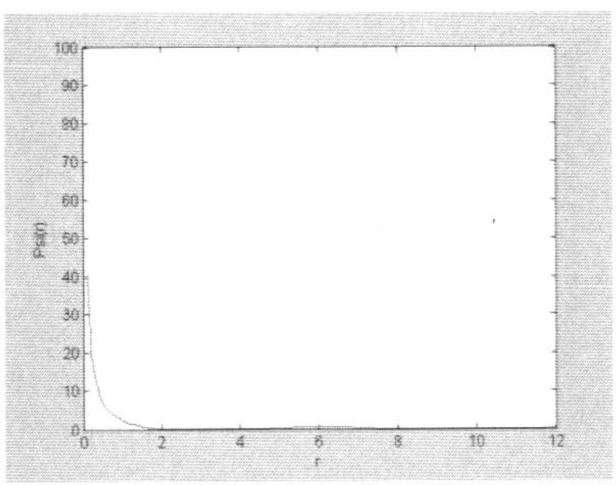

Figure 2. Ground State Radial Wave Function on Ground State for Scarf Trigonometry Potential (a = 2 and b = 2) in 3 Dimensions.

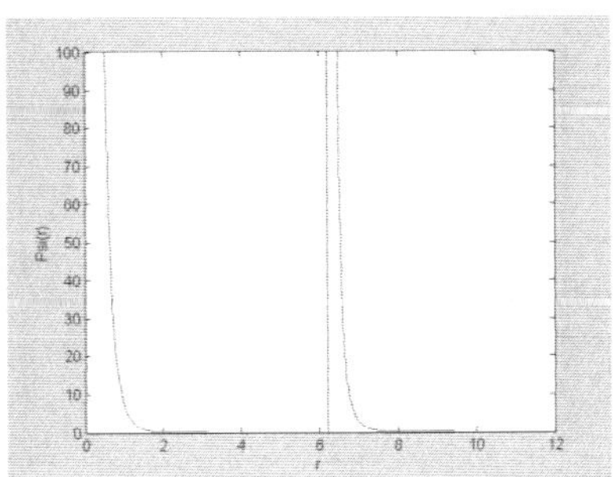

Figure 3. Ground State Radial Wave Function on Ground State for Scarf Trigonometry plus Non Central Rosen-Morse Potential (a = 2, b = 2, v = 2, and q = 2) in 3 Dimensions.

From Figure 2 and 3, we have the effect of each potential parameter. The effect of parameter Non-Central Rosen-Morse will not appear if the parameters of Scarf trigonometry are zero. It causes the radial wave function vanishes. If the Scarf trigonometry parameters are not zero, the Non-Central Poschl-Teller parameters duplicate the radial wave function as shown in Figure 3.

Solution of Angular part divided into 2 parts, that is:

$$Y_{l_1}^{(l)}\left(\Omega_D=\theta_1,\theta_2\right)=\Phi\left(\theta_1=\varphi\right)H\left(\theta_2\right) \tag{43}$$

Solution of azimuth part equation:

$$\Phi = Ae^{im\varphi} \tag{44}$$

$$A=\frac{1}{\sqrt{2\pi}} \tag{45}$$

Solution of polar part equation:

$$H\left(\cot\theta_2\right)=C_{n_{\theta_2}}\frac{\left(1+\cot^2\theta_2\right)^{\frac{1}{2}p_{\theta_2}}}{\left(\sin\theta_2\right)^{-\frac{1}{2}}}e^{-\frac{q}{p_{\theta_2}}\arc\tan\left(\cot\theta_2\right)}$$
$$\frac{d^n}{d\cot\theta_2^{\,n}}\left\{\left(1+\cot^2\theta_2\right)^{n_{\theta_2}-p_{\theta_2}}e^{\frac{2q}{p_{\theta_2}}\arc\tan\left(\cot\theta_2\right)}\right\} \tag{46}$$

with

$$p_{\theta_2}=\sqrt{\frac{\left[\frac{1}{4}+A_2\right]+\sqrt{\left[\frac{1}{4}+A_2\right]^2+4q^2}}{2}} \tag{47}$$

$$A_2=\left(\sqrt{A_1+\upsilon\left(\upsilon-1\right)}+n_{\theta_2}+\tfrac{1}{2}\right)^2$$
$$-\frac{q^2}{\left(\sqrt{A_1+\upsilon\left(\upsilon-1\right)}+n_{\theta_1}+\tfrac{1}{2}\right)^2}-\tfrac{1}{4} \tag{48}$$

Using substitution of all equation (46) to (48), we get unnormalized solution of 3 dimensional Schrodinger equations for Scarf hyperbolic plus non central Poschl-Teller potential, for example:

Table 1. Visualization of angular part wave function for Non-Central Rosen Morse-Teller Potential.

$H\left(\cos\theta\right)_{n_l m v q}$	3 Dimensions	2 Dimensions Projection
$H\left(\cos\theta_2\right)_{0000}$	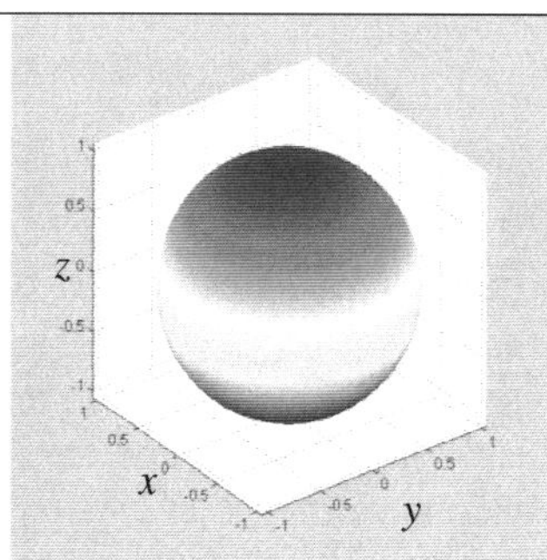	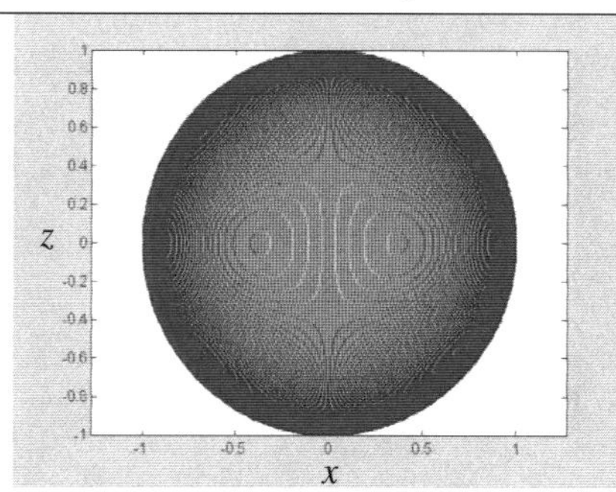

IOP Conf. Series: Journal of Physics: Conf. Series **997** (2018) 012014 doi:10.1088/1742-6596/997/1/012014

$H\left(\cos\theta\right)_{n_l m v q}$	3 Dimensions	2 Dimensions Projection
$H\left(\cos\theta\right)_{0033}$	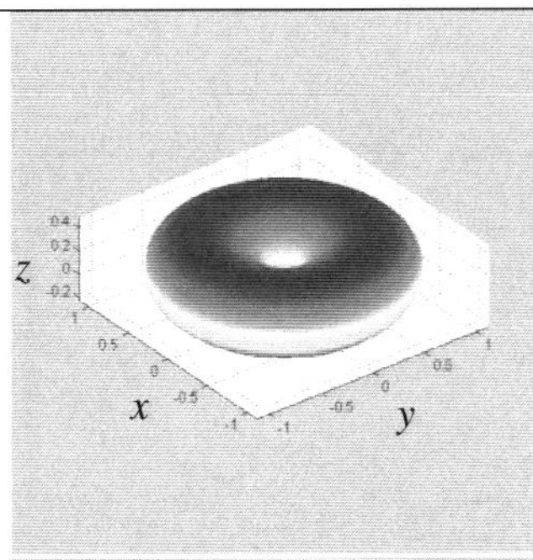	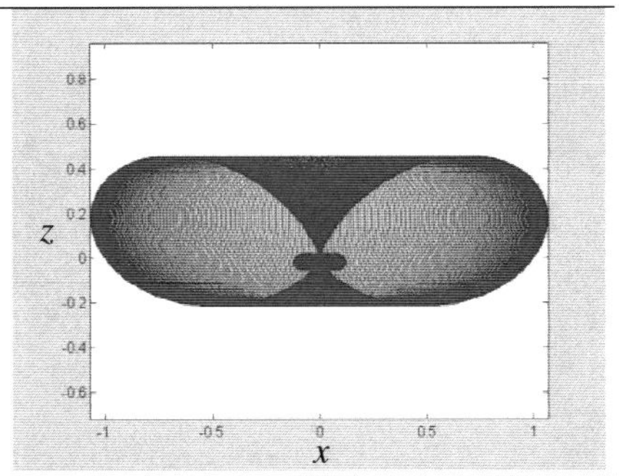
$H\left(\cos\theta\right)_{0100}$	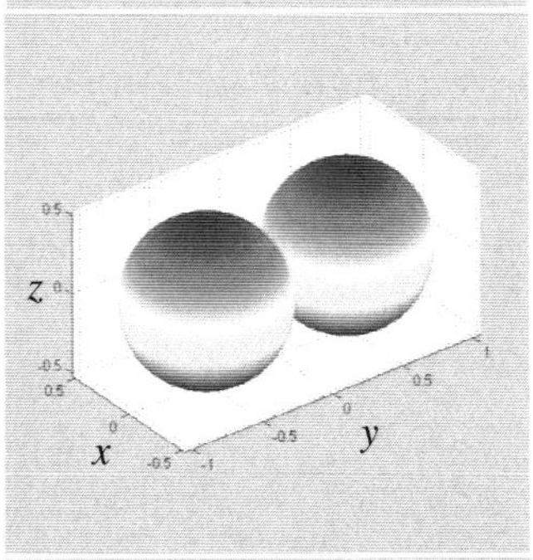	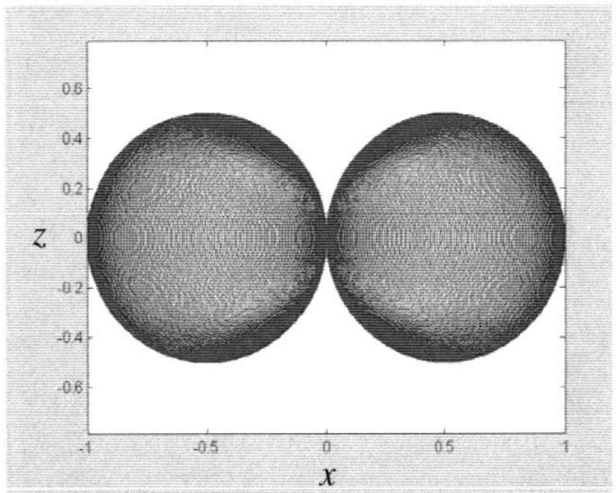
$H\left(\cos\theta\right)_{0133}$	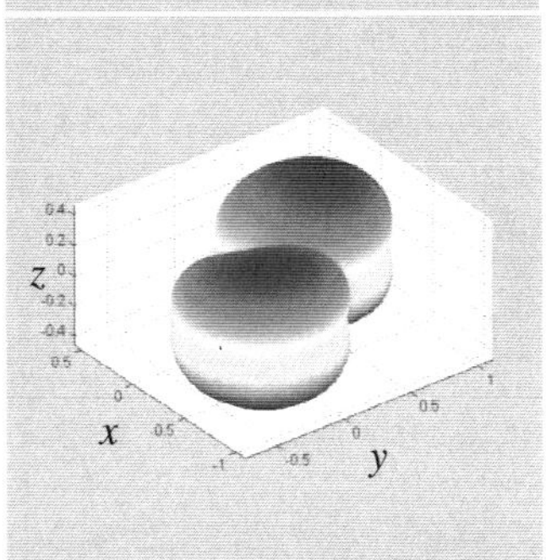	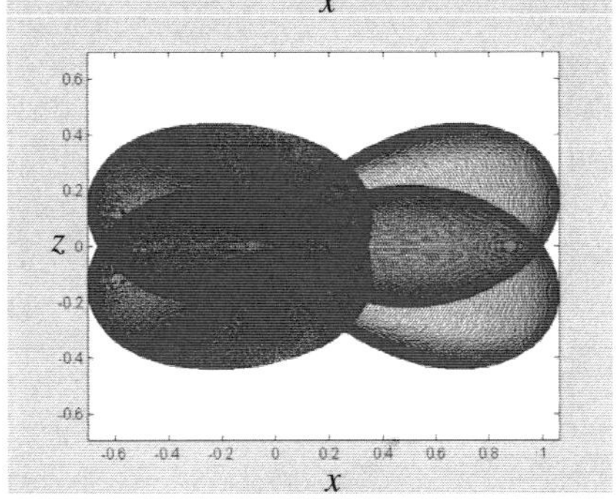
$H\left(\cos\theta\right)_{1000}$	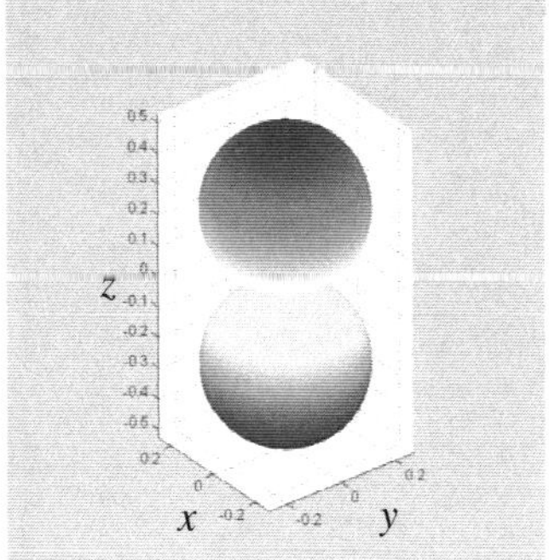	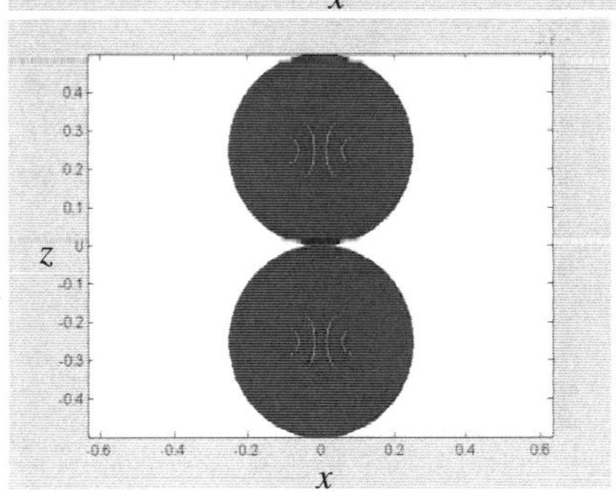

$H(\cos\theta)_{n_l mvq}$	3 Dimensions	2 Dimensions Projection
$H(\cos\theta)_{1033}$	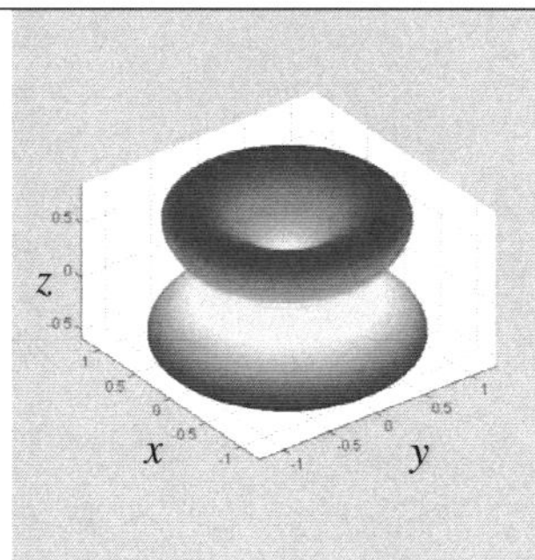	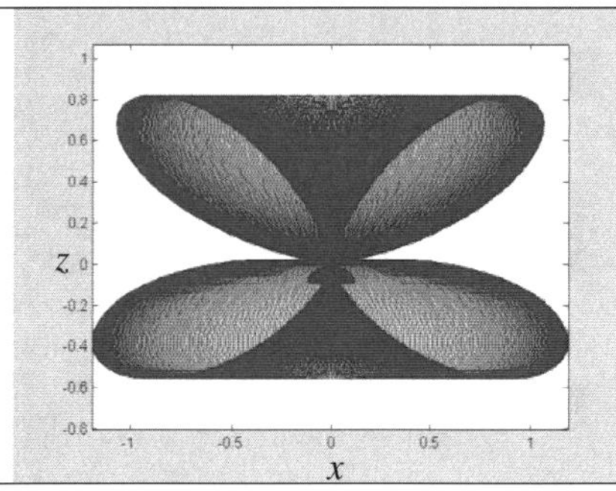

The effect of polar potential parameters in 3-dimensions of Non-Central Rosen-Morse Potential on the orbital (subshell) particle can be described in Table 1. Without any polar potential (v and q are zero), the orbital electron will be changed into orbitals in spherical harmonics or hydrogen-like atom. These phenomena are figured in the number 1, 3, and 7, which correspond to an orbital in spherical harmonics in the same parameters [23].

5. Conclusion

In this paper, we present the solutions of Schrodinger equation in D-dimension for Scarf Trigonometry plus non-central Rosen-Morse Potentials within the framework of an approximation to the centrifugal and high dimension term. The bound state energy is obtained in D-dimensions using the Nikiforov-Uvarov method, and it was found to agree with previous works [6, 18]. The corresponding wave function of the Scarf trigonometry plus non-central Rosen-Morse potentials was obtained in terms of the Jacobi polynomials. The example of bound state energy and wave function in 3 dimensions presented in the condition of the ground state. The existence of arbitrary dimensions increases the amplitude of wave function.

Acknowledgement
This research is partly supported by Hibah Peneliti Utama (PUT UNS) 2014 and DIKTI (Grant No. 165a/UN27.11/PN2013).

References

[1] Chen G 2004 *Z. Naturforsch* **59a** 875
[2] Dong S H and Sun G H 2003 *Phys Lett. A* **314** 261
[3] Agboola D 2010 *Chin. Phys. Lett.* **27** 040301
[4] Agboola D 2009 *Phys. Scr.* **80** 065304
[5] Deta U A, Suparmi and Cari 2013 *Adv. Studies Theor. Phys.* **13** 647
[6] Deta U A, Suparmi and Cari 2013 *AIP Conf. Proc.* **1554** 190
[7] Luqman H, Suparmi, Cari and Deta U A 2014 *J. Phys. Conf. Ser.* **539** 012019
[8] Alemgadmi K I K, Deta U A, Cari and Suparmi 2015 *AIP Conf. Proc.* **1677** 030004
[9] Deta U A, Suparmi and Cari 2014 *J. Phys. Conf. Ser.* **539** 012018
[10] Greene R I and Aldrich C 1976 *Phys. Rev. A* **14** 2363
[11] Xu Y, He S and Jia C S 2010 *Phys. Scr.* **81** 045001
[12] Suparmi A, Cari C, Deta U A and Handhika J 2016 *J. Phys. Conf. Ser.* **776** 012077

[13] Nikiforov A F and Uvarov V B 1988 *Special functions of Mathematical Physics* (Boston: Birkhauser Verlag Basel) p 317-318
[14] Negro J, Nieto L M and Rosas-Ortiz O 2004 *J. Phys. A: Math. Gen.* **43** 10079
[15] Castillo 2007 *Rev. Mex. Fis.* **E53/2** 143
[16] Jasso CBC and Kirchbach M 2006 *AIP Conf. Proc.* **857** 275
[17] Jasso C B C and Kirchbach M 2006 *J. Phys. A: Math. Gen.* **39** 547
[18] Deta U A, Suparmi, Cari, Husein A S, Yuliani H, Alemgadmi K I K, Hakim L and Supriyanto 2014 *AIP Conf. Proc.* **1615** 121
[19] Suparmi 2011 *Mekanika kuantum II* (Surakarta: Sebelas Maret University) p 166-169
[20] Ikot A N, Akpabio L E and Uwah E J 2011 *Electronic J. Theor. Phys.* **25** 225
[21] Ikot A N, Antia A D, Akpabio L E and Obu J A 2011 *J. Vestib. Res.* **6** 65
[22] Wray K 2011 *An introduction to string theory* http://math.berkeley.edu/~kwray/papers/string_theory.pdf [accessed 11 Juli 2013]
[23] Orchin M, Macomber R S, Pinhas A R and Wilson R M 2005 *The vocabulary and concepts of organic Chemistry - Second Edition* (Hoboken: John Wiley & Sons, Inc.) p 13-16

IOP Conf. Series: Journal of Physics: Conf. Series **997** (2018) 012015 doi:10.1088/1742-6596/997/1/012015

Spatial distribution of topsoil magnetic susceptibility in Sawahlunto City, West Sumatera

Afdal[1] and E S Wahyuni[1]

[1]Department of Physics, Universitas Andalas, Kampus UNAND, Limau Manis, Padang 25163, Indonesia

E-mail: afdal@fmipa.unand.ac.id

Abstract. A research to determine the spatial distribution of top soil magnetic suceptibility at Sawahlunto City, West Sumatra has been conducted. The top soil samples were taken at four locations ie the downtown area, the steam power plant area, the agricultural area, and coal mine area. At each location, the soil samples were taken at 10 points at a depth of 20 cm. Magnetic susceptibility were measured using Bartington MS2B Magnetic Susceptibility Meter. The topsoil samples from Sawahlunto city have relatively low average value of the magnetic susceptibility that is 67.0×10^{-8} m^3/kg. The magnetic susceptibility of topsoil samples from downtown area have the average and the highest value of magnetic susceptibility (100.6×10^{-8} and 259.9×10^{-8} m^3/kg), and followed by sample from the steam power plant area (98.4×10^{-8} and 258.0×10^{-8} m^3/kg), the agricultural area (56.2×10^{-8} and 83.7×10^{-8} m^3/kg), and coal mine area (12.9×10^{-8} and 26.8×10^{-8} m^3/kg). Soil samples from the steam power plant area have the widest range of magnetic susceptibility value range from 0.3×10^{-8} to 258.0×10^{-8} m^3/kg.

1. Introduction

Mining and industrial activity is involved in the drastic disruption of regional ecosystems and soil properties [1]. Industrial and mining activities has produced a significant release of heavy metals to the environment from processes such as energy production, mining, metal processing, manufacturing, burning of waste, and burning of fossil fuels [2]. Heavy metals can accumulate in the soil around the industrial area through atmospheric deposition, waterlogging of waste piles, and this causes serious pollution and degradation of soil quality [3]. The soil is the accumulation of toxic materials such as heavy metals. This non-biodegradable material deposited on surface soil can affect its environment for decades which has a strong negative impact on the environment [4]. The surface layer is the recipient of various pollutants, especially heavy metals and can be used as an indicator to determine the quality of the environment. The topsoil is the part that is most affected by industrial deposition and not depend on the type of industry [5].

Sawahlunto is one of the cities that has significant mineral resources. The city has a coal mining that has been operating since 1891, initially operated by the Dutch East Indies government. After Indonesia became independent, it was managed by the state through Ombilin Coal Mining Company. The company was then liquidated as a subsidiary of PTBA located in Tanjung Enim, South Sumatera. Coal from Sawahlunto is used to meet the needs of coal of PT Semen Padang and Ombilin power plant in the city of Sawahlunto. This mining site has not been operating since 2002. In addition, in the city of Sawahlunto also there are many traditional gold mines. Sawahlunto also has a coal fired power plant with coal consumption of 2000 tons per day. This activity is a source of soil contamination in the city of Sawahlunto besides from transportation and agricultural activities. There are many studies about the influence of mining on air and water quality, but few that focus on land quality issues by pollution of mines. So, it needs to know the contamination level of topsoil in city of Sawahlunto through the mapping its magnetic susceptibility value.

To estimate the concentration of heavy metals in the soil is usually used geochemical methods, but this method is expensive and time consuming, so it needs a time-efficient and cost-effective, non-destructive method. Magnetic methods have been successfully applied in several European countries [6]. The magnetic method has also conducted to determine the heavy metal distribution in Bucharest [7]. From these studies it is known that magnetic susceptibility can be used as an indicator of high concentrations of heavy metals in topsoil. Magnetic susceptibility is a measurable and concentration-dependent geophysical parameter that describes a material's ability to obtain magnetization when given an external magnetic field [8].

The application of magnetic susceptibility measurement for characterization and pollution checking is possible due to the presence of technogenic magnetic particles (TMP) in soils and sediments that are influenced by industrial deposition and urban dust derived from high-temperature engineering processes [9]. Areas which are closer to industrial areas have higher magnetic susceptibility values. The magnetic concentration parameters show significantly different magnetic responses to different sources of pollution. Magnetic proxies provide a fast way of detecting heavy metal contamination caused by anthropogenic pollution sources in large areas, although their sensitivity is controlled by pollution sources [10].

2. Methodology

The research was conducted at the Earth Physics Laboratory, the Physics Department of Andalas University. Sampling was conducted at four locations is in the Sawahlunto city, ie at the coal mine area, power plant area, agricultural area, and downtown (see Figure 1). At each of the locations, topsoil samples were collected at 10 points. The position of the sampling points in the locations are presented in Figure 2 to 5. At each of the 40 points, 50 g topsoil samples were collected at a depth of 20 cm. The distance between the sample points was 500 m.

Figure 1. Location of the research area in Sawahlunto City.

Equipments that were used in the research are MS2B magnetic susceptibility meter (to measure the magnetic susceptibility of the sample), core drill, and other supporting equipment such as balance, hammer, meter, and sample container. Measurement of the magnetic susceptibility of the samples were conducted in the laboratory. Each of the topsoil samples was mixed, air dried for 5 days to reduce the effect of water, and then sieved with 100-mesh sieve. Then a 12 g of each samples was stored into plastic container (sample holder), and their mass magnetic susceptibility was measured in the laboratory using MS2 magnetic susceptibility meter with MS2B sensor. The measurements were conducted in 15 directions and then its average was calculated.

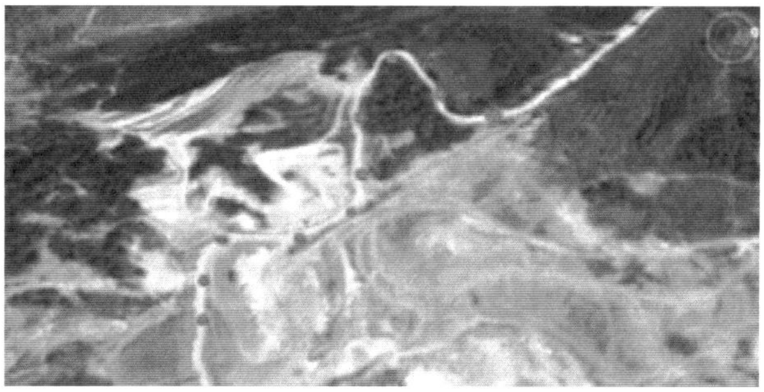

Figure 2. Sampling points in the coal mine area.

Figure 3. Sampling points near coal fired power plant area.

Figure 4. Sampling points in the downtown area.

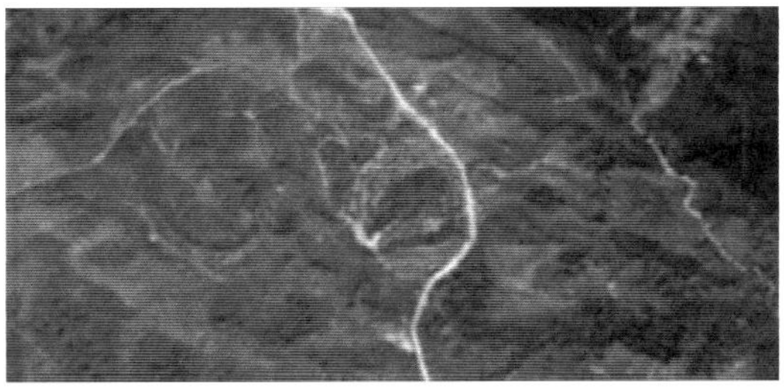

Figure 5. Sampling points in the agricultural area.

3. Results and Discussion

The value of the magnetic susceptibility of topsoil samples at four locations in the research area and theirs statistics are presented in Table 1. The topsoil samples from Sawahlunto city have relatively low (< 150×10^{-8} m^3/kg) average value of the magnetic susceptibility that is 67.0×10^{-8} m^3/kg. The value is ranging from 0.3 to 259.9×10^{-8} m^3/kg with only 3 samples have the value of magnetic susceptibility larger than 150×10^{-8} m^3/kg. For comparison with mean value (74.34×10^{-8} m^3/kg) found in urban topsoil of Ishafan in Central Iran [11] and median value (107×10^{-8} m^3/kg) found in urban topsoil of Xuzhou in China [12], the mean value in Sawahlunto city topsoil is lower. The mean value of the topsoil magnetic susceptibility in the research area is also lower than that in three mining cities in China those are Jinchang (327.55×10^{-8} m^3/kg), Baiyin (191.89×10^{-8} m^3/kg), and Jiayuguan (930.92×10^{-8} m^3/kg) [10].

Topsoil samples from the downtown area have the average and the highest value of magnetic susceptibility (100.6×10^{-8} and 259.9×10^{-8} m^3/kg), and followed by the sample from around the steam power plant (98.4×10^{-8} and 258.0×10^{-8} m^3/kg), the agricultural area (56.2×10^{-8} and 83.7×10^{-8} m^3/kg), and the coal mine area (12.9×10^{-8} and 26.8×10^{-8} m^3/kg). The relatively higher value of the magnetic susceptibility of the downtown area suggest that vehicle exhaust emissions is probably a dominant factor in increasing the magnetic susceptibility value of topsoil in the Sawahlunto city. The similar result found in agricultural soil of Tadla plain near Beni Mellal (Marocco) that area near to the road had higher value of magnetic suscpetibility (121×10^{-5} SI) than in area away from the road (53×10^{-5} SI) which was concluded as a consequence of road gasoline emissions [13]. Soil samples from the steam power plant area have the widest range of magnetic susceptibility from 0.3×10^{-8} to 258.0×10^{-8} m^3/kg.

Table 1. The value of magnetic susceptibility of topsoil samples at four locations in the research area.

Points	Magnetic Susceptibility ($\times10^{-8}$ m^3/kg)			
	Downtown	Agricultural Area	Power Plant	Coal Mine
1	44.3	36.5	123.0	13.7
2	138.1	20.1	66.7	14.7
3	36.5	82.7	13.7	5.8
4	259.9	77.3	11.3	6.6
5	94.3	69.3	22.5	14.5
6	96.6	51.9	174.9	24.2
7	64.4	39.8	126.9	11.9
8	58.3	52.2	258.0	26.8
9	86.5	83.7	186.9	8.0
10	126.8	48.9	0.3	2.9
Average	100.6	56.2	98.4	12.9
Maximum	259.9	83.7	258.0	26.8

Minimum	36.5	20.1	0.3	2.9

Figure 6 represents a comparison of the magnetic susceptibility value of samples for four locations in the Sawahlunto city. The topsoil samples from the coal mine area have the lowest value of magnetic susceptibility, then followed by samples from the agricultural and downtown area. Samples from the power plant area have the widest range of magnetic susceptibility value. Four samples from this area have higher value than samples from other locations, while the rest samples are not.

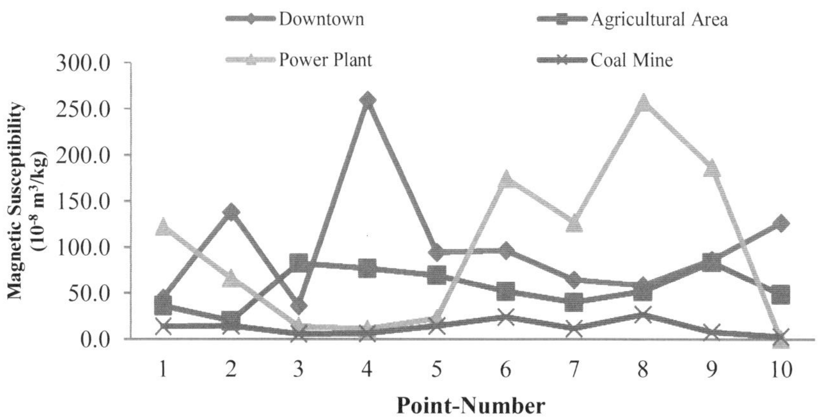

Figure 6. Graph of magnetic susceptibility value for four research locations in Sawahlunto city.

4. Conclusions

The topsoil samples from Sawahlunto city have relatively low ($< 150\times10^{-8}$ m^3/kg) average value of the magnetic susceptibility that is 67.0×10^{-8} m^3/kg.The magnetic susceptibility of the topsoil samples from downtown area have the average and the highest value of magnetic susceptibility (100.6×10^{-8} and 259.9×10^{-8} m^3/kg), and followed by the sample from around the steam power plant (98.4×10^{-8} and 258.0×10^{-8} m^3/kg), the agricultural area (56.2×10^{-8} and 83.7×10^{-8} m^3/kg), and surrounding coal mine sites (12.9×10^{-8} and 26.8×10^{-8} m^3/kg). The topsoil samples from the coal mine area have the lowest value of magnetic susceptibility, then followed by samples from the agricultural and downtown area. Samples from the power plant area have the widest range of magnetic susceptibility value. Four samples from this area have higher value than samples from other locations, while the rest samples are not.

Acknowledgments

We thank Program Bantuan Seminar Dalam Negeri, Universitas Andalas for transportation and accomodation financial support during the conference.

References

[1] Liu X, Baia Z, Zhoua W, Cao Y and Zhang G 2017 *J. Ecol. Eng.* **98** 228
[2] El Khalil H, El Hamiani O, Bitton G, Ouazzani N and Boularbah A 2008 *Environ. Monit. Assess.* **136**
[3] Li Z Y, Ma ZW, Kuijp TJ, Yuan ZW and Huang L 2014 *J. Sci. Total Environ.* **468** 843
[4] Adamczyk Z and Nowińska K 2016 *J. Environ. Earth Sci.* **75** 956
[5] Jordanova D, Goddu S R, Kotsev T and Jordanova N 2013 *J. Geoderma* **192** 237
[6] Kľučiarová D and Gregorová D 2007 *J. Geochem.: Explor. Environ. Anal.* **66** 313

[7] Panaiotu C G, Necula C, Panaiotu C E and Axente V 2005 *A Magnetic Investigation of Heavy Metals Pollution In Bucharest* (South Eastern Europe: Editura Politehnica-Timisoara)
[8] Evans M E and Heller F 2003 *Environmental Magnetism: Principles and Applications of Enviromagnetics* (California: Academic Press)
[9] Magiera T, Jabłońska M, Strzyszcz Z and Rachwał M 2011 *J. Atmos. Environ* **45**
[10] Wang B, Xia D, Yu Y, Jia J, Nie Y and Wang X 2015 *J. Environ. Pollut.* **207** 288
[11] Karimi R, Ayoubi S, Jalalian A, Sheikh-Hosseini A R and Afyuni M *J. Appl. Geophys.* **74** 1
[12] Wang XS and Xin Y 2005 *J. Environ. Geol.* **49** 10
[13] El Baghdadi M, Jakani K, Barakat A and Bay Y 2011 *J. Mater. Environ. Sci.* **2 (S1)** 513

IOP Conf. Series: Journal of Physics: Conf. Series **997** (2018) 012016 doi:10.1088/1742-6596/997/1/012016

Analytical solution of Schrödinger equation in minimal length formalism for trigonometric potential using hypergeometry method

I Nurhidayati[1], A Suparmi[1] and C Cari[1]

[1]Physic Departement, Graduate Program, Sebelas Maret University, Surakarta 57126, Indonesia

E-mail :inafisika03@gmail.com

Abstract. The Schrödinger equation has been extended by applying the minimal length formalism for trigonometric potential. The wave function and energy spectra were used to describe the behavior of subatomic particle. The wave function and energy spectra were obtained by using hypergeometry method. The result showed that the energy increased by the increasing both of minimal length parameter and the potential parameter. The energy were calculated numerically using *MatLab*.

1. Introduction

One of the main objectives in quantum mechanics is to obtain exact solutions of the relativistic and non-relativistic equations for some potentials [1]. The Schrödinger equation (SE) is one of non-relativistic quantum mechanics systems that give the exact solutions for shape invariant potential group. The most popular system in SE is a hydrogen atom (Coloumb) and a harmonic oscillator represents two typical examples in quantum mechanics [2-5]. For studying the quantum mechanical systems, it is necessary to study bound states to take the necessary information about the system and also solving scattering states for a system under the effect of a potential [6].

Since many type of research observation, with the other potential and providing analytic solution of the SE. The main problem in this SE is how to solve the SE to obtaining the wave function and the energy spectrum. The wave function and energy spectra were used to describe the behavior of sub-atomic particle.

Solution of the SE were obtained by inserting the potential into SE, then we used variable separation and substituting the suitable variable to reduce the SE become second-order differential equations such as Hermit function, Legendre, Laguerre, hypergeometric or confluent hypergeometric [7].

The quantum mechanical problems implying a generalized modified, there more piece of evidence for a minimal length (ML) of the order of Planck length [8]. The modified relation includes an MLF or Generalized Uncertainty Principle (GUP) [9]. In this paper we study the SE within the minimal length formalism using trigonometric potential.

The organization of the paper is as follows: in the next section we present an introduction to the MLF which is followed by the description of the SE within the MLF in subsection. In section 3, we explain the hypergeometric method. We use the standard hypergeometric method to obtain accurate numerical approximations to the energy. Finally in section 4, we investigate the result and discussion. The SE has been extended by applying the MLF for trigonometric cotangent potential. The wave function and energy spectra were obtained by using hypergeometric method. The energy was calculated numerically using *MatLab*.

2. Minimal Length Formalism

The theory of minimal length formalism (MLF) was motivated by Heisenberg algebra [8]. In the framework this formalism, the principal uncertainty can describe as:

$$[X, P] = [x, P] \tag{1}$$

With

$$X = x$$
$$P = [\left(1 + 2\alpha' p^2 + p^2\alpha'^2\right) p^2]^{1/2} \tag{2}$$

and parameter α' is minimal length parameter which is determined from a fundamental theory and varies in the interval $0 \leq \alpha' \leq 1$. For the $\alpha'^2 = 0$ then equation (2) becomes:

$$P^2 = \left(-\hbar\nabla^2 + 2\alpha'\hbar^2\nabla^4\right) \tag{3}$$

2.1. The Schrodinger Equation in MLF

The original schrodinger equation is described as

$$E\psi = -\frac{\hbar^2}{2m}\nabla^2\psi + V(r)\psi \tag{4}$$

The Schrodinger equation in minimal length formalism is obtained by inserting the minimal length formalism in equation (3) into equation (4) which is,

$$E\psi = -\frac{\hbar^2}{2m}(\nabla^2 - 2\alpha'\hbar^2\nabla^4)\psi + V(r)\psi \tag{5}$$

without the presence of minimal length equation (5) reduces to,

$$E^0\psi = -\frac{\hbar^2}{2m}\nabla_0^2\psi + V(r)\psi \tag{6}$$

In this condition, eq (6) is just the usual SE with the spectrum energy is given as $\boldsymbol{E^0}$. From the equation (6), we obtained $\boldsymbol{\nabla_0^2}$ as

$$\nabla_0^2 = -\frac{2m}{\hbar^2}(E^0 - V(r)) \tag{7}$$

Then by putting equation (7) into equation (8) so we get

$$E\psi - \left[-\frac{\hbar^2}{2m}\nabla^2 + 4\alpha' m(E^0)^2 - 8\alpha' m E^0 V(r) + 4\alpha' m(V(r))^2\right]\psi + V(r)\psi \tag{8}$$

with

$$\nabla^2 = \frac{1}{r^2}\frac{\partial}{\partial r}\left(r^2\frac{\partial}{\partial r}\right) + \frac{1}{r^2}\left(\frac{1}{\sin\theta}\frac{\partial}{\partial\theta}\sin\theta\frac{\partial}{\partial\theta} + \frac{1}{\sin^2\theta}\frac{\partial^2}{\partial\varphi^2}\right) \tag{9}$$

Now, equation (8) is expressed SE with Veff = $4\alpha' m\left(V(r)\right)^2 - 8\alpha' m E^0 V(r) + V(r)$.

3. Hypergeometric Method

Hypergeometric Differential equation was introduced by C.F.Gauβ [7] in term

$$z(1-z)\frac{\partial^2\Phi}{\partial z^2} + (c - (a+b+1)z)\frac{\partial\Phi}{\partial z} - ab\Phi = 0 \tag{10}$$

the solution of hypergeometric is described as,

$$_2F_1(a,b;c;z) = \Phi_1(z) = \sum_{n=0} \frac{(a)_n (b)_n}{(1)_n (c)_n} z^n = \sum_{n=0} \frac{(a)_n (b)_n}{n!(c)_n} z^n \tag{11}$$

To obtain the spectrum energy we set,

$$a = -n \; or \; b = -n \tag{12}$$

by appropriate subtitution of variable eq (8) is reduced to the standard hypergeometric equation, and so we get the spectrum energy and the corresponding wave function.

4. Result

4.1. Solution of SE in MLF

In this section we discussed the trigonometric potential model. By substitution of variable $\psi = \frac{\varphi}{r}$ to equation (8) we get the corresponding wave function

$$\left[\frac{\partial^2}{\partial r^2} - (\frac{L(L+1)}{r^2}) + \frac{16\alpha' m^2}{\hbar^2} E^0 V(r) - \frac{8\alpha' m^2}{\hbar^2} (V(r))^2 - \frac{2m}{\hbar^2} V(r) - \frac{8\alpha' m^2}{\hbar^2} (E^0)^2 + \frac{2m}{\hbar^2} E\right] \varphi = 0 \tag{13}$$

Equation (13) is the corresponding wave function for SE in MLF.
with the trigonometric potential cotangent,

$$V_{(r)} = V_0 \cot \gamma r \tag{14a}$$

$$V_{(r)}{}^2 = V_0{}^2 cot^2 \gamma r \tag{14b}$$

then by putting eq (14a) and (14b) into eq (13) so we get

$$\left[\frac{\partial^2}{\partial r^2} - (\frac{L(L+1)}{r^2}) + \frac{16\alpha' m^2}{\hbar^2} E^0 V_0 \cot \gamma r - \frac{8\alpha' m^2}{\hbar^2} (V_0 \cot \gamma r)^2 - \frac{2\,m}{\hbar^2} V_0 \cot \gamma r - \frac{8\alpha' m^2}{\hbar^2} (E^0)^2 + \frac{2\,m}{\hbar^2} E\right] \varphi = 0 \tag{15}$$

To obtain the exact solution of equation (15) we used centrifugal approximation

$$\frac{1}{r^2} = \frac{\gamma^2}{\sin^2 \gamma r} = \gamma^2 \csc^2 \gamma r \tag{16}$$

With suitable substitution variable then insert equation (16) to the equation (15) were reduced to differential hypergeometric is given as,

$$\left[\left(z(z-1)\frac{d^2}{dz^2} + (1-2z)\frac{d}{dz} - \left(L(L+1) + \frac{8\alpha' m^2}{\hbar^2 \gamma^2}(V_0)^2\right)\right) + \left(\frac{V_0\,(1-2z)(16E^0 \alpha' m^2 - 2\,m) - 8\alpha' m^2 V_0{}^2}{4\gamma^2 z(1-z)\;\;\hbar^2}\right) + \left(\frac{2\,mE - 8\alpha\prime\, m^2 E^{0\,2}}{4\gamma^2 z(1-z)\;\;\hbar^2}\right)\right] \varphi = 0 \tag{17}$$

then we set

$$\delta(\delta - 1) = \left(L(L+1) + \frac{8\alpha' m^2}{\hbar^2 \gamma^2}(V_0)^2\right) \tag{18}$$

$$2q = \frac{V_0\,(16E^0 \alpha' m^2 - 2\,m) - 8\alpha' m^2 V_0{}^2}{\hbar^2 \gamma^2} \tag{19}$$

$$-E' = \frac{2\,mE - 8\alpha' m^2 E^{0\,2}}{\hbar^2 \gamma^2} \tag{20}$$

by inserting eq (18), (19), and (20) into the eq (17) we obtained

$$\left[z(1-z)\frac{d^2}{dz^2}+(1-2z)\frac{d}{dz}+\delta(\delta-1)+\frac{2q(1-2z)}{4z(1-z)}-\frac{E'}{4z(1-z)}\right]\varphi=0 \quad (21)$$

The solution of wave function in eq (21) described as

$$\varphi=z^{\alpha}(1-z)^{\beta}f(z) \quad (22)$$

if we put eq (22) into eq (21) we obtained

$$z(1-z)\; z^{\alpha}(1-z)^{\beta}\left[\frac{\alpha(\alpha-1)}{z^2}f(z)-\frac{2\alpha\beta}{z(1-z)}f(z)+\frac{\beta(\beta-1)}{(1-z)^2}\;f(z)+\frac{2\alpha}{z}f'(z)-\frac{2\beta}{(1-z)}\;f'(z)+f''(z)\right]+(1-2z)z^{\alpha}(1-z)^{\beta}\left[\frac{\alpha}{z}f(z)-\frac{\beta}{(1-z)}f(z)+f'(z)\right]+\left(\delta(\delta-1)-\frac{4\alpha^2}{4z}-\frac{4\beta^2}{4(1-z)}\right)z^{\alpha}(1-z)^{\beta}f(z)=0 \quad (23)$$

by dividing Eq (23) with $z^{\alpha}(1-z)^{\beta}$then eq (23) can reduce as which is

$$z(1-z)\;f''(z)+[2\alpha+1-(2\alpha+2\beta+2)z\;]f'(z)-\left[(\alpha+\beta+\frac{1}{2})^2-\left(\delta-\frac{1}{2}\right)^2\right]f(z)=0 \quad (24)$$

Now, we can see from eq (24) the parameter hypergeometry as,

$$a'=\left(\delta-\frac{1}{2}\right)+(\alpha+\beta+\frac{1}{2})=(\alpha+\beta+\delta)$$
$$b'=-\left(\delta-\frac{1}{2}\right)+\left(\alpha+\beta+\frac{1}{2}\right)=(\alpha+\beta+\delta-1) \quad (25)$$
$$c'=2\alpha+1$$

where

$$4\alpha^2=(-2q+E') \quad (26)$$
$$4\beta^2=(2q+E') \quad (27)$$

The energy of SE in MLF is obtained by inserting eq (18), (19), and (20) into eq (26) and eq (27)

$$E=\frac{\hbar^2\gamma^2}{2\,m}\{L(L+1)+n^2\}-\;V_0\,(8\alpha' B_m E^0+1)-4\alpha' B_m {E^0}^2 \quad (28)$$

E is energy of the SE in MLF (eV) and E^0 is ground state energy of the SE in MLF (eV) which is given as

$$E^0=\left[\frac{\hbar^2}{2\,B_m}\{L(L+1)\gamma^2+\gamma^2 n^2\}+\;V_0\right] \quad (29)$$

4.2. Discussion

The SE which is based in MLF for a trigonometric potential cotangent was solved by using the hypergeometry method. The non-relativistic energy spectrum and the wave function were obtained by using hypergeometry method. In Table 1, we can see the result for sub-atomic particle (electron) by inserting m = 9.11 x 10^{-31}kg and a variation of *n*-number. For natural units, h = 1 and Vo = 5 eV, and α is minimal length parameter which was determined from interval $0 \leq \alpha' \leq 1$, so the value of energy was given in Table 1.

Table 1. Energy of SE in MLF.

	ENERGI (eV)										
(n.l)	0	0.1	0.2	0.3	0.4	0.5	0.6	0.7	0.8	0.9	1
(1.0)	1E+10	2E+10	2E+10	2.4E+10	2.9E+10	3.3E+10	3.8E+10	4.2E+10	4.7E+10	5.1E+10	5.6E+10
(2.0)	4E+10	1E+11	2E+11	2.6E+11	3.3E+11	4E+11	4.7E+11	5.4E+11	6.1E+11	6.8E+11	7.6E+11
(3.0)	1E+11	5E+11	8E+11	1.2E+12	1.5E+12	1.9E+12	2.3E+12	2.6E+12	3E+12	3.3E+12	3.7E+12
(4.0)	2E+11	1E+12	2E+12	3.6E+12	4.7E+12	5.9E+12	7E+12	8.1E+12	9.3E+12	1E+13	1.2E+13
(5.0)	3E+11	3E+12	6E+12	8.6E+12	1.1E+13	1.4E+13	1.7E+13	2E+13	2.3E+13	2.5E+13	2.8E+13

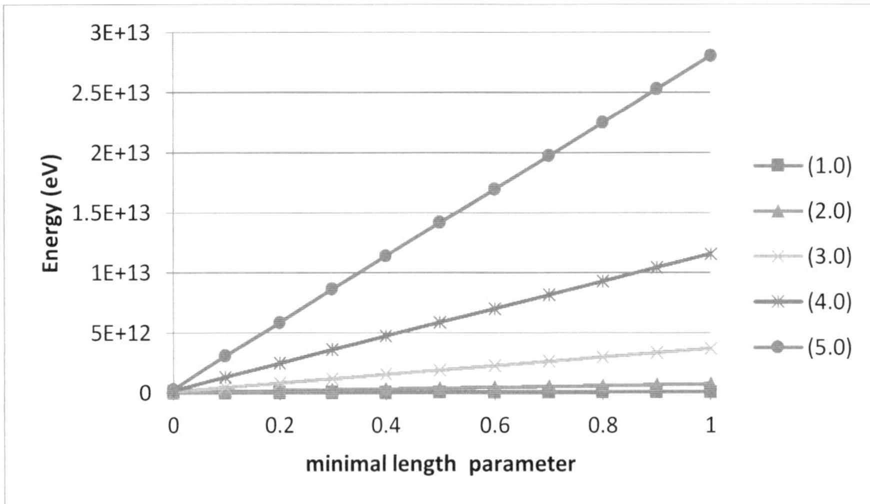

Figure 1. The graph of Energy spectrum SE in MLF.

5. Conclusion

We have studied the SE in MLF and have obtained the energy of this problem. The energy is calculated numerically using *MatLab*. The graph of energy for (n,l) with various of n number was described in Figure 1. The x-axis is parameter MLF and the y-axis is the Energy spectrum of SE in MLF (eV). The result showed that the energy increased by the increasing of both the minimal length parameter and the potential parameter. Based on the result of nonrelativistic energy is increased due to the increasing quantum number value and the increasing potential constant.

Acknowledgment

This research was partly supported by Sebelas Maret University Higher Education Project Grant Hibah Penelitian Berbasis Kompetensi 2018.

References

[1] Antia A D, Umo E A and Cyril C U 2015 *IJTPC* **10**
[2] Hamzavi M, Movahedi M, Thylwe K and Rajabi A A 2012 *Chin Phys Lett.* **29** 8
[3] Alemgadmi K, Suparmi, Cari and Deta U 2015 *AIP Conf. Proc.*1677
[4] Deta U, Alemgadmi K, Suparmi, Cari, Husein A, Yuliani H, Luqman H and Supriyanto 2014 *AIP Conf. Proc.* 1615 121127
[5] Liboff R L 2003 *Introductory Quantum Mechanics 4th ed* (San Francisco, CA: Addison Wesley)
[6] Pakdel F, Rajabi A and Hamzavi 2014 *Hindawi* **2014** 867483
[7] Suparmi 2011 *Mekanika Kuantum II* (Surakarta: UNS)
[8] Alimohammadi M and Hassanabadi H 2017 *Nucl. Phys. A.* **957** 439
[9] Chabab M, El Batoul A, Lahbas A and Oulne M 2016 *Phys. Lett. B.* **758** 212
[10] Suparmi A, Cari C, Handhika J, Yanuarief C and Marini H 2012 *IOSR-JAP* **2** 43
[11] Yi L Z, Diao Y F, Liu J Y and Jia C S 2014 *Phys. Lett. A.* **333** 212
[12] Ikot A N 2011 *J. Phys.* **6** 22
[13] Hassanabadi H, Zarrinkamar S and Rajabi A A 2011 *Theor. Phys.* **55** 541
[14] Qiang W C and Dong S H 2009 *Phys. Scr.* **79**
[15] Wei G F, Long C Y, Duan X Y and Dong S H 2008 *Phys. Scr.* **77**
[16] Hassanabadi H, Yazarloo B H, Ikot A N and Zarrinkamar S 2013 *Indian J. Phys.* **87** 1219
[17] Jia C S, Sun Y and Li Y 2002 *Phys. Lett. A.* **305** 231
[18] Budaca R 2014 *Phys. Lett. B.* **739** 56

Analysis of minimal length effect on the non-relativistic energy and wave function for the exponential type potential using Supersymmetric Quantum Mechanics

H Azizah[1], A Suparmi[2] and C Cari[2]

[1]Physics Department, Graduate Program, Sebelas Maret University, Indonesia
[2] Physics Department, Faculty of Mathematics and Fundamental Science, Sebelas Maret University, Indonesia

E-mail: husnunhimawan@gmail.com

Abstract. In this research, the minimal length effect was analysed on the energy and wave functions of the non-relativistic mechanical quantum system by studying the Schrodinger equation. The three dimensional Schrodinger equation with minimal length for the exponential type potential was solved by using supersymmetric quantum mechanics. To get the solution, the exponential type potential was reduced to the Morse potential. The non-relativistic energy was calculated numerically for diatomic molecules, LiH and HCl by using the Matlab software. From the results were shown that the presence of minimal length gives effect to the non-relativistic energy, it was seen clearly in the higher value of angular momentum l and quantum number n.

1. Introduction

The Schrodinger equation plays the main role in studying the non-relativistic system [1] that is used to analyse the non-relativistic energy and non-relativistic wave function which give the information of the quantum system [2]. Analytic solutions of the Schrodinger equation have been extensively studied with various potentials, as Eckart and Manning-Rosen [3], Morse [4, 5], and Scarf Trigonometry [6]. In this research, the Schrodinger equation with minimal length is solved for the exponential type potential, where the exponential type potential is reduced to the Morse potential.

The Schrodinger equation with minimal length for the exponential type potential can be solved by using the various methods, such as Nikiforov–Uvarov (NU) [7], Asymptotic Iteration Method (AIM) [8, 9], supersymmetry quantum mechanics [10, 11]. Supersymmetry quantum mechanics is one of the most well-liked methods which based on the shape-invariance concept. Supersymmetry appears based on the theory case in studying quantum mechanics at a higher dimension to unite the four basic interaction in nature, that are electro, weak, strong and gravitational interactions [11].

In recent years, the Schrodinger equation with minimal length has interested many researchers. The effect of minimal length has modified the Heisenberg uncertainty principle into generalised uncertainty principle (GUP) [12-14]. By applying minimal length term in the Schrodinger equation, the Schrodinger equation is modified with an additional form of p^4 [15,16] and become complex. For this reason, the modified Schrodinger equation is solved without minimal length. To get the usual non-relativistic energy, the linear momentum quadratic is only considered a function of r. The studies of the Schrodinger equation with the minimal length effect have been solved to scattering states of

Woods-Saxon interaction [14], Square Well Potential [5], hydrogen atom [16], and Bouncing Particle Spectrum [17].

In this paper, the Schrodinger equation is solved to obtain the non-relativistic energy and non-relativistic wave functions with the minimal length effect. Numerical non-relativistic energy is calculated to LiH and HCl molecule by using Matlab. The paper is structured as follows: The second and third section, the supersymmetry quantum mechanic and shape invariant are reviewed respectively. Section 4 gives the introduction of the Schrodinger equation with minimal length. In Section 5, the solution of the Schrodinger equation with minimal length for the exponential type potential using supersymmetric quantum mechanics is presented. Numerical result and discussion are presented in Section 6. Finally, the conclusions are given in Section 7.

2. Supersymmetry Quantum Mechanics

The main character in supersymmetric is a supercharge operator that commute with a Hamiltonian. For N=2, the supercharge operators are Q and $Q+$. Commutation relation of the operators with supersymmetric Hamiltonian Hss is satisfied the equation [11, 18].

$$\{Q,Q^+\}=Q^+Q+QQ^+=H_{ss} \tag{1}$$

Equation (1) shows that supercharges operators commute with supersymmetric Hamiltonian

$$\{Q^+,H_{ss}\}=\{Q,H_{ss}\}=0 \tag{2}$$

And Fermionic operators anti-commute with itself

$$(Q^+,Q^+)=(Q,Q)=0 \tag{3}$$

The supercharge operators are [11]

$$Q=\begin{pmatrix}0 & 0\\ A^- & 0\end{pmatrix},\qquad Q^+=\begin{pmatrix}0 & A^+\\ 0 & 0\end{pmatrix} \tag{4}$$

$A^{\pm}$ is Bosonic operator that written

$$A^{\pm}=\mp\frac{\hbar}{\sqrt{2m}}\frac{d}{dx}+\phi(x) \tag{5}$$

Furthermore, supersymmetric Hamiltonian is defined by

$$H_{SS}=\begin{pmatrix}A^-A^+ & 0\\ 0 & A^+A^-\end{pmatrix}=\begin{pmatrix}H^+ & 0\\ 0 & H^-\end{pmatrix} \tag{6}$$

Where supersymmetric partner Hamiltonian $H^{\pm}$,

$$H_{\pm}=-\frac{\hbar^2}{2m}\frac{d^2}{dx^2}+V_{\pm}(x)\,\text{with}\,V_{\pm}=\phi^2\pm\frac{\hbar}{\sqrt{2m}}\phi' \tag{7}$$

3. Shape invariance

Shape invariant potential system is two potentials that have the same shape. Two potential is said to be invariant if satisfies the following equation [19].

$$V_+(x,a_j)=V_-(x,a_{j+1})+R(a_{j+1}) \tag{8}$$

with $j=0,1,2,3....$ Supersymmetric partner potentials are defined by

$$V_+(x,a_j)=\phi^2(x,a_j)+\frac{\hbar}{\sqrt{2m}}\phi'(x,a_j) \tag{9}$$

and

$$V_-(x,a_j)=\phi^2(x,a_j)-\frac{\hbar}{\sqrt{2m}}\phi'(x,a_j) \tag{10}$$

By using super-algebra a given Hamiltonian is written as

$$H_{\pm} = H^{-} + E_0 = -\frac{\hbar^2}{2m}\frac{d^2}{dx^2} + V_{\pm}\left(x, a_j\right) + E_0 \tag{11}$$

Based on equation (8), equation (11) is reduced to

$$V\left(x\right) = V_{-}\left(x, a_j\right) + E_0 = \phi^2\left(x, a_j\right) - \frac{\hbar}{\sqrt{2m}}\phi'\left(x, a_j\right) + E_0 \tag{12}$$

V(x) is an effective potential of the system. In general, k^{th} Hamiltonian is written as

$$H_k = -\frac{\hbar^2}{2m}\frac{d^2}{dx^2} + V_{-}\left(x; \mathrm{a}_k\right) + \sum_{i=1}^{k} R\left(a_i\right), \quad k = 0, 1, 2... \tag{13}$$

where

$$R\left(a_i\right) = V_{+}(\mathrm{x}; \mathrm{a}_{i-1}) - V_{-}(\mathrm{x}; \mathrm{a}_i) \tag{14}$$

So, the general non-relativistic energy from $H\_$ is

$$E_n^{(-)} = \sum_{k=1}^{n} R(\mathrm{a}_k) \tag{15}$$

Then, the Hamiltonian from equation (11) is changed to the form

$$E_n = E_n^{(-)} + E_0 \tag{16}$$

En is total non-relativistic energy, and E_o is ground state non-relativistic energy

To get non-relativistic wave function, the ground state non-relativistic wave function is operated by lowering operators and the result must be zero [2]. The equation is written as

$$A\psi_0^{(-)} = 0 \tag{17}$$

The ground state non-relativistic wave function is obtained as

$$\psi_0^{(-)}\left(x, a_0\right) = N\exp\left[-\frac{\sqrt{2m}}{\hbar}\int^x \phi\left(x, a_0\right)dx\right] \tag{18}$$

N is the normalisation constant. The upper-level of non-relativistic wave function $\psi_n^{-}\left(n; a_0\right)$ can be determined by using a rising operator $\left(A^{+}\right)$ to the non-relativistic ground state wave function. In general, the excited non-relativistic wave equation of the nth level becomes

$$\psi_n^{-}\left(x; a_0\right) \sim A^{+}\left(x; a_0\right)\psi_{n-1}^{-}\left(x; a_1\right) \tag{19}$$

4. The Schrodinger Equation with minimal length

Generalised modified canonical commutation relation with minimal length is given as[12-13]

$$\left[X, P\right] = i\hbar\left(1 + \alpha P^2\right) \tag{20}$$

where

$$X_i = x_i \,, \quad P_i = \left(1 + \alpha p^2\right)p_i \tag{21}$$

α a very small positive parameter of the minimal length, in an interval $0 \le \alpha \le 1$. Here, p_i is the momentum operator at low energies and P_i is the momentum operator at high energies. At low energies, energies much smaller than the Planck mass and the second term of the right-hand side of equation (20) vanishes. The equation was well-known as Heisenberg uncertainty principle [12,14]. According to the equation (21), energy kinetic operator is written as

$$T = \frac{P^2}{2m} = \frac{\left(1 + 2\alpha p^2\right)p^2}{2m} \,, \quad p^2 = -\hbar^2\nabla^2 \tag{22}$$

The Schrodinger equation with the minimal length effect $(\alpha \neq 0)$ leads to

$$\left\{\frac{-\hbar^2}{2m}\nabla^2 - \frac{\alpha\hbar^4}{m}\Delta^2 + V(r)\right\}\chi = E\chi \tag{23}$$

To obtain the analytic solution of the second order differential, equation (23) is solved for $\alpha = 0$

$$H^0 = \frac{-\hbar^2}{2m}\Delta + V(r) \tag{24}$$

From the equation (24) we get

$$\Delta^2 = \left(\frac{2m}{\hbar^2}\right)^2\left[V(r) - E^0\right]^2 \tag{25}$$

Known that, electron revolute to the nuclei on polar shaped shall. It causes the non-relativistic energy and non-relativistic wave functions of an electron are employed by a solution of the Schrodinger equation in polar coordinate [20]. The Laplacian operator in polar coordinate is explained as follows

$$\nabla^2 = \frac{1}{r^2}\frac{\partial}{\partial r}\left(r^2\frac{\partial}{\partial r}\right) + \frac{1}{r^2}\Delta_\Omega \tag{26}$$

where

$$\Delta_\Omega = \frac{1}{\sin\theta}\frac{\partial}{\partial\theta}\left(\sin\theta\frac{\partial}{\partial\theta}\right) + \frac{1}{\sin^2\theta}\frac{\partial^2}{\partial\varphi^2} \tag{27}$$

Since the used potential only contains the variable r then the solution is only solved on the radial form. Where Δ_Ω is the angular part of the Laplacian operator. Then, the angular part changed to

$$\Delta_\Omega Y_{\ell m}(\theta,\varphi) = -l(l+1)Y_{\ell m}(\theta,\varphi) \tag{28}$$

by substituting equation (25) and (26) to equation (23), the Schrodinger equation is rewritten as

$$\left\{-\frac{\hbar^2}{2m}\left[\frac{1}{r^2}\frac{\partial}{\partial r}\left(r^2\frac{\partial}{\partial r}\right) - \frac{1}{r^2}l(l+1)\right] + 4\alpha m\left[V(r) - E^0\right]^2 + \left(V(r) - E\right)\right\}\chi(r) = 0 \tag{29}$$

Using the new non-relativistic wave function

$$\chi(r) = F(r)Y_{\ell m}(\theta,\varphi) \tag{30}$$

Equation (29) becomes

$$\left\{\frac{\hbar^2}{2m}\frac{1}{r^2}\frac{\partial}{\partial r}\left(r^2\frac{\partial}{\partial r}\right) - \frac{\hbar^2}{2m}\frac{1}{r^2}l(l+1) - \left[4\alpha m\left(V(r) - E^0\right)^2 + \left(V(r) - E\right)\right]\right\}F(r) = 0 \tag{31}$$

with $F(r) = \frac{\Psi(r)}{r}$, the Schrodinger equation is reduced to one dimensional only in radial function

$$\left\{\frac{\hbar^2}{2m}\frac{d^2}{dr^2} - \frac{\hbar^2}{2m}\frac{1}{r^2}l(l+1) + \left[-4\alpha m\left(V(r) - E^0\right)^2 - \left(V(r) - E\right)\right]\right\}\Psi(r) = 0 \tag{32}$$

Equation (32) is the Schrodinger equation in the presence of minimal length.

5. Solution of the Schrodinger equation with minimal length for the exponential type potential using supersymmetric quantum mechanics

The used exponential type potential is a part of the Morse potential. The potential is written as

$$V = De^{-\delta x} \tag{33}$$

with $x = \frac{r - r_0}{r_0}$ stand $\delta = ar_e$. D is dissociation energy, δ is the Morse parameter, r_0 is the equilibrium inter-nuclear distance. The exponential type potential in equation (33) is substituted into equation (32)

$$\frac{\hbar^2}{2m}\frac{d^2\Psi(r)}{dr^2} - \frac{\hbar^2}{2r}\frac{l(l+1)}{r^2}\Psi(r) + \begin{bmatrix} \left(-4\alpha mD^2\right)e^{-2\delta x} + \left(8\alpha mDE^o - D\right)e^{-\delta x} \\ +\left(E - 4\alpha m\left(E^o\right)^2\right) \end{bmatrix}\Psi(r) = 0 \quad (34)$$

To get the exact solution of the Schrodinger equation for $l \neq 0$, we need a suitable approach. In this research, we used the Pakeris approximation. The approximation is selected because of the centrifugal term in the exponential series base on the internuclear distance. To get the solution, the centrifugal term is simplified around $x=0$. Then, the second term in equation (34) becomes [7].

$$\frac{\hbar^2}{2m}\frac{l(l+1)}{r^2} = \frac{\hbar^2}{2m}\frac{l(l+1)}{r_0^2}\frac{1}{(1+x)^2} = \gamma\left(1 - 2x + 3x^2 + ...\right) \quad (35)$$

where

$$\gamma = \frac{\hbar^2}{2m}\frac{l(l+1)}{r_0^2} \quad (36)$$

With assumption fourth term and so on are very small, the approximation in equation (35) is only used until the third term. Then, the rotational term is replaced to

$$1 - 2x + 3x^2 + ... = D_0 + D_1 e^{-\delta x} + D_2 e^{-2\delta x} \quad (37)$$

Furthermore, equation (37) is rewritten in

$$1 - 2x + 3x^2 + ... = D_0 + D_1\left(1 - \delta x + \frac{\delta^2 x^2}{2!}\right) + D_2\left(1 - 2\delta x + \frac{4\delta^2 x^2}{2!}\right) \quad (38)$$

By combining the right and left term, the relation between the coefficient and the parameter δ is obtained as

$$D_0 = 1 - \frac{3}{\delta} + \frac{3}{\delta^2}, \quad D_1 = \frac{4}{\delta} - \frac{6}{\delta^2}, \quad D_2 = -\frac{1}{\delta} + \frac{3}{\delta^2} \quad (39)$$

So, equation (35) becomes

$$\frac{\hbar^2}{2m}\frac{l(l+1)}{r^2} = \gamma\left(D_0 + D_1 e^{-\delta x} + D_2 e^{-2\delta x}\right) \quad (40)$$

By substituting equation(40) into equation (34), the equation (34) is rearranged as

$$\begin{aligned} &-\frac{\hbar^2}{2m}\frac{d^2\Psi(r)}{dr^2} + \gamma\left[D_0 - D_1 e^{-\delta x} + D_2 e^{-2\delta x} + ...\right]\Psi(r) \\ &+\left[\left(+4\alpha mD^2\right)e^{-2\delta x} - \left(8\alpha mDE^o - D\right)e^{-\delta x} + 4\alpha m\left(E^o\right)^2\right]\Psi(r) = E\Psi(r) \end{aligned} \quad (41)$$

By introducing the following parameter

$$A_0 = \gamma D_0 + 4\alpha m\left(E^o\right)^2, \quad A_1 = \gamma D_1 + \left(8\alpha mDE^o - D\right), \quad A_2 = \gamma D_2 + \left(4\alpha mD^2\right) \quad (42)$$

Equation (41) was changed to the form

$$-\frac{\hbar^2}{2m}\frac{d^2\Psi(r)}{dr^2} + \left(A_0 - A_1 e^{-\delta x} + A_2 e^{-2\delta x}\right)\Psi(r) = E\Psi(r) \quad (43)$$

Finally, the effective potential is found as

$$V_{eff} = A_0 - A_1 e^{-\delta x} + A_2 e^{-2\delta x} \quad (44)$$

Based on the effective potential equation on equation (44), the corresponding superpotential is

$$\phi = C_1 - C_2 e^{-\delta x} \tag{45}$$

Using equation (45), equation (12) becomes

$$V_{eff} = C_1^2 - \left(2C_1C_2 + \frac{\hbar}{\sqrt{2m}}\delta C_2\right)e^{-\delta x} + C_2^{\,2} e^{-2\delta x} + E_0 \tag{46}$$

By combining equation (44) and (46), the value of *c1*, *c2* and E_0 are defined by

$$C_2 = \sqrt{A_2}\ ,\quad C_1 = \frac{A_1}{2\sqrt{A_2}} - \frac{\hbar}{2\sqrt{2m}}\delta \tag{47}$$

and

$$E_0 = A_0 - \left(\frac{A_1}{2\sqrt{A_2}} - \frac{\hbar}{2\sqrt{2m}}\delta\right)^2 \tag{48}$$

By inserting equation (47) into equation (45), the new superpotential becomes

$$\phi = \frac{A_1}{2\sqrt{A_2}} - \frac{\hbar}{2\sqrt{2m}}\delta - \sqrt{A_2}e^{-\delta x} \tag{49}$$

Using equation (49), equation (9) and (10) is written as

$$V_+(\mathrm{x};\mathrm{a}_0) = A_2 e^{-2\delta x} - \left(A_1 - \frac{2\hbar\sqrt{A_2}}{\sqrt{2m}}\delta\right)e^{-\delta x} + \left(\frac{A_1}{2\sqrt{A_2}} - \frac{\hbar}{2\sqrt{2m}}\delta\right)^2 \tag{50}$$

$$V_-(\mathrm{x};\mathrm{a}_0) = A_2 e^{-2\delta x} - A_1 e^{-\delta x} + \left(\frac{A_1}{2\sqrt{A_2}} - \frac{\hbar}{2\sqrt{2m}}\delta\right)^2 \tag{51}$$

By comparing equation (50) and (51), we get

$$a_0 = A_1\ ,\quad a_1 = A_1 - \frac{2\hbar\delta\sqrt{A_2}}{\sqrt{2m}}\ ,\quad a_2 = A_1 - \frac{4\hbar\delta\sqrt{A_2}}{\sqrt{2m}} \tag{52}$$

From equations (14), (50), (51), and (52), we get the constant value as

$$R(a_1) = \left(\frac{A_1}{2\sqrt{A_2}} - \frac{\hbar\delta}{2\sqrt{2m}}\right)^2 - \left(\frac{A_1}{2\sqrt{A_2}} - \frac{3\hbar\delta}{2\sqrt{2m}}\right)^2 \tag{53}$$

$$R(a_2) = \left(\frac{A_1}{2\sqrt{A_2}} - \frac{3\hbar\delta}{2\sqrt{2m}}\right)^2 - \left(\frac{A_1}{2\sqrt{A_2}} - \frac{5\hbar\delta}{2\sqrt{2m}}\right)^2 \tag{54}$$

Then the general form of constant value *R(an)* is

$$R(a_n) = \left(\frac{A_1}{2\sqrt{A_2}} - \frac{(2n-1)\hbar a}{2\sqrt{2m}}\right)^2 - \left(\frac{A_1}{2\sqrt{A_2}} - \frac{(2n+1)\hbar\delta}{2\sqrt{2m}}\right)^2 \tag{55}$$

So equation (15) becomes

$$E_n^{(-)} = \sum_{k=1}^{n} R(a_k) = \left(\frac{A_1}{2\sqrt{A_2}} - \frac{\hbar\delta}{2\sqrt{2m}}\right)^2 + \left(\frac{A_1}{2\sqrt{A_2}} - \frac{(2n+1)\hbar\delta}{2\sqrt{2m}}\right)^2 \tag{56}$$

$E_n^{(-)}$ is ground state non-relativistic energy from n[th] Hamiltonian. By inserting equations (56) and (48) into equation (16), the total non-relativistic energy of the system is

$$E_n = A_0 + \left(\frac{A_1}{2\sqrt{A_2}} - \frac{(2n+1)\hbar\delta}{2\sqrt{2m}} \right)^2 \tag{57}$$

Equation (57) shows that total non-relativistic energy is influenced by quantum number (n), momentum angular (l), minimal length parameter (α). The analytic solution in equation (57) is used to get the numerical calculation of non-relativistic energy. The calculation is not allowed to having the value A_2 in equals zero, so we have to set the value both l and α cannot be zero at the same time.

To calculate the non-relativistic wave function, equation (49) is substituted into equation (18) and we obtain

$$\psi_0^{(-)} = N\ Exp\left(-\frac{\sqrt{2m}}{\hbar} \int \left(\left(\frac{A_1}{2\sqrt{A_2}} - \frac{\hbar}{2\sqrt{2m}}\delta \right) - \sqrt{A_2} e^{-\delta x} \right) dx \right) \tag{58}$$

From equation (58), the ground state non-relativistic wave function is found as

$$\psi_0^{(-)} = N \exp\left(-\frac{\sqrt{2m}}{\hbar} \left(\frac{A_1}{2\sqrt{A_2}} - \frac{\hbar}{2\sqrt{2m}}\delta \right) x \right) \exp\left(-\frac{\sqrt{2m}}{\hbar} \frac{\sqrt{A_2}}{\delta} e^{-\delta x} \right) \tag{59}$$

With N is normalisation factor. Equation (59) shows that the non-relativistic wave function is influenced by momentum angular (l) and minimal length parameter (α).

6. Numeric Result

The numerical results of non-relativistic energy for the diatomic molecules were obtained by Matlab. The diatomic molecules were used because the Morse potential describes the vibration of the molecules with two atoms. The used molecules in the calculation were LiH and HCl as the example of diatomic molecules. In this paper, we used the natural unit for calculating the value of Non-relativistic energy. The result for the various quantum numbers and the minimal length parameter could be seen in Table 1 and 2. Table 1 and Table 2 showed the non-relativistic energy of LiH and HCl respectively.

Table 1 and 2 informed that the minimal length parameter affects the non-relativistic energy value both LiH and HCl. For the higher quantum number, the higher minimum length parameter increased the non-relativistic energy. The results showed that the non-relativistic energy with minimal length has the higher value than non-relativistic energy without minimal length. When α=0, the non-relativistic system was not influenced by minimal length and has very small energy than the system with minimal length. So, we know that the presence of minimal length which investigated by the Schrodinger equation upgraded the relativistic energy system significantly.

Table 1. The non-relativistic energy of LiH for various quantum numbers (n) (D = 20287 (cm^{-1}), m = 0.8801221 (amu), r_0 = 1.5956 (Å), a = 1.7998368).

	E(eV)		
n,l	α = 0	α = 0.005	α = 0.01
2,1	4.423E+02	3.100E+06	6.039E+06
3,1	4.429E+02	3.108E+06	6.056E+06
4,1	4.434E+02	3.116E+06	6.072E+06
5,1	4.440E+02	3.125E+06	6.088E+06
6,1	4.446E+02	3.133E+06	6.105E+06

Table 2. The non-relativistic energy of HCl for various quantum numbers (n) (D = 37255 (cm^{-1}), m = 0.9801045 (amu), r_0 = 1.2746 (Å), a = 2.38057042).

	E(eV)		
n,l	α = 0	α = 0.005	α = 0.01
2,1	1.367E+02	2.857E+05	5.495E+05
3,1	1.394E+02	2.991E+05	5.764E+05
4,1	1.420E+02	3.127E+05	6.036E+05
5,1	1.446E+02	3.266E+05	6.314E+05
6,1	1.473E+02	3.407E+05	6.595E+05

The non-relativistic energy also depended on the value of angular momentum. figure 1 and 2 showed the relation of non-relativistic energy H2 and LiH as a function of minimal length parameter for various angular momentum. The figure showed the higher value of minimal length increases the non-relativistic energy both molecules. The effect becomes sensible for higher angular momentum. These results were appropriate with ref. [12,13], that the non-relativistic energy increased with increasing angular momentum, and at higher minimal length values the non-relativistic energy increased visibly.

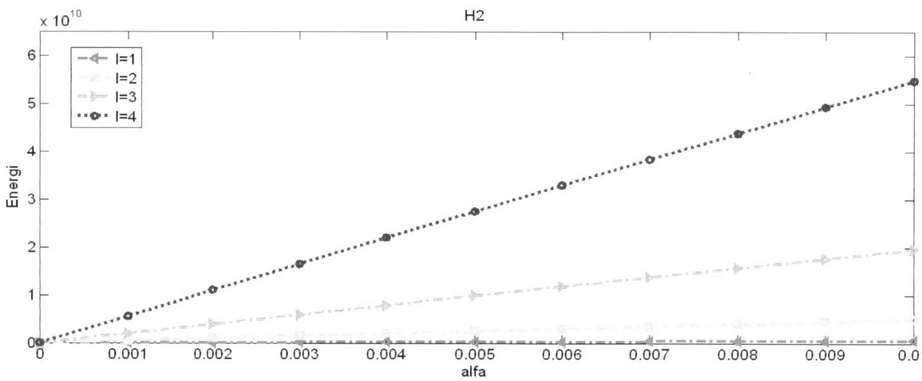

Figure 1. The non-relativistic energy of H_2 as the function of minimal length parameter for arbitrary angular momentum (l).

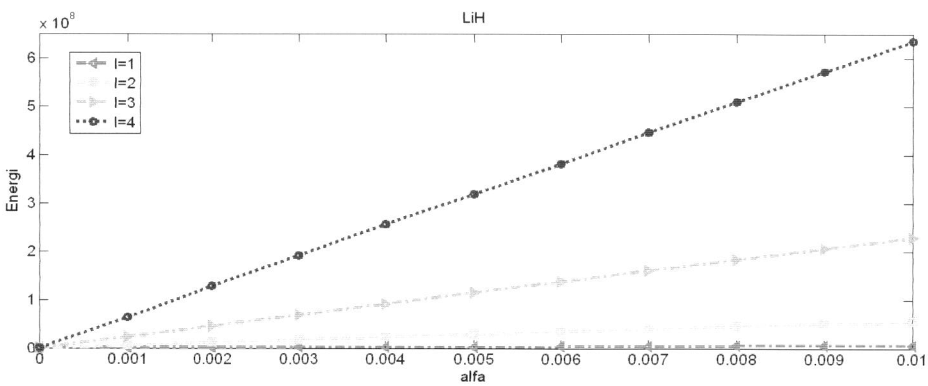

Figure 2. The non-relativistic energy of LiH as the function of minimal length parameter for arbitrary angular momentum (l).

The results showed that there was the difference value of non-relativistic energy both LiH and HCl. Table 1 and 2 informed that non-relativistic energy of LiH was higher than HCl, and figure 1 showed the different molecules have a slightly different slope. The difference occurred because of the input of molecules parameter. Where, dissociation energy (D), Morse parameter (δ), and mass (m) of HCl were higher than LiH while the equilibrium inter-nuclear distance (r_0) of HCl was lower then LiH.

7. Conclusion

In this research, the Schrodinger equation in the presence of minimal length was soluted. The solutions were obtained for the exponential type potential by using the supersymmetric quantum mechanics. The research was aimed to analyse the minimal length effect on the non-relativistic energy and non-relativistic wave function for diatomic molecules, LiH and HCl. The numerical results were calculated for molecules LiH and HCl by using Matlab. The results showed that the presence of the minimal length influenced the non-relativistic energy system for the various quantum number and angular momentum.

Acknowledgement
This research was partly supported by Sebelas Maret University Higher Education Project Grant Hibah Penelitian Berbasis Kompetensi 2018.

References

[1] Feizi H, Rajabi A A and Shojaei M R 2004 *J. Act. Phys. Polonica B* **42** 2143
[2] Cari, Suparmi and Marini H 2012 *Indonesian J. App. Phys.* **2** 112
[3] Farizky N M, Suparmi A, Cari C and Yunianto M 2016 *J. Phys. Conf. Series* **776** 1
[4] Chen Gang 2004 *J. Phys. Lett. A* **326** 55
[5] Roy A K 2013 *J. Results in Phys.* **3** 103
[6] Deta U A, Suparmi and Cari 2013 *Padjadjaran Int. Phys. Symp.* (Bandung: AIP Publishing LLC) 190-193
[7] Berkdemir C and Han J 2005 *Int. At. Energy Agency (IAEA)* **37** 10
[8] Bayrak O and Boztosun I 2007 *J. Mol. Struc. TEOCHEM* **802** 17
[9] Bayrak O and Boztosun I 2006 *J. Phys. A: Math. Gen.* **39** 6955
[10] Aktas M and Sever R 2004 *Mod. Phys. Lett. A* **19**, 2871
[11] Filho E D and Ricotta M R 2000 *Phys. Lett. B* **269** 269
[12] Alimohammadi M and Hassanabadi H 2017 *J. Nucl. Phys. A* **957** 439
[13] Chabab M, El Batoul A, Lahbas A and Oulne M 2016 *J. Phys. Lett B* **758** 212
[14] Hassanabadi H, Zarrinkamar S and Maghsoodi E 2012 *J. Phys. Lett. B* **718** 678
[15] Haouat S 2014 *J. Phys Lett B* **727** 33
[16] Brau F 1999 *J. Phys. A: Math Gen.* **32** 44
[17] Nozari K and Pedram P 2010 *Europhysics Lett.* **92** 5
[18] Suparmi and Cari 2011 *Mekanika Kuantum II* (Surakarta: Universitas Sebelas Maret)
[19] Saregar A, Suparmi, Cari and Yuliani H 2013 *Int. J. Eng. Sci.* **2** 14
[20] Suparmi and Cari 2011 *Mekanika Kuantum I* (Surakarta: Universitas Sebelas Maret)

IOP Conf. Series: Journal of Physics: Conf. Series **997** (2018) 012018 doi:10.1088/1742-6596/997/1/012018

The design of a PC-based real-time system for monitoring Methane and Oxygen concentration in biogas production

M Yantidewi[1], M S Muntini[2], U A Deta[1], and N A Lestari[1]

[1]Department of Physics, Universitas Negeri Surabaya, Gedung C3 Kampus Ketintang Unesa Surabaya 60231, Indonesia
[2]Department of Physics, Institut Teknologi Sepuluh Nopember, Kampus ITS Sukolilo-Surabaya, Indonesia

E-mail: metayantidewi@unesa.ac.id

Abstract. Limited fossil fuels nowadays trigger the development of alternative energy, one of which is biogas. Biogas is one type of bioenergy in the form of fermented gases of organic materials such as animal waste. The components of gases present in biogas and affect the biogas production are various, such as methane and oxygen. The biogas utilization will be more optimal if both gases concentration (in this case is methane and oxygen concentration) can be monitored. Therefore, this research focused on designing the monitoring system of methane and oxygen concentration in biogas production in real-time. The results showed that the instrument system was capable of monitoring and recording the data of gases (methane and oxygen) concentration in biogas production in every second.

1. Introduction

Energy has been a part of daily human life. Almost all sectors in human life require energy. Fossil fuels are the main source of energy that is widely utilized to meet their needs. Unfortunately, the human necessity for energy, especially fossil fuels, creates other problems. The increment in energy necessity is not followed by the fossil fuels reserves. Fossil fuels, just like another non-renewable energy, have a limit number of reserves. Furthermore, fossil fuels cause greenhouse-gas emissions. Based on this situation, a series of alternative energy development has been conducted.

A review by Adelekan in 2012 [1] said that Brazil and US had used ethanol as a biofuel alternative to gasoline since 2008. The review also said that several researchers had improved alternative energy production from various organic sources such as biogas from pretreated water hyacinth by Ofoefule et al [2], biogas from cassava peels mixed with poultry, piggery, and cattle wastes types by Adelekan and Bamgboye [3], and biogas from the peels of sweet potato and wild cocoyam by Adeyosoye et al [4].

According to the review, biogas is one of the most widely developed alternative energy sources and categorized as renewable energy. Biogas is a versatile renewable energy source, which can be used as a substitute for fossil fuels to generate power and heat, and it can be used also as gaseous vehicle fuel. Biogas production through anaerobic digestion process offers significant advantages over other forms of alternative energy production. Fehrenbach et al in Weiland stated that biogas is one of the most energy-efficient and environmentally friendly technology for alternative energy production. By making use of locally available resources, it is able to reduce greenhouse gases emissions dramatically compared to fossil fuels [5]. Biogas is a flammable gas comprised of organic wastes. Biogas mostly contains methane about 54-70%, carbon dioxide (20-30%), oxygen (0,1%), nitrogen (0,5-3%), and hydrogen sulfide (less than 1%) [6]. Methane gas –the main component in biogas- has a heat value between 4800-6700 kcal/m^3, meanwhile pure methane gas has a more great heat value around 8900 kcal/m^3 [7]. Daru in Roosganda and Rusdiana [8] also revealed that compared to coal gas and water gas, biogas has higher heat value. One m^3 of biogas is equal to 0,5 kg of liquid petroleum gas (LPG) and able to generate electricity around 1,25-1,50 kWh. This fact makes biogas can be utilized as alternative energy.

Biogas is produced through a process called anaerobic digestion. The process of anaerobic digestion is carried out by methanogen bacteria under the oxygen-depleted condition to produce a large amount of methane gas. If the conditions are less appropriate or inappropriate, the methanogen bacteria will not work optimally or even die. The optimum condition in biogas production process will determine the maximum production of biogas. One of those optimum conditions in biogas production process is the concentration of oxygen available in the biogas reactor. The existence of oxygen concentration in a biogas production process will obstruct methanogen bacteria [9].

However, the concentration of methane as a result of anaerobic digestion and also the concentration of oxygen as a factor that contributed to methane production cannot be known for sure. The estimation of oxygen concentration in material for biogas is needed to know if there is an oxygen methanogenic inhibition appears during the production without removing oxygen in the initial process [9]. To find out whether the biogas production is already optimum or not, a real-time monitoring system is needed to monitor it. Monitoring is an important strategy to achieve a better stability and higher conversion efficiencies in the anaerobic digestion process. The less suitable process indicators/conditions can affect the optimization of biogas production process (anaerobic digestion) [9].

The monitoring system is an instrument system or an instrument device that intertwined and serves as a means of measuring and collecting the data of a certain object. In this research, the instrument system consisted of gas sensors system and a data acquisition system which was able to carry out a real-time monitoring of methane and oxygen concentration in biogas production.

2. Materials and Methods

In general, the real-time monitoring system built in this research consisted of the biogas and the data acquisition system (Figure 1).

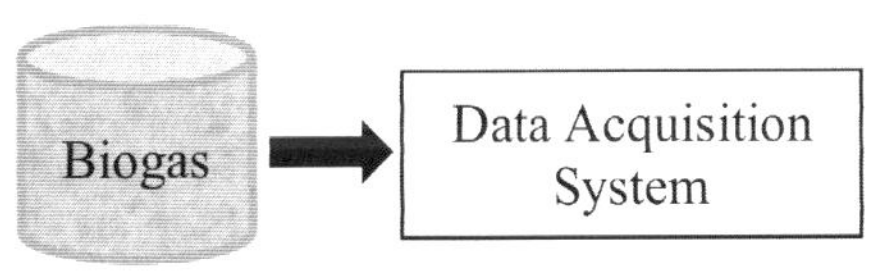

Figure 1. Block diagram of the instrument system in this research.

The procedure of the PC-based real-time system for monitoring the concentration of methane and oxygen gases in biogas process production is shown in Figure 2 below. The methane and oxygen concentration were detected by the gas sensors system then being interpreted by the data acquisition devices to be presented on PC.

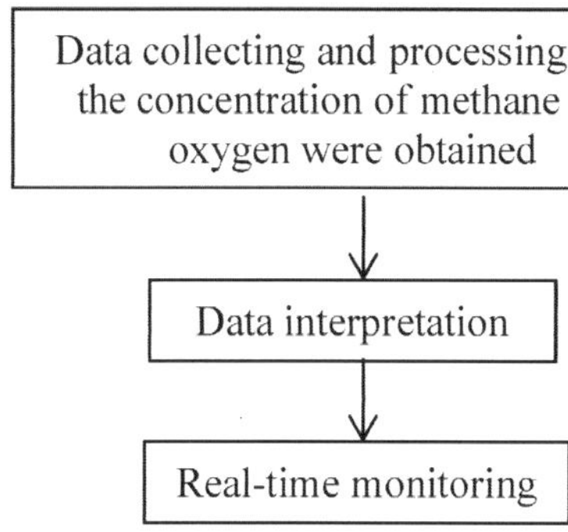

Figure 2. The procedure of a PC-based real-time system for monitoring methane and oxygen concentration.

2.1. Biogas Production

The biogas produced in this research was made on a laboratory scale as illustrated in Figure 3. The biogas was produced by mixing the cow dung with an amount of water. The mixture composition had a ratio of 1:1 and should be mixed well. The mixture was then placed into an 8 dm^3 biogas reactor or biogas digester connected to an airtight container before it connected to a 6 dm^3 gas reservoir. The biogas reactor serves to accommodate the mixture of cow dung and water as well as in the process of fermentation by bacteria in anaerobic (anaerobic digestion). Whilst, the gas reservoir is to gather the resulted gas/biogas.

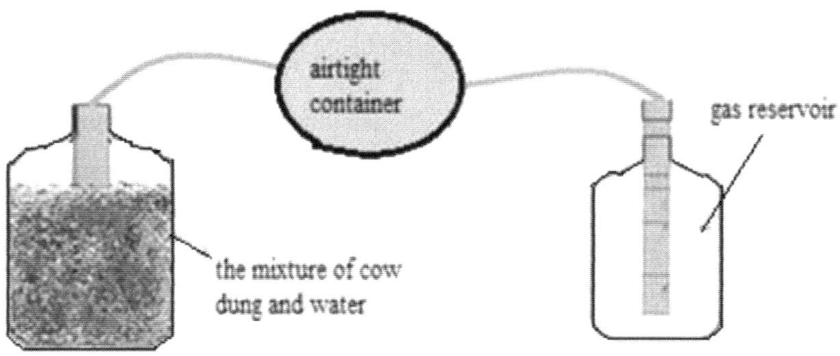

Figure 3. Illustration of biogas production.

2.2. The Data Acquisition System (DAS)

A data acquisition system (DAS) can be defined as a system that serves to retrieve, collect, and prepare data, to process it into the desired data [6]. Data acquisition system consists of a number of elements or components that are interconnected with each other formed in such a way that the system is able to retrieve, collect and store data quickly, accurately, and in real-time so that the data is ready for further processing [7].

A data acquisition system usually consists of a sensor or transducer, data acquisition devices, and a personal computer (PC). The figure below (Figure 4) shows the data acquisition system applied in this research.

Figure 4. Block diagram of data acquisition system.

The sensors system used in the research is gas sensors. The gas sensors detected the existence of methane and oxygen gases then transmitted them to data acquisition devices before continuing the data to the PC. The PC functioned to record, interpret, and display the information which retrieved from the data acquisition devices. The data acquisition devices converted the analog signal, produced by the gas sensors system attached to the measured objects, to its digital signal then relaying it to the PC.

2.2.1. The Gas Sensors System

The gas sensor system used TGS 2611 as methane gas sensor and KE-50 as oxygen gas sensor. TGS 2611 is a metal oxide semiconductor gas sensor which used PdO_2 as its sensing element. TGS 2611 requires two input voltage, heater voltage, and circuit voltage. KE-50 is a galvanic-cell type gas sensor. Compared to TGS 2611, KE-50 requires no heater voltage. Both TGS 2611 and KE-50 data output are still in the form of the analog data, therefore an analog-to-digital converter (ADC) is required. TGS 2611 has measurement range between 500-10000 ppm, while KE-50 has a measurement range from 0% to 100%. Note that this sensor system must be calibrated to obtain a comparable data as a conversion reference of sensor output unit.

2.2.2. Data Acquisition Devices

Generally, data acquisition devices consist of a signal conditioning and an analog-to-digital converter (ADC). Signal conditioning is a device works to condition the sensor output signal to be read by signal processing elements or an analog-to-digital converter. Although there were two gas sensors that were applied in the instrument system, the signal conditioning used in the data acquisition devices was non-inverting amplifier only. Both methane gas sensor –TGS 2611- and oxygen gas sensor -KE-50- used the ADC inside the AVR ATMEGA 8 microcontroller. Here, the microcontroller would process the input data, namely methane and oxygen level, derived from the measurement by the gas sensors system. After processing both methane and oxygen level, the microcontroller relayed them to the PC.

2.3. The PC-Based Real-Time System

A real-time system is a system that capable to produce an appropriate response in specific time arrangement. A real-time system is possible to send data to the computer centre, process the received data in the computer centre and then send them back to the sender at that time [11].

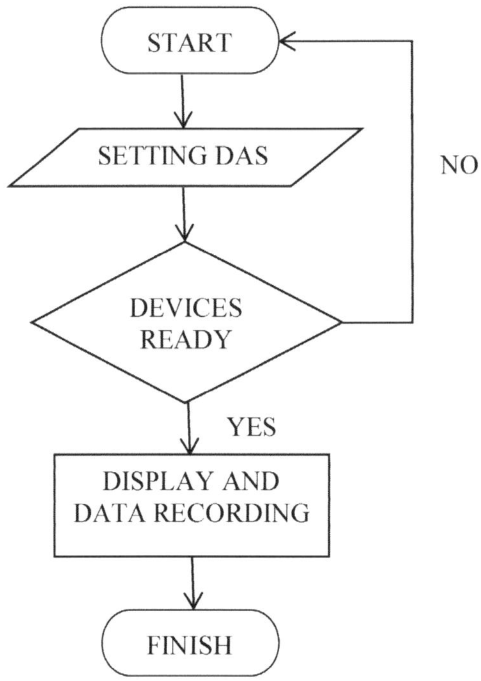

Figure 5. Flowchart of PC-based real-time system.

Figure 5 depicts a flowchart of the PC-based real-time system in this study. If the data acquisition devices are in the ready state, the devices will transfer the information they get to the PC. Later, the PC would display and perform the data recording. On the contrary, if the data acquisition devices are not in the ready condition then the devices will return to their original condition i.e. START.

3. Results and Discussion

The first stage in this study was producing biogas in laboratory scale, then continued by monitoring the concentration of methane and oxygen in biogas production. The monitoring system would be held in real-time to find out whether the production process had already optimum or not.

3.1. Biogas Production

As mentioned before in the introduction part, besides methane, biogas also contains a little hydrogen sulfide. Even though the amount of hydrogen sulfide is just little, this gas is corrosive. If there is corrosion during the production, the biogas will stop producing. Therefore, the selection of materials used for biogas reactor and gas reservoir plays an important role. This study used glass material as the biogas reactor and plastic material as the gas reservoir to prevent corrosion occurred.

3.2. Data Acquisition in Gas Sensors System

Biogas production is the most commonly monitored indicator because it represents the overall process performance and can be measured by some sensor [10]. In this study, the methane gas sensor was put into the gas reservoir because according to the theory, biogas mostly contains methane gas. Meanwhile, the oxygen gas sensor was placed inside the biogas reactor to know whether there is leakage inside the reactor or not. Figure 6 below illustrates the data acquisition in gas sensors system.

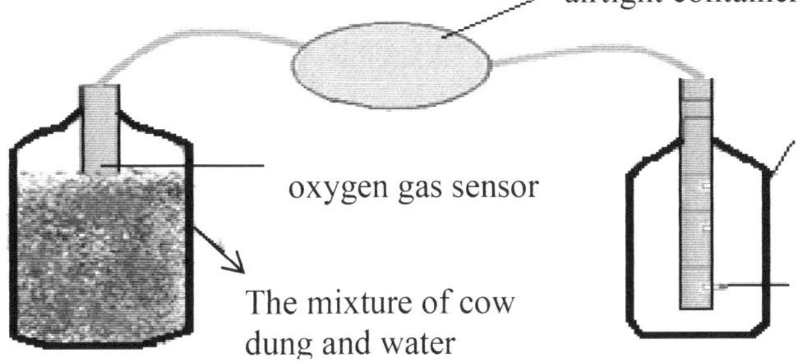

Figure 6. Illustration of data acquisition in gas sensors system.

3.3. The PC-Based Real-Time System

A real-time system is a data communication with speed and accuracy. The necessity for information has to be fulfilled in recent time or the same treatise immediately [6]. In this system, the process is performed in a matter of a few seconds only. Therefore, it takes a fast-communication path and fast-processing system, as well as a huge memory and storage system.

Methane and oxygen concentration data -detected by the gas sensors system- were forwarded to the computer (PC) by a serial communication in turn after being processed by the microcontroller. The data sent earlier was then received by computer serial receiver and displayed on the screen in real-time. In addition to being displayed on computer screen, the measured methane and oxygen concentration data were also stored in the database.

The PC-based real-time system in this study comprised of three main pages, among others: the user page, main-monitoring page, and searching page. This monitoring system utilized Delphi as its data viewer software of methane and oxygen gas concentration in biogas production. User-friendly design for the beginner programmer, own a large number of components, also the resulting application is quite complex and realistic are the reasons behind the selection of Delphi as data viewer software. In addition, Delphi has a fast compilation speed making it easier to modify the software also it has the ease of creating User Interface [12, 13]. Before entering the main monitoring page, the user would meet the user page as the initial display in the PC-based real-time system. Meanwhile, Figure 7 presents the main-monitoring page in the instrument system. The main-monitoring page would begin to show and record the concentration of methane and oxygen gases detected along with the measurement time if we clicked the 'START' button.

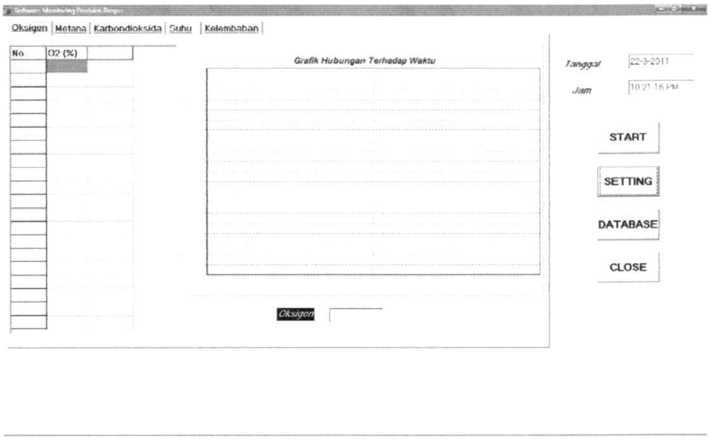

Figure 7. Main-monitoring page

The searching page (Figure 8) contained the record of measured physical parameters (methane and oxygen concentration) on specific dates. The measurement data recording was divided into two types namely graph and table, whereas each type was displayed on different tabs.

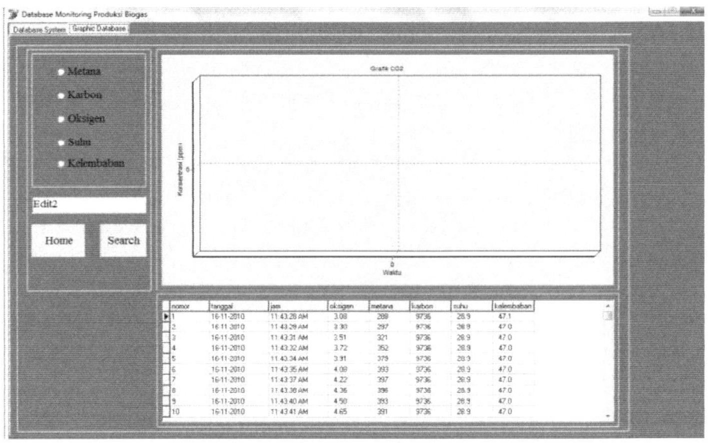

Figure 8. Searching page.

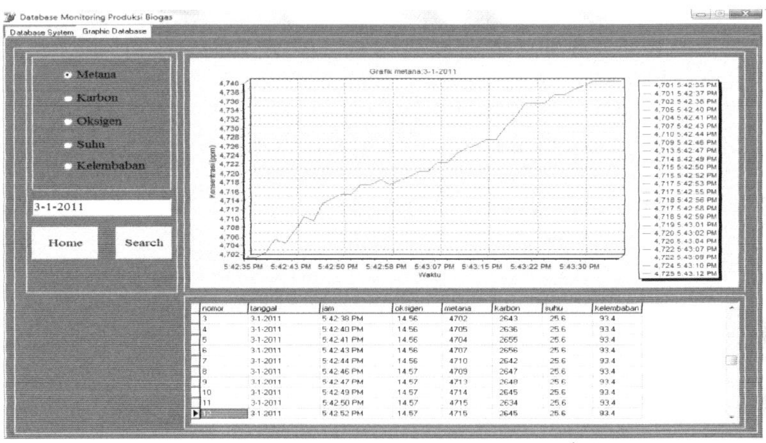

Figure 9. Methane production on the 13th day.

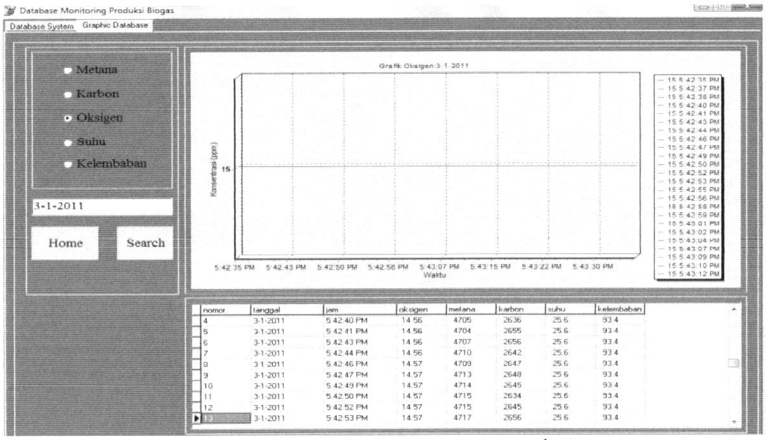

Figure 10. Oxygen level on the 13th day.

The two figures above (Figure 9 and10) depict methane and oxygen concentration found on the 13th day of biogas production. As seen in Figure 9, the level of methane concentration has a drastic increase trend. According to Putri et al [14], the increase in biogas production occurs on day 4 to day 25. Putri et al [14] also stated that the ratio of cow dung to water namely 1:1 makes the methane production becomes inconstant because the water level is not comparable with the level of cow manure in the biogas reactor. Meanwhile, Figure 10 depicts stability of oxygen concentration in biogas reactor. Although the oxygen existence in biogas reactor was detected, it did not affect the methane concentration.

4. Conclusions

The instrument system built in this research had the capability to record and display the methane and oxygen gases concentration of biogas production based on the real-time system. This PC-based real-time system used Delphi as monitoring software, and there were three main pages in its appearance menu.

References

[1] Adelekan BA 2012 *Phys. Rev. Res. Int.* **2** 1
[2] Ofoefule A U, Uzondima E O and Onukwuli O D 2009 *Int. J. Phys. Sci.* **4** 535
[3] Adelekan B A and Bamgboye B I 2009 *Afr. J. Agric. Res.* **4** 571
[4] Adeyosoye O I, Adesokan I A, Afolabi K D and Ekeocha A H 2010 *Afr. J. Environ. Sci. Technol.* **4** 388
[5] Weiland P 2010 *Appl. Microbiol. Biotechnol.* **85** 849
[6] Yantidewi M 2011 *Pemantauan Produksi Biogas Secara Waktu Nyata* (Surabaya: Institut Teknologi Sepuluh Nopember)
[7] Alfanz R, Nurhadi A and Laksmono J A March 2016 *J. Nas. Teknik Elektro* **5** 128
[8] Roosganda E and S Rusdiana 2011 *Pros. Semin. Nas. Era Baru Pembangunan Pertanian: Strategi Mengatasi Masalah Pangan, Bioenergi, dan Perubahan Iklim* (Bogor: Pusat Sosial Ekonomi dan Kebijakan Pertanian Badan Penelitian dan Pengembangan Pertanian Kementerian Pertanian) pp 220-234
[9] Wresta A and Saepudin A 2013 *Energy Procedia* **32** 115
[10] Kanokwan B, Bastone D J, Steyer J-P and Angelidaki I 2010 *Water Res.* **44** 5973
[11] Aldarmi S A 1998 *Real-Time Database Systems: Concepts And Designs* (York: University of York)
[12] Nagara N and Yazid P I 2012 *J. Oto. Ktrl. Inst. (J. Auto. Ctrl. Inst)* **4** pp 17-24
[13] Sihombing M A, Harmawan A A and Manik T N 2017 *Jur. Fisika Flux* **14** 53
[14] Putri D A, R R Saputro and Budiyono 2017 *Int. J. Renew. Energy Dev. (IJRED)* **1** 61

Microseismicity of Blawan hydrothermal complex, Bondowoso, East Java, Indonesia

S Maryanto[1, 2, 3]

[1] Geophysics Laboratory, Physics Department, Faculty of Sciences, Brawijaya University, Malang, Indonesia
[2] Centre of Energy and Natural Resources, Institute of Research and Community Services, Brawijaya University, Veteran Street 65145, Malang, Indonesia
[3]Brawijaya Volcanology and Geothermal Research Centre, Faculty of Sciences, Brawijaya University, Malang, Indonesia

E-mail: sukir@ub.ac.id

Abstract. Peak Ground Acceleration (PGA), hypocentre, and epicentre of Blawan hydrothermal complex have been analysed in order to investigate its seismicity. PGA has been determined based on Fukushima-Tanaka method and the source location of microseismic estimated using particle motion method. PGA ranged between 0.095-0.323 g and tends to be higher in the formation that containing not compacted rocks. The seismic vulnerability index region indicated that the zone with high PGA also has a high seismic vulnerability index. This was because the rocks making up these zones were inclined soft and low-density rocks. For seismic sources around the area, epicentre and hypocentre, have estimated base on seismic particle motion method of single station. The stations used in this study were mobile stations identified as BL01, BL02, BL03, BL05, BL06, BL07 and BL08. The results of the analysis particle motion obtained 44 points epicentre and the depth of the sources about 15 – 110 meters below ground surface.

1. Introduction

Geothermal is one of the biggest potentials of the renewable energy sector in Indonesia. It was a natural source of energy derived from the rock interaction and heat flow beneath the earth's surface. In Indonesia, there were 256 geothermal areas and have 27,441 MV of potential energy distributed from Nangroe Aceh Darussalam to Irian Jaya [1]. The geothermal system in Indonesia dominated by volcanic geothermal system (volcano hosted geothermal) that associated with a volcano quarter. It was commonly located in the Quaternary volcanic arc that stretching in Indonesia as well as for Blawan-Ijen Complex in East Java [2-6].

Some non-seismic geophysical methods have been applied to observe the existing of geothermal Blawan area. Based on the geo-magnetic result indicated the geothermal source volume estimated about 133.16 million m^3 which has temperature around 70.2°C [5].Geo-electric survey at Blawan identified the underground seepage of hot water which following the fault direction and river flow with some hot spring that founded which has temperature about 50°C [3]. Furthermore, the natural waters have been also identified with geo-electric survey and results it is in about 9-25 meters depth [7]. However, a seismic method had not been applied at Blawan volcano hosted geothermal area.

IOP Conf. Series: Journal of Physics: Conf. Series **997** (2018) 012019 doi:10.1088/1742-6596/997/1/012019

Basically, the geothermal system with hydrothermal type was formed as a result of heat transfer from source to its surroundings by conduction and convection [8]. The three-component seismic method is very effective used for exploration in the geothermal area. The seismic wave events as measurement results and the time function consist of three components i.e. two horizontal components (North-South and East-West), and one vertical component (Up-Down) [9].

The geothermal systems generally located in tectonic areas or high volcanic activity [10]. Almost all geothermal areas affected by regional geological structures. Microtremor as the passive seismic method record all motions generated by earth or human activities. Moreover, using the particle motion analysis in horizontal and vertical components could determine the distribution of the location of earthquake sources position (epicentre and hypocentre).

2. Methods

Seismic data recorded in Blawan-Ijen Complex, East Java at 10 mobile stations (BL01 – BL10). The station selected according to their event characteristics i.e. BL01, BL02, BL03, BL05, BL06, BL07, and BL08. The data recorded by TDS 3 components seismometer i.e. North-South (NS), East-West (EW) and Up-Down (UD). These three components will be used for seismic particle motion analysis. Time domain data was transformed into frequency domain using FFT (Fast Fourier Transform) to obtain the frequency spectrum of each component. In addition, spectrogram analysis was needed to know the variation of the signal frequency distribution. Then cut-off frequency determined for the filtering process and using band-pass Butterworth filter.

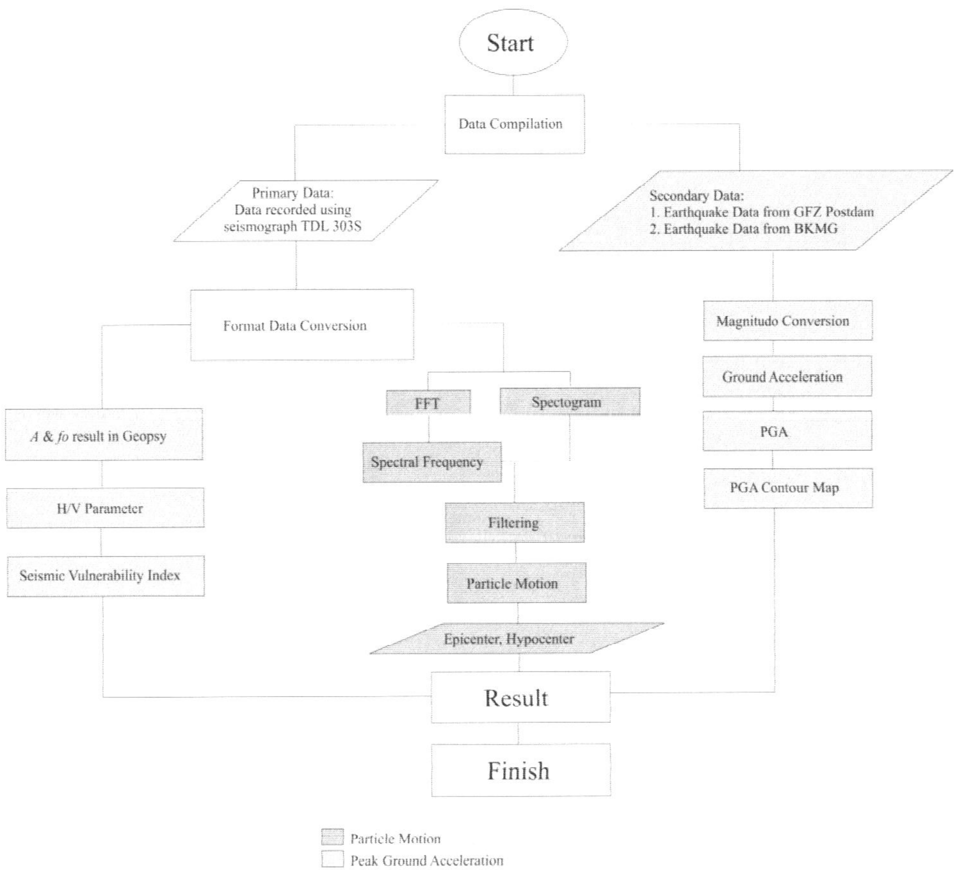

Figure 1. Flowchart of methods used in this research.

Particle motion of horizontal and vertical components plotted to estimate the epicentre and hypocentre of seismic sources. The amplitudes of EW and NS components plotted to obtain horizontal P-wave particle motions. The epicentres estimated by calculating the azimuth direction of horizontal particle motions. The amplitude resultant of both horizontal components plotted with the vertical components for predicting the hypocentres. Additional data such geological structure data at Blawan-Ijen Complex used for assuming the source positions influenced by the hydrothermal activities. The flowchart of the methods can be seen in figure 1.

3. Results and Discussion

3.1. Peak Ground Acceleration (PGA)

The acceleration of maximum ground vibration due to earthquake events was estimated by the attenuation function that formulated by scientist. Until now in Indonesia did not have its own attenuation function because the historical data of the earthquake was not enough to formulate the attenuation function. The Fukushima–Tanaka attenuation [11] was used in this research because the acceleration of land in the hills and mountains was assumed to have almost the same morphological conditions. In addition, this attenuation function had chosen because there were several fractures and fault indications in the research area so it was possible to use the function. Furthermore, the attenuation function by Young *et al* (1997) [12] also used because this function was considered appropriate for the research area as it was suitable for earthquakes with more than 200 km depths and was also suitable for subduction earthquake sources type.

From the calculation of the two functions, they were obtained different PGA results but has a similar plot contours. The value obtained from the calculation of Fukushima–Tanaka was smaller than the Young method. The result of PGA with Fukushima–Tanaka method ranged from 0.0953–0.3227 g while the method of Young attenuation obtained results between 1.7664–1.9464 g. Figures 2 and 3 were results of PGA calculation with attenuation function of Fukushima–Tanaka and Young.

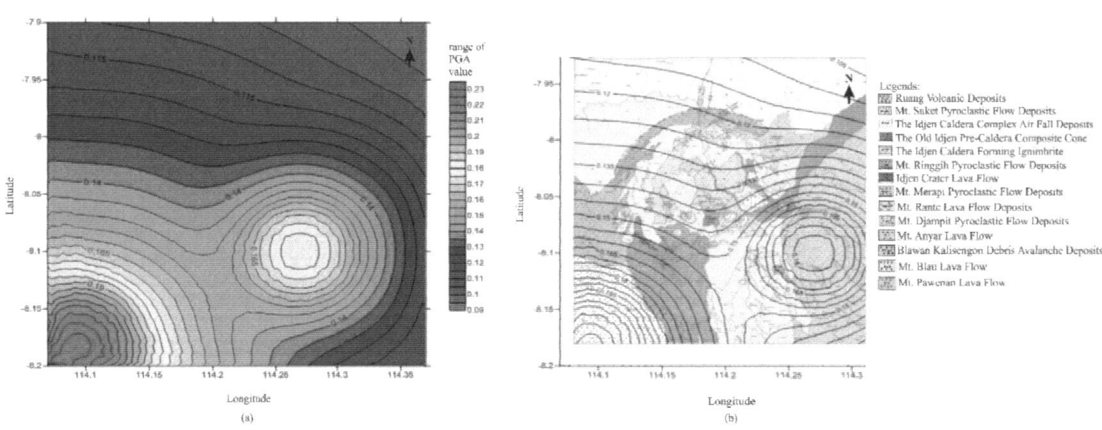

Figure 2. Contour of Peak Ground Acceleration of Blawan area using Fukushima-Tanaka attenuation function.

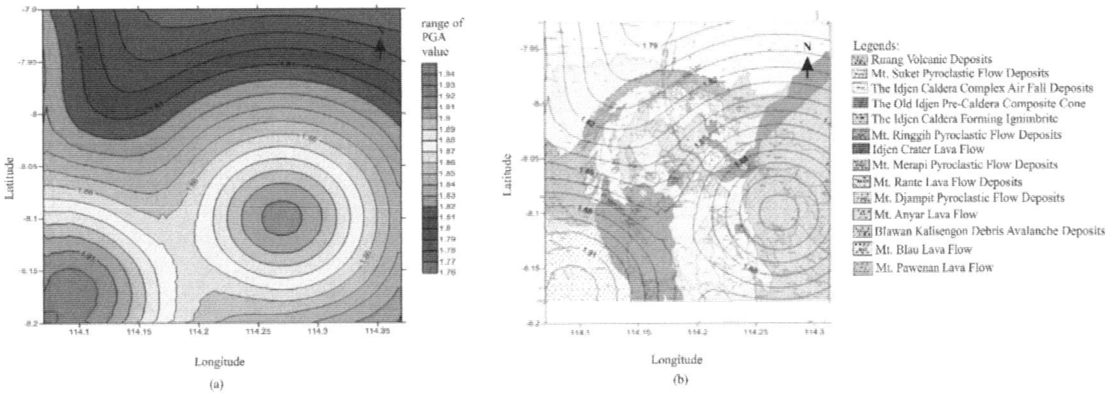

Figure 3. Contour of Peak Ground Acceleration of Blawan area using Young attenuation function.

Based on figures 2 and 3, the values of the Peak Ground Acceleration were indicated by colour bar with the highest value shown in red colour. The areas with PGA values tend to be high at coordinates -8.15° LS 114.1° BT and -8.1° LS 114.28° BT. At coordinates -8.15° LS 114.1° BT has a PGA value of 0.3227 g for Fukushima–Tanaka and 1.946 g for Young attenuation. This large PGA by using Young attenuation tends to be high when compared to the Peak Ground Acceleration of Indonesian bedrock map [13] which only ranges from 0.15–0.25 g, but for Fukushima–Tanaka show the results which not different significantly. The variation of soil acceleration values influenced by the distance of hypocentre (*H*) and the magnitude of the earthquake which occurred around the area. The closer distance of the earthquake hypocentre to the point of observation showed a greater value of PGA. Similarly, the magnitude of the earthquake also provides a large PGA as well.

Regarding to geology of the region, there are faults that affects the earthquakes. In Blawan region, there are about 4 faults and graben which affect the acceleration value of ground vibration. The highest PGA values found in southwest of research area. This area was formed by predominantly eruption of Mount Raung which composed of andesitic lava, lava, pyroclastic flows, and pyroclastic fall. The physical properties of rocks which not too hard and compact indicated a high seismic vulnerability, like andesitic & breccias. Tuff sandstone (pyroclastic fall) which contain lapilli tuff and pumice stone also have less physical characteristics so it has an appreciable seismicity. The lava, which consists of breccias, tuff sand, lava block, andesitic and basalt, has a higher susceptibility, as well as alluvium deposits.

3.2. Particle Motion

Based on seismic particle motion analysis in Blawan Complex, Ijen Mountain, East Java, there are 44 spots of epicentre suspected as hydrothermal activity below earth surface. They are supported by the distribution of geothermal manifestations around the area. The dominant rocks and layers in Blawan Complex area, according to geoelectric and geomagnetic methods, are sedimentary layer which was deposit of Blawan Lake in the form of clay tuff, limestone sediments, gravel, water contained water, and layers of volcanic rocks which are breccias tuff of volcanic rock and basaltic lava. Lava rocks contain many cracks which become spaces for fluid flow (water). Cracks may occur due to volcanic as well as tectonic activities around Mount Ijen. Hot rocks will heat the fluids beneath the earth's surface, it increased the hot fluid activity and cause earthquakes. Figure 4 explained the seismic epicentre points suspected to be an undiscovered hot spring.

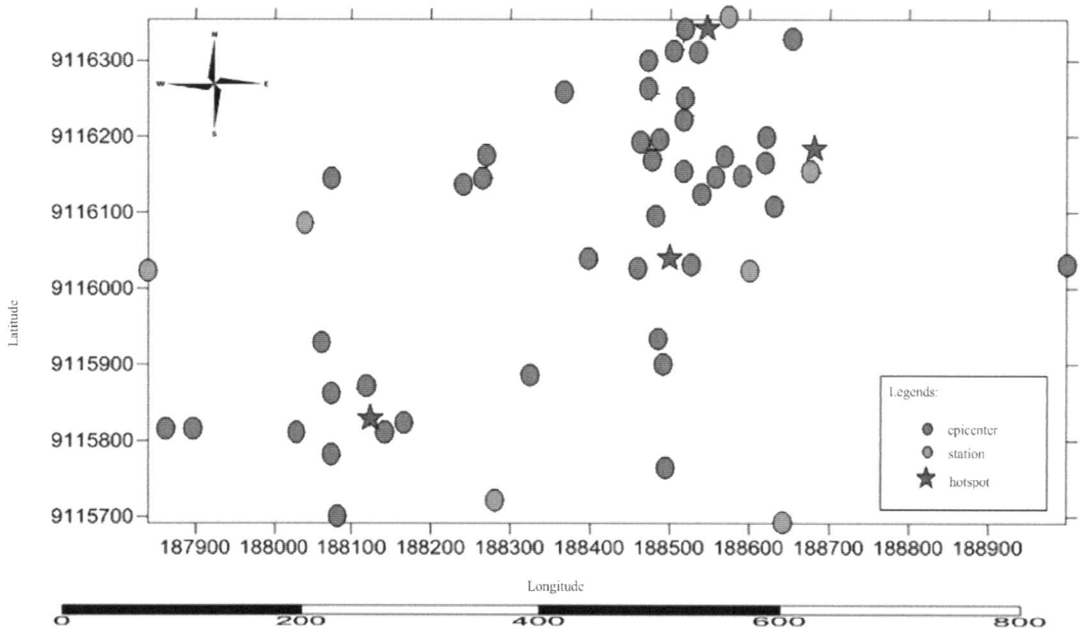

Figure 4. Map of geothermal epicentre distribution at Blawan Complex Mount Ijen, East Java.

Figure 4 shows that the epicentre spread near the hotspots around study area. From the calculation using vertical component of particle motion shows 31 points of hypocentre below the study area. Some of them can be seen in table 1. From table 1, the hypocentres of events in study area are around 15 – 110 meters below the surface. Figures 5 show the distribution of hypocentre in study area. Figure 5a was hypocentre distribution in South-North section, while figure 5b was West-East section. From figures 5a and b shows that hypocentre distribution each events flow through every hot spots on earth surface.

Table 1. Some coordinates of hypocentre location around study area.

No	Longitude (UTM)	Latitude (UTM)	Depth (m)
1	188295	9115927	65
2	188252	9115973	80
3	188301	9115961	60
4	188372	9115916	20
5	188269	9115873	30
6	187876	9115882	40
7	188536	9116108	110
8	188480	9116276	15
9	188010	9116041	70

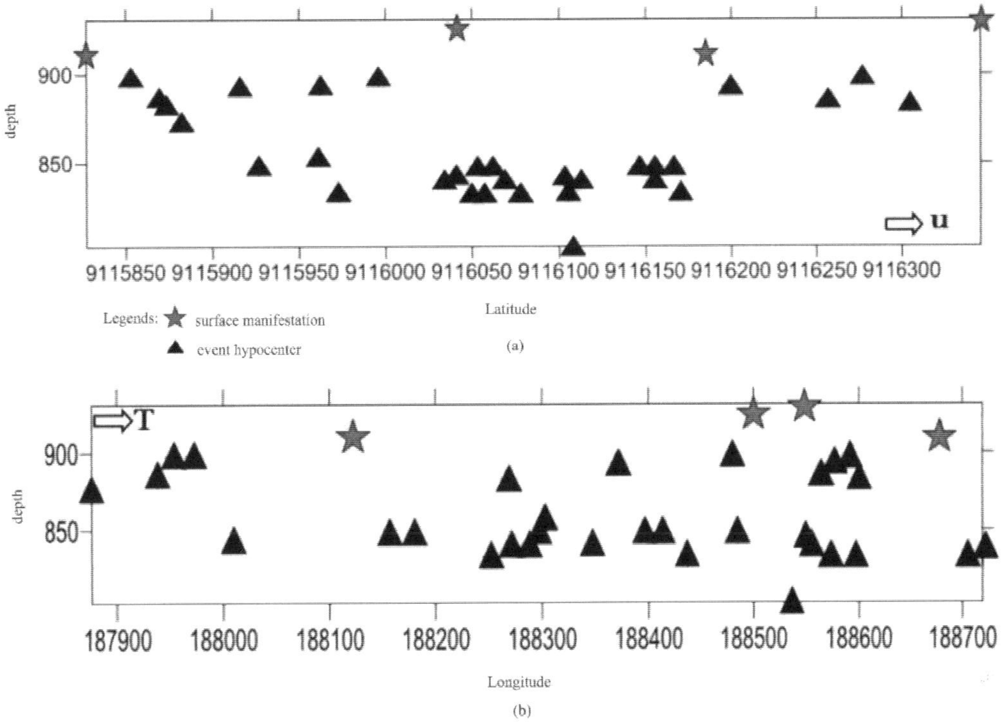

Figure 5. Hypocentre distribution of seismic events, (a) South-North section, (b) West-East section.

3.3. Seismic Vulnerability Index

Seismic vulnerability index was a value that can describe the level of surface layer susceptibility to deformation that occurs during earthquake. The seismic index obtained from the calculation of H/V parameters in the form of A and f_0. Besides, to calculate seismic vulnerability of research area, H/V parameter can also used to know the dominant period of recorded microtremor. The dominant period can illustrate the geological condition of the research area so it can complete the information as the basis for the analysis of vulnerability and PGA values obtained. The results obtained from the HVSR method are in table 2.

Table 2. The results of seismic vulnerability index calculation.

No	Point Name	A	f_0	Dominant Period (s)	Seismic Vulnerability Index
1	BL01	0.526	2.49	0.401606	0.111115
2	BL02	0.849	4.65	0.215054	0.155011
3	BL03	10.749	4.16	0.240385	27.77428
4	BL04	1.806	7.68	0.130208	0.424692
5	BL05	6.134	1.522	0.65703	24.72139
6	BL06	0.747	4.01	0.249377	0.139154
7	BL07	0.554	2.67	0.374532	0.11495
8	BL08	0.722	2.52	0.396825	0.206859
9	BL09	2.812	0.836	1.196172	9.458545
10	BL10	0.982	1.79	0.558659	0.538728

The value of dominant period can be used to determine the geological condition of the study area. Based on the results, the dominant period at point BL01 was 0.401606. According to the soil classification of Kanai & Tanaka [14] and Omote-Nakajima [15] conversion in Table 3, the dominant period value of BL01 was 0.40, which is III/B type. This type of soil indicates the presence of alluvial rocks consisting of gravel, hard clay sand, and clay, but they were still unknown formations. Points BL02, BL03 and BL04 obtain different dominant period values but all three include in the same type of soil, i.e. II/A type. The soil conditions of II/A type were almost identical to the III/B type soil comprising gravel, hard clay sand, and clay with a thickness of 5 meters. At point BL05, the dominant period value that tend to be high, amount to 0.65703. The high dominant period value indicate the classification of IV/C type soils composed of alluvial rocks formed from delta, top soil, and mud sediment with a depth of 30 meters. The BL06 point indicates the II/A type with the dominant period value 0.249377. The points BL07 and BL08 have the dominant period values 0.374532 and 0.396825, both of them were classified in soil III/B type with soil conditions similar to II/A type, but they were still unknown formation. The points BL09 and BL10 included on the soil IV/C type because it has the value of the dominant period 1.196172 and 0.558659. Therefore, it was indicated alluvial rocks due to sedimentation delta, top soil and mud at a depth of 30 meters.

Table 3. Soil classification by Kanai–Tanaka and Omote–Nakajima modified from Parwatiningtyas et al (2013) [16].

Soil Classification		Dominant Period (s)	Specification
Kanai-Tanaka	Omote-Nakajima		
I	A	0.05–0.15	Tertiary rocks or older. Consists of hard sandy gravel
II	A	0.10–0.25	Alluvial rocks with 5 meters width. Consists of sandy gravel, sandy hard clay, clay, and loam
III	B	0.25	Alluvial rocks similar with II type rocks
IV	C	>0.40	Alluvial rocks formed from delta sedimentation, top soil, mud, etc with 30 meters depth

The result of Seismic Vulnerability Index in table 2 also shows the higher values. When Seismic Vulnerability Index was larger, the risk of damage due to earthquake becomes greater. The high values of Seismic Vulnerability Index were affected by geological condition around the area. They related to wave amplification which propagated in soil layers or rocks. If the soil thickness relatively large associated with large amplification, the Seismic Vulnerability Index was relatively large too. Figure 6 presents the contour of Blawan Seismic Vulnerability Index. High Seismic Vulnerability Index was shown on red colour. Based on geological map (picture 2 and 3), the area with high Seismic Vulnerability Index such as BL03 and BL05 which shown on red colour in contour map were in sedimentary deposits of Blawan lake. It was characterized by the shrinkage between the shale and sand with a minimum thickness about 10 meters and consists of fossil, lime sediment, and hard. The rocks consist of andesitic fragment, basalt, pumice stone, quartz, and granodiorite. These rocks have less coherent structure so it has a high vulnerability. The materials of these rocks also affect the wave amplification, while the rocks which have low density (less coherent structure) will also have relatively high risk of damage.

IOP Conf. Series: Journal of Physics: Conf. Series **997** (2018) 012019 doi:10.1088/1742-6596/997/1/012019

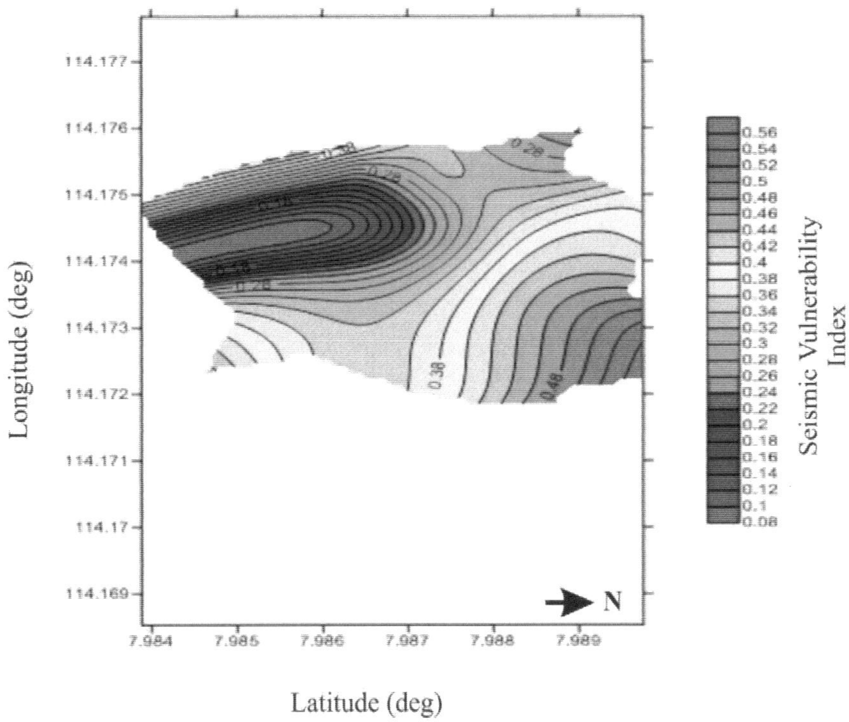

Figure 6. Contour of Seismic Vulnerability Index in Blawan area.

4. Conclusion

Based on the results, the hypocentre located about 15 to 110 meters below the subsurface. While epicentre distributed around the hot springs of Blawan area. The existing of epicentre and hypocentre can relate to the existence of hydrothermal activity beneath the earth's surface. In addition, the maximum land acceleration value of the geothermal potential area of Blawan-Ijen East Java ranges from 0.093 to 0.322 for the Fukushima-Tanaka attenuation function and by 1.766 to 1.946 for Young attenuation function. Region with the largest PGA values located at coordinates -8.15° LS 114.1° BT and -8.1° LS 114.28° BT. The region was a zone that also has a high seismic susceptibility index due to soft-tending constituents and low rock densities.

Acknowledgement

We thank to PEER Science grant PGA 2000005053 for partially financial support, Wurung-Ijen crater team and all members of Brawijaya Volcano Geothermal Research Center (BRAVO GRC) Brawijaya University that joined during data acquisition at Ijen volcano complex.

References

[1] Maryanto S 2017 *Proc. 8th. Int. Conf. on Global Resource Conservation* **19** **8**
[2] Maryanto S, Dewi C N, Syahra V, Rachmansyah A, Nadhir A and Santoso D R 2017 *J. Geosci.* **7** 12
[3] Maryanto S, Suciningtyas I K L N, Dewi C N and Rachmansyah A 2016 *Int. J. of Geophys.* **2016** 14
[4] Suciningtyas I K L N, Maryanto S and Rachmansyah A 2013 *J. Natural-B* **2** 164
[5] Afandi A, Maryanto S and Rachmansyah A 2013 *J. Neutrino* **6** 1
[6] Nuha D Y U, Maryanto S and Santoso D R 2017 *J. Penelit. Fis. Apl.* **7** 2

[7] Wiyono, Soemarno, Maryanto S and Rachmansyah A 2017 *American J. Env. Eng.* **7** 10
[8] Gupta H and Roy S 2007 *Geothermal Energy: An Alternative Resource for the 21th Century* (Amsterdam: Elsevier) p 31
[9] Ihsan A B 2011 *Characterization of Microtremor at Porong River, Kebon Agung Sidoarjo* (Malang: Brawijaya University)
[10] Iguchi M *et al* 2012 *J. Disaster Res.* **7** 26
[11] Fukushima Y and Tanaka T 1990 *Bull. Seism. Soc. Am.* **80** 757
[12] Young R R, Chiou S J, Silva W J and Humphrey J R 1997 *Seism. Res. Lett.* **68** 58
[13] Irsyam M *et al* 2010 *Study Result Summary of Indonesia Earthquake Revision Team 2010* (Bandung: Bandung Technology Institute) p 29
[14] Kanai K and Tanaka T 1961 On Microtremors VIII *Bull. Earthq. Res. Inst.* **39** 99
[15] Omote S and Nakajima N 1966 *Proc. of Japan Earthq. Eng. Symp.* **1** 21
[16] Parwatiningtyas D, Ambarsari E W, Marlina D and Wiratomo Y 2013 *Proc. The 3rd. Int. Symp. for Suistainable Humanosphere (ISSH)* **3** 46

IOP Conf. Series: Journal of Physics: Conf. Series **997** (2018) 012020 doi:10.1088/1742-6596/997/1/012020

Analysis of graphical representation among freshmen in undergraduate physics laboratory

A S Adam[1], S Anggrayni[1], A Kholiq[1], N P Putri[1] and N Suprapto[1,a]

[1]Department of Physics, Faculty of Mathematics and Natural Science, Universitas Negeri Surabaya, Jl. Ketintang, Surabaya 60231, Indonesia

E-mail: [a]nadisuprapto@unesa.ac.id

Abstract. Physics concept understanding is the importance of the physics laboratory among freshmen in the undergraduate program. These include the ability to interpret the meaning of the graph to make an appropriate conclusion. This particular study analyses the graphical representation among freshmen in an undergraduate physics laboratory. This study uses empirical study with quantitative approach. The graphical representation covers 3 physics topics: velocity of sound, simple pendulum and spring system. The result of this study shows most of the freshmen (90% of the sample) make a graph based on the data from physics laboratory. It means the transferring process of raw data which illustrated in the table to physics graph can be categorised. Most of the Freshmen use the proportional principle of the variable in graph analysis. However, Freshmen can't make the graph in an appropriate variable to gain more information and can't analyse the graph to obtain the useful information from the slope.

1. Introduction

Transferring process of raw data is an important part of physics laboratory to make analysis process easier. It can transform in table or graph to make an appropriate conclusion. The proportional principle easy to describe when the raw data transferred into the graph [1]. It shows clearly the relation between variables [2]. In the experiment, there are several types of variables. The independent, dependent and control variables. The independent variable is the one condition that changes in an experiment. The dependent variable is the variable that measures or observe. A controlled variable or constant variable is a variable that does not change during an experiment [3]. The tendency of one variable to other not necessarily show a straight-line trend [4]. A graph can show the specific constant according to some physical law and aberrant data in sequence. It shows up in the graph clearly. A mistake from plotting graph in variable or experiment can detect from graphical representation [5].

The graph should show enough data because a lot of data make an appropriate decision to take better conclusion. Clearly, it needs more than two points for data plotting in the graph, more data should be taken to confirm that the data are real [5] such as the region around the velocity of sound. If there are too few data, the velocity of sound will fail to see in the long range. It never knows the velocity in any range as long as the resonance tube. In a spring system, a lot of data will show the elastic limit of the specific spring. The elastic limit of spring depends on the material of spring and it is the condition when spring permanently stretched so that the material doesn't return to its original length when the force is removed. In a simple pendulum, a lot of range of angle will show that the simple pendulum equation unvalued. That equation doesn't express for any angles, but in a few angles (less than about 10°) [6].

If the data is uncorrelated and no relationship between variables, further analysis isn't necessary. But in quantitative data such as the velocity of sound, simple pendulum and spring system topics need to establish the mathematical relationship between the variables. The graph should transform the linear graph to analyse it easier [7]. In the linear graph, the gradient of the graph can use to determine the specific constant according to some physical law [8]. In the velocity of sound, constant gradient use to determine the velocity of sound. In simple pendulum topic, constant gradient use to determine the gravitational acceleration. In the spring system, gradient used to determine the spring constant [6,9].

The representation such as diagram and graph play a critical role in the effectiveness of the interactive engagement between students in these physics learning an environment [10]. The graph is very important because there is a proportional principle. This proportional principle gives the dependent and independent variable for the experiment describes due to the aim [11].The graph depicting a physical proportional which cannot easily be recognized in the table of the same data [12]. There is a tendency among freshmen and graph proportional view and a tendency among incorrect existing knowledge and graphing problem to gain more information from the graph [13].

Physics laboratory for freshmen is the activities in a basic physics course. It consists of mechanics and thermodynamics topics [14]. After getting the related information in class, freshmen apply the physics' concept in the laboratory. That concept isn't only about the physics' principle but also how that principle gets rationally [15]. Physics concept understanding is the importance of the physics laboratory, especially in mechanics and thermodynamics concept. So, the appropriate analysis of data takes to the correct conclusion. For that reason, a graphical representation is required. Based on the importance of graph to make an appropriate conclusion for freshmen, so it is necessary to analyse the graphical representation among freshmen in undergraduate programs.

2. Method

2.1. Context and participants

This particular research uses the experiment data of 20 freshmen's laboratory report in basic physics class at the State University of Surabaya (Universitas Negeri Surabaya), Indonesia. Freshmen are typically first-year students the undergraduate program, all taking the basic physics course as part of physics education and the physics study program. The course is taken in the first semester of 2017. Freshmen have 2,5 hours per week in the laboratory, with 2,5 hours lecture per week in class, and meet for a pre-laboratory session before taking laboratory activities. The pre-laboratory sessions are generally traditional, with freshmen answering questions related to the topic with laboratory assistant.

2.2. Data and analysis

The sources of data for these studies are freshmen's experimental report in the velocity of sound, simple pendulum and spring system topics. It chooses 20 randomly from a group and from all classes. It analyses in the indicator adapted from Christopher Deacon [5]. More particularly, the indicator items are shown in the Table. 1. In this research, the indicators are slightly modified to questions form. The result of this analysis is recapitulated and the graphs of the freshmen are studied in detail.

Table 1. Indicator of graphical representation.

	Yes	No
Do freshmen make a graph?		
Do freshmen use appropriate variable in its axis?		
Do freshmen use proportional principle?		
Do freshmen use gradient analysis?		
Do freshmen use aberrant data analysis?		

3. Results

This section starts by presenting the detailed result of the graphical representation of freshmen experimental report in the velocity of sound, simple pendulum and spring system topics. The second part is the analysis of the whole data. These studies are based on modification in the original indicator from Christopher Deacon [5].

3.1. Initial study: Graphical representation findings

In the first study, the graph took from the experimental report in the data and analysis chapter and focus on information from graph analysis.

3.1.1. Velocity of sound. This topic aims to know the proportionality between the frequency of sound and length on first harmonic. Moreover, this topic aims to determine the velocity of sound in the air. The appropriate graph for this topic shown in figure 1. Period (1/frequency) is proportional to the length on the first harmonic and the velocity of sound *(v)* obtains with an equation 1 with m is a gradient of the slope in figure 1.

$$v = 4m \tag{1}$$

Based on the graph shown in figure 1, the speed of sound in the air can be determined as follows:

$$v = 4m = 4 \times 84{,}046 = 336.18\ m/s$$

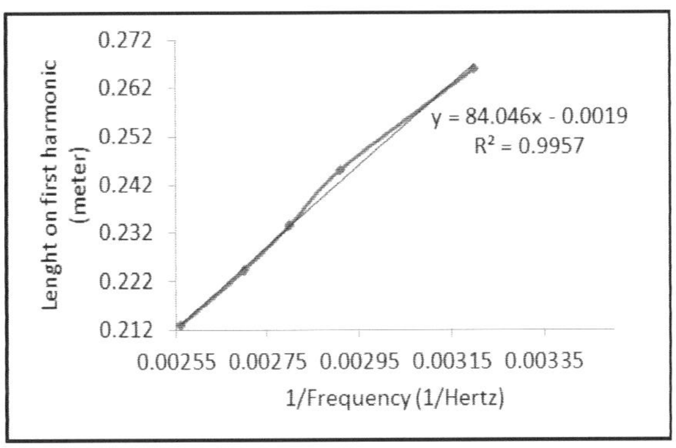

Figure 1. Graph plotted to obtain the velocity of sound.

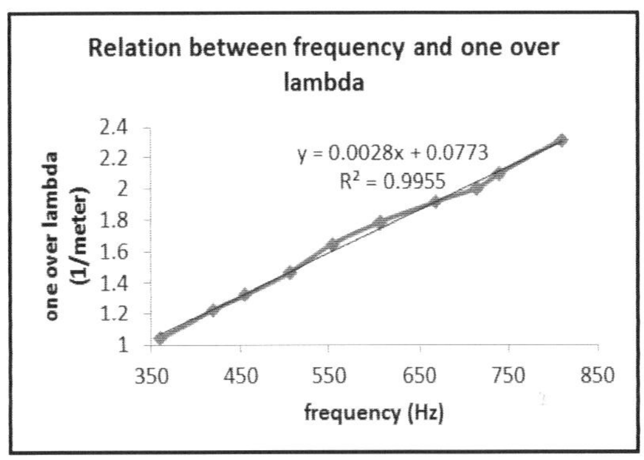

Figure 2. Freshman's graph interpretation of the experiment report.

In this topic, the independent variable is plotted in x-axis as one over frequency and dependent variable plotted in y-axis as a length of first harmonic. Both of x-axis and y-axis didn't contain control variable. From the graph, an increase of the period causes the increase of length in the first harmonic (period and length of first harmonic have linear proportionality) or an increase of the frequency cause decrease of the length at the first harmonic (frequency and length in first harmonic have inverse proportionality). From this variable in an axis, the slope of the graph is used to determining the velocity of sound. Besides this model, there is another graph to determine the velocity of sound using a length of the second harmonic in *the* y-axis. It will show the same value of the velocity of sound.

In freshman's graph shown in Figure 2 titled as 'relation between frequency and one over wavelength', frequency directly proportional to one over the length of harmonic. This graph shows the mistake. It doesn't explain the number of harmonic in that graph. This topic uses closed-end air columns and there are two harmonics, first and second harmonic. This graph can't use to determine the speed of sound and freshmen didn't analyse that. Freshmen just analyse proportionality of this graph. Most of the freshmen use the proportionality by writing the linear proportionality or the increase of frequency causes the increase of variable in *y*-axis such as wavelength and length of first or second harmonic.

At figure 2 shown, freshmen use one over a wavelength in *y*-axis used to make the slope increase. This is the miss-understanding among freshmen that slope didn't use to determine any constant due to determine the velocity of sound. This miss-understanding also found at no one freshmen use a gradient to determine the velocity of sound. The gradient just is shown in the graph without any analysis.

3.1.2. Simple pendulum. This topic aims to know the proportionality between the length of string and period. Moreover, this topic aims to determine the gravity acceleration of earth. An appropriate graph for this topic shown in figure 3. From the graph in figure 3, the length of the string is proportional to the square of the period and gravity acceleration of earth *(g)* obtains with the equation 2 with *m* is a gradient of the slope in figure 3 and *l* is the length of the string.

$$g = \frac{4\pi^2 l}{m} \tag{2}$$

Based on the graph shown in figure 3, the gravity acceleration of earth *(g)* can be determined as follows:

$$g = \frac{4\pi^2}{m} = \frac{4 \times 3{,}14^2}{4{,}0152} = 9{,}84\ m/s^2$$

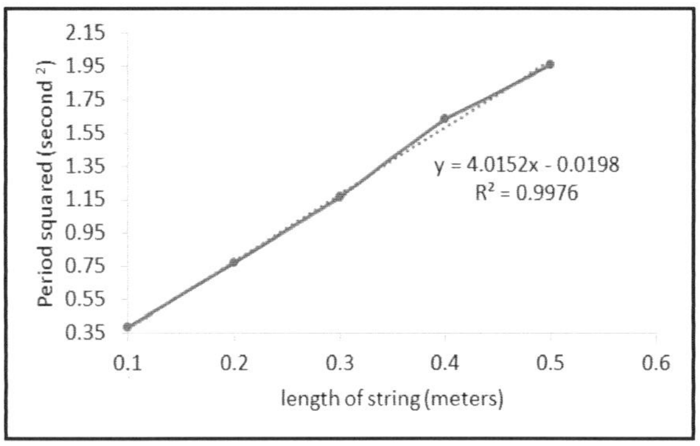

Figure 3. Graph plotted to obtain gravity acceleration of earth.

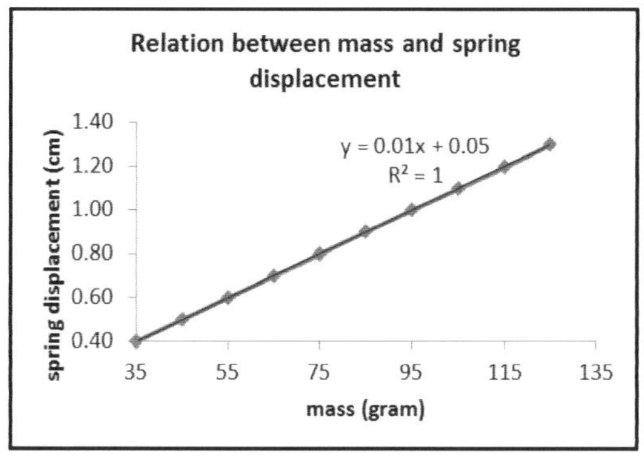

Figure 4. Freshman's graph interpretation of the experiment report.

In this topic, the independent variable is plotted in *x*-axis as the length of string and dependent variable plotted in *y*-axis as the square of the period. Both of *x*-axis and *y*-axis didn't contain control variable. From the graph, raising length of string cause the increase of the square of the period (the length of string and square of the period have linear proportionality). From this variable in the axis, the slope of the graph is used to determining the gravity acceleration of earth.

In freshman's graph shown in figure 4 titled as 'relation between string length and period', the length of the string directly proportional to the period. This topic uses small angle (less than 10°) in the pendulum system. This graph can't use to determine the gravity acceleration of earth and freshmen didn't analyse that. Freshmen just analyse proportionality of this graph. Most of the freshmen use the proportionality and write linear proportionality or the increase of string length causes the increase of period.

3.1.3. Spring system. This topic aims to know the proportionality between mass and spring displacement. Moreover, this topic aims to determine spring constant. The appropriate graph for this topic shown figure 5. From the graph in figure 5, the mass is proportional to spring displacement and

IOP Conf. Series: Journal of Physics: Conf. Series **997** (2018) 012020 doi:10.1088/1742-6596/997/1/012020

spring constant *(k)* obtains with the equation 3 with *m* is a gradient of the slope in figure 5 and *g* is gravity acceleration of earth *($g = 9,8\ m/s^2$)*.

$$\boldsymbol{k} = \frac{\boldsymbol{g}}{\boldsymbol{m}} \tag{3}$$

Based on the graph shown in figure 5, the spring constant *(k)* can be determined as follows:

$$k = \frac{g}{m} = \frac{9.8}{0{,}1933} = 50{,}7$$

In this topic, the independent variable is plotted in *x*-axis as mass and dependent variable plotted in *y*-axis as spring displacement. Both of *x*-axis and *y*-axis didn't contain control variable. From the graph, an increase of mass causes the increase of spring displacement (mass and spring displacement have linear proportionality). From this variable in the axis, the slope of the graph is used to determining the spring constant.

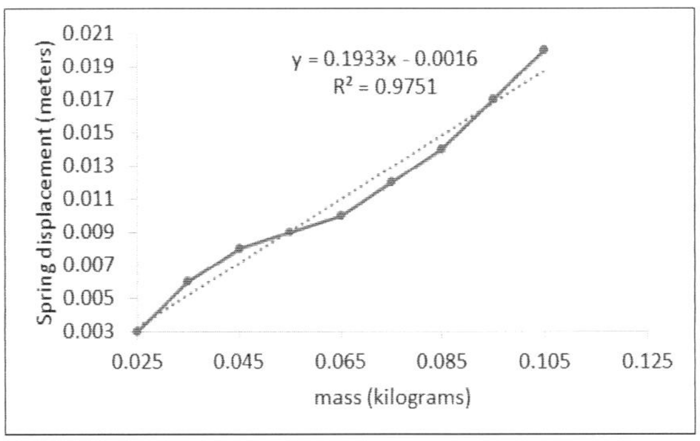

Figure 5. Graph plotted to obtain spring constant.

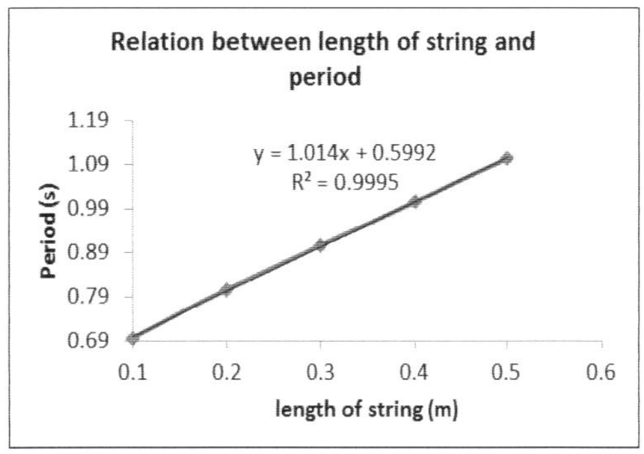

Figure 6. Freshman's graph interpretation of the experiment report.

In the freshman's graph shown in figure 6 titled as 'relation between mass and spring displacement', mass directly proportional to spring displacement. The graph in figure 6 shows that the variable in x-axis and y-axis are appropriate, but freshmen didn't use to determine the spring constant. Freshmen just analyse proportionality of this graph. Most of the freshmen use the proportionality and write linear proportionality or the increase of mass causes the increase of spring displacement.

3.2. Analysis

Graph representation on freshmen's experiment report is compiled to table 2 below. This table will show the graphical representation among freshmen.

Table 2. Graphical representation among freshmen

	Yes		No	
Do freshmen make a graph?	18	90%	2	10%
Do freshmen use appropriate variable in its axis?	10	50%	10	50%
Do freshmen use proportional principle?	18	90%	2	10%
Do freshmen use gradient analysis?	0	0%	20	100%
Do freshmen use aberrant data analysis?	0	0%	20	100%

Based on the table above, most of the freshmen (90% of the sample) can make a graph derived from the data of the physics laboratory. It means the transferring process of raw data which illustrated in the table physics graph can be categorised. Freshmen who made graph didn't make a mistake in transferring the raw data into a graph. Most of the freshmen also use proportional principle on their graph. It means, all of the freshmen who make graph use proportional principle, although half of sample use an inappropriate variable in x-axis and y-axis. They have understood that graph is used to show proportionality between variables easily.

In graph axis, half of the freshmen use an appropriate variable like describe at the previous part and half of the freshmen make inappropriate variable. All of the freshmen who make inappropriate variable use an independent variable in *the* x-axis and dependent variable in y-axis without any consideration to gain more information from the graph.

In gradient using, freshmen didn't analyse the gradient from the graph. Actually, the gradient important to determine specific value in every topic. From this procedure, a specific value shows the universal value in every independent value. Freshmen also didn't analyse the aberrant data from the graph. Aberrant data in a sequence of measurements will usually show up in the graph more readily than in a table of data.

Freshmen use *Microsoft Excel* TM to make their graph. McKenzie and Padilla state that graphs are an important tool in enabling students to predict relationships between variables and to substantiate the nature of these relationships. They suggest a relationship between the student's graphing skills and their ability to understand scientific relationships [16]. At this research, the relation between the computer-based graph and the ability to understand scientific relationships aren't noticed. This also notices the reason for freshmen didn't the ability to use an appropriate variable, gradient analysis and aberrant analysis.

A graph is a tool for data analysis. As the other tools, the effectiveness of the tools will depend on the user's proficiency. A competence freshman will be able to make and interpret the graph. Physics concern for natural phenomena and mathematics describes these phenomena in the term of the equation and predict the behaviour of the physical system as conditions change. A well-drawn graph provides the combination between physics and mathematics.

A good data is needed to make a good graph. A good understanding in order to get more information from the graph. Based on five fundamental steps to produce a good graph [5], before plotting the data, the aim of the experiment should be clearly set and can be answered. After that, plotting the data to the easy visualize the unit, linear function graph and choose useful variable in axis due to gain more information (use any software to produce graph easily). Then use the slope and interpret it to obtain physical specific's constant as well as describe the relation with experiment. For physics laboratory activity, the good graph producing must be explained to the freshmen due to experiment activity. This explanation must be well-understanding for freshmen and show the whole process from transferring process of the raw data to the graph analysis.

4. Conclusion

Most of the freshmen (90% of the sample) make a graph based on the data from physics laboratory and use the proportional principle of the variable in graph analysis. It means the transferring process of raw data which illustrated in the table to physics graph can be categorised and freshmen understand the main function of the graph as tools. However, some freshmen (50% of the sample) can't make the graph in an appropriate variable to gain more information and can't analyse the graph to obtain useful information from the slope and no one freshmen use the gradient and aberrant data analysis. A good graph produces from five fundamental mentioned and need the whole graphical process explanation.

References

[1] McKenzie D L and Padilla M J 1986 *J. Res. Sci. Teach.* **23** 571
[2] Wainer H and Thissen D 1981 *Annu. Rev. Psychol.* **32** 191
[3] Shukla R K and Srivastava A 2006 *Practical Physics* (New Delhi: New Age International

Limited Publisher) p 10
[4] Lichten W 1989 *Am. J. Phys.* **57** 1112
[5] Deacon C 1999 *Phys. Teach.* **37** 270
[6] Serway and Jewett 2006 *Physics for Scientists and Engineers (sixth edition)* (California: Thomson Brook/Cole) p 468
[7] de Cock M 2012 *Phys. Rev. ST Phys. Educ. Res.* **8** 1
[8] Planinic M, Milin-Sipus Z, Katic H, Susac A and Ivanjek L 2012 *Int. J. Sci. Math. Educ.* **10** 1393
[9] Nowikow I and Heimbecker B 2001 *Physics Concept and Connections* (Canada: Irwin Publishing) p 174
[10] Fredlund T, Airey J and Linder C 2012 *Eur. J. Phys.* **33** 657
[11] Bell S 1999 *A Beginner's Gide to Uncertainty of Measurement* (United Kingdom: Crown) p 62
[12] Beichner R J 1994 *Am. J. Phys.* **62** 750
[13] Mokros J R and Tinker R F 1987 *J. Res. Sci. Teach.* **24** 369
[14] Team 2017 *Buku Pedoman Universitas Negeri Surabaya Fakultas Matematika dan Ilmu Pengetahuan Alam* [Guide Book of The Faculty of Mathematics and Natural Sciences Universitas Negeri Surabaya] (Surabaya: Unesa University Press) p 157
[15] Admoko S and Supriyono 2016 *J. Penelit. Fis. Apl.* **6** 34
[16] McKenzie D L and Padilla M J 1984 *Paper presented at the meeting of the Nat. Assoc. Res. Sci. Teach.* (New Orleans: LA, Louisiana)

IOP Conf. Series: Journal of Physics: Conf. Series **997** (2018) 012021 doi:10.1088/1742-6596/997/1/012021

Optimization of thermal conductivity lightweight brick type AAC (Autoclaved Aerated Concrete) effect of Si & Ca composition by using Artificial Neural Network (ANN)

Zulkifli[1] and G P Wiryawan[1]
[1]Engineering Physics Department, Sepuluh Nopember Institute of Technology, Surabaya,60111 Indonesia

E-mail : zulab@ep.its.ac.id

Abstract. Lightweight brick is the most important component of building construction, therefore it is necessary to have lightweight thermal, mechanical and aqustic thermal properties that meet the standard, in this paper which is discussed is the domain of light brick thermal conductivity properties. The advantage of lightweight brick has a low density (500-650 kg/m^3), more economical, can reduce the load 30-40% compared to conventional brick (clay brick). In this research, Artificial Neural Network (ANN) is used to predict the thermal conductivity of lightweight brick type Autoclaved Aerated Concrete (AAC). Based on the training and evaluation that have been done on 10 model of ANN with number of hidden node 1 to 10, obtained that ANN with 3 hidden node have the best performance. It is known from the mean value of MSE (Mean Square Error) validation for three training times of 0.003269. This ANN was further used to predict the thermal conductivity of four light brick samples. The predicted results for each of the AAC1, AAC2, AAC3 and AAC4 light brick samples were 0.243 W/m.K, respectively; 0.29 W/m.K; 0.32 W/m.K; and 0.32 W/m.K. Furthermore, ANN is used to determine the effect of silicon composition (Si), Calcium (Ca), to light brick thermal conductivity. ANN simulation results show that the thermal conductivity increases with increasing Si composition. Si content is allowed maximum of 26.57%, while the Ca content in the range 20.32% - 30.35%.

1. Introduction

Energy consumption of a building is influenced by various factors, one of which relates to the thermal insulation capability of the material used in the construction of the building. Use of materials with good thermal insulation capability can reduce energy consumption for indoor cooling requirements. One material that is currently widely used because it has good thermal insulation capability is a lightweight brick. The lightweight brick has a porous structure so it has lower density and better thermal insulation ability than conventional brick. The material's thermal insulation capability, including the lightweight brick is determined by its thermal conductivity where the smaller the thermal conductivity value, the better its thermal insulation capacity. In the case of light bricks, factors affecting thermal conductivity include : density, moisture content, temperature, porosity, and lightweight brick mineral composition [1-5]. Research conducted by Stuharova (2016) shows that there is a nonlinear relationship between thermal conductivity, density and the content of light brick vapor. This relationship is expressed in terms of mathematical equations, developed from the experimental data. In some cases, experimental data are not possible to be used in modeling the relationship between the properties of a material and the factors that influence it in terms of mathematical equations. This can be accomplished by modeling artificial neural networks. Artificial Neural Network (ANN) is an information processing system with the ability to learn, remember, and solve problems based on the learning process (training). ANN has been widely used in various studies to predict material properties including thermal properties and mechanical properties [4-7]. In this study, using Backpropagation Neural Networks to predict lightweight brick thermal conductivity based on Si and Ca composition variations.

2. Methodology

2.1 Materials and Methods

The first step in this research is data collection. In this research, there are two types of data that is secondary data for training of ANN and primary data used for testing. Based on the results of the literature study, 18 pairs of data were composed of Si, Ca, Al, O and density from three research papers [8-10]. This data is collected in a database as input for the training needs of ANN. Meanwhile, the next variable is the thermal conductivity paired with each of the above variables collected as the target data. The summary of input and target data can be shown in table 1 and table 2 below:

Table 1. Input Training Data.

Variables	*N	Maximum Composition (wt.%)	Minimum Composition (wt.%)	Mean Composition (wt.%)	Standard Deviation
Si	18	32.56	12.18	20.67	6.93
Ca	18	31.96	18.3	23.69	3.51
Al	18	6.67	1.11	4.09	2.22
O	18	46.28	38.87	41.66	1.86
ρ	18	1210	504	712	180.13

** ρ – density (kg/m³) ; * N - the amount of data*

Table 2. Targeted Training Data.

Variables	*N	Maximum Composition (wt.%)	Minimum Composition (wt.%)	Mean Composition (wt.%)	Standard Deviation
*ρ	18	0.376	0.123	0.23	0.09

** ρ – density (kg/m³) ; * N - the amount of data*

Primary data for the purposes of ANN testing were obtained from thermal conductivity testing on four light brick samples with each AAC1, AAC2, AAC3 and AAC4 samples.

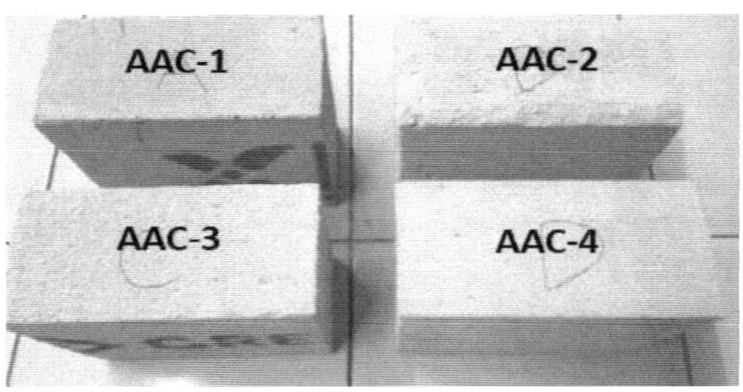

Figure 1. Four light brick sample models.

IOP Conf. Series: Journal of Physics: Conf. Series **997** (2018) 012021 doi:10.1088/1742-6596/997/1/012021

In the four light brick samples, density measurements, composition tests with EDX shown in figure (2), and thermal conductivity tests were performed. The summary of test input and target data is shown in table 3 below.

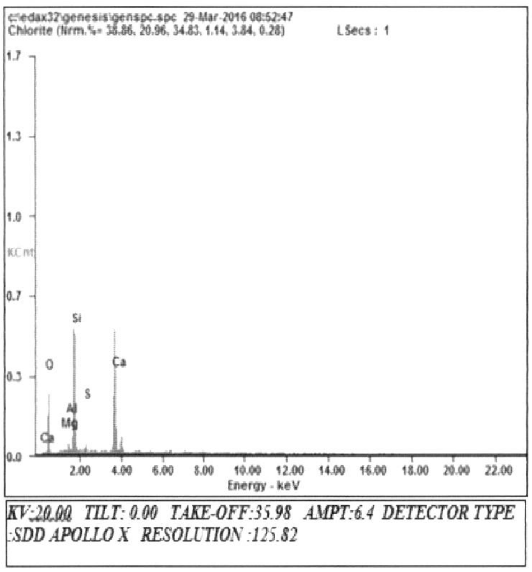

Element	*Wt %*	*At %*
O K	44.76	62.28
MgK	00.94	00.86
AlK	02.02	01.67
SiK	25.41	20.14
S K	00.92	00.64
CaK	25.94	14.41

EDAX ZAF QUANTIFICATION STANDARDLESS SEC TABLE : DEFAULT

Figure 2. EDX composition test results of sample AAC1.

Table 3. Summary of Testing Input Data.

Variables	*N	Maximum Composition	Minimum Composition	Mean Composition	Standard Deviation
Si	4	5.56	27.37	17.24	11.02
Ca	4	29.62	51.5	40.43	11.87
Al	4	0	1.69	0.84	0.97
O	4	38.71	41.8	40.12	1.61
***ρ**	4	509	632	585	53.33

** ρ – density (kg/m³) ; * N - the amount of data*

2.2 . Making Artificial Neural Networks (ANN)

The making of ANN program is done by using Matlab R2015a software. Neural Network Toolbox in Matlab provides various functions that support the creation and development of ANN. Making Artificial Neural Networks is done through the following stages:

- Loading training data (input and target)
- Normalization of input and target data into ranges 1 and -1.
- Function equations of data normalization used are as follows:

$$y = \frac{2(x - x_{min})}{x_{max} - x_{min}} - 1 \quad (1)$$

Where: x is initial value, y is values after normalization, x_{max} is maximum value in the data, x_{min} is the minimum value in the data

1. Initialization of Artificial Neural Networks. The activation function used is tansig and purelin
2. Training parameter determination (three important parameters specified are epochs = 100, training goal = 1e-5; validation checks = 10, while other parameters are left to default)
3. Initiate training and display the weight of training outcomes and validation.
4. Save on the network.

In making this Artificial Neural Network conducted variations of hidden node is 1-10. For each variation of hidden node, do the training 3 times with different initial weight. In this study, 85% of the total 18 data were used for training purposes, while the remaining 15% were used for validation purposes. Artificial Neural Network Architecture Backpropagation used as in figure (3) below:

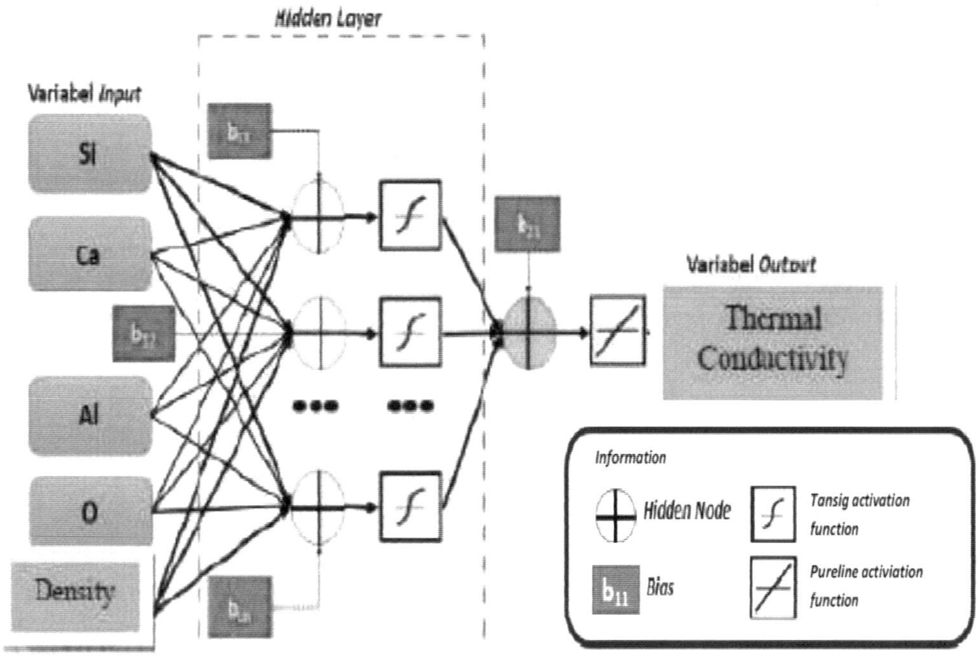

Figure 3. Backpropagation ANN architecture.

2.3 Artificial Neural Network (ANN) Selection

To determine the best ANN model of 10 different hidden node types depends on the mean value of MSE (Mean Squared Error) validation of each ANN. MSE is calculated by the following equation (2):

$$MSE = \frac{1}{N}\sum_{i=1}^{N}(T_i - Y_i)^2 \tag{2}$$

Where : N is amount of training or validation data, Ti is target value of ANN, Yi is the output value of ANN

2.4 Artificial Neural Network (ANN) Test

At this stage, ANN testing is performed by looking at network output (prediction) if new input data is fed to the network. The first step that needs to be done before testing is to call the data and load it into the Matlab workspace. The network used in the testing phase is the network selected from the evaluation results [11]. Next will be compared the value of target and output of ANN which are both lightweight thermal conductivity

2.5 Analysis of the Effect of Composition on Thermal Brick Thermal Conductivity

Si and Ca, and is the main composition that affects the formation of dominant phase contained in the light brick of calcium silicate hydrate. The effect of such composition variation on thermal conductivity can be known by simulating using ANN, where one of the composition values is varied while the other composition is kept constant. The boundary value of the composition of an element found in the light brick can be found using the equation below:

$$Vf = \frac{\lambda_f(\lambda_m - \lambda)}{\lambda(\lambda_m - \lambda_f)} \quad (3)$$

Where : V_f is fraction volume, λ is thermal conductivity of simulation, λ_f is thermal Conductivity at atmosfer pressure and T = 300 K (0.026 W/m.K), λ_m is thermal conductivity of tobermorite (0.2 W/m.K).

3. Result and Discussion

3.1 Lightweight Brick Composition

Table 4 below shows the composition contained in lightweight bricks (four samples AAC1, AAC2, AAC3 and AAC4). The high concentrations of calcium and oxygen and the presence of sulfur in all light brick samples indicate that the area on the surface of the test sample has been carbonated due to environmental influences. Carbonation of one type of light brick degradation that occurs when $Ca(OH)_2$ and C-S-H phase found in lightweight brick react with CO_2 in the air to form $CaCO_3$. and degradation can also occur when carbonated $CaCO_3$ reacts with sulfuric acid in the air and produces gypsum. This mechanism leads to the emergence of sulfur content in the four light brick samples [12].

Table 4. EDX composition test results on four light brick samples.

Sample Lightweight brick	The composition (wt%) is average					
	Si	**Ca**	**Al**	**O**	**Mg**	**S**
AAC1	27.37	30.74	1.67	38.79	0.61	0.83
AAC2	23.89	29.62	1.69	41.19	0.61	1.00
AAC3	5.56	51.50	0	41.80	0	1.14
AAC4	1.26	49.88	0	38.71	0	1.26

3.2 Results of Training and Evaluation of Artificial Neural Network (ANN)

The training of ten models of ANN with variations of the number of hidden nodes gives the results as shown in Table 5 below. The ANN with 3 hidden nodes yields the MSE average validation of the smallest value of 0.003269.

Table 5. Comparison of MSE training and validation scores ANN with hidden.

Amount hidden node	MSE (Average)	
	Training	Validation
1	0.002855	0.005908
2	0.002668	0.005516
3	0.002434	0.003269
4	0.004751	0.003833
5	0.002383	0.006464
6	0.002625	0.007059
7	0.002716	0.008955
8	0.002250	0.003891
9	0.002897	0.008206
10	0.002974	0.004768

Table 6 below shows a comparison of MSE training and validation between 3 ANNs with 3 hidden nodes trained with different initial weights. ANN2 resulted in MSE validation of the smallest value of 0.002252.

Table 6. MSE Value Validation of ANN with one hidden node for three training.

Filename of ANN	MSE Training	MSE Validation
ANN1	0.002694	0.003716
ANN2	0.002764	0.002252
ANN3	0.001845	0.003840

The comparison between MSE training and ANN2 validation is shown in Figure 4 below. The results show the best validation performance or MSE validation reaches the minium value in the 11th epoch. In the next epoch, it shows that MSE validation has increased until training ends at the 21st epoch

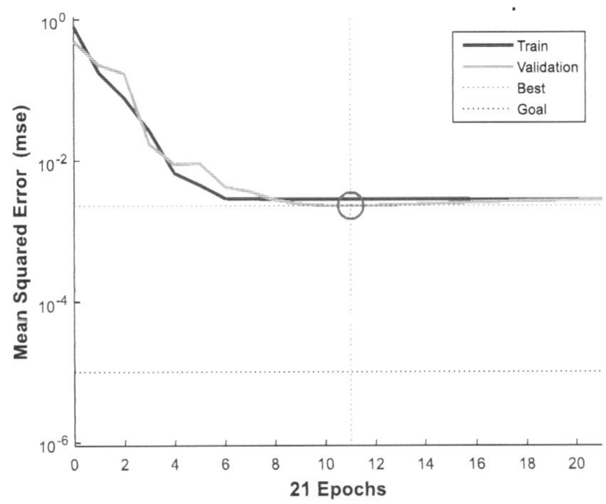

Figure 4. Training performance and validation curve of ANN2.

Based on the training function with the Levenberg-Marquardt (trainlm) method, training will cease during one of the following six conditions: Epoch, *Performance Goal, Gradient, Mu, Validation Checks, and Time are reached*. In MSE validation reaches a minimum, training is still on going to find out whether the value of MSE can still be reduced or not. If, in this case, in the next ten epoch fails to decrease the MSE and the five cessation conditions (*Epoch, Performance Goal, Gradient, Mu, Time*) have not been reached, the training will cease with a termination condition of validation checks or max_fail achieved, as shown in figure (5). The stored network weight refers to the weights when the MSE validation is of minimum value or when the training reaches the 11th epoch, then the weighting of the ANN used refers to the value of the validated MSE, Figure 5 below:

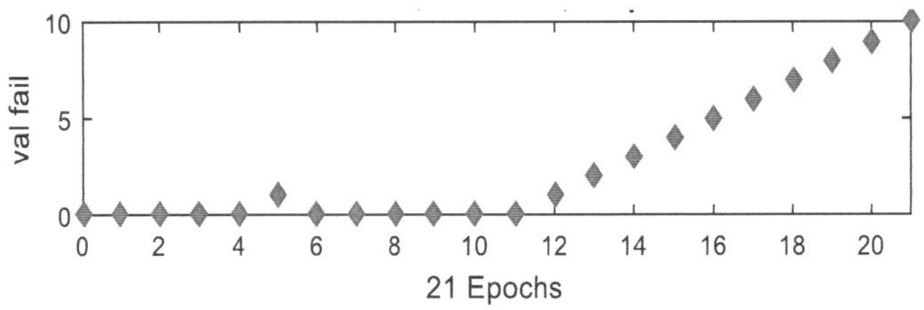

Figure 5. Validation check curve of ANN2.

The next step taken to evaluate the artificial neural network linearity is with the regression curve. The regression curve shows the relationship between the output of network (output), regression of training result and validation of ANN2 obtained value of training correlation coefficient and validation reach 0.99. This value of 0.99 shows the oprimal correlation between the target value and the output of the ANN2 training results, shown in figure (6) below:

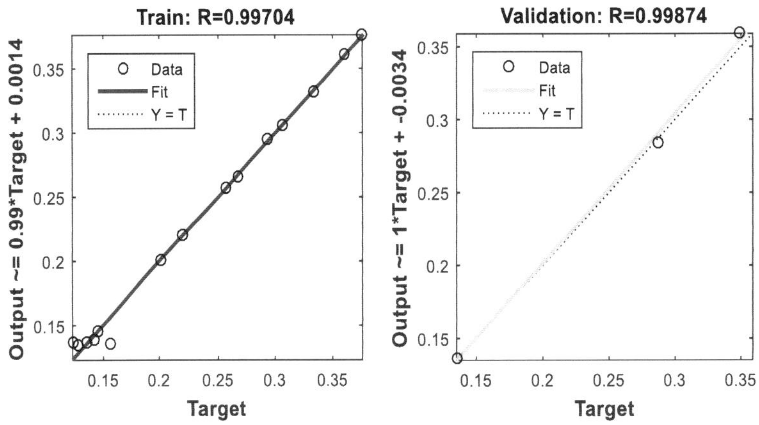

Figure 6. Training regression curve and validation of ANN2.

3.3 Artificial Neural Network Testing

The ANN test with composition input data from the EDX test results, and light brick density yields the output value (prediction) as shown in Table 7 below. Based on the training MSE curve and the validation of the ANN as shown in Figure (4), it appears that the validation and training curves are almost coincidental. This indicates that ANN has a good generalization. This generalization is related to the prediction ability of ANN if given new input data that is not part of the training data.

Table 7. The value of thermal conductivity between the target and the predicted ANN.

Sample Lightweight brick	Thermal conductivity (W/m.K) **Target**	Thermal conductivity (W/m.K) **Prediction**
AAC1	0.214	0.243
AAC2	0.212	0.29
AAC3	0.214	0.32
AAC4	0.214	0.32

3.4. Effect of Silicon Variation (Si) and Ca Calcium

Si, and Ca, is the main composition that affects the formation of dominant phase found in light brick of Calcium Silicate Hydrate (C-S-H). In addition, the density also affects the lightweight thermal brick conductivities. Simulation of ANN using ANN2 is done by varying the composition to its thermal conductivity.

3.4.1 Effect of Silicon (Si) variation on thermal conductivity

Figure 7 below shows the relationship between thermal conductivity and the variation of Si composition. The result shows that the larger the Si composition on the lighter brick the higher the thermal conductivity, this is due to the quartz-SiO_2 content in the lightweight brick one of the constituent elements is Si. The matrix phase in the AAC light brick dominated by the Calcium Silicate Hydrate phase (C-S-H) is the tobermorite phase. When the autoclave process takes place most SiO_2 does not react to produce quartz residue (SiO_2).

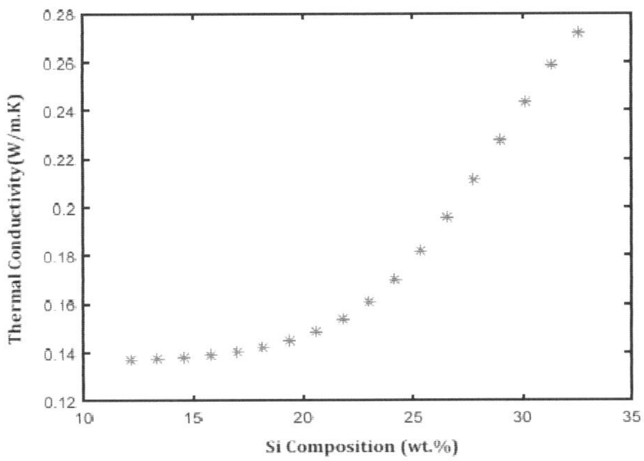

Figure 7. Thermal conductivity curves with variations of Si composition.

Table 8. Thermal conductivity of minerals on light brick [13].

Mineral	Thermal conductivity (W/mK)
Quartz (kristal tunggal)	7.2 – 13.6
Tobermorite	0.18 – 0.2

Table 8 above shows the thermal conductivity value of quartz and tobermorite. It appears that quartz has a thermal conductivity that is worth 35 times higher than the lightest thermal brick conductivity in general. Therefore, to produce lightweight bricks with low thermal conductivity values, one of the things to note is to try to keep the quartz residue present in the matrix phase much lower. This can be done with regard to the ratio of cement and sand composition to sufficient mineral minerals to be reacted with silica. Another thing that can be done to reduce the quartz residue contained in the lightweight brick matrix is to use a more reactive silica source so it can more easily react when the autoclave process takes place. To determine the maximum limit of Si content in light brick, calculated the value of V_f using equation (3). V_f denotes the fraction of air volume in the lightweight brick. Based on this simulation result, the limit of Si content in light brick is 26.57%, where there is composition of Si worth more than 26.57%, V_f will be negative value.

3.4.2.Effect of Ca variation on thermal conductivity

Ca composition is related to the content of CaO which is one of the main minerals of light brick minerals that is calcium silicate hydrate. Based on the result of simulation of Ca variation shown in figure (8) and calculation of V_f value using equation (3), it is known that the limitation of Ca composition in light brick is 22.32% - 30.35%. Beyond this limitation, V_f is negatively negative which is not allowed.

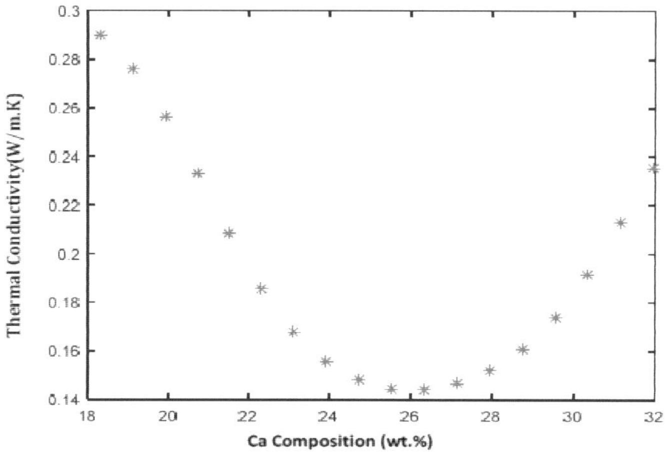

Figure 8. Thermal conductivity curves with variations of Ca composition.

4. Conclusion

From the results of this study can be drawn conclusion as follows: (1) Based on ANN simulation having performance with MSE validation of 0.002252; (2) A light brick ANN simulation found that the maximum allowable composition for Silicon (Si) is 26.57%, while the composition of Calcium (Ca) is 20.32% - 30.35%; and (3) ANN tests performed for the prediction of thermal conductivity in four light brick samples were obtained: AAC1 = 0.243 W/m.K, AAC2 = 0.29 W/m.K, AAC3 = 0.32 W/m.K, AAC4 = 0.32 W/m.K, or thermal conductivity average is 0.2933 W/m.K.

References

[1] Struharova A 2016 *App. Mech. Mater.* 824 100
[2] Narayanan N and Ramamurthy K 2000 *Cem. Concr. Compos.* 22 321
[3] K Wakili, E Hugi, L Karvonen, P Schnewlin and F Winnefeld 2015 *Cem. Concr. Compos.* 62 52
[4] Siang J 2009 Artificial Neural Network and Programming using Matlab Edition 2 (Yogyakarta: Andi Yogyakarta)
[5] Bal L and Bodin F 2013 *Constr. Build. Mater.* 38 248
[6] Lee J, Lee J and C B S 2012 *Int. J. Concr. Struct. Mater.* 6 177
[7] Singh R, Bhoopal R and Kumar S 2011 *Build. Environ.* 46 2063
[8] Baspinar M S, Demir I, Kahraman E and Gorhan G 2014 *KSCE J. Civ. Eng.* 18 47
[9] Kuramaa H, Topcub I and Karakurt C 2009 *J. Mat. Process. Technol.* 209 767
[10] Kunchariyakun K, Asavapisit S and Sombatsompop K 2015 *Cem. Concr. Compos.* 55 11
[11] Ghiormez L, Panoiu M, Panoiu C and Tirian O 2018 *IOP Conf. Ser. Mater. Sci.* 294 012026
[12] M Grutzeck 2005 Cellular Ceramics: Structure, Manufacturing, Properties, and Application (Weinheim: Wiley) 193-221
[13] B Straube and H Walther 2011 *AAC with Low Thermal Conductivity* (Online) Available on https://gazobeton.org/sites/default/files/sites/all/uploads/acc_with_low_termal_conductivity.pdf Accessed February 17, 2017

The effectiveness of Concept Mapping Content Representation Lesson Study (ComCoReLS) model to improve skills of Creating Physics Lesson Plan (CPLP) for pre-service physics teacher

E Purwaningsih[1], Suyatno[2], Wasis[3] and B K Prahani[3,a]

[1]Department of Physics, Faculty of Mathematics and Natural Science, State University of Malang, Indonesia
[2]Department of Chemistry, Faculty of Mathematics and Natural Science, Universitas Negeri Surabaya, Indonesia
[3]Department of Physics, Faculty of Mathematics and Natural Science, Universitas Negeri Surabaya, Indonesia

E-mail: [a]binarprahani@gmail.com

Abstract. This research is aimed to analyse the effectiveness of ComCoReLS (Concept Mapping Content Representation Lesson Study) model towards the improvement skills of Creating Physics Lesson Plan (CPLP) for pre-service physics teacher. This research used one group pre-test and post-test design on 12 pre-service physics teacher at University of Malang State (Indonesia) in academic year 2016/2017. Data collection was conducted through test and interview. Skills of creating physics lesson plan for pre-service physics teacher measurement were conducted through Physics Lesson Plan Evaluation Sheet (PLPES). The data analysis technique was done by using paired t-test and n-gain. The CoMCoReLS model consists of 5 phases, including (1) Preparation, (2) Coaching, (3) Guided Practice, (4) Independent Practice, and (5) Evaluation. In the first, second, third and fifth phases are done at University of Malang State, while the fourth phase (Independent Practice) is done in SMAN 1 Singosari, SMAN 2 Malang, SMA Lab UM, MAN 3 Malang. The results showed that there was a significant increase in skills of creating physics lesson plan for pre-service physics teacher at $\alpha = 5\%$ and n-gain average of high category. Thus, the ComCoReLS model is effective for improving skills of creating physics lesson plan for pre-service physics teacher.

1. Introduction

Teachers are the most important factor for learners especially during the learning process [1]. Teachers are the determinants of what is taught in the classroom and how to teach it [2]. The quality of education in schools is determined and accounted for by teachers [3]. The quality of teachers will have a direct impact on student's competence [4,5]. Lesson planning is very important as the main guide or determinant of activities that teachers will do in the classroom during the learning process [6,7]. In accordance with the teacher's understanding of the PCK, the teacher can determine the direction of learning done and can determine the best and fun actions that can be done to help learners achieve learning objectives [8]. In this study, the indicators of creating physics lesson plan skills include: objectives, learning strategies, physical materials, assessment, instructional media [9-13].

The importance of creating physics lesson plan skills in Indonesia is not realized. The results of previous research studies are as follows. (1) The result of evaluation on education quality trend in Indonesia, it can be concluded that the target of lesson plan has not been reached [14]. (2) Basic concepts (essential material) that learners should learn as vehicles to achieve basic competence demands, taught by teachers are not essential. Consequently the basic competencies that should be achieved by learners cannot be achieved optimally [15]. (3) Teachers tend to teach in the same way as has been taught to them [9]. (4) The quality of education in Indonesia is low [3]. Findings from previous studies of research literature studies are strengthened by preliminary studies by researchers in Physics Education State University of Malang as follows. The difficulties experienced by pre-service physics teacher include: (a) determining the essential material to be taught, (b) creating a learning experience that is varied and meaningful for learners; (c) creating challenging activities to better enable learners; (d) integrating technology in learning, (e) compile an authentic HOTs assessment, (f) develop worksheet that can trace the skills of the science process [15,16]. The results of the study literature and preliminary studies above can be used as the basis that still need innovative physics learning to improve the creating physics lesson plan skills of pre-service physics teacher in Indonesia. The results of this research prove that the awareness (responsibility) of teachers creating physics lesson plan skills is low.

In the previous research, ComCoReLS (Concept Mapping Content Representation Lesson Study) model has been developed to improve Pedagogic Content Knowledge (PCK) for pre-service physics teacher. The CoMCoReLS model has been specially designed to increase the skills of creating physics lesson plan for pre-service physics teacher. The CoMCoReLS model consists of 5 phases, including (1) Preparation, (2) Coaching, (3) Guided Practice, (4) Independent Practice, and (5) Evaluation. The previous research developed a device of learning physics as an operational form of ComCoReLS model developed. Actually, the implementation of the ComCoReLS model that has been developed quality of pre-service physics teacher.

A model is said to be of high quality when it fulfill the criteria of validity, practicality, and effectiveness [17]. Indicators of creating physics lesson plan skills include: objectives, learning strategies, physical materials, assessment, instructional media [9-13]. The purpose of this research is to analyse the effectiveness of CoMCoReLS model to improve the skills of creating physics lesson plan skills for pre-service physics teacher. The focus of the problem in this study included: (1) whether there was a significant increase (statistically) of skills of creating physics lesson plan skills for pre-service physics teacher before and after the CoMCoReLS model was applied, (2) how much level of skills of creating physics lesson plan skills for pre-service physics teacher increased before and after applied CoMCoReLS model.

2. Methodology of Research

2.1. *General Background of Research*

This research was conducted at Malang State University (Indonesia), SMAN 1 Singosari, SMAN 2 Malang, SMA Lab UM, MAN 3 Malang. The scope of this research is on pre-service physics teacher in academic year 2016/2017. This research is emphasized on the analysis of the fulfillment of the effectiveness of CoMCoReLS model by analyzing the improvement of creating physics lesson plan skills for pre-service physics teacher before and after following the CoMCoReLS model. The effectiveness of the CoMCoReLS model was determined based on a significant increase in scores (statistically) between pre-test and post-test of creating physics lesson plan skills for pre-service physics teacher, as well as the mean of n-gain determined by criteria: low, medium and high.

2.2 Sample of Research
The samples in this study were 12 pre-service physics teacher of Malang State University, Indonesia. Pre-service physics teacher in academic year 2016/2017 take the Study and Field Practice course in physics education.

2.3 Instrument and Procedures
Creating physics lesson plan for pre-service physics teacher measurement were conducted through Physics Lesson Plan Evaluation Sheet (PLPES) [10]. PLPES includes objectives, learning strategies, physical materials, assessment, instructional media [9-13]. In this study, the indicators of Creating Physics Lesson Plan (CPLP) skills include: objectives, learning strategies, physical materials, assessment, instructional media [9-13]. Study and Field Practice course used in physics learning. This research uses one group pretest-posttest design, which is O1 X O2 [18]. The learning process begins by giving pre-test (O1). Each pre-service physics teacher is required to complete the PLPES. After the pre-test, the lecturer applies the CoMCoReLS model and learning tool in each group (X). Physics learning is consisting of: CPLP skills include: objectives, learning strategies, physical materials, assessment, and instructional media at each learning phase. The process of physics learning ends with post-test (O2). Every pre-service physics teacher is required to post-test of CPLP skills by PLPES.

2.4 Data Analysis
Creating physics lesson plan skills of pre-service physics teacher is analyzed based on the assessments obtained by pre-service physics teacher before and after learning using the CoMCoReLS model. The pre-test, post-test, and n-gain data of CPLP skills of pre-service physics teacher were further analyzed using inferential statistical tests with the help of SPSS and supported by qualitative descriptive analysis. The score level for CPLP is based on indicators of objectives, learning strategies, physical materials, assessment, and instructional media [9-13]. The n-gain value is determined by the equation: n-gain = (score post-test - score pre-test) / (maximum score - pre-test score) [19]. According to the following criteria: (1) if n-gain $\geq .7$ (high), (2) if $.3 <$n-gain $< .7$ (moderate), and (3) if n-gain $\leq .3$ (low).

3. Result of Research
The learning outcomes of all groups related to the creating physics lesson plan skills (CPLP) of pre-service physics teacher are presented in Figures 1 and Table 1. Shape bar represent the mean of pre-test, black bars scores represent the mean post-test scores, and Vertical bar scores represent the n-gain scores. Figure 1 shows the average post-test scores of CPLP skills of pre-service physics teacher for all groups is greater than the pre-test score. The average pre-test, post-test, and n-gain scores associated with CPLP skills of pre-service physics teacher indicators for all groups are presented in detail in Table 1. Figure 1 show the average n-gain value of CPLP skills for SMAN 1 Singosari; SMAN 2 Malang; SMA Lab UM; MAN 3 Malang is respectively .72; .76; .73; and .72. The average n-gain value of CPLP skills of pre-service physics teacher for all groups is in the high category.

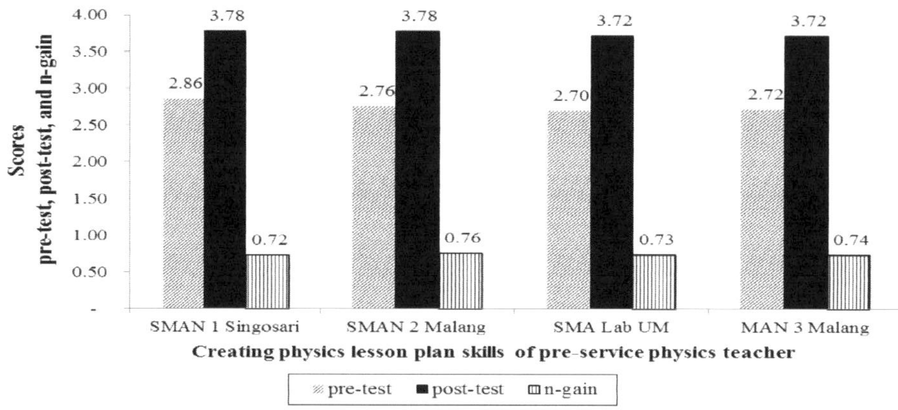

Figure 1. The average pre-test, post-test, and n-gain scores of creating physics lesson plan skills of pre-service physics teacher in all groups.

Table 1. The average score of pre-test, post-test and n-gain of creating physics lesson plan skills of pre-service physics teacher in all groups.

Groups	**Scores**	**Indicator of Creating Physics Lesson Plan (CPLP) Skills**				
		Objectives	**Learning Strategies**	**Physical Materials**	**Assessment**	**Instructional Media**
SMAN 1 Singosari	Pre-test	2.40	2.60	3.50	2.30	3.50
	Post-test	3.70	3.80	3.80	3.90	3.70
	n-gain	**.81**	**.86**	**.60**	**.94**	**.40**
SMAN 2 Malang	Pre-test	2.30	2.50	3.40	2.20	3.40
	Post-test	3.70	3.80	3.80	3.90	3.70
	n-gain	**.82**	**.87**	**.67**	**.94**	**.50**
SMA Lab UM	Pre-test	2.40	2.40	3.30	2.10	3.30
	Post-test	3.70	3.70	3.70	3.80	3.70
	n-gain	**.81**	**.81**	**.57**	**.89**	**.57**
MAN 3 Malang	Pre-test	2.30	2.60	3.20	2.30	3.20
	Post-test	3.60	3.80	3.60	3.90	3.70
	n-gain	**.76**	**.86**	**.50**	**.94**	**.63**

Table 1 shows that the CPLP skills test scores of each indicator include: objectives, learning strategies, physical materials, assessment, and instructional media are moderate for all groups and the post-test score CPLP skills in each indicator is high for all groups. The n-gain of CPLP skills score of each indicator includes: objectives, learning strategies, physical materials, assessment, and instructional media in all groups are moderate/high.

Table 2. Paired test result of creating physics lesson plan skills for all groups.

Implementation	**Inferential Statistics Test**	**Data**	**Asymp. Sig. (p)**
SMAN 1 Singosari; SMAN 2 Malang; SMA Lab UM; MAN 3 Malang	Paired t-test	*Pretest Posttest*	.00

$p < .05$ (2-tailed)

Table 2 shows the improving in pre-test and post-test of CPLP skills tested using Paired t-test. The p score gives a value of .01 for SMAN 1 Singosari; SMAN 2 Malang; SMA Lab UM; MAN 3 Malang. Each score is considered significant, because $p < .05$. Because Z the result of the calculation is negative so it shows that there is an increase of creating physics lesson plan skills of pre-service physics teacher after applied learning with CoMCoReLS model for all groups.

4. Discussions

The effectiveness of the CoMCoReLS model on increasing Creating Physics Lesson Plan (CPLP) skills of pre-service physics teacher included (1) an increase in pre-test and post-test scores and (2) n-gain CPLP skills value, as shown in Figure 1, Table 1, and Table 2. Before the CoMCoReLS model is applied; pre-service physics teacher moderate mastery the CPLP skills, average score of pre-service physics teacher is under the standard score (minimum score 2.70 in score range 1-4), that is the average score of CPLP skills for SMAN 1 Singosari; SMAN 2 Malang; SMA Lab UM; MAN 3 Malang respectively are 2.86; 2.76; 2.70; 2.72. All this time, pre-service physics teacher are not used to objectives, learning strategies, physical materials, assessment, and instructional media. The result of evaluation on education quality trend in Indonesia, it can be concluded that the target of lesson plan has not been reached [14]. (3) Teachers tend to teach in the same way as has been taught to them (Etkina, 2010). This is supported by [15,16] found that most pre-service physics teacher of physics education program, Malang State University include: (1) determining the essential material to be taught, (2) creating a learning experience that is varied and meaningful for learners; (3) creating challenging activities to better enable learners; (4) integrating technology in learning, (5) compile an authentic HOTs assessment, (6) develop worksheet that can trace the skills of the science process.

After the CoMCoReLS model was applied to the pre-service physics teacher, the mastery of CPLP skills rises above average and becomes high; the average score of creating physics lesson plan skills for SMAN 1 Singosari; SMAN 2 Malang; SMA Lab UM; MAN 3 Malang respectively to 3.78; 3.78; 3.72; and 3.72 (well beyond the minimum score of 3.72 in the 1-4 score range). The increase in CPLP skills is allegedly influenced by design of scenario in phase of the CoMCoReLS model. Each phase emphasizes CPLP skills of the CoMCoReLS model is a combination of Concept Mapping, CoRe and Pap-eR as PCK representations that will be trained on potential physics teachers through Lesson Study. CoRe and PaP-eR can be used as valid methodologies for assessing science teacher's PCK [23]. The CoMCoReLS model will be implemented in the Study and Field Practice courses. The CoMCoReLS model consists of 5 phases, including (1) Preparation, (2) Coaching, (3) Guided Practice, (4) Independent Practice, and (5) Evaluation. Preparing pre-service physics teacher focused on improving the ability to plan and implement learning and self-regulated learning. The ability of pre-service physics teacher in planning and implementing learning is low, experiencing some difficulties especially in determining essential material, creating challenging activities, integrating technology [20,21].

The increase of CPLP skills of pre-service physics teacher score in all groups was significant at the 5% significance level with n-gain of .75 for group-1; .73 for group-2; .73 for group-3; and .72 for group-4. This means that there is an increase in CPLP of pre-service physics teacher after the implemented CoMCoReLS model. Increased CPLP skills is supported by the availability of CPLP skills of pre-service physics teacher learning needs as suggested in phase of the CoMCoReLS model. The principle of reaction of the CoMCoReLS model relates to attitudes, and the response of lecturers in observing and treating pre-service physics teacher including to questions or attitudes proposed by pre-service physics teacher. The ComCoReLS model, the way teachers' pay attention and treat students should: (1) Lecturers motivate and remind students of pre-service physics teacher to always take responsibility. (2) Lecturers give praise, opportunities for pre-service physics teacher to ask questions, to argue, to criticize the process of physics learning in order to skill planning and skills to implement PCK learning increases. (3) Motivate, guide by guiding it through questions, until the pre-service physics teacher believes that the verdict is correct. (4) Accommodate the suggestion and opinion of the pre-service physics teacher, immediately provide feedback. It is supported by social constructivist theory: pre-service physics teacher will build their knowledge through social interaction with other pre-service physics teacher and with lecturers [22,6]. Cognitive apprenticeship theory, an expert (lecturer) gives an example of problem solving, can be a tutoring [22]. Modelling theory by Bandura: prospective physics teachers can learn through observation and explanation from others [23]. An effective teacher starts his lesson by reviewing, explaining the purpose of learning in a language that is easy to understand, and showing how learning is now related to learning in the past [6].

Feedback given by the teacher can help the development of learning process and the increase of learners' learning performance [23,24]. The results showed that CoRe and Pap-eR can be used as a valid methodology for assessing teacher science PCK [25]. The introduction of CoRe in lectures strengthened awareness about PCK components [26]. Based on the above description, the implementation of CoMCoReLS model is able to develop the role of physics lesson plan skills of pre-service physics teacher.

5. Conclusions

The CoMCoReLS model consists of 5 phases, including (1) Preparation, (2) Coaching, (3) Guided Practice, (4) Independent Practice, and (5) Evaluation. In the first, second, third and fifth phases are done at University of Malang State, while the fourth phase (Independent Practice) is done in SMAN 1 Singosari, SMAN 2 Malang, SMA Lab UM, MAN 3 Malang. The results showed that there was a significant increase in skills of creating physics lesson plan for pre-service physics teacher at $\alpha = 5\%$ and n-gain average of high category. Thus, the ComCoReLS model is effective for improving skills of creating physics lesson plan for pre-service physics teacher. The implication of this research is that CoMCoReLS model can be used as an alternative to overcome the low of creating physics lesson plan skills for pre-service physics teacher. To improve the result of this research, it is necessary to do further generalizations in various education levels and countries.

6. Acknowledgements

The author's gratitude goes to the Ministry of Research, Technology and Higher Education who has funded the Research. Likewise, the author's gratitude goes to the Malang State University, SMAN 1 Singosari, SMAN 2 Malang, SMA Lab UM, and MAN 3 Malang that has provided research opportunities.

References

[1] Blanton L P, Sindelar P T and Correa V I 2006 *J. Spec. Educ.* **40** 115
[2] Abell S K 2007 *Handbook of Research on Science Education* (NJ: Lawrence Erlbaum)
[3] Hendayana S, Asep S and Imansyah H 2010 *J. Int. Educ. Coop.* **2** 41
[4] Hammond D L Holtzman D J Gatlin S J and Heilig J V 2005 *Educ. Policy Anal. Arch.* **13** 1
[5] Rivkin G S, Hanushek E A and Kain J F 2005 *Econometrica* **73** 417
[6] Arends R I 2012 *Learning to Teach* (New York: McGraw-Hill Companies, Inc)
[7] Reigeluth C M and Stein F S 1983 *Instructional Design Theories and Models. Hillsdale* (Mahwah, NJ: Erlbaum Associates)
[8] Rozenszajn R and Yarden A 2014 *Res. Sci. Educ* **44** 189
[9] Etkina E 2010 *Phys. Rev. ST Phys. Educ. Res.* **6** 020110
[10] Loughran J J, Mulhal P and Berry A 2011 *Int. J. of Ed.* **30** 1301
[11] Magnusson S, Krajcik J and Borko H 1999 *Examining Pedagogical Content Knowledge* (Dordrecht: Kluwer Academic Publishers)
[12] Ertmer P A and Timothy J N 2013 *Performance Improvement Quarterly* **26** 65
[13] Williams J, Eames C, Hume A and Lockley J 2012 *Res. Sci. Technol. Educ.* **30** 327
[14] Raharjo S and Budi 2012 *J. Penelit. Eval. Pend.* **16** 298
[15] Purwaningsih E 2015 *Indonesian J. Applied Phys.* **5** 9
[16] Purwaningsih E, Nur M and Wasis 2015 The TPACK Model of Prospective Physics Teacher in Universitas Negeri Malang *Proc. Int. Conf. Math. Sci. and Educ.* (Mataram Unram)
[17] Plomp T and Nieveen N 2013 *Educational Design Research* (Enschede: SLO)
[18] Fraenkel J R, Wallen N E and Hyun H H 2012 *How to Design and Evaluate Research in Education* (New York: McGraw-Hill)
[19] Hake R R 1998 *Am. J. Phys.* **66** 64
[20] Purwaningsih E 2015 Penguatan Pengetahuan Materi Fisika dengan *Content Representation (CoRes)* pada Calon guru Fisika *Semin. Nas. Pend. Sains Pascasarjana Universitas Negeri*

Surabaya (Surabaya: Universitas Negeri Surabaya)
[21] Purwaningsih E and Yuliati L 2015 Prospective Physics Teacher Ability on Designing Lesson Plan At Senior High School in Terms the TPACK Framework *Proc. Int. Conf. Math. Sci. and Educ.* (Mataram: Unram)
[22] Santrock J W 2011 *Educational Psychology* (New York: McGraw-Hill)
[23] Wilbert J, Grosche M Gerdes and Heiki 2010 *Learn. Disabilities: A Contemporary J.* **8** 43
[24] Lina A and Sunarti T 2017 *J. Penelit. Fis.Apl.* **7** 68
[25] Bertram A and Loughran 2012 *J. Res. Sci. Educ.* **42** 1027
[26] Hume A C and Berry A 2011 *Res. Sci. Educ.* **41** 341

The use of Papuan iron sand and river sand for fine aggregate in mortar for nuclear radiation shield application

K Dahlan[1], E Haryati[1] and Y. S. Aninam[1]

[1]Department of Physics, Cenderawasih University, Indonesia

E-mail: khaeriahd@gmail.com

Abstract. This study explores the effect of fine aggregate on mortar properties and its application as a nuclear shield. This study was based on a hypothesis that the types of aggregate applied as radiation shield determined the level of its effectiveness on preventing nuclear radiation. There are two types and sources of fine aggregate that was used as main ingredients for mortar production in this research, namely iron sand and river sand. Both types of sand were derived from the respective regions of Sarmi and Jayapura, Papua. The results showed that the mortar materials that were produced with the iron sand provided better results in dispelling radiation than that of river sand. The compressive strength of fine aggregate from the iron sand was 21.62 MPa, while the compressive strength of the river sand was 16.8 MPa. Measuring the attenuation coefficient of material, we found that the largest aggregated value of mortar with fine iron sand reached 0.0863 / cm. On the other hand, the smallest HVT (Half Value Thickness) was obtained from the iron sand mortar, at 8.03 cm.

1. Introduction

Nuclear technology has been widely applied in many areas of daily life. However, despite having many advantages, this technology also possesses harmful effects on health and the environment. In order to reduce the risk of radiation hazards, there is a need for a protective shield that can absorb nuclear radiation and minimize its intensity. We expect the shield to reduce the radiation into a quantity that can be accepted by the human body.

This study proposes an application of iron sand as basic material for radiation shields. The experiment in this study consisted of placing a radiation shield between the ionized radiation source and its surrounding environment. The effectiveness of the shield depends on the type of radiation, radiation energy, and materials the shield is composed of [1]. The radiation shield is based on the principle of attenuation: the ability to block or to reduce the intensity of radioactive through photoemission and scattering performed by barrier materials. This study suggested the application of the iron sand into concrete or mortar, as a basic material for the radiation barrier [2]. When the radiation enters the shield, a portion of the radioactive will be absorbed by the shield. As a result, the intensity of radiation after entering the absorbent material is reduced significantly. In this case, the intensity of radiation after being absorbed by the protective material will decrease without diminishing the energy from radiation passing through the material. The resulting absorption value of the interaction is expressed quantitatively as a linear coefficient, or attenuation coefficient (μ). This coefficient represents the probability of one interaction per unit of distance. The process of attenuation of both gamma-ray radiation in a shield material is exponential according to the following equation:

$$I = I_0 \, e^{-\mu x} \tag{1}$$

where I in this calculation means the intensity of radioactive after going through the radiation shield, Io as the intensity of radiation before it passes through the shield, and x as the thickness of the shield material [3]. Therefore, the attenuation coefficient can be calculated using the following equation:

$$\mu = \frac{ln\left(\frac{I_0}{I}\right)}{x} \tag{2}$$

Previous studies have investigated the manufacturing process of mortar or concrete for use in nuclear radiation shields, especially in the medical sector, by replacing aggregates with materials that can block radioactive. Natural resources and materials such as arc furnace slag aggregate [4], steel balls [5], lead oxide [6], hematite [7], and other concrete mixtures [8] have been widely developed for medical purposes, including applications for nuclear radiation shields.

Principally, to shield radiation effectively, the materials should have high atomic number and density. According to Raju [9], high density can be found by using high density aggregates. In this research, mortars were created with a mixture of iron sand and river sand as fine aggregates with high density. The compositions of the iron sand and river sand were 4.311 [10] and 2.55, respectively. Whereas the density for normal aggregates were about 2.5-2.7.

Thus, we can conclude that both the iron sand and river sand fall under the category of high-density aggregates, suitable to produce high density mortar. It is also safe to conclude that the mortar can provide high levels of radiation shield capability by incorporating different types of aggregate. As in this study, two different types of aggregate were assessed from these two samples. The materials for the samples were obtained locally in Jayapura, Papua. The primary locations to retrieve the sands were Sarmi and Doyo. Sarmi is a regency in Papua Province, located in the outskirts of Jayapura. It is one of the prime locations where massive stocks of iron sand are found along rivers and beaches. Doyo, on the other hand, is a district in Jayapura regency with a substantial amount of sand on its rivers. The local people use sand for many purposes, such as building constructions.

The purpose of this research is to determine the value of mortar compressive strength and the coefficient value of material, to engage nuclear radiation. We developed the mortar by mixing pure iron sand and river sand. We then determined the Half Value Thickness (HVT) of the two types of mortar constructed from iron sand and river sand.

2. Experimental Method

2.1 Fine Aggregate

Two different types of fine aggregate with particle sizes smaller than 0.15 mm were treated as follows: 1) the iron sand obtained from Sarmi regency was treated using a permanent magnet to obtain pure iron sand, while 2) the river sand from Jayapura regency was not treated.

2.2 Development of Mortar

Preparing waterproof molds with varying thickness precedes the preparation of mortar. The result was in the form of paste that consisted of cement, fine aggregate, water, and viscocrete-10, with a general ratio of cement : fine aggregate : viscocrete-10 = 1 : 1.5 : 0.015. After this process, the paste was inserted into the mold to mark volume height for each mold. Variations created in the manufacturing process of mortar are variations in the ratio of river sand and iron sand as fine aggregate, as seen in the following table:

Table 1. Sample Code.

Sample Code	Information
Mortar I	Comparison of fine aggregates between iron sand & river sand (1 : 0)
Mortar II	Comparison of fine aggregates between iron sand & river sand (0 : 1)
Mortar III	Comparison of fine aggregates between iron sand & river sand (0.5 : 0.5)

After the mortars were finished, they were then printed and dried, and afterwards tested to determine their compressive strengths by using a press machine. After that, the samples were tested for their capability for use in radiation shields.

2.3 Testing Mortar through press and strength
The strength of the mortar load was the magnitude of the load per unit area, which causes the mortar assay to be destroyed when burdened with a certain compressive force produced by the compressor [11].

2.4 Testing of Mortar Capability as a Radiation Shield
This test was performed with a set of Geiger counting/Geiger Muller series using ^{137}Cs as the source of radiation.

3. Result and Discussion

3.1. Compressive Strength of Mortar
The mortar compressive strength values for each type of mortars after 14 days can be seen in the following table:

Table 2. Mortar pressing values.

No.	Sample Code	Pressing Strength Value (MPa)
1.	Mortar I	21.62
2.	Mortar II	16.68
3.	Mortar III	17.97

According to Table 2, it can be seen that mortar with iron sand fine aggregate had the highest compressive strength value, at 21.62 MPa; while the mortar with river sand fine aggregate had the smallest compressive strength value of 16.68 MPa.

3.2. The Mortars' Attenuation Coefficient to Radiation
The material attenuation coefficient values are displayed in the graphs below. The values represent a relationship between the ratio of initial radiation and the radioactive that passed through the mortar to the thickness of the mortars; as shown in Figure 1.

IOP Conf. Series: Journal of Physics: Conf. Series **997** (2018) 012023 doi:10.1088/1742-6596/997/1/012023

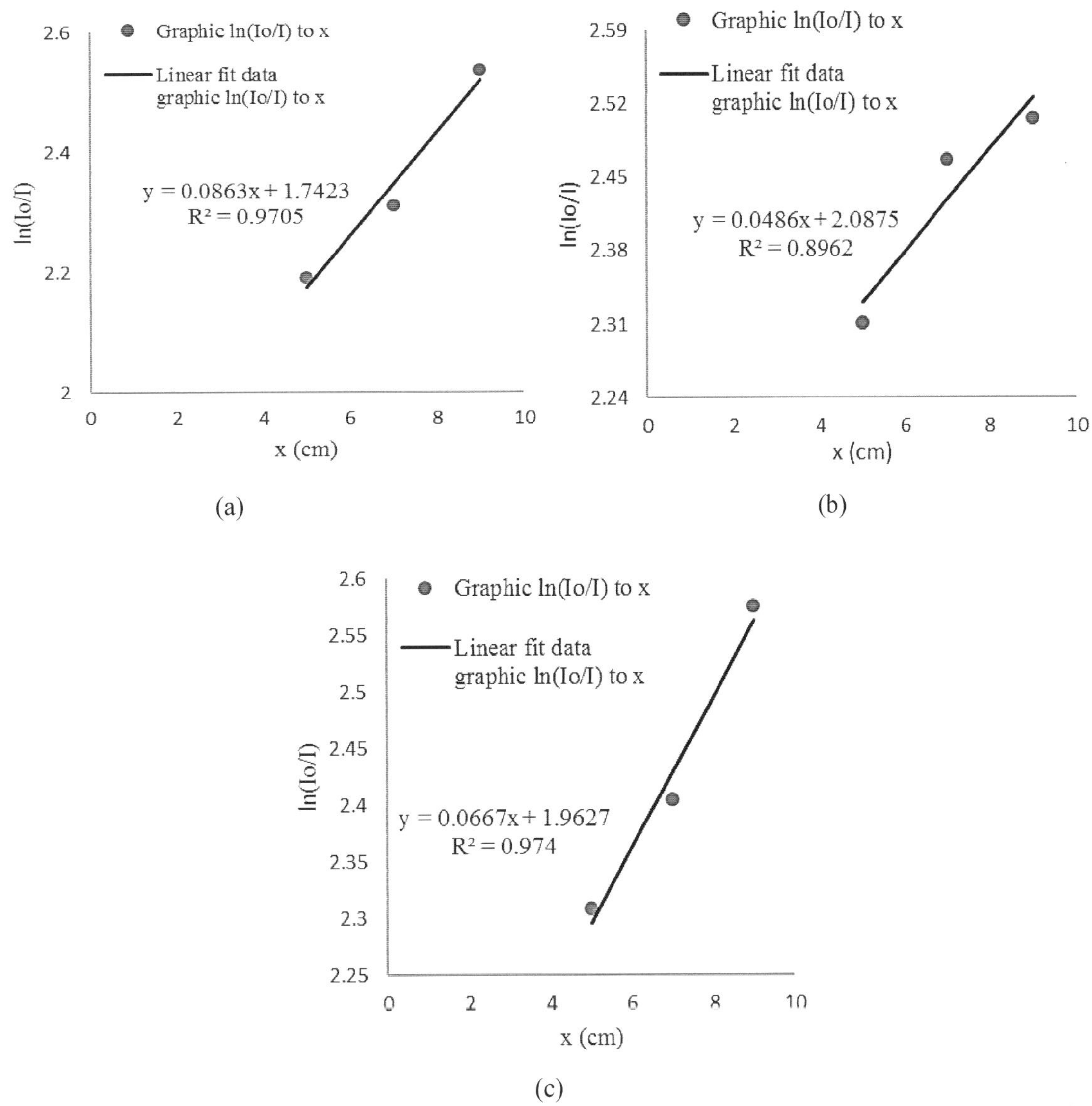

Figure 1. Graph of the relationship between ratio of the initial radiation and the radiation that passed through the mortar to the thickness of the mortar for (a) Mortar I, (b) Mortar II & (c) Mortar III.

Data processing using graph (a) gives us an equation of $y = 0.0863x + 1.7423$, where the straight line equation equals the material's attenuation factor, hence $\mu = 0.0863$/cm. Graph (b) gives us an equation of $y = 0.0486x + 2.0875$, and graph (c) gives us $y = 0.0667x + 1.9627$. Therefore, the attenuation coefficients are as seen in Table 3 below.

Table 3. The mortars' attenuation coefficient values.

No.	Sample Code	Attenuation Coefficient (μ) (per cm)
1.	Mortar I	0.0863
2.	Mortar II	0.0486
3.	Mortar III	0.0667

Table 2 shows that the highest material attenuation coefficient value against nuclear radiation occurs on mortar with purified iron sand fine aggregate, at 0.0863/cm. It is then followed by the mortar with fine aggregate composed of 50% iron sand and 50% river sand, at 0.0667 /cm. The lowest attenuation coefficient against nuclear radiation, at 0.0486/cm, occurred on the mortar with river sand fine aggregate.

3.3. Half Value Thickness (HVT)
Half Value Thickness (HVT) is the density of the mortar that allows it to effectively absorb 50% of the nuclear radiation. Half Value Thickness for each type of mortar depends on the material's attenuation coefficient value against the nuclear radioactive. The HVT values for each mortar can be seen in Table 4.

Table 4. Half Value Thickness (HVT) for each mortar.

No.	Sample Code	Half Value Thickness (HVT) (cm)
1.	Mortar I	8.03
2.	Mortar II	14.26
3.	Mortar III	10.39

Table 4 shows that Mortar I had the lowest HVT value of 8.03 cm and Mortar II had the highest HVT value of 14.26 cm. This proves that the lower a mortar's Half Value Thickness, the better its quality as a nuclear radiation shield.

In addition to being influenced by fine aggregate materials, the mortar's attenuation coefficient value against nuclear radioactive was also influenced by its compressive strength value. The higher compressive strength, the better its capacity in absorbing value to the radiation. The results of this study therefore indicate that by using iron sand as a fine aggregate in mortar, the mortar's ability in absorbing radiation was increased significantly.

4. Conclusion
We have successfully developed mortars using Papuan iron sand and river sand as base materials. We have also clearly deduced from our observation that the type of fine aggregate plays a key role in applying concrete as a material in a nuclear radiation shield. From the measurement of compressive strength, the mortar composed of iron sand fine aggregate had a higher compressive strength value. In testing the mortars' capabilities as radiation shields, we have also measured their material attenuation coefficient and HVT values. Consequently, the values in mortars composed of iron sand fine aggregate are higher compared to those composed of river sand, as well as to those composed of a mixture of iron sand and river sand. The findings of this study also suggests that more research on potential applications of natural resources for radioactive shields should be well considered in the future.

References
[1] Elmahroug Y, Tellili B and Souga C 2013 *Int. J. Phys. Res.* **3** 7
[2] Tung L, Chi P, Wai L, Tai C and Karl F 2013 *Mater. Struct.* **46** 1133
[3] Ahmed O 2015 *Prog. Nucl. Energy* **79** 48
[4] Maslehuddin M, Naqvi A, Ibrahim M and Kalakada Z 2013 *Ann. Nucl. Energy* **53** 192
[5] Ali A, Kahtan M, Mohd M A, Kamarudin H, Andrei S and Rafiza R 2013 *Mat.* **6** 4836
[6] Yasinta S 2013 *Civ. Eng. Forum* **22** 1349
[7] Fugaru V, Bercea S, Postolache C,and Manea S 2015 *Acta. Phys. Pol., A* **127** 1427
[8] Raju N K 1993 *Design of Concrete Mixes 3rd Edition* (New Delhi: CBS Publishers & Distributors)

[9] Gencel O, Brostow W, Ozel C and Filiz M 2010 *Mater. Sci.* **16** 249
[10] Putra H, Satyarno I and Wijatno A 2008 *Forum Teknik Sipil* **18** 909
[11] Badan Standar Nasional 1990 *Standar Nasional Indonesia* 03-1974-1990 (Jakarta: Badan Standar Nasional)

IOP Conf. Series: Journal of Physics: Conf. Series **997** (2018) 012024 doi:10.1088/1742-6596/997/1/012024

Correction factors in determining speed of sound among freshmen in undergraduate physics laboratory

A Lutfiyah[1], A S Adam[1], N Suprapto[1], A Kholiq[1] and N P Putri [1,a]

[1]Department of Physics, Faculty of Mathematics and Natural Science, Universitas Negeri Surabaya, Surabaya 60231, Indonesia

E-mail: [a]nugrahaniprimary@unesa.ac.id

Abstract. This paper deals to identify the correction factor in determining speed of sound that have been done by freshmen in undergraduate physics laboratory. Then, the result will be compared with speed of sound that determining by senior student. Both of them used the similar instrument, namely resonance tube with apparatus. The speed of sound indicated by senior was 333.38 ms^{-1} with deviation to the theory about 3.98%. Meanwhile, for freshmen, the speed of sound experiment was categorised into three parts: accurate value (52.63%), middle value (31.58%) and lower value (15.79%). Based on analysis, some correction factors were suggested: human error in determining first and second harmonic, end correction of tube diameter, and another factors from environment, such as temperature, humidity, density, and pressure.

1. Introduction

Basic physics course on the speed of sound topic can be supported by laboratory-based experiment. Physics concept is not only about the concept but also about how the concept is obtained rationally [1]. Because of that, experiment activity allows students to dig more knowledge about sound waves, because they obtain hands-on experience in the lecture process. It will help them to be more competence as a pre-service teacher. A traditional experimental method can be used to determine the speed of sound in the air, studying sound waves harmonic in closed-tube [2]. Its instrument is called resonance tube with apparatus. It is consist of loudspeaker, amplifier, closed-tube with scales and a piston. Piston in the tube is moved to compress the gas. Before it is moved, the gas is undisturbed and having uniform density. The gas just in front of it is compressed when the piston is suddenly pushed, so the pressure and density in this region are now higher than before [3].

Changes of pressure because of a rapid movement of the piston, gives the sound as a result. It will be traveling pressure that is reflected at the bottom of the tube. The superposition of an incident and reflected waves give standing wave will have a node at one end and antinode at the other end. This is the fundamental frequency or first harmonic of the air column in the tube. This oscillation has the longest wavelength (λ), four times greater than the length of the air column ($\lambda = 4L_1$, i.e. the length of air column) [4].

Speed of sound can be determined using the equation,

$$v = \lambda f \qquad (1)$$

Where f is fundamental frequency and λ is the wavelength. $\lambda = 4L_1$, where L_1 is the length of the air column in first harmonic.

An end correction factor must be added to the length because the antinode forms above the top of the tube. Based on theory, it is found that correction factor e is equal to $0.4d$ for closed-tube [4], where d is the inner diameter of the tube.
Hence,

$$\lambda = 4(L_1 + e) \tag{2}$$

$$v = 4f(L_1 + 0.4d) \tag{3}$$

In closed-tube, the length of an air column for second harmonic is $L_2 = \frac{3}{4}\lambda$ [5,6]. So that, end correction can be calculated by subtituting equation (2) into $L_2 + e = \frac{3}{4}\lambda$ because there is an end correction factor for second harmonic. Therefore,

$$e = \frac{L_2 + 3L_1}{2} \tag{4}$$

Sound wave propagates in the air depend on ambient temperature [1]. In air 0°C and 1 atm, sound travels at speed 331 ms^{-1}. If v_0 refers to the speed of sound in the air at 0°C and T_c to the medium temperature in degrees Celcius, then the speed of sound in any temperature is given the following formula [7,8],

$$v = v_0 + (0.6T_c) \tag{5}$$

2. Method

The research design is by using quantitative descriptive method. The sample is taken from freshmen in Universitas Negeri Surabaya, Indonesia. The research subjects are 19 experiment groups in physics laboratory. Each of the groups is consist of two or three members.

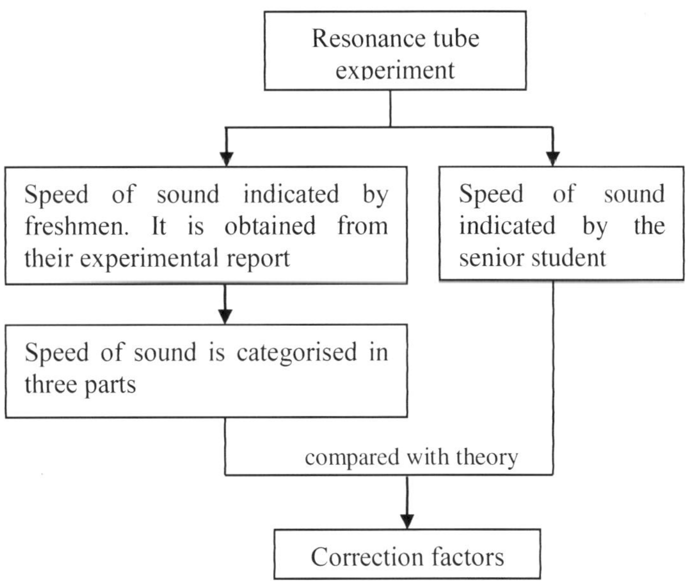

Figure 1. Flowchart of the research methods.

Data is obtained from the result of the senior student's experiment and freshmen's experiment report. Both of them used similar instrument, namely resonance tube with apparatus. It is conducted by using loudspeaker as a sound source and set amplifier for particular frequency. Piston movement in a tube make different tone from the source. At the time of recording it, the scale of loudest tone is a length of air column that resonance occurs. It is observed to get first and second harmonic. Senior experiment activity is selected five frequencies in the audible range, 311-395 Hz. Every frequency is repeated three times for getting accurate data. The data of waves are calculated to determine the speed of sound from that experiment. Afterwards, it is compared to the speed of sound in theory. Difference result both of the speed of sound in experiment and theory are analysed to figure out the correction factors.

One of parts in freshmen's experiment report is discussion part. They write their prediction and discussion about the factors that affect to their experiment result on it. Then it will be analysed to find the precise reason. The correction factors of them are presented in percentage form in a table.

3. Results and Discussion

Explanation of the result will be showed in two parts. Experimental data by the senior student is explained first in the table 1. The speed of sound is calculated with adding end correction factor.

Table 1. Experimental data performed by the senior students.

No.	Frequency (Hz)	L_1 (x 10^{-2} m)	L_2 (x 10^{-2} m)	Speed of sound (ms^{-1})
1.	311.4	29.97	83.17	331.33
2.	338.0	28.90	77.90	331.24
3.	354.6	26.47	73.17	331.20
4.	375.9	22.47	67.33	337.31
5.	394.5	21.57	64.13	335.85
	Speed of sound average			333.38

Determining speed of sound that indicated by the senior student is 333.38 ms^{-1} at room temperature. It has deviation to the theory about 3.98%. Then the data is represented in a graph to get speed of sound using the gradient of graph in figure 1.

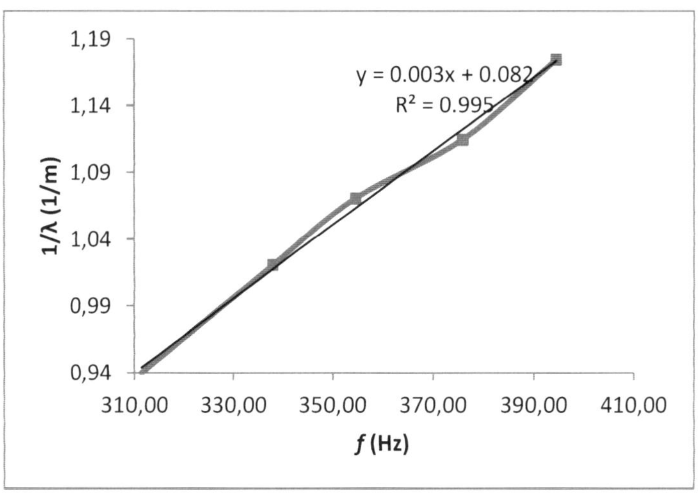

Figure 1. Plotting of $1/\lambda$ as a function of f to determine v, indicated by the senior student.

Based on the graph shown in Fig. 1, the speed of sound in the air can be obtained as follows:

$$v = \frac{1}{m} = \frac{1}{0.003} = 333.33\ ms^{-1}$$

The speed of sound that indicated by freshmen is categorised in three parts: accurate, middle and lower. It base on the difference between the speed of sound in the air that indicated by freshmen and theory. For dry air condition with 1 atm pressure, the speed of sound that calculated by using equation (5) at room temperature 27°C is 347.2 ms^{-1} and at 20°C is 343 ms^{-1}. Some previous research found different values of speed of sound such as: (343±3) ms^{-1} at 25.5 °C [10], 342.96 ms^{-1} at 23°C [6], (344±3) ms^{-1} at 23°C [11], and the average speed of sound value of (336±4) ms^{-1} at various temperature from 19.1°C to 26.8°C [9]. The categories are presented in the table 2.

Table 2. The speed of sound categories in air indicated by freshmen.

No.	Speed of Sound Range (m/s)	Categories	Percentage (%)
1	343 – 359	accurate	52.63
2	298 – 337	middle	31.58
3	117 – 125	lower	15.79

The largest percentage is in the accurate category where the value is approximate to the speed of sound in theory. Meanwhile, there is still a speed of sound in the lower category. The result in other categories is caused by factors as a correction of the experimental results. The correction factors that mentioned freshmen are presented in the table 3.

Table 3. Correction factors based on freshmen's experiment report.

No.	Correction Factors	Percentage (%)
1.	Error in determining first and second harmonic	21.28
2.	Temperature	19.15
3.	Noisy	17.02
4.	Pressure	12.77
5.	Too low frequency	10.64
6.	Density	6.38
7.	End-correction of tube diameter	6.38
8.	Uncertainty of measure tool	4.26
9.	Moving air condition	2.13

A correction factor about 4.5 $ms^{-1}{}^{\circ}C^{-1}$, can be used as a scale between temperature and speed of sound in cold water. The important role to respect the speed of sound in water is temperature and salinity [12]. The speed of sound depends on temperature and does not depend on pressure in the air [13]. The speed of sound is increase as the equation (5). The increase of density is affected by the increasing of pressure. There is no speed of sound change caused by pressure. But it is true if the temperature remain constant. Density change is caused by temperature change. It does not affect pressure. Pressure changes may affect density but do not vice versa [13]. Other studies have suggested that temporal or spatial variations in pressure (e.g., atmospheric influences) have a small effect on the speed of sound so that the pressure attenuation can be ignored [12].

Another factor that affects the density is humidity. Dry air is more dense than moist air (not particularly visible). This causes an increase in the speed of sound. The decrease of specific-heat ratio is caused by moisture, which cause speed of sound to decrease. However, increasing moisture will increase the speed of sound because the decrease in the density is dominant [13]. Sound waves have a strong influence on temperature and relative humidity [14] because it affects to the amount of absorption by the atmosphere. It also deals with the frequency which is the high frequencies will be absorbed more than low frequencies [14,15].

The end correction of tube diameter is one of the reasons for the difference between the speed of sound and theory. It should be calculated based on the equation (4). While the instrument that is used to measure the length of air column has uncertainty about 0.05 cm.

Moving air or wind conditions have an effect on the speed of the sound. If the medium moves, the sound will be transported further [15,16]. The high horizontal wind speed will produce sound refraction because it causes the average speed of sound to vary [17]. However, in determining speed of sound was not done in outdoor so no moving air factors that affect.

There are many predictions of correction factors that have been determined by freshmen. The percentage of error when determining the first and second harmonic is the greatest of all. One of the reasons expressed by the freshmen is when pulling the piston is not slowly. Longitudinal waves located in the compressed area by the piston move at a certain velocity v, but the piston does not move with velocity v because the speed of sound is greater than the speed of the piston [3]. The error of determining the first and second harmonic affects to the length of the air column so give different wavelength as the results. It has a major effect on the speed of sound, especially in the lower category.

According to table 2, the correction factors corresponding to the experiments are error in determining first and second harmonic, uncertainty of the instrument, end correction of tube diameter and environmental factors such as temperature, humidity, density, and pressure.

From the experiment result and explanation above, some suggestion to the freshmen is doing experiment in the noiseless place so that the sound coming out can be heard clearly. It will reduce the error of obtaining the first and second harmonic. Experimental report, especially in discussion part should be analysed deeply for determining correction factors precisely. Analyse ability will give benefit for student to get one of skills in 21 century and improve conception of learning physics [18].

4. Conclusion

Speed of sound indicated by freshmen is different. They are categorized into three groups: accurate (52.63%), middle (31.58%) and lower (15.79%). The most results are accurate, the value approximate to theoretical. Speed of sound indicated by senior students is 333.38 ms^{-1} with a deviation to the theory about 3.98 %. The correction factors are suggested: error of determining first and second harmonic, uncertainty of the measuring instrument, end correction of the tube diameter, and environmental factors such as temperature, humidity, density, and pressure.

References

[1] Admoko S and Supriyono 2016 *J. Penelit. Fis. Apl.* **6** 34
[2] Yavuz A and Temiz B K 2016 *Phys. Educ.* **51** 015009
[3] Serway R A and Jewett J W 2004 *Physics for Scientists and Engineers* (California : Thomson Brooks/cole) 513-514
[4] Amrani D 2013 *Phys. Educ.* **48** 65
[5] Purjiyanta P, Handayani L and Marwoto P 2016 *J. Phys. Conf. Ser.* **739** 012032
[6] Yavuz A 2015 *Phys. Educ.* **50** 281
[7] Giancoli D C 2008 *Physics for Scientists and Engineers with Modern Physics* (New Jersey : Pearson Education) p 425
[8] LoPresto M C 2014 *Phys. Educ.* **46** 437
[9] Silva A A 2017 *Eur. J. Phys.* **38** 045802
[10] Gómez-Tejedor J A, Castro-Pallacio J C and Monsoriu J A 2014 *Phys. Educ.* **49** 310
[11] Da Silva W P, Precker J W, e Silva D D P S and e Silva C D P S 2005 *Phys. Teach.* **43** 219
[12] Dushaw B D, Sagen H and Beszczynska-Möller A 2016 *J. Accoust. Soc. Am.* **140** 622
[13] Bohn D A 1988 *J. Audio Eng. Soc.* **36** 223

[14] Mahdi Z S and Al-Jumaily K 2012 *Iraqi J. of Sci.* **53** 689
[15] Ogunsote O O 2007 *Propagation of Sound* (Nigeria: Ondo State) 6-8
[16] Chen Z and Maher R C 2004 *J. Montana State Univ. Propr.* 1-10
[17] Wiener F M and Keast D N 1959 *J. Accoust. Soc. Am.* **31** 724
[18] Suprapto N, Chang T-S and Ku C-H 2017 *J. Baltic Sci. Educ.* **16** 7

The scientific argumentation profile of physics teacher candidate in Surabaya

T N Ain[1], H A C Wibowo[1], A Rohman[1] and U A Deta[2]

[1]Department of Physics Education, STKIP Al Hikmah Surabaya
[2]Department of Physics, Universitas Negeri Surabaya

E-mail: trisenurulain@gmail.com

Abstract. The ability of scientific argumentation is an essential factor that must be mastered by physics teacher candidate as a requirement in explaining good and accurate scientific concepts. In the process of arguing, students develop explanations or persuade colleagues to support their hypotheses, express doubts, ask questions, relate alternative answers, and confirm what is unknown to develop the ability to provide rational and scientific explanations. The design of this research is descriptive qualitative with the subject of research is 20 undergraduate students of Physics Education Department in Surabaya. The research instrument consists of four casuistic questions related to the concept of kinematics. The argumentation pattern of physics teacher candidate is coded using Toulmin's argumentation pattern. The results show that the student's ability in providing scientific argument is at the level of providing claims with the support of a weak warrant. The students are not able to provide excellent rebuttals. In each case given, the student can give a good claim statement in answering the questions. However, the concept used to support the claim is not correct. This case causes the warrant used to support the claim is weak. Students also do not analyse other facts that affect the system. Students have not reached a higher level because the understanding of physics is not deep enough.

1. Introduction

As a branch of the natural sciences, physics has an important role in explaining natural behavior and predict the possibility of changes in the universe. Therefore, the study of physics cannot be solved only through a set of mathematical formulas, but also rather requires a scientific explanation. This explanation includes supporting the empirical facts, analyzing the fact, and verifying the truth [1]. To explain a natural phenomenon with an accurate and reliable scientific explanation, the physics teacher must have good scientific argumentation ability. The ability of the argument must be based on scientific thought and must be supported by empirical evidence.

The ability of physics teachers to explain the concept of physics is needed because many students are still having difficulty in learning physics. From previous research, the greatest difficulty of students in learning physics is understanding the physical concept of the topic[2]. Mentioned in other research [3], one of the problems in learning physics is the students just memorise the formula without deep understanding concepts. Also in the same paper mentioned that most of the students feel challenged to learn more about physics. That, because they know concepts of physics are easy to find in daily life.

Educating a good teacher who has an excellent scientific competence requires good critical and reflective thinking skills. One way to accelerate this ability is by equipping the power of good scientific argumentation. As a prospective teacher, physics education program students also should be

IOP Conf. Series: Journal of Physics: Conf. Series **997** (2018) 012025 doi:10.1088/1742-6596/997/1/012025

able to explain the physics topics to learners. The ability to provide rational and scientific explanation can be developed through the habituation of arguing [4]. While the scientific argumentation ability can be improved in various ways. The physics teachers candidate who is weak in scientific argumentation ability must learn more and deeply to improve their ability[5].

One of material physics that requires good scientific argumentation from the students is kinematics concept. The subject of kinematics is quite easy to apply because its physical phenomena are often encountered in everyday life. The mastery of a kinematics concept that can be reflected in the physical implementation is needed to be owned by physics teacher candidate students so that their knowledge is not limited to using the formula to solve mathematical problems only [6]. The ability to solve kinematics problems correctly is not enough for physics teacher candidates [7]. It needs precision analysis through scientific argumentation in explaining the concept. Therefore, research on the ability of physics teacher candidates in scientific argument needs to be done, especially on kinematics material.

2. Related Studies

2.1. Scientific Argumentation of Physics Teacher Candidate

Arguments are a strategy for explaining the concept. Argumentation comes from rational thinking and critical thinking. The argumentation process enables the student to process the explanation to support their opinion, to invite or persuade colleagues to support their opinion, to express doubts, to ask questions, to relate alternative answers, and to analyses the interrelationships between concepts that affect a particular physical system [8]. Also, through the process of argumentation, there is a social interaction between the personal and the social dimensions that can ultimately lead to a wider and deeper development of knowledge [9].

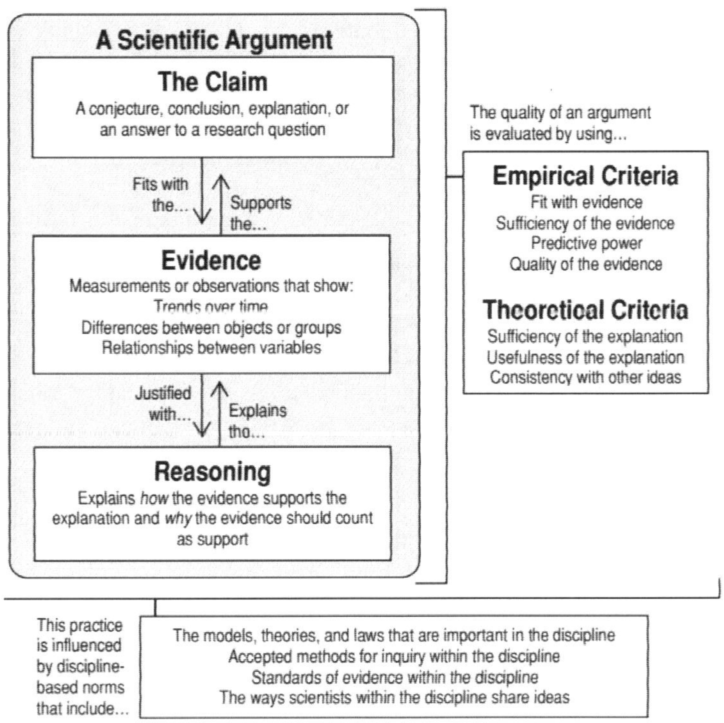

Figure 1. Illustration of the components that build scientific argumentation [11].

In the era of technology and ASEAN Economic Community challenges, physics-minded students must take on the role of being actively involved in the latest issues related to science. One of the skills a physics teacher should have is to argue scientifically "well" and appropriately using evidence in explaining and testing their opinion [10]. The scientific argument requires the analysis of critical thinking based on logical evidence and reason to substantiate a claim. The illustration of the components that explain a scientific argument well processed is presented in Figure 1.

2.2. Toulmin Argumentation Pattern (TAP)

There are researches on the argument in the learning process have been done. One of the studies was Toulmin's (1958) research on Okumus [1] which produced the following model of argumentation capability.

Table 1. Toulmin Argumentation Pattern [1].

Term	Description
Data	Facts, based on reality, used to prove claims.
Claim	The statement used to answer a problem (hypothesis).
Warrants	Describing the relationship between data and claim. Warrants explain how data support claims.
Qualifiers	The expression of the degree of certainty and uncertainty of an argument, such as the use of the word "possibility", "dependent", and "certain".
Rebuttals	Counter-arguments or statements used when a claim is not accepted.
Backing	Assumptions that support the acceptance of warrants. The backing is needed if the warrant is not acceptable.

Osborne [12] mention that the claim given by the student is the most important factor in argument construction. This is because the claim is the first foundation given by students in building great argumentation. The simplest argument is an argument that consists of data and claims only, whereas complex arguments are arguments that contain rebuttals. Arguments that have rebuttals are the best arguments. Students have to use complex analytical skills which are relating concepts that influence the claims given to form rebuttals. This causes the use of arguments consisting of data, claims, warrants, backing, and rebuttals have a positive impact on learning [1]. The division of argument levels based on Erduran, Simon, and Osborne [1] is presented in the Table 2.

Table 2. The TAP Argument Level was developed by Erduran, Simon and Osborne [1].

Level	Characteristic of the Argument
Level 5	Extensive arguments with more than one rebuttal
Level 4	Arguments with a claim with a clearly identifiable rebuttal. This argument may also have multiple claim and counter-claims.
Level 3	Arguments in the series of claims or counter-claims with data, warrant, or backing are sometimes accompanied by weak rebuttal
Level 2	Arguments consist of data, claims, warrant, or backing, but without rebuttal Osborne advocated further distinction at this level: Level 2B - arguments consist of claims supported by more than one data, warrant, or backing, but without rebuttals Level 2A - arguments consist of claims supported by data, warrant, or backing, but without rebuttals
Level 1	Simple arguments in the form of claim and counterclaim or claim versus claim

3. Method

This research used survey research method. Surveys are research methods to check and to measure the empirical symptoms that take place at the location of study. The survey method was conducted on the sample units faced as respondents and not the entire target population.

3.1. Subject of the Research

The subject of this research is 20 students of the Physics education program in Surabaya. The first-year students are chosen because in the first semester they have studied kinematics material. They are students who are prepared to be prospective teachers with adequate pedagogy and scientific skills.

3.2. Research Design

The implementation of this research is done gradually by the research flow in Figure 2.

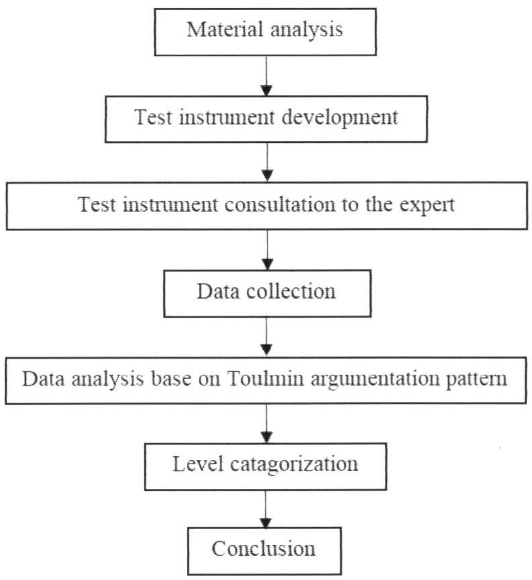

Figure 2. The scheme of research implementation.

The first step that researchers do is analyse the concept of kinematics material that became a topic in this study. Furthermore, researchers made a test instrument adapted from physics book [13]. The test instrument is then consulted to the expert who explores the field of assessment. Collecting data in this research is by test instrument and interview. The results of the data then analysed and categorised based on Toulmin's argumentation pattern.

3.3. Data Collection Technique

Data collection technique in this study used the test instrument in the form of four description questions adapted from [9]. The four questions of the description contain matters of kinematics concepts related to the free fall motion chapter, upward vertical motion, and parabolic motion which requires good argument in solving them. Besides that, the researcher also conducts direct questioning to the students to ensuring the answers and conceptual knowledge of students.

3.4. Data Analyze Technique

Data analysis technique used in this research is quantitative descriptive analysis technique. Data related to the students' scientific argument are analyzed by using an observer rubric adapted of Toulmin. The data that had been analyzed categorized into the level of argumentation developed by Erduran, Simon, and Osborne.

4. Results and Discussion

Student's scientific argument profile is analysed based on the answers given by 20 physics education program students in the Surabaya area on four questions about casuistic essay related to kinematics material. The pattern of argumentation is analysed by Toulmin's argumentation pattern which is then categorized into levels as shown in Table 2. Toulmin's argumentation pattern consists of Claim (C), Data (D), Warrants (W), Qualifiers (Q), Rebuttals (Q) and Backing (B). Student argumentation patterns consisting of Data (D) and Claim (C) are marked C-D, whereas if student argumentation pattern is Data (D), Claim (C), Warrants (W), and Rebuttals (R) then marked D-C-W-R.

The following is a description of the mapping of ability profile of physics teacher candidate in giving scientific argumentation. The first question discusses the concept of instantaneous velocity and average velocity on three kinds of motion. The first question used in this study is:

Question 1

Based on the following phenomenon, give your argument and analysis. Is there an object that has an instantaneous velocity equal to the average velocity at a particular point as long as the object moves?

A. A ball is thrown vertically upward until it reaches its highest point then falls back into the hands of the thrower.

B. A racing car moves from the rest to achieve a speed of 100 m / s.

C. Spacecraft in space that travels at constant speed.

Student responses to the above questions are very diverse. Most students provide clear claims that three phenomena have the same average speed as the instantaneous speed at a certain point along the moving object. However, the claim is not supported by the correct data, so the warrants become weak. The majority of students are only able to reach a level 2B; the argument consists of claims, data and a simple warrant (C-D-W). Others are only able to provide arguments at level 1 that contains claims and data (C-D). The simple argument at level one presents data with a claim without a clear relationship to support the data already mentioned in the claim given.

The weak argumentation given by the students for the above case is due to the concept of instantaneous velocity, and average speed are often given separately. Students are accustomed to being given a matter of instantaneous velocity or on average speed only. So when the students are given a phenomenon in which both concepts involve, students are confused to identify data that support the given opinion. Through simple interviews of some students who became subjects in this study, the authors can provide conclusions that students are accustomed to the problem related to the average speed and instantaneous speed in the form of mathematical calculations without adequate physical explanation.

The second question is related to the concept of free falling motion of two skydivers as follows:

Question 2

In the case of a skydiver who flies from the plane, the first skydiver jumps from the plane. Later, the second skydiver jumps so that both of them move freely in the vertical direction. Give the argument and analysis whether:

A. During free-fall motion, their speed at all times is the same.

B. During free-fall motion, their speed is different at any time.

C. During free-fall motion, both skydivers always have the same distance until landing on the ground.

The argumentation pattern given by the student for this case is better than the argument given in the first case. In this phenomenon, the student has been able to explain the concept that although the motion experienced by skydivers are uniformly accelerated motion, the acceleration owned by each skydiver are equal that is the earth gravitational acceleration. So, the skydiver who jumped first always has greater speed. This is then increasing separation between the two skydivers. Students have also been able to provide rebuttals in the form of factors that may affect the system such as a negligible air friction force, the mass of objects is assumed to be the same and skydivers jumping are assumed to have no significant effect.

The argumentation ability given by the students, in this case, has reached level 3 with the C-D-D-W-R argumentation pattern. The others are still at level 2B with the C-W and C-D-W argumentation patterns. Students who give level 1 of argumentation pattern in the form of claim only (C) is the student who is analysing this problem using the unsuitable concepts of air pressure, the uniform linear motion concept, and gravity. This proves that the students' concept of understanding is the basic requirement for giving argumentation.

The third question discusses the concept of parabolic motion as follows:

Question 3

A stone is thrown at a certain speed, so it moves with a parabolic path. According to your understanding, is there a point along the path taken by the rock where speed and acceleration that:

A. Parallel or in line each other.

B. Perpendicular each other.

In this case, the students are divided into two categories namely level 1 and level 2B category. There is only one student who gives the argumentation pattern at level 3. This problem is because the students can not describe velocity vector and acceleration vector on parabolic motion correctly and precisely. The student has not understood that in parabolic motion there is a uniform linear motion in the x-axis and non-uniform linear motion in the y-axis. Through an interview to the students who can not describe the parabolic motion with vector analysis, the student argued that the concept of parabolic motion they received so far did not explain vector analysis and concept of parabolic motion correctly. The material they receive is mostly about the formula of the maximum distance, the maximum height, and the travel time of the object in parabolic motion. One student who gives the argumentation pattern at level 3 mentions rebuttal that is the conditions affected the system, for example, the angle which affects the parabolic motion.

The last question is:

Question 4

Three balls are on the roof of the house with the same height. The first ball (A) is thrown vertically upward with initial speed (x m/s), the second ball (B) is thrown horizontally at speed (x m/s), the third ball (C) is released vertically down.

A. Will all balls reach the ground at the same time? If it is not, mention the order of the balls reached on the ground. Give your analysis of your answer.

B. Do all balls have the same speed when touching the ground? If not, specify the order of the speed with the relevant analysis.

The fourth question is about the velocity of the objects which is thrown vertically upwards, and horizontally to form parabolic motion, and also the velocity of the objects which being free thrown to form free fall motion. In this case, there are three different levels of argumentation given by the students: level 1, level 2B, and level 2A. The student can state that form the three of the motions, the vertical upward motion takes the longest time for the object to touch the ground compared to the others. However, students have not been able to analyse the parabolic motion with a free fall motion will give the same time for the ball for moving to the ground. This has an implication when the student answers the point B about the speed of three objects. Students' weakness in giving the argument, in this case, is because students are not used the critical thinking to analyse the physical phenomena. When interviewing a student who writes the C-D-W-C-W pattern (level 2), the researchers gradually and deeply guide the analysis that students write until the student finds that the time and velocity of B and C are equal. The majority of students argue that the fastest ball reaches the ground is ball C because it is affected by the acceleration of gravity and the shortest distance assumption without scientific proof of accurate data.

Based on the whole analysis of the argumentation pattern that the students are given, it can be seen that the argumentation has an important role in building students' concept of understanding. As a prospective teacher, students of physics education should have a good understanding of concepts and scientific argumentation to explain the physical phenomenon to their students later. This research is the first step for further research for improving students ability in the pedagogical and conceptual knowledge.

5. Conclusion

Based on the results and discussion, the students' scientific argument ability is at the level of providing claims with weak warrant. It can be concluded that most students are unable to provide appropriate rebuttals. Students need to deepen the physical concept, not only paying attention to the mathematical aspects but also other possibilities that have been described separately including enriching the field's direct observation of the phenomenon that has been studied. As a Physics teacher candidate, it is bad if the understanding of problems is limited to the mathematical solution that has often been obtained.

References

[1] Okumus S and Unal S 2012 *Proc. Soc. Behav. Sci.* **46** 457
[2] Karim S and Saepuzaman D 2016 Analisis Kesulitan Mahasiswa Calon Guru Fisika dalam Memahami Konsep Gerak Parabola *Pros. SNF.* **5** (Surabaya: Unesa) 51
[3] Azizah R, Yuliati L and Latifah E 2015 *J. Penelit. Fis. Apl.* **5** 44
[4] Kuhn D, Zillmer N, Crowell A and Zavala J 2013 *Cognit. Instr.* **31** 456
[5] Hsu C C, Chiu C H, Lin C H and Wang T I 2015 *Comput. Educ.* **91** 46
[6] Datur I S, Yuliati L and Mufti N 2016 Eksplorasi Kemampuan Pemecahan Masalah Siswa Fisika pada Materi Fluida Statis *Pros. Semin. Pend. IPA Pascasarjana UM.* **1** (Malang: UM) 294
[7] Afwa I L, Sutopo and Latifah E 2015 Perlunya Program *Deep Learning Question* untuk Meningkatkan Penguasaan Konsep Kinematika Mahasiswa Fisika *Pros. Semin. Fisika dan Pembelajarannya Pascasarjana Universitas Negeri Malang* (Malang: UM) 66
[8] Eskin H and Bekiroglu F O 2009 *Eur. J. Math. Sci. Technol. Ed.* **5** 63
[9] Simon S, Erduran S and Osborne 2002 Enhancing the Quality of Argumentation in School Science *Paper Presented at the Annual Meeting of the National Association for Research in Science Teaching*
[10] Wardani A D, Yuliati L and Taufiq A 2016 Kemampuan Argumentasi Ilmiah dan Pemecahan Masalah Fisika Siswa SMA pada Materi Gaya dan Gerak *Pros. Semin. Pend. IPA Pascasarjana UM.* **1** (Malang: UM) 13
[11] Sampson V, Grooms J and Walker J P 2010 *Explan. Study Wiley Period. Ed.* **95** 217
[12] Osborne J, Erduran S, Simon S and Monk M 2001 *Sch. Sci. Rev.* **82** 301
[13] Serway R A and Jewett J W 2014 *Fisika untuk Sains dan Teknik* (Jakarta: Salemba Teknika)

The differences in analysing strategy of viscosity experiment between freshmen and laboratory assistant

S Anggrayni[1], H Mubarok[1], N P Putri[1], N Suprapto[1] and A Kholiq[1,a]

[1]Department of Physics, Faculty of Mathematics and Natural Science, Universitas Negeri Surabaya, Surabaya 60231, Indonesia

[a]Email: kholiq@unesa.ac.id

Abstract. The viscosity is defined by dimension of a fluid that resists the force tending to motive the fluid to flow. The aim of viscosity experiment is to determine the fluid viscosity coefficient value. By using graphical analysis, the result of oil viscosity coefficient value which performed by laboratory assistant showed: (1) 0.20 Pa.s using solid ball with accuracy 99.64% and (2) 0.21 Pa.s using smaller solid ball with accuracy 99.17%. Meanwhile, the result of oil viscosity coefficient value which performed by freshmen showed: (1) 0.44 Pa.s using solid ball with accuracy 87.85% and (2) 0.32 Pa.s using smaller solid ball with accuracy 89.84%. The differences result of the freshmen and assistant laboratory viscosity experiment are caused by the freshmen calculated the coefficient viscosity value without velocity correction factor and they used small range fluid so the times are not identified well.

1. Introduction

The viscosity of a fluid reflects the property of being able to transmit friction forces between particles or to exhibit resistance to shearing flows. For a quantitative characterisation of viscosity of a fluid, the laminar flow is considered when the relative motion between two adjacent layers generates shear stresses [1]. The viscosity is defined by a dimension of a fluid that resistant force tending to motive the fluid to flow [2]. Liquids with a high viscosity flow slowly and vice versa [3]. For example, we are easier to pour water from a bottle than pour honey. This is due to the viscosity of honey greater than the water viscosity [4]. The viscosity coefficient value shows the degree of viscosity of a fluid. In physics teaching, the conventional method for measuring the viscosity of liquids, called the falling ball method, is based on the fall of an object through a liquid. The experiment is realized using a setup called the falling ball viscometer, which typically measures the viscosity [3]. There is important to determine the viscosity coefficient value because the value shows the ability of the current fluid [9], and developing the scientific process for us. The scientific process are attitudes include curiosity, respect for proof, critical thinking, creative thinking, speaking based on concrete evidence or data, and care about the environment [10]. Additionally, this scientific process triggers conception of learning physics [11].

For determining the fluid viscosity coefficient one can drop a ball which known the density and then measure the terminal velocity of the ball [5]. The object which is moving in a fluid always gets a friction force in the opposite direction of the motion of the object (Figure 1). The magnitude of the friction force depends on the relative velocity of the object in the fluid and the shape of the object. For ball-shaped objects, the magnitude of the frictional force satisfies Stokes's Law.

$$F = 6\pi\eta r v \tag{1}$$

where, r is the radius of the ball; and v is the ball velocity

Figure 1 (a) The ball which is moving in a fluid always gets a friction force in the opposite direction of the motion of the object, (b) The force which is acting on the ball when falls into the fluid. (adaptation from [5])

If the ball is dropped in the fluid then the ball moves down with an increasing velocity due to the acceleration of gravity. But, at a point the velocity does not change anymore. This velocity called the terminal velocity [3]. The forces which act on the falling object are the gravity force, the Archimedes force, and the Stokes force. When the ball reaches in the terminal velocity, the three forces are balance. Based on the ball terminal velocity, we can determine the coefficient viscosity value of fluid. Equation (2), (3), and (4) describes the third of the forces.

The Gravity Force

$$w = m_b g = \rho_b V g = \rho_b \left(\frac{4}{3}\pi r^3\right) g \tag{2}$$

The Archimedes Force

$$F_A = \rho_f g V = \rho g \left(\frac{4}{3}\pi r^3\right) \tag{3}$$

The Stokes Force

$$F_S = 6\pi\eta r v \tag{4}$$

Applying Newton's second law, we can write the following equation of motion for the falling ball:

$$m\frac{dv}{dt} = w - F_A + F_S \tag{5}$$

where v is the instantaneous velocity of the falling ball at a time t [3].

When the object reaches in the terminal velocity, the three forces satisfy

$$w = F_A + F_S$$

$$\rho_b \left(\frac{4}{3}\pi r^3\right) g = \rho_f g \left(\frac{4}{3}\pi r^3\right) + 6\pi\eta r v$$

$$\eta = \frac{2r^2 g(\rho_b - \rho_f)}{9v} \tag{6}$$

The measured viscosity of a suspension appears to depend upon the diameter of the tube of the capillary viscometer used in determining it [6]. The metal ball is dropped into a cylinder which is containing the fluid will tend to be forced to move away from the cylinder wall. This condition is called the wall effect [6, 7]. The terminal velocity will be achieved if the ball is far from the initial location when it's released in the fluid. Therefore, we need the velocity correction factor because the tube which is used in this experiment has limited diameter and length.

The velocity correction factor based [8]

$$\boldsymbol{v} = v'\left(1 + 2.4\frac{r}{R}\right)\left(1 + 3.3\frac{r}{L}\right) \tag{7}$$

Where, v' is the ball terminal velocity; R is the radius of the tube; and L is the length of fluid. The unit of coefficient viscosity value is $Pa.s$ with $\mathbf{1\ poise = 1\frac{dyne.s}{cm^3} = 10^{-1}\ \frac{N.s}{m^3} = 10^{-1}\ Pa.s.}$

The errors that can reduce the accuracy of measurements in this experiment are the inaccuracy in determining the constant point and the inaccuracy in measuring travel time of the ball falling [7]. From the results of viscosity experiments that have been done in basic physics laboratories, there are significant results differences between physics major and non-physics major. Therefore, it is necessary to do research on the viscosity experiment to find the cause of the differences result.

In this article, we discuss about the differences of analysis strategy of the viscosity experiment result by freshmen and laboratory assistant. The purpose of the viscosity experiment is to determine the value of oil viscosity coefficient.

2. Method

Based on the research aim, this research is a descriptive research using the quantitative descriptive method in explaining the differences of analytical strategies on viscosity experiment between freshmen and laboratory assistant. In this case, the subject research is the freshmen in majoring mathematic, physics, chemistry, and biology of the faculty of mathematics and natural sciences of The State University of Surabaya. The method of this experiment is falling a ball which known the density and then measure the terminal velocity of the ball. For measuring the ball terminal velocity is manipulating the travel of the ball and then measuring the times. The fluid of this experiment is oil with two methods, they are using a solid ball and using a smaller solid ball [8].

3. Result and Discussion

The viscosity experiment we got some data, there are the mass of the ball, the diameter of the ball, the diameter of the tube, the density of the fluid, the length of fluid, and the terminal velocity. For calculating the terminal velocity is dividing between the long travel of the ball and the times. In this experiment, we manipulated the long travel of the ball and measuring the times. The long travel manipulation used distance after constant point, the meaning is the measuring the terminal velocity when the ball move with constant velocity in the fluid. After we got the data, we can calculate use the viscosity equation. Based the viscosity experiment which performed by freshmen, the oil viscosity coefficient value as follows

Table 1. The result of oil viscosity coefficient value which performed by freshmen.

Major	Department	The Coefficient Viscosity value, η (Pa.s)	
		Solid Ball	Smaller Solid Ball
Physics Major	Physics	0.0760	0.0285
Non Physics Major	Mathematic	0.5153	0.4870
	Chemistry	0.9424	0.6398
	Biology	0.2130	0.1140

The result of the viscosity coefficient value above is obtained from calculate using equation 5 at each distance manipulation then determine the viscosity coefficient value average. The result of oil viscosity coefficient value which performed by freshmen showed: (1) 0.44 Pa.s using solid ball with accuracy 87.85% and (2) 0.32 Pa.s using smaller solid ball with accuracy 89.84%. Meanwhile, the viscosity experiment which performed by laboratory assistant, the oil viscosity coefficient value using graphical analysis as follows

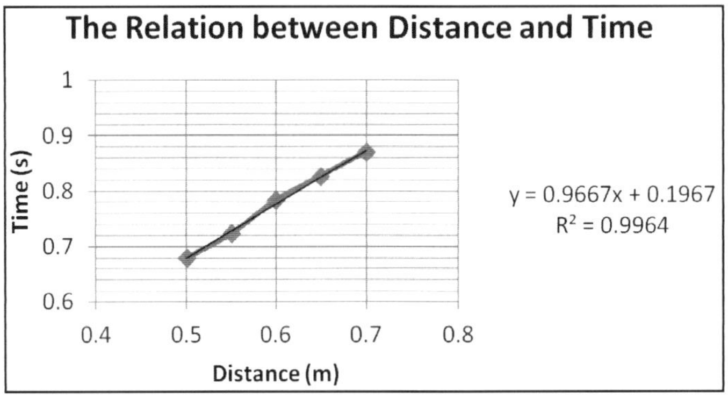

Figure 2. Graphic analysis of the viscosity experiment result using solid ball.

The graph line equation in Figure 2 shows the relation between distance and time. Y-axis shows the times, X-axis shows the times, and the gradient of the line equation is 0.9667. From the gradient we can determine the coefficient viscosity value use equation (7) and R^2 shows the accuracy of this experiment. The accuracy of this experiment is gotten 99.64%.

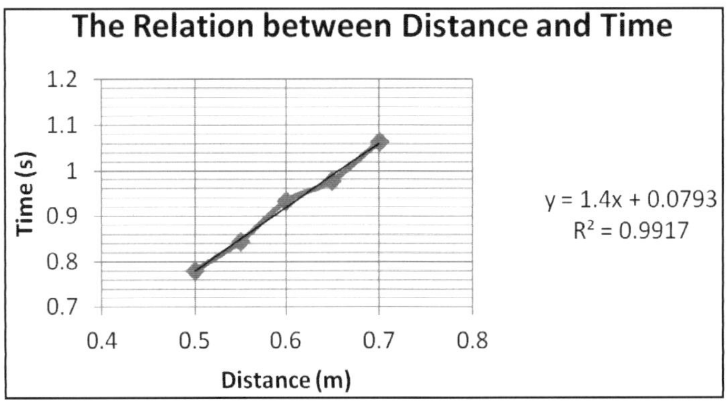

Figure 3. Graphic analysis of the viscosity experiment result using smaller solid ball.

The graph line equation in Figure 3 shows the relation between distance and time. *Y*-axis shows the times, *x*-axis shows the times, and the gradient of the line equation is 1.4. From the gradient we can determine the coefficient viscosity value use equation (7). And R^2 shows the accuracy of this experiment. The accuracy of this experiment is gotten 99.17%.

Based the graph of the relation between distance and time above, we can determine the value of the fluid viscosity coefficient through the gradient value of the graph. Using the value of a velocity correction factor in the equation (6), obtained the value of viscosity coefficient using graphical analysis as follows

$$\eta = \frac{2r^2 g\, m\,(\rho_b - \rho_f)}{9\left(1+2.4\frac{r}{R}\right)\left(1+3.3\frac{r}{L}\right)} \qquad (8)$$

where, m is the graph gradient [7].

The result of the oil viscosity coefficient value which performed by a laboratory assistant showed: (1) 0.20 Pa.s using solid ball with accuracy 99.64% and (2) 0.21 Pa.s using smaller solid ball with accuracy 99.17%. The result of the oil viscosity value which performed by freshmen is different with the result of oil viscosity coefficient value which performed by laboratory assistant due to the freshmen calculated the viscosity coefficient value without using the velocity correction factor, and they used relatively small distance manipulation so that they couldn't identify the times well.

Based two analytical strategies above, we know that the result analysis of this experiment using graphical analysis based on the equation (7) is better than analysing use equation (5). This is caused by velocity correction factor equation (6) is neglected by the equation (5). In this case, we can't neglect the velocity correction factor because the tube which is used in this experiment has limited diameter and length.

4. Conclusion

The result of this experiment which performed by freshmen got the oil viscosity coefficient value using solid ball and smaller solid ball show different values. It is also showing the different result if it is compared with the viscosity coefficient value of laboratory assistant's result. The differences are caused by the freshmen calculated the viscosity coefficient value without using the velocity correction factor, and they used relatively small distance manipulation so that they couldn't identify the times well. So, if we want to get accuracy result in analysing our data in our experiment, more better if we use graphical analysing, and attention the correction factor the experiment.

References

[1] Ciornei F C, Alaci S, Amarandei D, Irimescu L, Romanu I C and Acsinte L I 2017 *IOP Conf. Series: Mat. Sci. Eng.* **174** 012041
[2] Suprapto N, Zamroni A and Yudianto E A 2017 *Chem.* **26** 615
[3] Houari A 2011 *Phys. Educ.* **46** 688
[4] Serway R A and Jewet J W J 2004 *Physics for Scientists and Engineers* 6th edn (California: Thomson Brooks/Cole) p 1926
[5] Abdullah M 2016 *Fisika Dasar 1* (Jakarta: Kampus Ganesa) p 1068
[6] Maude A D 1967 *Brit. J. Appl. Phys.* **18** 1193
[7] Chusni M M and Toifur M 2012 *Pros. Pertemuan Ilmiah XXVI HFI Jateng & DIY* 4
[8] Tim Dosen Pembina Praktikum Fisika Dasar 1 2014 *Panduan Praktikum Fisika Dasar 1* (Surabaya: Unipress) 37
[9] Widdicombe A T, Ravindrarajah P, Sapelkin A, Phillips A E, Dunstan D, Dove M T, Brazhkin V V and Trachenko K 2014 *Phys. Educ.* **49** 406
[10] Admoko S and Supriyono 2016 *J. Penelit. Fis. Apl.* **6** 34
[11] Suprapto N, Chang T-S and Ku C-H 2017 *J. Baltic Sci. Educ.* **16** 7

IOP Conf. Series: Journal of Physics: Conf. Series **997** (2018) 012027 doi:10.1088/1742-6596/997/1/012027

The estimation of occupational dose in 15 MV varian clinac iX room by Argon-41 as an activation product of photoneutron

R Latifah[1,a], Bunawas[2] and J A E Noor[3]

[1] Department of Health, Faculty of Vocational Education, Universitas Airlangga, Jalan Airlangga, Surabaya, 60115, Indonesia
[2] Center for Safety Technology and Radiation Metrology, National Nuclear Energy Agency of Indonesia (BATAN), Jalan Kuningan Barat, Jakarta, 12710, Indonesia
[3] Department of Physics, Universitas Brawijaya, Jalan Veteran, Malang, 65145, Indonesia

E-mail: [a]risalatul.latifah@vokasi.unair.ac.id

Abstract. Linear accelerator (linac) becomes the most commonly used treatment to damage and kill cancer cell. Photon and electron as the radiation beam are produced by accelerating electrons to very high energy. Neutrons are generated when incident high photon energy interacts with component of linac such as target, flattering filter and collimator via photoneutrons reaction. The neutrons can also produce activation of materials in treatment room to generate radioactive materials. We have estimated the concentration of Argon-41 as activated product from argon-40 in the linac room using foil activation. The results show that the Argon-41 concentration in linac room which is operated 15 MV for 1 treatment (1 minute) is 1440 Bq/m^3. Accordingly that concentration, the occupational dose is 6.4 mSv per year.

1. Introduction

In modern radiotherapy, both high energy X-ray and electron beam are often used in the treatment of cancers. These beams mainly are generated by linear accelerator (linac). However, when high-energy accelerators operated roughly at 10 MV or higher, a significant number of neutrons are generated via photoneutron reactions (n, γ). The photoneutron are produced when high energy photon collides with the high atomic number of material in linac system such as the accelerator head, all collimators, the target and flattening filter [1-4]. This reaction consequently brought an undesirable neutron production. Moreover, the neutrons can activate the material in treatment rooms to generate radioactive materials which raise a concern about irradiation safety. Argon is one of the gaseous in the air with stable nuclide in the form of ^{40}Ar. Although the argon concentration in air is relatively small, but the chances for activation by neutrons are relatively large because argon has a fairly high cross section value of 162 barn [5]. When the stable Argon ^{40}Ar interact with neutron, Argon will be activated as a radioactive in the form ^{41}Ar. Recently, little attention has been emphasized to gaseous radionuclide ^{41}Ar which can be generated by thermal neutron activation of stable ^{40}Ar in air.

Numerous papers were focusedon the study of neutron field and its production in linac, but it mostly concerns to the patients. This study observed that the impact on occupational radiation dose due to activation product contributed by Argon-41 [6]. The neutron activation approach is implemented to estimate the concentration of Argon-41 in linac room. Indium foils are used for neutron activation because it has a high cross section for thermal neutron [7-8].

2. Materials and Methods

2.1. Activation of Indium Foil

The measurements of thermal neutron flux were carried out with the use of induced activity method [7-8]. The indium foils (^{115}In isotop) in the shape of circle with the diameter of 2.3 cm were applied as activation detector. Isotope ^{115} In has a relatively high cross-section for thermal neutron capture (162 barn). The thermal neutron induces the activity of the foil according to the following reaction:

$$^{115}_{49}In + ^{1}_{0}n \rightarrow ^{116}_{49}In^{*}$$

$$^{116}_{49}\mathrm{In}^{*} \rightarrow \beta\_ + ^{116}_{49}\mathrm{Sn}^{*} \rightarrow ^{116}_{49}\mathrm{Sn} + \mathrm{gammas}$$

The activities of indium foils were determined on the basis of measurement of gamma energy spectrum from deexcitation of 116Sn*. The spectral measurements were performed using HPGe detector connected to a multichannel analyzer. The HPGe detector was calibrated with the filtered source multiradionuclides including 241Am, 133Ba, 137Cs and ^{60}Co. The activities of the foil were determined by counting the gammas of 1293.54 keV. From counting, the activity of indium can be calculated with [9]:

$$A = \frac{\frac{area}{t_c}}{\varepsilon \cdot yield\ (E_\gamma)} \tag{1}$$

Other The thermal neutron flux is defined and the induced radioactivities of indium foils, as follows:

$$\phi_{th} = \frac{A}{\sigma_{In} \times N_A \times \frac{m \times a}{M} \times (1 - e^{-\lambda t_i}) \times e^{-\lambda t_d}} \tag{2}$$

where ϕ_{th} denotes the thermal flux (cm^{-2} s^{-1}) which represents an activity of indium that is counted by gamma spectrometer, λ is the decay constant (2.135×10^{-4} s^{-1}), σ_{In} is microscopic cross-section (162 barn), N_A is Avogadro's number, *m* is the mass of indium foil (g), *M* is the atomic weight of indium (114.82), *a* is the isotopic abundance of ^{115}In (95.7%), t_i is irradiation time and t_d is the time duration between irradiation and measurement [6].

2.2. Irradiation Treatment

Indium foils (purity 99.993%) with an averaged mass of (1.4884 ± 0.0004 g) placed on and distributed around medical accelerator room (Varian Clinac iX), such as on Figure 1. The accelerator was operated at 15MV for 1 minute and collimator was opened at a field size of 20×20 cm^2.

The irradiated foils were immediately counted using gamma spectrometer with HPGe detector. The measured gamma-ray spectra was collected with a multichannel analyzer and was further analyzed by gamma-spectrum software (Gammavision-32). Foils were placed on the face of detector for counting; the efficiency was determined to be 3.2% at the characteristic gamma-ray energy of 1294 keV which is emitted from activated nuclide ^{116m}In for 5 minutes.

2.3. Estimation of The Argon-41 Concentration and Corresponding Dose

Argon-41 is produced by photoneutron reaction between high energy photon and material component in linac. The concentration is difficult to measure directly [8], so activation foil approach is applied. The value of ϕ_{th} from irradiated foil in linac room is used to determine the accumulation of argon-41 concentration (C_{Ar}) is derived as [6]:

$$C_{Ar}(Bqm^{-3}) = F \times \bar{\phi}_{th} \times \sigma_{Ar} \times N_{Ar} \times \left(1 - e^{-\lambda_{Ar} t_i}\right) \tag{3}$$

where *F* correction factor (9.2), $\bar{\phi}_{th}$ is the averaged thermal neutron flux in lianc room, σ_{Ar} is the neutron capture cross section of argon (0.64 barn) [10], N_{Ar} is concentration argon-40 in air (3.18× $10^{23} m^{-3}$), λ_{Ar} is the disintegration rate of Argon-41 (1.05× $10^{-4} s^{-1}$) and t_i is irradiation time.

The corresponding dose of Argon-41 are calculated from three kind of doses. They are external dose, internal dose and immersion dose. The calculation using dose conversion facror as follows

$$D^m_{i,r} = X^m_i \cdot F^m_{i,r} \tag{4}$$

where $D^m_{i,r}$ is received dose of organ *r* from radionuclide *i* on track *m*, X^m_i is the concentration of radionuclide *i* on track *m* and $F^m_{i,r}$ conversion factor of radionuclide *i* for organ *r* on track *m* [9].

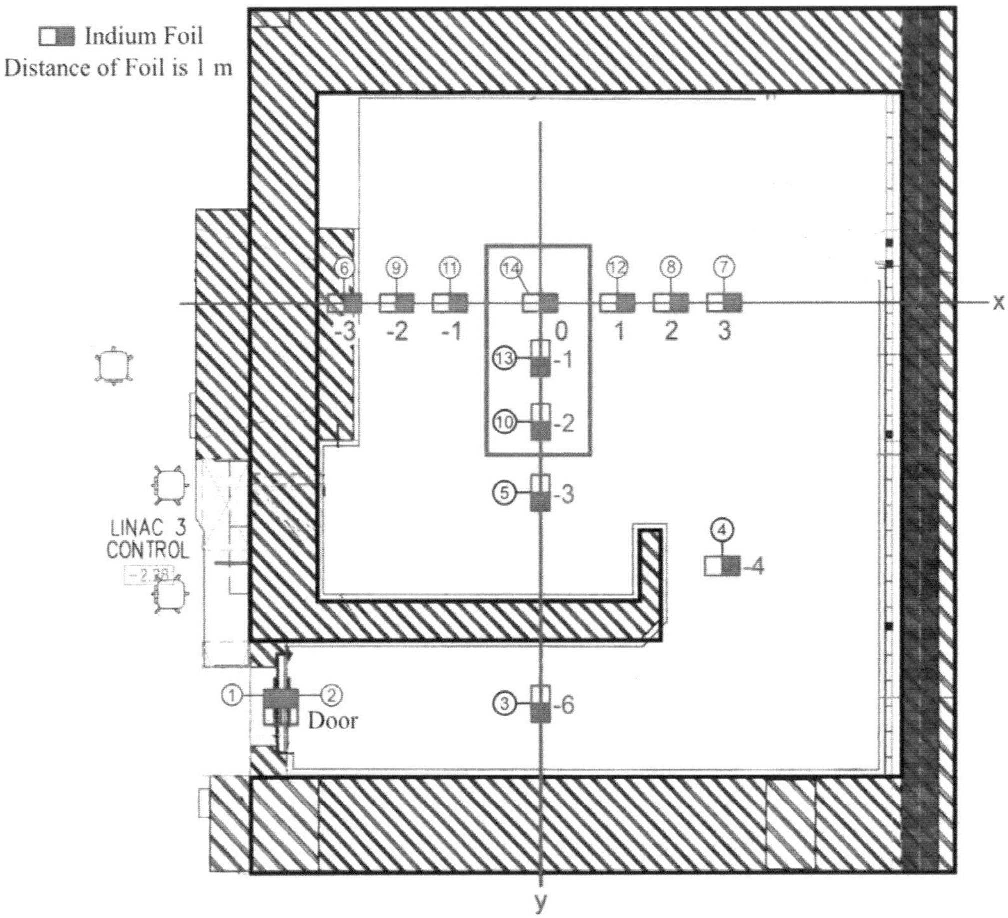

Figure 1. Distribution of indium foils in linac room.

3. Result and Discussion

3.1. Spatial Distribution of Thermal Neutron in Linac Room.

Table 1 shows the thermal neutron flux ϕ_{th} from irradiated foil around in linac room. Based on measurement of 14 indium foils, the average of thermal neutron flux in linac room varied from $1.4 \times 10^2 - 9.7 \times 10^3$ $cm^{-2} \cdot s^{-1}$. The highest value is founded at the linac isocenter (9.6×10^3 $cm^{-2} \cdot s^{-1}$) and the lowest is at labirin. Figure 1 shows that area of x-negative-axis nearer to concrete than x-positive-axis and the result show that the distribution of ϕ_{th} x-negative-axis is higher than x-positive-axis because there are reflected neutron from concrete and contributed from thermalization of fast neutron. Thermal neutron flux has declined gradually with distance from the isocenter and has not detected on the outer door in operator room. It indicates that linac control room has been safely from thermal neutron as activation product of photoneutron reaction.

Table 1. Measurement Result of Thermal Neutron Flux in Linac Room from Indium Foil Placed to around Linac Room.

Num Foil	*Foil coordinate*	$\bar{\Phi}_{th}$ ($cm^{-2} \cdot s^{-1}$)
1	(-3,-6.2)	0
2	(-3,-6)	142
3	(0,-6)	304
4	(3,-4)	3429
5	(0,-3)	6802
6	(-3,0)	8234
7	(3,0)	5804
8	(2,0)	6753
9	(-2,0)	9280
10	(0,-2)	7974
11	(-1,0)	9619
12	(1,0)	8507
13	(0,-1)	9619
14	(0,0)	9660

Measurement of thermal neutron flux in linac room has been done. It is listed to table II. Medical linear accelerator operating above 10 MV produce photoneutron via interaction high-energy photons with high atomic number (Z) materials. The material of linac will become main factor for releasing secondary neutron to treatment room. Table 2 shows the comparison some of linac type, Varian Clinac iX produce neutron lower than the other type of linac.

Table 2. List of Study Thermal Neutron Flux in Linac Room.

Reference	*Beam*	*linac*	*Area (cm²)*	Φ_{th} ($n \cdot cm^{-2} \cdot Gy^{-1}$)
This study	15 MV	Varian C*linac* iX	20 × 20	$9.7 \cdot 10^3$
Liu, 2011 [2]	15 MV	Varian C*linac* 21EX	20 × 20	$2.0 \cdot 10^5$
Konefal, 2005 [10]	15 MV	Primus Siemens	20 × 20	$2.9 \cdot 10^5$
Gur, 1978 [11]	18 MV	Philips SL/75-20	10 × 10	$1.4 \cdot 10^6$
Palta, 1984 [12]	18 MV	Mevatron 77	15 × 15	$2.3 \cdot 10^5$
Lin, 2001 [14]	20 MV	Varian C*linac* 2300	10 × 10	$1.5 \cdot 10^6$
Uwamino, 1985 [15]	21 MV	Microtron, MM22	10 × 10	$3.2 \cdot 10^5$

3.2. The Concentration of Argon

Argon-41 is product by activation Argon-40 in air with neutron as a product of photoneutron reaction. The average thermal neutron flux in linac room can be related to estimate the concentration Argon-41. The activated Argon-41 was assumed to diffuse rapidly and to be uniformly to distribute in the treatment room. As a whole, the average of Argon-41 concentration in linac room calculated by equation (3) is 1440 Bq/m³.

Remembering Argon-41 is a gaseous radionuclide, the Argon-41 concentration in room will be depend on radioactive decay and ventilation. This study has measured the ventilation rate of air on both the intake vent (inlet) and the exhaust vent (outlet). The measurement resulted ventilation rate is 37 m³/h. Refer to the result, we can calculate its ventilation coefficient $T_{\frac{1}{2}v}$ is 7.2/min. The half-life of Argon-41 as time function can be expressed as follows [6]:

$$\frac{1}{T_{\frac{1}{2}total}} = \frac{1}{T_{\frac{1}{2}Ar}} + \frac{1}{T_{\frac{1}{2}v}} \quad (5)$$

$$\frac{1}{T_{\frac{1}{2}total}} = \frac{1}{110} + \frac{1}{7.2} \quad (6)$$

$$T_{\frac{1}{2}tot} = 6.75 \text{ minutes}$$

The equation 6 can be used to determine the concentration dynamics of argon-41 in the linac room during treatment activities. Figure 2 shows the concentration dynamics of argon-41 during weekdays or treatment activities (8 hours) by taking the assumption that every 20 minutes there is irradiation using high energy 15 MV.

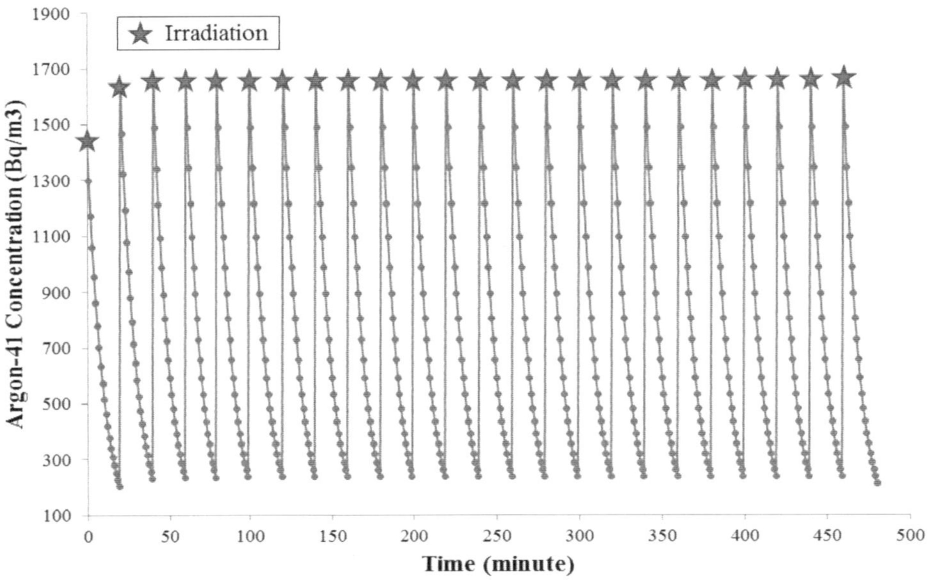

Figure 2. Argon-41 Concentration with time during a treatment.

Figure 2 shows that in the 20 minutes, concentrations of Argon-41 is about 184 Bq/m^3, it decreases 87% from the initial concentration when treatment irradiation with the energy of 15 MV was done. The concentration of argon-41 in the linac room can be used as information for determining a safe time to entry in linac room for nurses or medical personnel. The safe time of entry is time in which the concentration of argon-41 in the room has been fully exhausted. By knowing the safetime of entry, it can be used to radiation protection efforts on the medical staff.

3.3. Radiation Dose of Argon-41

The corresponding dose of Argon-41 in linac room can be contributed from three ways. They are external dose (contributed from gamma irradiation when people are in irradiation field), internal dose (contributed from inhalation of Argon-41) and beta immersion dose (contributed from absorbed Argon-41to whole body surface). They are express in mathematically as follows:

$$D_{eks} = t \cdot \sum X_{Ar-41,air} \cdot F_{Ar-41}^{eks} \quad (7)$$

$$D_{imersi} = t \cdot \sum X_{Ar-41,air} \cdot F_{Ar-41}^{immersion} \quad (8)$$

$$D_{inhalasi} = t \cdot \sum X_{Ar-41,air} \cdot F_{Ar-41}^{int} \quad (9)$$

where D is dose (mSv) t is time exposure (s), X is concentration (Bq/m^3), F is conversion factor $\left(\frac{mSv}{Bq \cdot m^3 \cdot jam}\right)$.

Determination of the dose of the contribution of the activated argon-41 using dose conversion factors. Thus, it is necessary to determine the value of the dose conversion factors. Dose conversion factors referenced in the 1977 USNRC for external dose, internal and immersion in a row at 4.5×10^{-7}, 1.4×10^{-7} and 4.5×10^{-7} with unit (mSv/min)/(Bq/m^3) [15]. There are additional correction for immersion dose. This correction factor is caused each person's body is different. Mattar estimated the surface area of the skin in the function of mass and height [16], then refined by Wa'id which incorporates gender or gender function f, as in the following equation 10:

$$S = 0{,}0072B^{0{,}425} \times T^{0{,}725} \times f(m^2) \quad (10)$$

where S is the surface area of the skin (m^2), B is body mass (kg), T is the body height (cm) and f is the shape correction factor/ factors of gender, for men and 0.95 for women and 1.05 [18]. As standard, ICRP has published standard mass and height as a reference for men 73 kg; 176 cm (S = 1.61 m^2) and for women 60 kg; 163 cm (S = 1.59 m^2) [19]. So, the new conversion factor for immersion dose contributions from Argon-41 is

$$F_k = 4.5 \cdot 10^{-7} \times \frac{1{,}78}{surface\ body} \quad \text{(male)} \quad (11)$$

$$F_k = 4.5 \cdot 10^{-7} \times \frac{1{,}73}{surface\ body} \quad \text{(female)} \quad (12)$$

By using immersion dose conversion factor in equation (11) and (12), we need the data of mass and height workers who are around in such as linac operator, medical physicists and nurses. This study has made generalization for using standard data of mass and height based on IAEA, mass and height standards for the Asian men and women respectively are 60 kg; 170 cm and 51 kg; 160cm [19]. Thus, the conversion factor for immersion dose are $4.97 \times 10^{-7}(\frac{mSv/_{menit}}{mSv/_{m^3}})$ for male and $4.90 \times 10^{-7}(\frac{mSv/_{menit}}{mSv/_{m^3}})$.

As a whole, the conversion factor for three doses can be tabulated in Table 3 as follow:

Table 3. Conversion Factor of Contributed Dose by Argon-41.

Type of dose	*Conversion factor* $\left(\frac{mSv/_{menit}}{Bq/_{m^3}}\right)$
External	4.5×10^{-7}
Internal	1.4×10^{-7}
Immersion	4.9×10^{-7}

By knowing these dose conversion factors, it can be predicted how many additional doses contributed by Argon-41 obtained if people are in the linac room during t time. If a worker is assumed in the linac room operated at energy 15MV for 5 min after irradiation and the daily working period of worker is seven hours and five days in a week, thus calculation dose can be listed at Table 4. It is still relatively small when related to the annual dose for occupational exposure.

Table 4. Total Contributed Dose by Argon-41 for Workers per Year.

Type of dose	**Calculation**
external	2.4 mSv
Internal	0.7 mSv
Immersion	2.6 mSv
Total	5.7 mSv

4. Conclusion

Argon-41 is an activation product due to secondary neutron which is created via photoneutron reaction with material of linac. This study aims to estimate the concentration of Argon-41 using neutron activation approach and predict the radiations dose received by worker. The result show that the annual occupational dose from Argon-41 is estimated equal to 6.4 mSv. This exposure can be significantly reduced with maintenance or ventilation and a delay of few minutes between the end of the irradiation and the entering inside the treatment room. Other paragraphs are indented

References

[1] Konefal A, Orlef A and Dybek M 2008 *Phys. Méd.* **24** 212
[2] Liu W S, Changlai S P and Pan L K 2011 *Phys. Chem.* **80** 917
[3] Konefal A, Orlef A and Laciak M 2012 *Rep. Pract. Oncol. Radiother.* **17** 339
[4] Tuo F, Zhou F and Yi Y 2006 *Appl. Radiat. Isot.* **64** 910
[5] IAEA 1974 *Handbook on Nuclear Activation Cross-Sections* (Vienna: International Atomic Energy Agency)
[6] Chao J H, Liu W S and Chen C Y 2007 *Radiat. Meas.* **42** 1538
[7] Chao J H and Chiang A C 2010 *Radiat. Meas.* **45** 1024
[8] Chao J H, Hsu P C and Liu H M 2001 *Radiat. Isot.* **55** 549
[9] Cahyana C, Umbara H and Lubis E 1998 *Pemodelan Isodosis dari Dispersi Radionuklida di Atmosfir di Daerah PPTN Serpong* (Online) Available on http://digilib.batan.go.id/e-prosiding/File%20Prosiding/Informatika/lkstn/LKSTN%20Ke%2012/Chepy.pdf?mainFrame = Accessed January 15, 2017
[10] Konefal A, Dybek M and Zipper W 2005 *Nukleonika* **50** 73
[11] Gur D, Rosen J C and Bukovitz A G 1978 *Méd. Phys.* **5** 221
[12] Palta J R, Hogstrom K R and Tannanonta C 1984 *Radiat. Meas.* **11** 498
[13] Lin J P, Chua T C and Lin S Y 2001 *Appl. Radiat. Isot.* **55** 397
[14] Uwamino Y, Nakamura T and Ohkubo T 1986 *Méd. Phys.* **13** 374
[15] USNCR 1977 *Calculation of Annual Doses to Man from Routine Releases of Reactor Effluents for the Purpose of Evaluating Compliance With 10 CFR Part 50, Appendix I* (USA)
[16] Mattar J A 1989 *Care Med.* **17** 846
[17] Abuhoza A A A 2007 *Comparison Study of Reflected and Transmitted Thermal Neutron Flux in Water and Other Moderators* (Saudi Arabia: King Saudi University Riyadh)
[18] ICRP 1995 *Basic Anatomical and physiological Data for use in radiological protection* (Vienna: International Commission of Radiological Protection)
[19] IAEA 1998 *Compilation of Anatomical, Physiological and Metabolic Characteristics for a Reference Asian Man* **2** (Vienna: International Atomic Energy Agency)

IOP Conf. Series: Journal of Physics: Conf. Series **997** (2018) 012028 doi:10.1088/1742-6596/997/1/012028

Horizontal Displacement Vector Analysis in Ujong Muloh GPS Station (UMLH) Sumatra Island on March 27 – April 25, 2012

S S Pamungkas[1], S Koesuma[1, a] and Budi Legowo[1]

[1]Department of Physics, Faculty of Mathematics and Natural Sciences, Sebelas Maret University, Surakarta, Indonesia

[a]E-mail: sorja@uns.ac.id

Abstract. Sumatra Island is an area that has high tectonic activities. This is because the island of Sumatra is located in two major plates of the world, the Indo-Australian plate and the Eurasia plate. The subduction zone causes Sumatra to deform from time to time. The deformation of Sumatra Island can be observed by continuous recording coordinates using the GPS Station. Continous-GPS (C-GPS) in Sumatra Island is named Sumatran GPS Array (SuGAr), one of them named UMLH. The UMLH GPS station used to observe the displacement in the Aceh City of Sumatra Island, is located in Ujung Muloh. The changes of GPS coordinate recording data can represent the deformation pattern that occurred in Sumatra. On April 11, 2012, according to USGS data, there had been an earthquake in the city of Aceh about 8.6 at coordinates of 2.433°N, 93.072°E. The purpose of this research is to analyze the horizontal displacement due to the occurrence of the earthquake. Data processing is carried out using software GAMIT/GLOBK. The magnitude of the displacement of Sumatra Island before the earthquake, during the earthquake, and after the quake on component X were respectively: 0.04 mm/day, 56.63 mm/day, and 8.28 mm/day; while on component Y were respectively: 0.03 mm/day, 23.78 mm/day, and 1.22 mm/day. The direction of displacement was 253.8° towards Southwest with the assumption that 0° was in the North.

1. Introduction

Sumatra Island is located near in subduction zone, which is a meeting zone of the Indo-Australian Plate and the Eurasian Plate. This is the cause of Sumatra has the potential to experience an earthquake [1]. A tectonic activity can be followed by a deformation. Deformation is a change in the shape of an object due to the transformation of the position of its constituent particles. Deformation can also be seen as a position from point to point. So that when tectonic activity occurs, an island has a certain direction displacement.

In Sumatra, there is a permanently installed GPS station to record coordinates continuously. The GPS station is named Sumatran GPS Array (SuGAR). SuGAr Station is operated by LIPI and Tectonic Observatory Caltech (USA) [2]. GPS observation data are stored in the Script Orbit and Permanent Array Center (SOPAC) which can be used for geodynamic measurement studies in Sumatra. Vector analysis is one method to observe deformation on an island [3]. Observation of the displacement is done using continuous GPS recording data [4-7].

Previous research on the displacement in North Sumatra had been done after the Aceh earthquake in 2004. One of them was at UMLH SuGAR Station (Ujong Muloh) on Day of Year (DOY) 19.2005-

19.2006. The horizontal motion showed the direction of the southwest with a velocity of 41.0 ± 0.1 cm/year prior to the outbreak of SAE 2004. The displacement rate then decreased to 18.0 ± 0.1 cm/year during the period 2006.19-2007.19 [8].

Based on USGS data dated April 11, 2012, there had been a tectonic earthquake with a magnitude of 8.6 in North Sumatra at 2.433°N, 93.072°E with a depth of 25 km. This research was conducted to observe the characteristics of plate movement horizontally. The GPS station used was UMLH in the Aceh area because the quake was felt in the city and the GPS station is the only one in Aceh. Observations were made by analyzing the magnitude of the displacement before and after the earthquake events.

2. Data and Methods

The research used continuous GPS observation data (SuGAr) which can be accessed through http://sopac.ucsd.edu/dataBrowser.shtml page. The range of data used was 30 DOY (87.2012-118.2012) in the event of an earthquake that laid in the mid-range of the DOY. The selected station was UMLH as deformation observation station and 7 reference GPS stations were BAKO, COCO, DGAR, IISC, PTAG, LHAZ. All data had been downloaded in RINEX data format. Then the data was processed using GAMIT/GLOBK 10.61 software because it had a better correction [9]. The design of deformation observation in this study is described in figure 1.

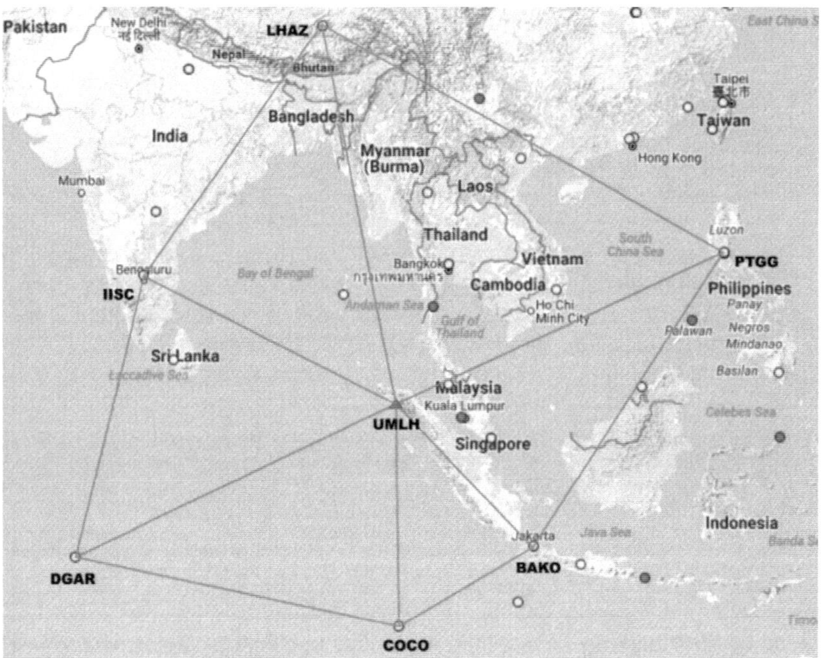

Figure 1. Design of deformation observation

The selection of these stations was based on the configuration of the connective nets. The quality of the data in the reference GPS station, as well as the availability of reference GPS data on the DOY corresponding to the amount of GPS deformation station data used [9]. RINEX data processing scheme is shown in the following figure 2.

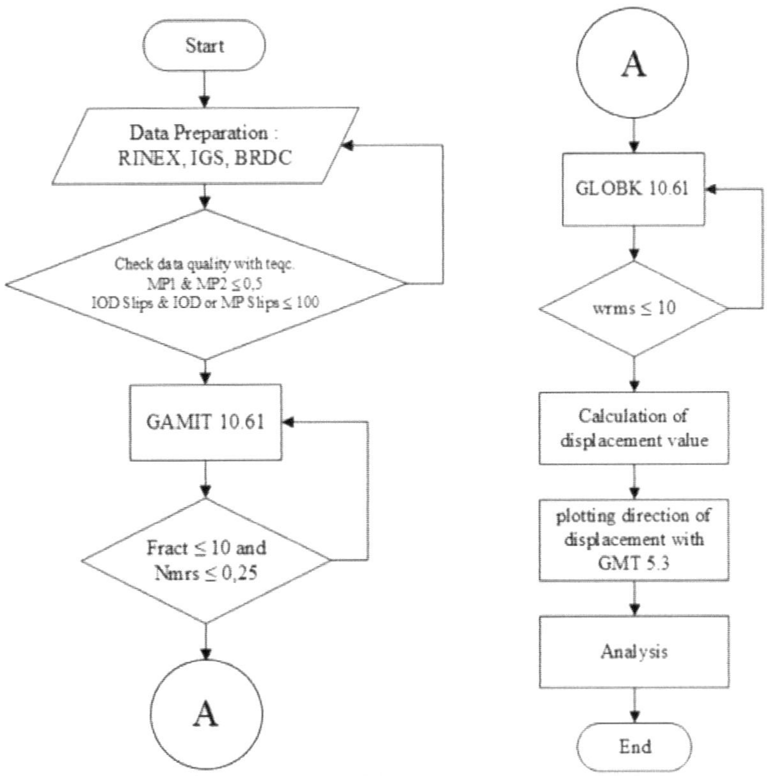

Figure 2. Processing scheme

3. Results and Discussion

3.1. Results of Data Quality Checks

The quality of data received for processing with GAMIT has the criteria of MP1 and MP2 values below 0.5 m [9]. The Rinex data used as a whole has good criteria as shown in figure 3 and figure 4. The smallest value of MP1 is shown in DOY 90, 102, and 116 with value 0.29 m; while the largest value found in 110 is 0.33 m. The average value of MP1 as a whole is 0.31 m. Day of year (DOY) is the order of the day in one year. Then the average value of MP2 is 0.25 m, with the smallest value is 0.23 m and the highest value is 0.27.

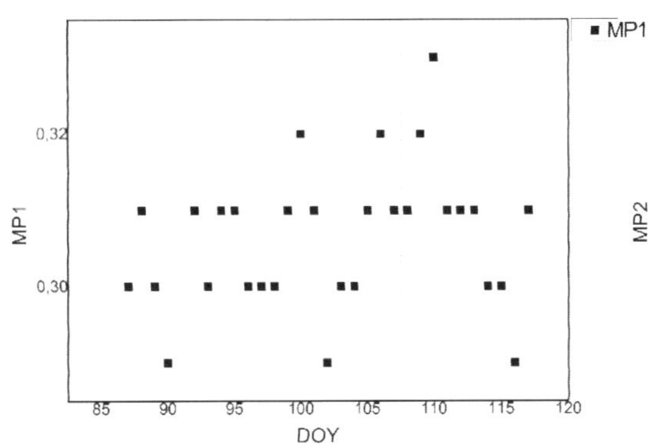

Figure 3. Graph of MP1 versus DOY

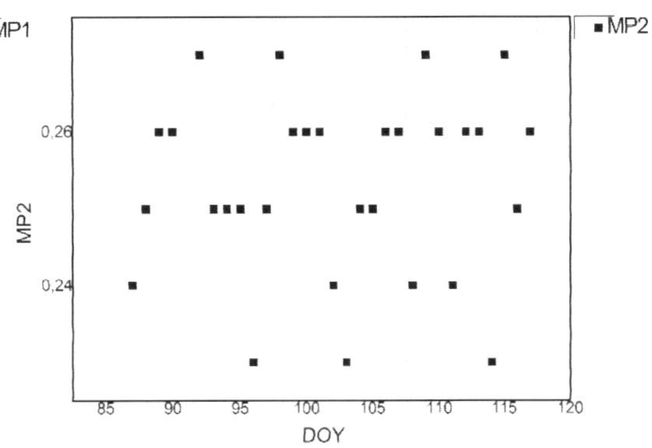

Figure 4. Graph of MP2 versus DOY

IOD Slips stands for Ionosphere Observable Delay (IOD), is a parameter showing the delay of GPS signal due to ionosphere influence [10]. The IOD value criteria that meet the processing standard is below 100 [10]. The IOD value of the slips has an average of 75.6 with the lowest value is 25 and the highest value is 97. The result of IOD or MP slips in Figure 4 has an average of 61.5 with the lowest value is 25 and the highest value is 99.

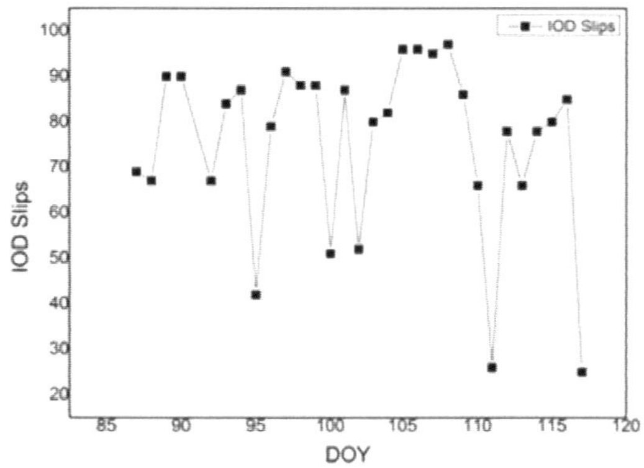

Figure 5. Graph of relationship IOD Slips versus DOY

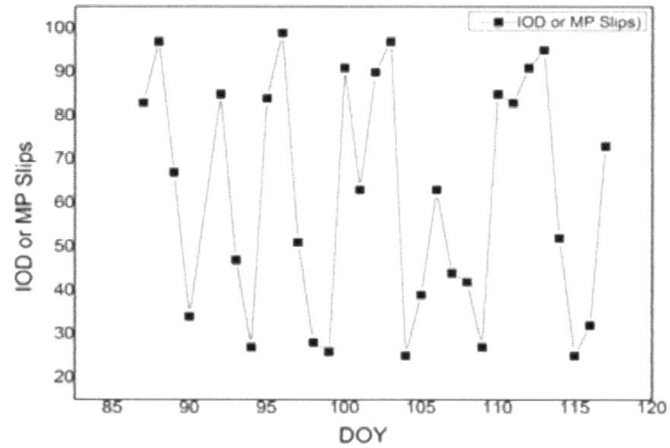

Figure 6. Graph of relationship IOD or MP slips versus DOY

3.2. Results of Gamit Processing

The fract value of the GAMIT processing result can be seen from the q-files generated in the respective DOY folder. Good fract value criterion is less than 10 [9].

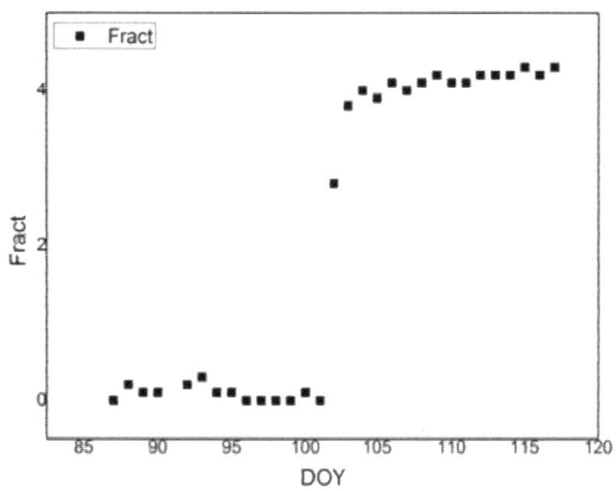

Figure 7. Graph of relationship longitude fract component versus DOY

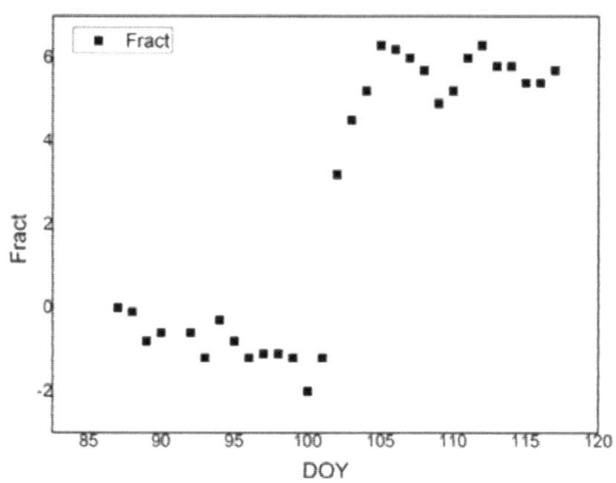

Figure 8. Graph of relationship longitude fract component versus DOY

The processing result shows that the fract value for the latitude component undergoes an average change after DOY 102 is 4.1; while the average value of fract before DOY 102 is 0.1. This significant change can be investigated further through GLOBK processing results. Because in general this fract value should not have a significant change in Figure 7 and Figure 8. This may indicate that there is an effect on the UMLH station so that the fract value undergoes significant changes in DOY 102. The fract value of the longitude component has a good quality for processing in which the lowest fract value is -2 at DOY 100 and the highest fract value is 6.3 at DOY 112. Overall fract value of the GAMIT processing result has fulfilled criterion with maximal value 10.

After looking at the quality of GAMIT processing using fract value, the quality test result of GAMIT process is done by analyzing the value of postfit nmrs. This value can be seen in q-file of each DOY after final process using GAMIT [9]. The result of this value is plotted in the relationship graph of postfit nmrs to the DOY to see the change in the value of postfit nmrs. Postfit nmrs results are divided into 2 solutions namely biases free and biases fixed.

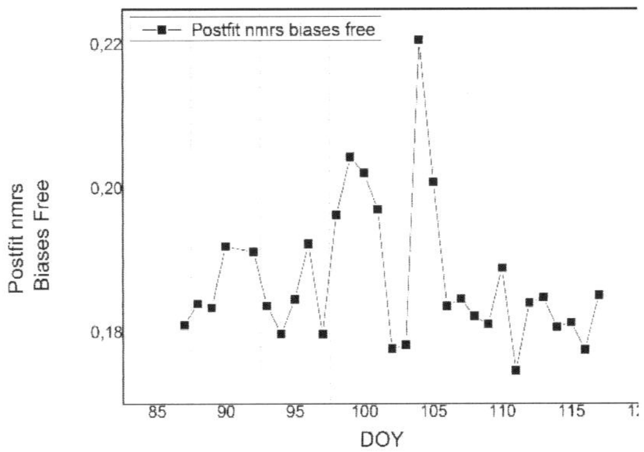

Figure 9. Graph of relationship biases free postfit nmrs value versus DOY

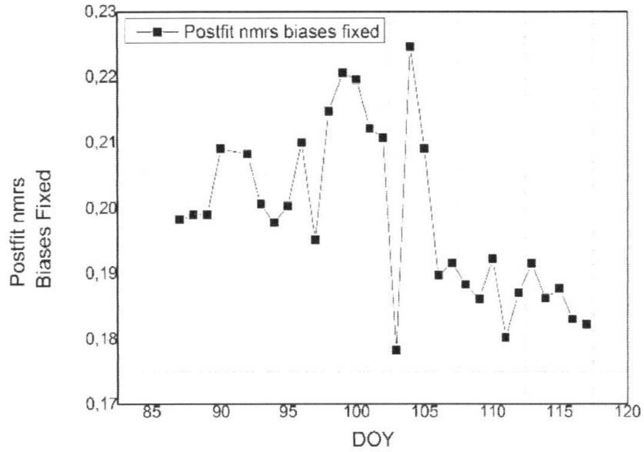

Figure 10. Graph of relationship biases fixed postfit nmrs value versus DOY

The postfit nmrs biases free value of 30 DOY fluctuated with the highest value of the DOY 104 of 0.22 m and the lowest value at the DOY 111 of 0.17 m. While for postfit nmrs biases fixed value also shows a pattern of postfit nmrs fluctuate value. The lowest postfit value in the fixed biases solution is on the DOY 103 of 0.178 m and the highest postfit value of nmrs at the DOY 104 of 0.224 m. Overall postfit nmrs value meets criteria with values below 0.5 m. This may indicate that in observation there is no effect of cycle slips that have not been removed.

3.3. *Results of Processing GLOBK*

The time series plot is the result of the evaluation process of GLOBK used to see the existence of outlier data in the range of data used as the observation [9]. Outlier data is data whose value is the most different from other values. The value may be caused by incomplete data recording, in which one of them can be caused by an instrument error.

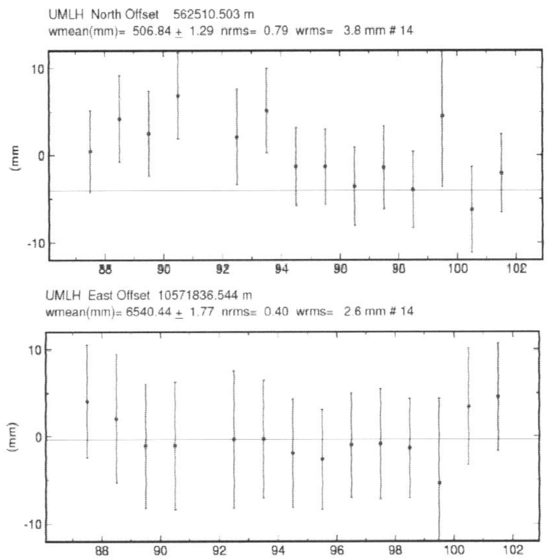

Figure 11. Plot time series before the quake

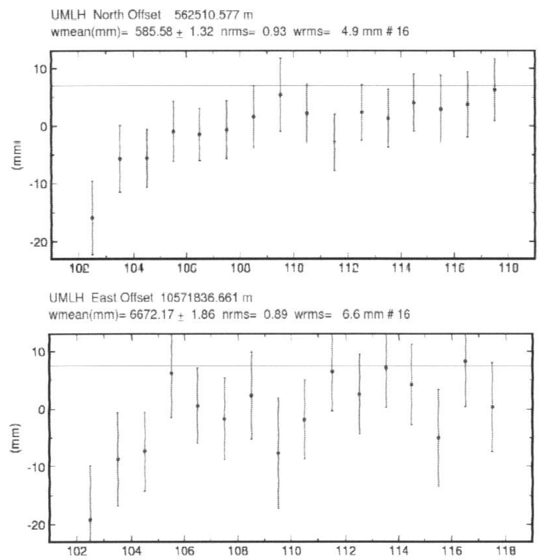

Figure 12. Plot time series after the quake

Based on figure 11 and figure 12 there is no outlier data, where the existence of outlier data can be shown with high wrms value. The maximum limit of wrms that can be tolerated for GLOBK processing is below 100.

3.4. Displacement of Station UMLH

Figure 13 and figure 14 is a plot of coordinate points from DOY 85 to 120 for X and Y components, respectively. It is clearly seen that in DOY 102 for both graphs shows a significant displacement. This is indicated by the earthquake on April 11, 2012, because DOY 102 is the same as April 11, 2012.

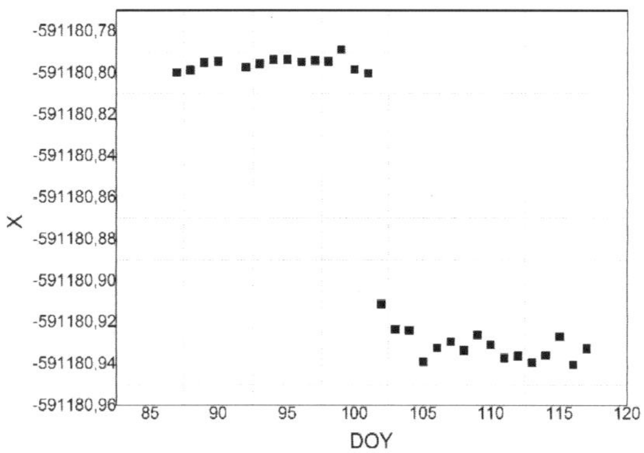

Figure 13. Displacement on component X

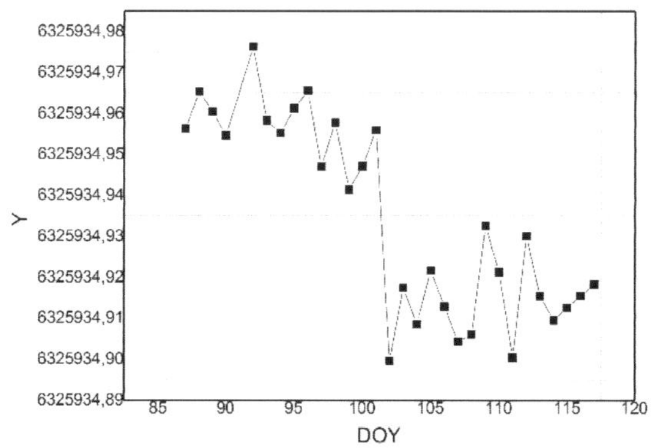

Figure 14. Displacement on component Y

The analysis of the displacement before and after the tectonic earthquake on April 11, 2012, based on Figure 13 and Figure 14 is obtained in X and Y components. The magnitude of the displacement values is summarized in table 1.

Table 1. Magnitude of displacement UMLH GPS Station

Event	Displacement Average	
	X (mm/day)	Y (mm/day)
Before	0.04	0.03
Earthquake	56.63	23.78
After	8.28	1.22

Overall at DOY 87.2012 - 117.2012, the shift occurred at UMLH GPS Station leads to the Southwest with a speed at component X is 4.58 mm/day and on the Y component is 1.31 mm/day. The displacement vector at the station is illustrated in Figure 15. Azimuth shifts that occured are calculated using the equation:

$$displacement\ of\ azimuth = arc.tan\ \frac{\Delta N}{\Delta E} \qquad (1)$$

Where the shift that occured is 253.8° with a reference angle is 0° in North.

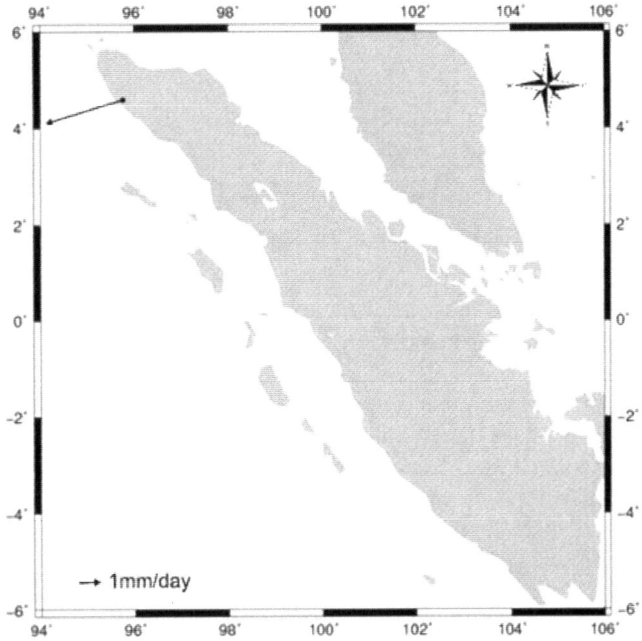

Figure 15. The direction of displacement UMLH GPS station

4. Conclusion

Analysis of the displacement caused by the earthquake on April 11, 2012, for 30 DOY with the observation of UMLH GPS station showed the direction of the displacement to the Southwest with Azimuth 253.8°. The magnitude of displacement before and after the earthquake as seen in table 1, shows that the displacement before the earthquake has a smaller value than after an earthquake. This shows that the tectonic movement forces in the magnitude of changes in the displacement of the island's plate.

References

[1] Bock Y, Prawirodirdjo L, Genrich J F, Stevens C W, Mccaffrey R, Subarya C, Puntodewo S S O and Calais E 2003 *J. Geophys. Res.* **108** 2367
[2] Kreemer C, Blewitt G, Hammond W C and Plag H P 2006 *Earth, planets* and *space* **58** 141
[3] Abidin H Z 2007 *Penentuan Posisi dengan GPS dan Aplikasinya* (Jakarta: PT. Pradnya Paramita)
[4] Satirapod C, Trisirisatayawong I, Fleitout L, Garaud G D and Simons W J F 2013 *Adv.Space Res.* **51** 1565
[5] Hanifa N R, Sagiya T, Kimata F, Efendi J, Abidin Z A and Meilano I 2014 *Earth and Planet. Sci. Lett.* **401** 159
[6] Ardika M, Meilano I and Gunawan E 2015 *J. Asian Earth Sci.* **8** 127
[7] Simons W J F, Socquet A, Vigny C, Ambrosius B A C, Haji Abu S, Promthong C, Subarya C, Sarsito D A, Matheussen S, Morgan P, Spakman W 2007 *J. Geophys. Res.* **112** 1978
[8] Gunawan E, Sagiya T, Ito T, Kimata F, Tabei T, Ohta Y and Sugiyanto D 2014 *J. Asian Earth Sci.* **88** 218
[9] Hearing T A, King R W, Floyd M A and McClusky S C 2015 *GAMIT Reference Manual* (Departement of Earth, Atmospheric, and Planetary Sciences: MIT)
[10] Estey L and Wier S 2014 *Teqc Tutorial: Basic of Teqc Use and teqc products* (UNAVCO: Boulder, Colorado USA)

Development of multimedia learning based inquiry on vibration and wave material

H Madeali[1] and B K Prahani[2,a]

[1]Department of Physics Education, Faculty of Teacher Training and Education
Haluoleo University, Indonesia
Jl. H.E.A. Mokodompit, Kendari, 93231, Indonesia
[2]Department of Physics, Faculty of Mathematics and Natural Sciences
Universitas Negeri Surabaya, Indonesia
Jl. Ketintang, Surabaya 60231, Indonesia

E-mail: [a]binarprahani@gmail.com

Abstract. This study aims to develop multimedia learning based inquiry that is interesting, easy to understand by students and streamline the time of teachers in bringing the teaching materials as well as feasible to be used in learning the physics subject matter of vibration and wave. This research is a Research and Development research with reference to ADDIE model that is Analysis, Design, Development, Implementation, and Evaluation. Multimedia based learning inquiry is packaged in hypertext form using Adobe Flash CS6 Software. The inquiry aspect is constructed by showing the animation of the concepts that the student wants to achieve and then followed by questions that will ask the students what is observable. Multimedia learning based inquiry is then validated by 2 learning experts, 3 material experts and 3 media experts and tested on 3 junior high school teachers and 23 students of state junior high school 5 of Kendari. The results of the study include: (1) Validation results by learning experts, material experts and media experts in valid categories; (2) The results of trials by teachers and students fall into the practical category. These results prove that the multimedia learning based inquiry on vibration and waves materials that have been developed feasible use in physics learning by students of junior high school class VIII.

1. Introduction

The rapid development of technology has a role that can be utilized for various purposes including for education and learning. This is indicated by the number of choices or ways that can be used in learning. One of them is the utilization of multimedia in learning [1-3]. The development of multimedia technology unlocks great potential in changing ways of learning and how to obtain information [4-6]. The development of multimedia has opened the opportunity for teachers to develop the learning system in order to get optimal results. Similarly for students, with multimedia learning will be easier to determine in what way and how to absorb the information delivered quickly and efficiently. The sources of information and knowledge that students acquire are no longer only fixated on books but more widely and diverse [4-7].

One of the classifications of multimedia learning is hypermedia [1]. Hypermedia is a medium that has the composition of materials that are not sequential. The term hypertext describes non-consecutive documents, consisting of text, audio and visual information stored on a computer. Hypertext is viewed as nonlinearly organized information and accessed in textual form including diagrams, tables and drawings [2]. Unlike textbooks, hypertext can be presented by combining it with other media such as animation, sound, images, and graphics. Hypertext makes it easier for us to find the materials we want because the subject matter is neatly sorted by using the help of links [3]. Links have a sense of the relationship of each node or part that becomes the infrastructure of each node or part of it to connect to text, graphics, audio, video and other animations. Hypertext can be used as one of the tools that allow learners to build their own knowledge. But there are also deficiencies in the use of hypertext, hypertext systems can be made in various ways, but there is enough reason to believe that most of these ways will not produce good learning results. This is because the hypertext can just make the reader confused. Readers have gone astray in links with so many branches. In addition, since nothing will deter the reader from linking the two nodes of information and continuing with other vertices, there is the possibility that navigation will end up with some incoherent information. The creation of multimedia using hypertext needs to be neatly compiled, so it will not cause the multimedia users to get lost in the [1-10] information link.

This finding is of particular concern to the improvement of multimedia implementation in physics learning. Implementation of multimedia can be applied in physics learning [7-10]. One of them is by inquiry learning [11-16]. The study of inquiry has been shown to improve the skills of the process of science, conceptualization, creativity, learning outcomes, problem solving, critical thinking, and student creativity [17-30]. The inquiry model is defined as one of the inquiry learning model that presents the problem, the questions and the material or the supporting material is determined by the teacher. Teachers given problem encourage students to investigate and determine the answers [32-34]. The inquiry model offers an integrated investigation planned and guided by librarians and teachers, enabling students to gain a better understanding of the subject matter of curriculum content and concept information. Develop the skills and abilities necessary for work and everyday life in the 21st century [21,27,31]. The social constructivism theory developed by Vygotsky has three major implications in the impression of the inquiry model: 1) through the social interaction of the students become conscious of their basic mental function and able to use them for growth; (2) the teacher provides the tasks within the range of students (zone of proximal development); (3) Providing learning with scaffolding [35-36]. According to Lev Vygotsky, language and thought will become increasingly independent in the development of the child [37]. Vygotsky states that children develop a more systematic, rational concept that is the result of dialogue with a skilled counsellor [38-40]. Language and thought will become increasingly independent in the development of the child when the multi-representational ability of the student develops optimally [37].

Inquiry learning can encourage learning toward the student centred, where the teacher's position shifts from the traditional instructor to the mentor [17-30]. In addition, student-cantered learning can lead to active, creative, interactive, and fun learning [32-34]. Multimedia utilization is expected to help students understand the material learned because by using multimedia students can obtain information that is difficult to be presented in the student's learning room even more so in the process of physics learning which most of the material is abstract [1-10]. Supported by the use of hypertext that is part of the multimedia element itself can make students more interested in learning because the use of hypertext is different from ordinary text books, hypertext can store text, animation in it that has been linked by the link that makes it more interesting [1-6] . In addition, the addition of aspects of inquiry makes students into learning centres that no longer make teachers as learning centres, so students are required to be able to find the concept of what he learned [1-6]. This makes the student more understand the material he learned because the concept of the material he found himself and make the science he studied longer stored in him. Researchers will have developed an inquiry-based hypertext multimedia to enhance the physics concept of vibration and wave matter. Stages of development research refer to the ADDIE development model by Dick and Carey, including Analysis, Design, Development, Implementation, and Evaluation [41]. The formulation of this research problem as follows.

a. How is the validity of multimedia learning based inquiry on vibration and wave materials developed?
b. How is the practicality of multimedia learning based inquiry on vibration and wave materials developed?

2. Methodology of Research

2.1. Time and Place of Study Implementation

The research was carried out in February 2016 to finish in Laboratory of Department of Physics Education Faculty of Teacher Training and Education University of Halu Oleo Kendari and tested in state junior high school of 5 Kendari.

2.2 Types of Research

This type of research is research development (Research & Development) to develop multimedia learning based inquiry on vibration and wave materials.

2.3 Research Design

This research uses Research and Development. The research stages are based on the ADDIE development model developed by Dick and Carey (1978), including Analysis, Design, Development, Implementation, and Evaluation.

2.4 Research Instruments

In the validation by experts consists of 3, which includes 2 learning experts, 3 material experts and 3 media experts While in the trial phase consists of 2 that includes teacher responses to the program and student responses to the program. The validation sheet given to the experts is to determine the level of validity and learning multimedia developed. While the questionnaire given to teachers and students aims to determine the practicality level of developed multimedia learning. The validation sheet is a questionnaire.

2.5 Data Analysis Techniques

Determination of the validity level and practicality of multimedia learning based inquiry on vibration and wave materials developed using assessment criteria by [42-44].

3. Result and Discussions of Research

3.1 Validity of Multimedia Learning Based Inquiry on Vibration and Waves Materials

Multimedia learning based inquiry on vibration and wave materials is packaged in hypertext form using Adobe Flash CS6 Software. The inquiry aspect is constructed by showing the animation of the concepts that the student wants to achieve and then followed by questions that will ask students what is observable from the animation. The validity of multimedia learning based inquiry on the vibration and wave materials that has been developed is then validated by 2 learning experts. The validation results are presented in Table 1. The validity of multimedia learning based inquiry on vibration and wave materials with an average of 86.10 with valid and reliable categories, so that it can be used in physics learning on vibration and wave materials.

Multimedia learning based inquiry on vibration and wave materials that has been developed is then validated by 3 content experts. The validation results are presented in Table 2. The validity of instructional media with the average percentage of 83.30 with valid and reliable categories, so that it can be used in physics learning on vibration and wave materials.

IOP Conf. Series: Journal of Physics: Conf. Series **997** (2018) 012029 doi:10.1088/1742-6596/997/1/012029

Table 1. Validity of multimedia learning based inquiry on vibration and wave materials by 2 learning experts.

Validity of multimedia learning based inquiry on vibration and wave materials by 2 learning experts	Validity		Reliability
1. Interesting multimedia is used	100	Valid	Reliable
2. Ease of multimedia users	100	Valid	Reliable
3. Material compatibility with student characteristics	75.00	Valid	Reliable
4. Clarity of feedback / response	75.00	Valid	Reliable
5. Use of Hypertext words	62.50	Valid	Reliable
6. The inquiry is seen in the presentation of matter	100	Valid	Reliable
7. The precision of the presentation sequence	87.50	Valid	Reliable
8. Ease of understanding the content (content) of the lesson	87.50	Valid	Reliable
9. Ease of understanding of sentence there is text / writing	75.00	Valid	Reliable
10. Clarity of learning instructions	100	Valid	Reliable
11. Material compatibility with learning objectives	87.50	Valid	Reliable
12. Adequacy of the given example	75.00	Valid	Reliable
13. Sufficiency of material description in clarity of concept of vibration and wave	75.00	Valid	Reliable
14. Clarity of target use of multimedia learning	100	Valid	Reliable
15. Conformity of material sequences contained in the media	75.00	Valid	Reliable
16. Clarity of learning indicators	100	Valid	Reliable
17. Conformity of the material with the basic competencies used	87.50	Valid	Reliable

Table 2. Validity of multimedia learning based inquiry on vibration and wave materials by 3 content experts.

Validity of multimedia learning based inquiry on vibration and wave materials by 3 content experts	Validity		Reliability
1. Conformity test with answer key	91.70	Valid	Reliable
2. Compliance test with basic competencies and indicators	75.00	Valid	Reliable
3. Response or answer key	91.70	Valid	Reliable
4. Exact precision	83.30	Valid	Reliable
5. Into the material suitability of the image	83.30	Valid	Reliable
6. Use of hypertext words	83.30	Valid	Reliable
7. Use of language	91.70	Valid	Reliable
8. The inquiry aspect is seen in the presentation of the material	66.70	Valid	Reliable
9. The truth of material substance	83.30	Valid	Reliable

Multimedia learning based inquiry on vibration and wave materials has been developed and then validated by 3 media experts. The validation results are presented in Table 3. The validity of instructional media with an average of 85.40 with valid and reliable categories, so that it can be used in physics learning on vibration and wave materials.

Table 3. Validity of multimedia learning based inquiry on vibration and wave materials by 3 media experts.

Validity of multimedia learning based inquiry on vibration and wave materials by 3 media experts	Validity		Reliability
1. Efficiency of text	83.30	Valid	Reliable
2. Efficiency of screen usage	83.30	Valid	Reliable
3. Clarity of instructions for use	83.30	Valid	Reliable
4. Consistency of the button	91.70	Valid	Reliable
5. Navigation	91.70	Valid	Reliable
6. Fascination and motivation	91.70	Valid	Reliable
7. Quality of interaction	83.30	Valid	Reliable
8. Ease of language understanding	83.30	Valid	Reliable
9. Quality of interaction	83.30	Valid	Reliable
10. Ease of understanding language	91.70	Valid	Reliable
11. Quality of animation	83.30	Valid	Reliable
12. Audio quality	83.30	Valid	Reliable
13. Quality of photos, pictures, graphics	83.30	Valid	Reliable
14. Colour composition	91.70	Valid	Reliable
15. Type and size of text	83.30	Valid	Reliable
16. Use of hypertext words	91.70	Valid	Reliable
17. Use of the button	83.30	Valid	Reliable
18. Menu view	75.00	Valid	Reliable

The creating of multimedia learning based inquiry on vibration and wave materials has passed the development stage using the ADDIE model. After passing through these stages a multimedia learning is produced. Multimedia learning is then validated by 2 learning experts, 3 material experts, and 3 media experts. Table 1, Table 2, and Table 3 show the results of validation by 2 experts obtained the final value of 86.10, then the validation results by 3 material experts obtained the final value 83.30 and the results of validation by the media expert obtained the final value of 85.40. The final grades obtained from the learning expert, the material expert and the media expert enter the valid percentage range based on the qualification of the validity rate.

Based on the data in Table 1, Table 2, and Table 3 indicates that multimedia learning based inquiry on vibration and wave materials has been declared valid and reliable by learning experts, materials experts, and media experts. Multimedia learning based inquiry on vibration and wave materials can be used in physics learning. The results of this study in accordance with the findings of previous research that the implementation of multimedia can be applied in the learning of physics [7-10], Hypertext can be used as one of the tools that give opportunity to the learners to build their own knowledge [1-3]. The integration of the inquiry model also received a positive assessment from the validator. This is supported by research [17-30] that inquiry learning can encourage learning toward the student centred, where the teacher position shifts from traditional instructor to mentor [17-30]. The development of multimedia learning based inquiry on vibration and wave materials that has been valid and can be used in physics learning is supported by social constructivism theory Vygotsky namely: 1) through social interaction students become aware of their basic mental function and are able to use it for growth; (2) the teacher provides the tasks within the range of students (zone of proximal development); (3) Providing learning with scaffolding [35-36].

3.2 The practicality of multimedia learning based inquiry on vibration and wave materials

The limited trial stage was conducted by giving questionnaires to 3 teachers of Junior High School and 23 students of State Junior High School of 5 Kendari. Limited trials are conducted in order to know the practicality level of multimedia learning in helping the learning process of students. Table 4 shows that the practicality of multimedia learning based inquiry on vibration and wave material by 3 teachers is 94.20 very good and reliable categories.

Table 4. Practicality of multimedia learning based inquiry on vibration and wave materials by 3 teachers.

Practicality of multimedia learning based inquiry on vibration and wave materials by 3 teachers	Practicality		Reliability
1. Interesting multimedia is used	100	Very good	Reliable
2. Ease of multimedia users	100	Very good	Reliable
3. Material compatibility with student characteristic	91.70	Very good	Reliable
4. Clarity of feedback / reason	83.30	Very good	Reliable
5. Use of Hypertext words	91.70	Very good	Reliable
6. Adequacy of the given example	91.70	Very good	Reliable
7. The precision of the presentation sequence	100	Very good	Reliable
8. Ease of use of the content (content) of the lesson	100	Very good	Reliable
9. Ease of understanding of sentences in the text	100	Very good	Reliable
10. Clarity of learning instructions	91.70	Very good	Reliable
11. The inquiry aspect is seen in the presentation of the material	100	Very good	Reliable
12. Conformity of material with purpose / indicator	91.70	Very good	Reliable
13. Achievement of basic indicators and competencies	83.30	Very good	Reliable

Table 5 shows that the practicality of multimedia learning based inquiry on vibration and wave materials by 23 students is 83.90 categories is very good and reliable.

Table 5. Practicality of multimedia learning based inquiry on vibration and wave materials by 23 students.

Practicality of multimedia learning based inquiry on vibration and wave materials by 23 students	Practicality		Reliability
1. Adequacy questions	81.70	Very good	Reliable
2. Quality of interaction	81.70	Very good	Reliable
3. Ease of understanding language	80.00	Very good	Reliable
4. Use of hypertext	80.90	Very good	Reliable
5. Use of images	80.90	Very good	Reliable
6. Matching of Colour	87.80	Very good	Reliable
7. Key Clarity	93.00	Very good	Reliable
8. Display	93.00	Very good	Reliable

Multimedia learning based inquiry on vibration and wave materials that has been valid and reliable is used in the implementation of physics learning. Multimedia learning based inquiry on vibration and wave materials was tested to 3 junior high school teachers and 23 students of State Junior High School of 5 Kendari. The final score obtained from the teacher's test is 94.20 (very good and reliable) and the test by the students is 83.90 (very good and reliable). The findings of this study are supported by [17-30] that the study of mercury has been shown to improve the skills of the science process, conceptual satisfaction, creativity, learning outcomes, problem solving, critical thinking, and student creativity. Student-centred inquiry can lead to active, creative and interactive learning, and fun 32-34. Multimedia utilization is expected to help students understand the material learned because by using multimedia students can obtain information that is difficult to be presented in the student's learning room even more so in the process of physics learning which most of the material is abstract [1-10]. Supported by the use of hypertext that is part of the multimedia element itself can make students more interested in learning because the use of hypertext is different from ordinary text books, hypertext can store text, animation in it that has been linked by the link that makes it more interesting [1-6]. In addition, the addition of aspects of inquiry makes students into learning centres that no longer make teachers as learning centres, so students are required to be able to find the concept of what he learned [1-6]. The results of this study indicated that multimedia based learning inquiry on vibration and wave materials was feasible use in physics learning at junior high school level because it has been declared valid and practical.

4. Conclusion

Multimedia based learning inquiry is packaged in hypertext form using Adobe Flash CS6 Software. The inquiry aspect is constructed by showing the animation of the concepts that the student wants to achieve and then followed by questions that will ask the students what is observable. Multimedia learning based inquiry is then validated by 2 learning experts, 3 material experts and 3 media experts and tested on 3 junior high school teachers and 23 students of state junior high school 5 of Kendari. The results of the study include: (1) Validation results by learning experts, material experts and media experts in valid categories; (2) The results of trials by teachers and students fall into the practical category. These results prove that the multimedia learning based inquiry on vibration and waves materials that have been developed feasible use in physics learning by students of junior high school class VIII.

5. Acknowledgements

The author's gratitude goes to the Ministry of Research, Technology and Higher Education who has funded the Research. Likewise, the author's gratitude goes to the Haluleo University and SMPN 5 Kendari that has provided research opportunities.

References

[1] Ismiasri T 2011 *Cakrawala Pend.* **13** 5
[2] Aliwar 2013 *Al-Izzah* **8** 2
[3] Putra I B G A 2013 *Kumpulan Artikel Mahasiswa Pend. Teknik Informatika* **2** 5
[4] Mayer R E 2001 *Multimedia Learning* (New York: Cambridge University Press)
[5] Mayer R E 2003 *Learn. Instr.* **13** 125
[6] Mayer R E 2005 *The Cambridge Handbook of Multimedia Learning* (Cambridge: Cambridge University Press)
[7] Schnotz W and Lowe R 2010 *Learn. Instr.* **13** 117
[8] Schnotz W and Lowe R 2003 *Learn. Instr.* **13** 116
[9] Kustijono R *J. Penelit. Fis. Apl.* **2** 10
[10] Habibbullh M, Jatmiko B and Widodo W 2017 *J. Penelit. Fis. Apl.* **7** 27
[11] Ariesta R and Supartono 2011 *J. Pend. Fis. Indonesia* **1** 62
[12] Ashiq H A, Muhammad A and Azra S 2011 *Int. J. Humanities Soc. Sci. 1* 269
[13] Astuti Y and Setiawan B 2013 *J. Pend. IPA Indonesia* **2** 88
[14] Ferlina I, Suyatna and Maharta *J. Pend. Fis.* **1** 175
[15] Wenning C J 2011 *J. Phys. Teach. Educ. Online* **6** 7
[16] Prahani B K, Limatahu I, Soegimin W W, Yuanita L and Nur M 2016 *Int. J. Educ. Res.* **4** 231
[17] Ambarsari W, Santosa S and Maridi 2012 *J. Pend. Biologi Universitas Sebelas Maret* **5** 81
[18] Berg C A R, Bergendahl V C B and Lundberg B K S 2003 *Int. J. Sci. Educ.* **25** 112
[19] Crawford B A 2000 *J. Res. Sci. Teach.* **37** 916
[20] Crockett M D 2002 *Teach. Teach. Educ.* **18** 609
[21] Gerald F L 2011 *Scan.* **30** 26
[22] Khan M S, Hussain A R, Majoka M I and Ramzan M 2011 *Int. J. Acad. Res.* **3** 955
[23] Kirschner P A, Sweller J and Clark R E 2006 *Educ. Psychol.* **41** 75
[24] Lee H, Linn M, Varma K and Liu O 2010 *J. Res. Sci. Teach.* **47** 71

[25] Luft J 2001 *Int. J. Sci. Educ.* **23** 517
[26] Michal Z 2007 *Res. Sci. Educ.* **3** 59
[27] Opara J A and Oguzor N S 2011 *Current Res. J. Soc. Sci.* **3** 188
[28] Pandey G K Nanda and Ranjan V 2011 *J. Innovative Res. Educ.* **1** 7
[29] Sadeh I and Zion M 2009 *J. Res. Sci. Teach.* **46** 1137
[30] Wilson C D, Taylor J A, Kowalski S M and Carlson J 2010 *J. Res. Sci. Teach.* **47** 276
[31] Benckert and Pettersson 2008 *Eur. J. Math. Sci. Techno. Educ.* **2** 121
[32] Abdurrahman, Liliasari, Rusli A and Bruce W 2011 *Cakrawala Pendidikan* **30** 30
[33] Bao L, Fang K, Cai T, Wang J, Yang L, Cui L, Han J, Ding L and Luo Y 2009 *Am. J. Phys.* **77** 1118
[34] Mercer N, Dawes L, Wegerif R and Sams C 2004 *British Educ.l Res. J.* **30** 359
[35] Arends R I 2012 *Learning to Teach* (New York: McGraw-Hill Companies)
[36] Slavin E R 2011 *Educational Psychology Theory and Practice* (USA: Pearson)
[37] Moreno R 2010 *Educational Psycology* (New York: John Wiley & Sons Inc)
[38] Fu Y, Han C T and Hui L W 2013 *Australasian J. of Educ. Techno.* **29** 89
[39] Zydney J M 2010 *Computers and Education* **54** 150
[40] Yu F Y 2009 *Computers in Human Behavior* **25** 1129
[41] Walter D and Carey L 1978 *The Systematic Design of Instruction* (USA: Scott, Foresman and Company)
[42] Arikunto S 2010 *Prosedur Penelitian Suatu Pendekatan Praktik* (Yogyakarta: Rineka Cipta)
[43] Fraenkel J R, Wallen N E and Hyun H H 2012 *How to Design and Evaluate Research in Education* (New York: McGraw-Hill)
[44] Plomp T and Nieveen N 2013 *Educational Design Research* (Enschede: SLO)

IOP Conf. Series: Journal of Physics: Conf. Series **997** (2018) 012030 doi:10.1088/1742-6596/997/1/012030

Students' understanding of forces: Force diagrams on horizontal and inclined plane

J Sirait[1], Hamdani[1] and S Mursyid[1]

[1]Department of Physics Education, Tanjungpura University, Pontianak, Indonesia

E-mail: judyantosirait@gmail.com

Abstract. This study aims to analyse students' difficulties in understanding force diagrams on horizontal surfaces and inclined planes. Physics education students (pre-service physics teachers) of Tanjungpura University, who had completed a Basic Physics course, took a Force concept test which has six questions covering three concepts: an object at rest, an object moving at constant speed, and an object moving at constant acceleration both on a horizontal surface and on an inclined plane. The test is in a multiple-choice format. It examines the ability of students to select appropriate force diagrams depending on the context. The results show that 44% of students have difficulties in solving the test (these students only could solve one or two items out of six items). About 50% of students faced difficulties finding the correct diagram of an object when it has constant speed and acceleration in both contexts. In general, students could only correctly identify 48% of the force diagrams on the test. The most difficult task for the students in terms was identifying the force diagram representing forces exerted on an object on in an inclined plane.

1. Introduction

Force is one of the most fundamental concepts in physics. It is used to describe and explain a variety of phenomena in mechanics and electromagnetism. Therefore, many physics education researchers conducted studies concerning conceptual understanding of forces involving students from high school to university students [1-3]. The findings show that many students still have alternative conceptions of forces and some students cannot not clearly explain the meaning of the term force. These difficulties might affect students' understanding of how to apply the force concept in other contexts.

One of the tools to help students understand force concepts is a force diagram (otherwise called a free body diagram-FBDs) – a representation that depicts an object of interest and the forces exerted on this object by other objects [4]. One can delineate a problem with different representations such as sketch or picture and diagrams. For example, a car is at rest on inclined plane. Students might visualize this verbal description using a sketch and draw a force diagram as shown in figure 1. In this diagram, the object of interest (the system, in this case, the car) is represented as a dot and forces exerted by other objects are shown with the vector arrows labelled with two subscripts. The first subscript signifies the system and the second-the object that exerts the force on the system.

A group of researchers in the United States carried out both quantitative and qualitative approaches to investigate whether university students who use force diagrams consistently during their course spontaneously utilize the diagrams on multiple choice tests to solve problems without getting credit for the solution [4]. They also compared performance of the students' who used forced diagrams to those who did not and the quality of students' representations. Rosengrant et al. found that students who learned free body diagrams tended to draw a diagram while solving problems on multiple-choice tests and obtained higher scores than students who did not. Also, they found that high achieving students harnessed FBD to understand the problem and evaluate the answer.

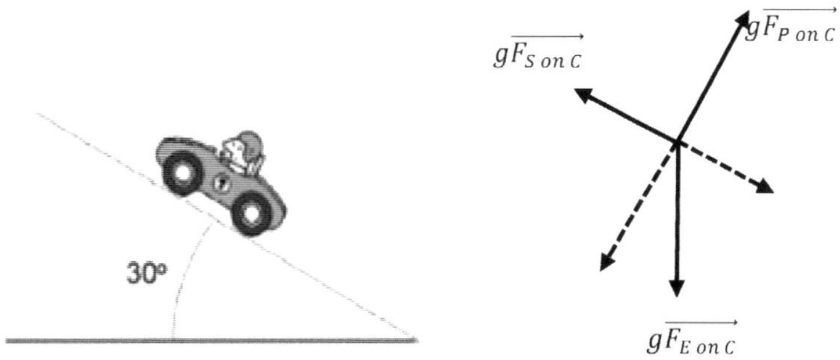

(a) (b)

Figure 1. Different types of representations: (a) sketch or picture and (b) force diagram.

In helping students to identify correctly forces and to draw correctly free body diagram, Savinainen et al. suggested to draw interaction diagram (ID) before constructing free body diagram (FBD) [5]. The purpose of ID is to identify all objects that interact with an object of interest and ID is depicted verbally (words). The researchers conducted a study of the impact of using ID by involving high school students in Finland. They divided students into those who were taught interaction diagrams and those who were not. Students were asked to solve eight questions which cover Newton's Law concepts by constructing both ID and FBD for each question. Their findings show that while there is a positive correlation between constructing ID and identifying forces, they still had students who correctly constructed interaction diagrams, but did not always correctly identify forces and did not always correctly draw FBD. This shows that using an interaction diagram is not sufficient to support students in drawing a free body diagram.

The use of free body diagrams (FBDs) does not always contribute positively on students' performance while solving physics problems. A study analysed the effect of encouraging students to draw force diagrams while solving problems [6]. The quantitative study involved university students in United States who were taking a calculus-based introductory physics course with two different classes: traditional, or regular class, and honours class. Students in the regular class learned with using force diagrams during the problem-solving exercise (common problem solving strategy) while the teaching strategy in the honours class focused on an explicit problem solving approach which did not follow common problem solving procedures, but in which students also practised constructing force diagrams. Students were asked to solve four open problems concerning Newton's laws. Within each class students received two distinct formats of the problems: some were asked to draw a diagram and some had no prompting for a diagram in the problems. Heckler found that students who were prompted to draw diagrams were less successful in determining the correct solution than those students who were not. Students who did not receive the prompt to draw a diagram employed intuitive solutions instead of formal strategies. However, students have different style in approaching physics problems; it depends on the type of questions. Students, for example, might have different strategies in solving Newton's laws problems in horizontal and inclined context. It means that intuitive knowledge is not always appropriate to all problems.

The goal of this study is to investigate students' difficulties in understanding force diagrams for horizontal surfaces and inclined planes which addresses three concepts: an object is at rest, at constant speed, and at constant acceleration.

2. Methodology

To acquire detailed information about students' ability to understand force concepts, a survey was applied by involving 212 pre-service physics teachers (freshmen, sophomore, junior students) who had taken a basic physics course at the Teacher Training and Education Faculty, Tanjungpura University. The participants in this study are dominantly by female students (153 females and 59 males). These students will become physics teachers in junior and senior high schools after completing four-year study in the department of physics education.

We adapted an existing test from Aviani et al. [7] as an instrument to measure students' understanding about forces in terms of force diagram. The translators who work at the science education department translated all items from English into Indonesian. We then chose items based on our needs for this study and designed a number of other items to complete the test. Afterwards, five faculty members of the physics department validated the items to measure (face validity) how well the test items addressed physics content. The test, which is multiple-choice, consists of six items which covers three situations: an object is at rest, at constant velocity and at constant acceleration. The purpose of this test is to examine students' ability to choose the appropriate force diagrams in different contexts-horizontal surface and inclined plane. An open question which covers inclined plane context, was also given to 60 first year students (freshmen students) to obtain more detail how students approach the question. Students were given 40 minutes to solve all questions. Before analysing statistically, students' works were firstly tabulated.

3. Result and Discussion

The percentage of students successfully answered more than or equal to three to six items ($\geq 50\%$) is 56%. Based on the analysis, students faced difficulties to solve inclined problems (whether an object is at rest or is moving). Students were only able to answer inclined plane questions less than 50%. Furthermore, in the context of horizontal surface students correctly solved the problems more than a half of the items. Students' answers for each item are displayed in Table 1.

Table 1. The description of items and the percentage of students' answers.

Context	Item	A	B	C	D	E
Horizontal Surface	1	12	2	83	2	1
	2	11	6	27	15	40
	3	52	16	10	13	9
Inclined Plane	4	4	40	18	7	31
	5	16	42	23	10	8
	6	55	15	10	7	13

Overall, the average percentage of problems that can be successfully solved by students is 48% (about three items). More than 80% of students successfully answered question #1: a block rests on a horizontal surface; students were asked to choose the correct free body diagrams. They chose correctly force diagrams. Students were able to identify all forces exerted on the block such as earth force on the block (weight force) and table force on the block (normal force). Only few students (about 20%) conceived that only a weight force is exerted on the object.

urthermore, more than 50% of students seemed to lack the ability to identify force diagrams, while an object moves with constant speed and constant acceleration (#2 and #3). For a block on constant speed, students did not completely identify all forces acting on the object. Besides, they couldn't successfully apply their conceptual knowledge (Newton's laws) to this situation to determine the resultant force (the net force is zero when an object moves at constant speed). Question #3 which shows that a block is accelerating on a horizontal surface, is presented in figure 1. The students' difficulty is they are not aware of friction force, weight force, and normal force.

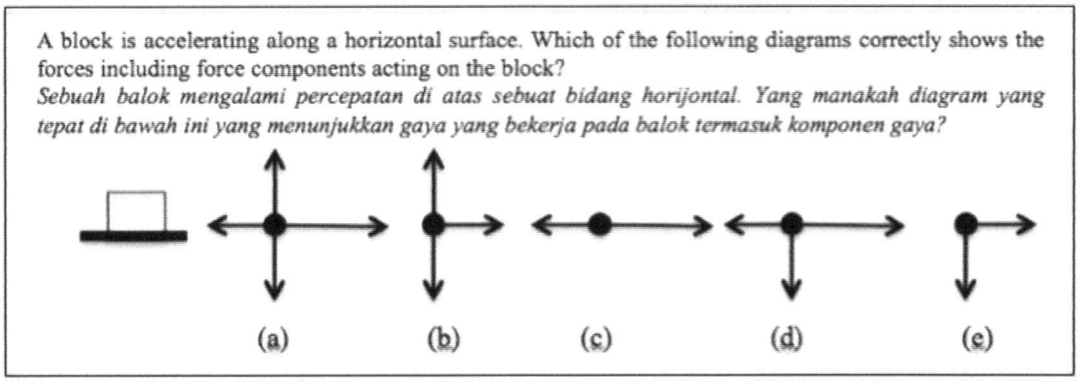

Figure 2. Question number 3.

On an inclined plane, on the other hand, the students were not aware of the situation of the object (whether the block is at rest or moving). Moreover, they just selected the free body diagram consisting of real forces without considering about the components of the forces. Also, the students could not successfully apply their conceptual understanding on the horizontal surface to an inclined plane. Palmer [8] stated that students are inconsistent in applying the same concept in different situations.

About 70% of students were not able to provide the correct answers when an object acts at rest. Figure 2 shows question #4 which represents force diagram of a block at rest on an inclined plane. Students only noticed the real forces on an object without thinking about vector components (why the object does not move on an inclined plane). Moreover, when an object moves with constant speed and constant acceleration, about 50% of students were not able to correctly distinguish free body diagrams for every situation.

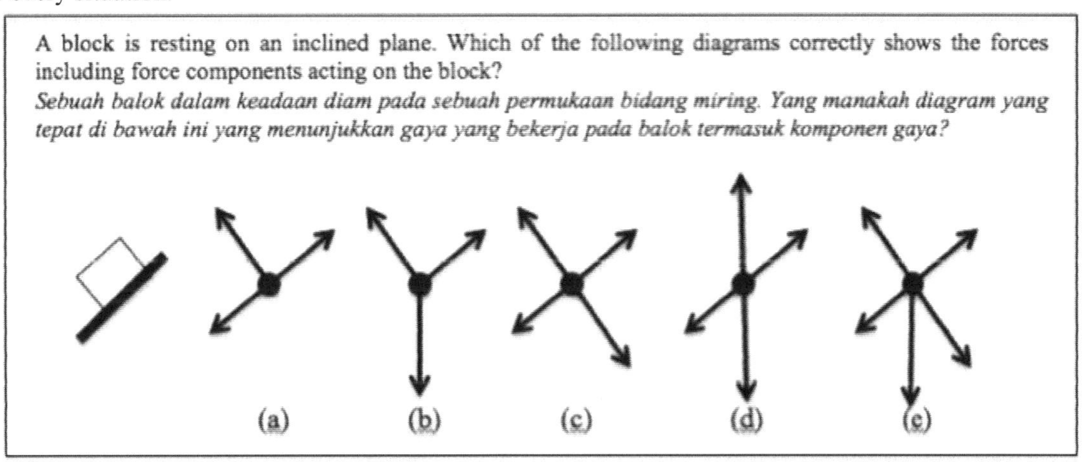

Figure 3. Question number 4.

Furthermore, for an object at constant speed, many students were not able to identify the correct force diagram. Some students did not fully identify all of the forces exerted on the object, and did not successfully apply trigonometry concepts in determining the resultant of the forces. Students also lack the ability to apply Newton's First Law in this situation in order to determine that the resultant force is zero when an object moves at constant speed. This indicates that the students could not completely represent all of the forces exerted on the block with diagrams by considering the law of inertia. Therefore, students should be encouraged to check the consistency of their understanding [9]. In this occasion, students should evaluate whether the concept of Newton's first law is consistent with the net force (force diagram).

Some of the students (about 50%) did not include the normal force or the component of weight force on the object while the block is accelerating along an inclined plane. In this context, they applied the concept of Newton's Second Law, but the lack understanding of force components. This suggests that a teacher should be careful in prompting students to use force diagram while solving physics problems [10,11]. In addition to utilize force diagrams, students 'views about representations including force diagrams has a positive correlation with students' grade on basic physics course [12].

Besides multiple-choice questions, an open-ended question (inclined context) was given to 60 students (first year students) to seek how students approach the problem. The question is "a car which has 15000 N weight is at rest on a 30o inclined plane. The coefficient of static friction between box and the surface is 0.9 and the coefficient of kinetic friction is 0.8. Find the magnitude of friction force on the car (note: sin 30° = 0.5; cos 30° = 0.86; gravitational acceleration in earth = 10 m/s²)." The physics concept of this problem is similar to question number 4 on multiple-choice problem.

Table 2. The list of students' difficulties.

Students' difficulties	N and percentage
To draw sketch/picture	3 (5%)
To identify forces acting on the car	45 (75%)
To draw correctly the direction of the forces	35 (58%)
To select correctly the appropriate equations	58 (96%)

Table 2 shows the list of students' difficulties in solving the inclined plane problem. Even though the question is not provided a sketch (visualisation), almost students (95%) drew correctly pictorial representation to represent the situation; only 3 students did not provide a picture and drew an incorrect sketch.

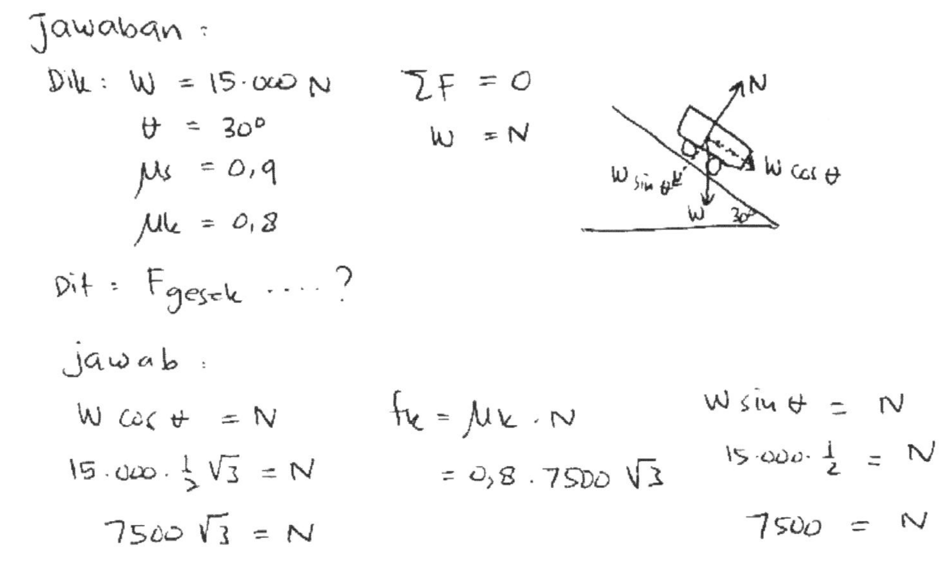

Figure 4. An incorrect student's answer.

Identifying forces exerted on the car is one of students' difficulties. They had lack ability to identify completely forces such as force of earth on car (weight force), force of surface on the car (normal force), force of surface on the car (static friction force), and force component of weight force.

Furthermore, 58% of students struggled to draw correctly the direction of the forces. Mostly students provided incorrect direction of the component of weight force. An example of incorrect answer is shown in figure 3. Due to drawing the incorrect directions of forces and not drawing free body diagrams, students did not select correctly the appropriate equation to solve the problem. Students jump directly to an equation $f_s = \mu_s N = \mu_s mg \cos\theta$. They did not notice that the car is at rest, so the static friction force is equal (but opposite direction to) the component of weight direction with x- axis.

A correct answer is presented in figure 4. The student starts by drawing sketch or picture then drawing free body diagram, and making label for each of force. The students completely identify all forces exerted on the car and also correctly draw the direction of forces. Furthermore, the student also draws a diagram to visualize all forces and component before choosing the appropriate equation to solve the problem. This diagram is very helpful to determine the net force in terms of Newton's Laws. Sirait et al. [13] found that most students agreed that using representations such as force diagrams help them in solving physics problems.

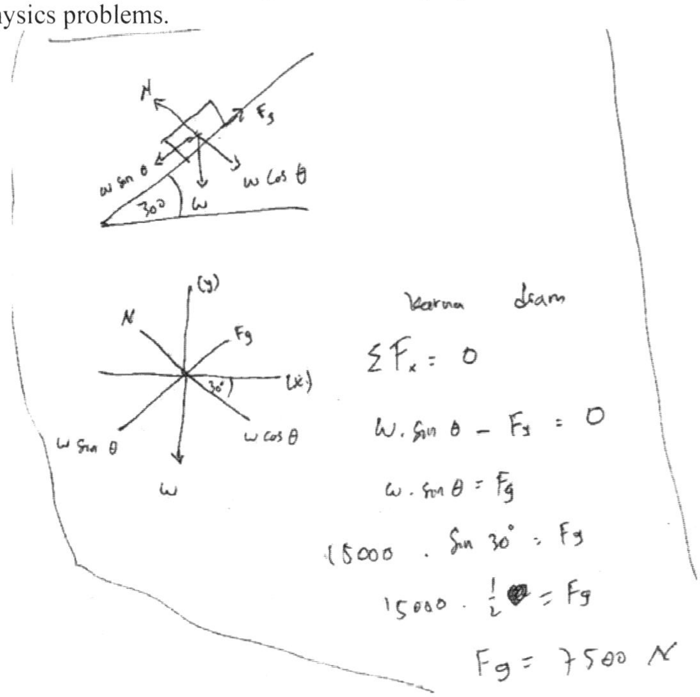

Figure 5. A correct student's answer.

4. Conclusion

In summary, the most difficult task for students in terms of the force concepts is identifying the force diagrams while an object is placed on an inclined plane whether it is at rest, moving with constant speed and constant acceleration. Students should be facilitated with mathematical tool such as vector and trigonometry. Thus force diagrams should be taught clearly and students are given opportunity to create or draw their own representations.

Acknowledgement

This study was supported by The Ministry of Research, Technology, and Higher Education – Indonesia grant No: 107/SP2H/LT/DRPM/IV/2017. We thank to lectures and students at department of physics education – Tanjungpura University who participated in this study.

References

[1] Hestenes D, Wells M and Swackhamer G 1992 *Phys.Teach.* **30** 141
[2] Savinainen A 2002 *Phys. Edu.* **37** 45

[3] Thornton R K and Sokoloff D R 1998 *Am. J. Phys.* **66** 338
[4] Rosengrant D, Van Heuvelen A and Etkina E 2009 *Phys. Rev. ST Phys. Educ. Res.* **5** 010108
[5] Savinainen A, Makynen A, Nieminen P and Viiri J 2013 *Phys. Rev. ST Phys. Educ. Res.* **9** 010104
[6] Heckler A F 2010 *Int. J. Sci. Educ.* **32** 1829
[7] Aviani I, Erceg N and Mešić V 2015 *Phys. Rev. ST Phys. Educ. Res.* **11** 020137
[8] Palmer D 1997 *Int. J. Sci. Educ* **19** 681
[9] Low D J and Wilson K F 2017 *Am. J. Phys.* **85** 54
[10] Lin S and Singh C 2015 *Phys. Rev. ST Phys. Educ. Res.* **11** 020105
[11] Sirait J, Hamdani and Oktavianty E 2017 *J. Turkish. Sci. Educ.* **14** 82
[12] Hamdani, Mursyid S, Sirait J and Etkina E 2017 *Jur. Penel. Pengem. Pend. Fis.* **3** 151
[13] Sirait J, Sutrisno L, Balta N and Mason A 2017 *Jur. Pend. Fis. Indo.* **13** 79

The Students' misconceptions profile on chapter gas kinetic theory

M N R Jauhariyah[1], N Suprapto[1], Suliyanah[1], S Admoko[1], W Setyarsih[1], Z Harizah[2] and I Zulfa[2]

[1] Physics Education Program Study, Physics Department, Faculty of Mathematics and Natural Sciences, Universitas Negeri Surabaya, Indonesia
[2]Physics Education Program, Postgraduate Program, Universitas Negeri Malang, Indonesia

E-mail: mukhayyarotinjauhariyah@unesa.ac.id and mukhayyarotin_87@yahoo.com

Abstract. Students have conception and misconceptions in the learning process. Misconceptions are caused by the teacher, students, and learning source. In the previous study, the researcher developed a misconception diagnosis instrument using three-tier on chapter gas kinetic theory. There are 14 items including 5 sub-chapters on gas kinetic theory. The profile of students' misconceptions shows that students have misconceptions in each sub-chapter. The cause of misconceptions came from preconceptions, associative thinking, reasoning, intuition, and false negative. The highest cause of misconception in this chapter is student's humanistic thinking.

1. Introduction

Humans have a structure of knowledge in the brain like boxes containing meaningful information that is different. The same experience for some people will be interpreted differently by each individual and stored in different boxes. Piaget said that each new experience is associated with the known structure in the brain's structure [1]. Different information owned by students sometimes causes some students to have a concept that is not in accordance with the scientific concepts. The mismatch of concepts that students possess with the concept of experts can be called using several terms such as "children's ideas" [2], "mental model" [3-5], "alternative conceptions" [6-10], "misconceptions" [11-16], and so on. The term "misconceptions" is used in this study.

The term "misconceptions" is going to be used for students' conceptions that contradict the scientific concept [16]. Vamvakoussi and Vosniadou used the "misconceptions" as a label for synthetic concepts that mismatch with the accepted view and that form student's attempt to integrate the existing knowledge with new information before a deeper conceptual change occurs [16].

The cause of misconception can come from various sources, including textbooks, teachers, and students themselves. In Physics subjects, students often experience misconceptions on the abstract material. Teachers will have more difficulty in explaining abstract concepts because there are no real examples in the daily life of the students. One example of the abstract material in Physics is the Gas Kinetic Theory. Based on the questionnaire given to the students of Driyorejo Senior High School, there are still many students who do not understand the material about gas kinetic theory, even though the material has been taught. This is corroborated by the students' test results on the gas kinetic theory material that shows more than 40% of students have values below minimum standards. According to one teacher at Driyorejo Senior High School statement, students are more focused on memorizing the formula so that students are less understanding of the concepts of gas kinetic theory. This leads to misconceptions in students.

It can't be denied that students build their own knowledge while learning. But, because students do not have a scientific frame of mind that fits to serve as a benchmark so that students are having errors in constructing their knowledge. This causes of the students' misconceptions. Misconceptions can happen because of the students' learning difficulties. There are external factors and internal factors in student learning difficulties. Internal factors can be a lack of interest and lack of motivation to learn, while external factors are lack of media variation and learning methods [17]. Further, Azizah et.al identified the problem-solving difficulties of high school students caused by several factors, partly because of the material being studied, the learning activities in the classroom, and the teacher's teaching style [18]. Suparno states some of the things that are causing misconceptions originating in students, among others, (1) preconceptions or early concepts of students, (2) student associative thinking, (3) humanistic thinking, (4) reasoning, (5) false intuition, (6) the stage of cognitive development of students, (7) the students' abilities, and (8) interest in student learning [19].

Misconceptions can be identified by providing diagnostic tests to students. Diagnostic tests are assessment tools which are concerned with the persistent or recurring learning difficulties that are left unresolved and are the causes of learning difficulties [20]. These diagnostic test instruments bring to light the disparity between what we want our students to know or learn and what they really know or learn. Habibbulloh et.al developed the Certainty of Response Index (CRI) as the diagnostic test instrument to reduce the students' misconceptions on Photoelectric Effect topic [21]. There are many kinds of methods to identify the student's misconceptions such as interviews, open-ended tests, multiple-choice tests, multiple-tier test [16]. The multiple-tier tests consist of two-tier, three-tier, and four-tier tools of the diagnostic test. The three-tier diagnostic test is a diagnostic test instrument consisting of three levels of questions, the first level of which contains the usual multiple-choice questions, the second level contains the choice of reason, and the third level contains questions relating to the belief in the answer chosen in two the previous level [22-23]. Compared to the two-tier diagnostic test, this three-tier diagnostic test is more effective to distinguish between nonconformist students and students who have the misconception with the added question of belief in choosing the answer [23]. Using three-tier diagnostic test, it can be known the profile of students who understand the concept, students who have experience misconception, and students who do not understand the concept.

Wiyono et.al used the three-tier diagnostic test instrument to identify the junior school students' misconceptions on motion topic [24]. The motion of three-tier diagnostic instrument effective to found and described the students' misconceptions on motion topic [24]. In the previous study, the three-tier diagnosis test developed on gas kinetic theory material. This instrument used to diagnose the students' misconceptions on chapter gas kinetic theory. This study describes the students' misconceptions profile on chapter gas kinetic theory.

2. Research Methods

The three-tier diagnostic test instrument for the gas kinetic theory (GKT) given to 34 students of Driyorejo Senior High School to identify the students' misconceptions and the cause of misconceptions. The GKT responses categorize into five categories based on the adoption and adaptation of Kutluay [25] research, showed by Table-1. Besides that, the answer of students reasoning analyses to know the cause of the students' misconceptions.

Table 1. The combination of three-tier test answer (adaptation from Kutluay research [26]).

Phenomena (P)	Reasoning (R)	Confidence	Category
First-tier	Second-tier	Third-tier	
True	True	Sure	Understand the Concept (UC)
True	True	Unsure	Less Understand the Concept (LUC)
False	False	Sure	Do Not Understand the Concept (NUC)
True	False	Unsure	Guessing (G)
True	False	Sure	
False	True	Sure	Misconceptions (M)
False	False	Sure	

3. Results and Discussion

The identification of students' misconceptions used the GKT three-tier diagnostic test consist of 14 test items shows that many students have the misconception. In addition, many students categorized into do not understand the concept. Tabulation of the student's combination of answers using the three-tier diagnostic test shown in Table 2.

Table 2. The combination of student's answer categories.

The number of item test	Combination of three-tier answer categories				
	UC	LUC	NUC	G	M
1	7 (22.58)	1 (3.23)	7 (22.58)	6 (19.35)	10 (32.26)
2	7 (21.88)	3 (9.38)	8 (25.00)	4 (12.5)	10 (31.25)
3	4 (12.9)	2 (6.45)	8 (25.81)	4 (12.9)	13 (41.94)
4	0 (0)	9 (0)	8 (25.00)	4 (12.5)	20 (62.5)
5	6 (20.7)	3 (10.3)	5 (17.24)	7 (24.24)	8 (27.6)
6	4 (12.9)	2 (6.45)	5 (16.13)	4 (12.9)	16 (51.61)
7	6 (17.65)	5 (14.7)	9 (26.47)	6 (17.65)	8 (23.53)
8	3(9.1)	4 (12.1)	7 (21.21)	7 (21.21)	12 (36.4)
9	0 (0)	0 (0)	8 (23.53)	7 (23.53)	19 (55.88)
10	2(6.25)	0 (0)	10 (31.25)	6 (31.25)	14 (43.75)
11	6 (18.75)	1 (3.13)	6 (18.75)	4 (18.75)	15 (46.88)
12	1 (3.22)	1 (3.23)	7 (22.58)	7 (22.58)	15 (48.39)
13	0 (0)	0 (0)	10 (30.3)	7 (21.22)	16 (48.50)
14	1 (3.12)	2 (6.25)	7 (21.88)	3 (9.37)	19 (59.38)

UC = Understand the Concept, LUC = Less Understand the Concept, NUC = do Not Understand the Concept, G = Guessing, M = Misconceptions

Based on Table 2, there were no students who understand the concept (UC) in test items 4, 9, and 13. Students' understanding of the material of ideal gas properties, gas laws, and kinetic energy of the gas is still very weak. While in the category of not understanding the concept (NUC), students most included in the category for test items 10 and 13. The test item contains the sub-material of the laws of the ideal gas. Students do not understand the concept (NUC) can happen because students quickly forget what he has learned. Not understanding the concept can also occur because students were not listening to students on going or because the methods used by teachers were boring so that students did not listen during the lesson so there are some concepts that are not understood by students. Item test no. 4,9,10, and 13 are not included in the LUC category. This indicates that there were no students who have not understood the concept. It also means that students are confidently choosing answers at the first and second level, but there were answers into other categories. Category guessing (G) can be caused by students who did not understand the concept of the concept of the item so that in addition to the students answer wrong on the first or second level, students are also not sure of the answer he chose. On the whole items there are a percentage of students in the guess category, but under 25% in the guessing category. The following describes the understanding of the student's concept in each sub-chapter.

3.1. Sub-chapter the ideal gas properties

Based on Figure 1, item test number 1 has the largest understanding concept percentage among other items, 22.58%. While the test item number 2 has a concept percentage of 21.88%. The percentage of misconceptions (M) in test items 1 and 2 respectively 32.26% and 31.25%. This shows that from a total of 34 students, about 11 students have misconceptions on the sub-chapter properties of the ideal gas.

In the first item test, the tube contains the ideal gas and the students are asked to choose a statement related to the characteristics of the ideal gas. The correct answer is "the gas particles will experience a change in direction of momentum when collided with the tube wall" with the choice of reason 'When colliding the wall, the particle changes the direction of the velocity so that the particle also changes in the direction of momentum'. Ideal gaseous particles experience complete tubing with walls. When subjected to collisions with walls, the particles change in the direction of speed so that the direction of momentum also changes.

Item test 2 is presented about the motion of the ideal gas particles and the student must determine the particle motion correctly. The correct answer to the test item number 2 is 'move randomly in all directions' with the choice of reason 'gas particles move randomly in accordance with Brownian motion'. A poor student's understanding of the properties of the ideal gas causes the student to be ambiguous in determining the characteristics of the characterizing the ideal gas and movement of ideal gas particles resulting in misconception.

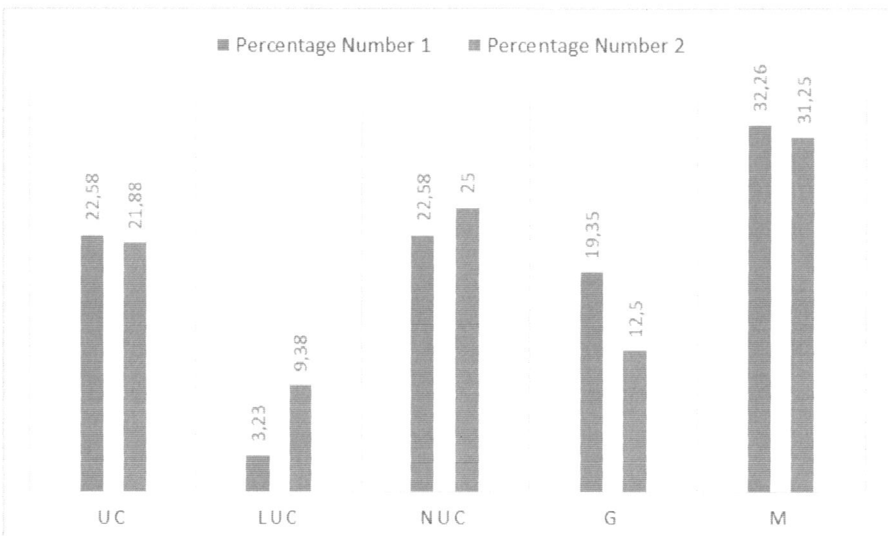

Figure 1. The percentage of Understand the Concept (UC), Less Understand the Concept (LUC), do Not Understand the Concept (NUC), Guessing (G), and Misconceptions (M) for item test number 1 and number 2.

3.2. Sub-chapter gas state equation

Test items that contain the equation of the gas state are on test items 3 and 6. Percentages for each category of students understanding the concept on test items 3 and 6 are shown in Figure 2.

Based on Figure 2, the percentage of students experiencing misconception is 41.94%. Misconceptions occur when the student incorrectly gives an answer at the first or the second level, Item test number 3 presented two images. The first picture is a pump in the normal state that the end is closed. The second image is a pump image with a closed end and pressed from the top. The student determines the movement of the gas molecule after the pump is pressed. The correct answer for this test item is "the molecule is more often subjected to collisions". While, the correct reason is "when pressed, the scope of the gas gets narrower." The high percentage of misconceptions on the item test shows that students' understanding is still wrong in determining the movement of gas particles in the pump when the pump is pressed.

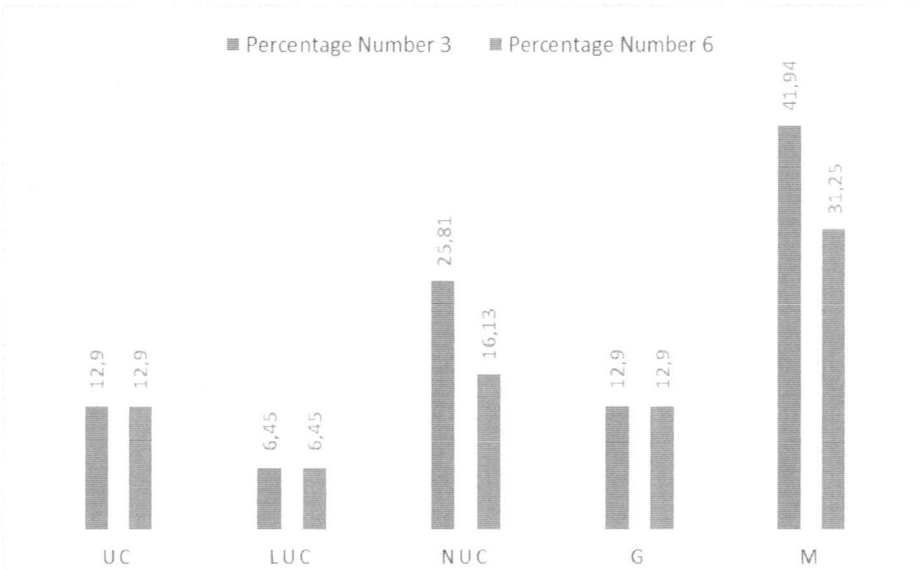

Figure 2. The percentage of Understand the Concept (UC), Less Understand the Concept (LUC), do Not Understand the Concept (NUC), Guessing (G), and Misconceptions (M) for item test number 3 and number 6.

For the percentage of student misconception on item test 6, is big enough, that is equal to 51.61%. On this point, there are four tubes which have the same number of molecules and are being heated to a certain temperature on each tube. Nearly half of students experience misconceptions because students confidently answer but misread the tubes that have the most pressure to the smallest. In addition, the category of misconception also comes from answers that students choose on the second level is also wrong. The answer the student should have chosen is "tube 4 > tube 3 > tube 2 > tube 1" on the reason "the greater the temperature, the faster the molecular movement increases the pressure".

3.3. Sub-chapter ideal gas laws

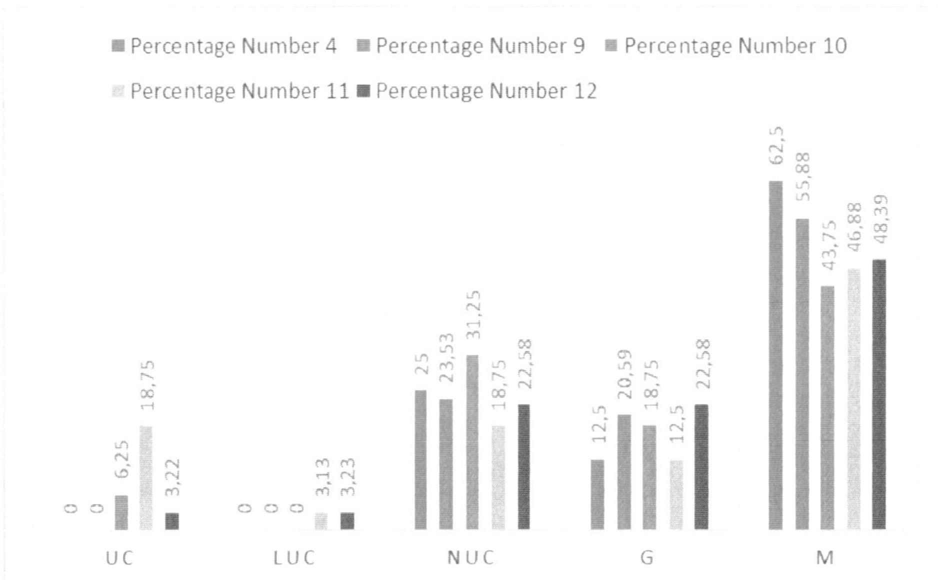

Figure 3. The percentage of Understand the Concept (UC), Less Understand the Concept (LUC), do Not Understand the Concept (NUC), Guessing (G), and Misconceptions (M) for item test number 4, 9, 10, 11, and 12.

The items containing the substance of the ideal gas laws are contained in item 4, 9, 10, 11, 12. On these five items are presented phenomena related to the laws of the ideal gas. The following graph of the percentage of misconceptions in each category for items 4, 9, 10, 11, and 12 are shown in Figure 3 Based on Figure 3, the students' misconceptions percentage for that five items are big, even though there are achieved 62.5%. This shows that students do not fully understand about ideal gas laws including the application of that law in everyday life, i.e item 12 when the balloon entering to the two liquids with different temperature, many students wrong to conclude what is happening in the balloon. In accordance with Charles' Law, in enclosed spaces, if the gas pressure is constant, the temperature is directly proportional to the volume. In A tube, with a liquid temperature of 100°C will make the balloon expand because the heat energy from outside the balloon is transferred into the balloon thus making the balloon expand. In tube B, the liquid temperature is smaller than the temperature of the balloon, the heat energy of the balloon is transferred out of the balloon. That's why the balloon on tube B shrinks.

3.4. Sub-chapter gas kinetic energy

The material about gas kinetic energy contains in items 5, 7, 8, and 13. The students understanding concept percentage of gas kinetic energy shows in Figure 4. Based on Figure 4, the student's misconceptions on sub-chapter gas kinetic theory less than that of on sub-chapter ideal gas laws. The highest misconceptions on sub-chapter gas kinetic energy there on test item 13 as much as 48.5%. In question number 13 students have not been able to distinguish the true statement about the kinetic energy of the gas. In addition, the second highest percentage of misconception was found in point 8 of 36.4% of students experiencing misconceptions. In this case, the students have not fully understood the effect of heating or of rising gas temperature in a confined space to the change of kinetic energy of gas particles.

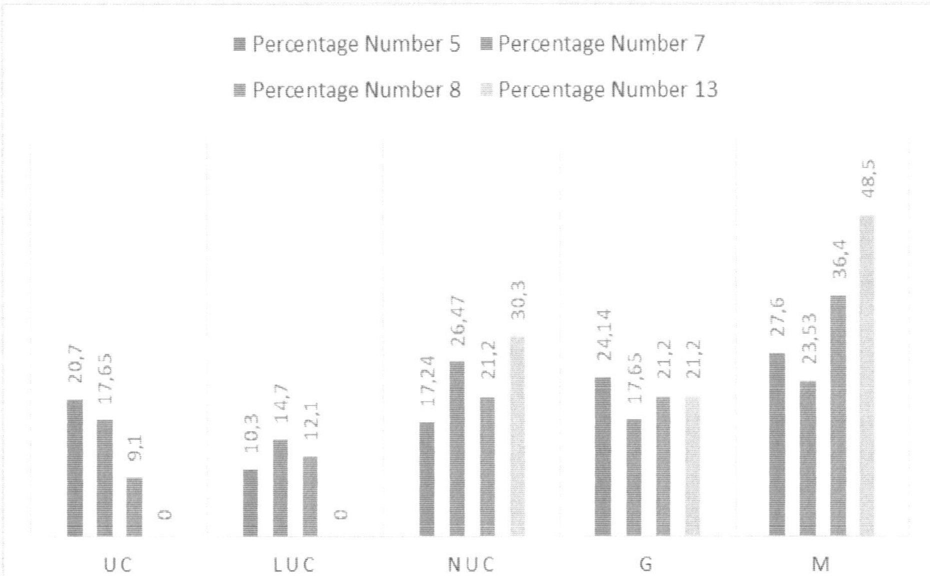

Figure 4. The percentage of Understand the Concept (UC), Less Understand the Concept (LUC), do Not Understand the Concept (NUC), Guessing (G), and Misconceptions (M) for item test number 5, 7, 8, and 13.

3.5. Sub-chapter equipartition of energy

The item that contains the energy equipartition sub is contained in item 14. In this case, three different tubes are presented, then the student selects the statement according to the students according to the inner energy of the three tubes. Here is the percentage of students' understanding of the sub-matter of equipartition of energy.

Based on Figure 5 on item 14, students who experienced misconception as much as 59.38%. This is much different with the percentage of students who understand the concept of 3.12%. Based on these data, more than half of students are still wrong in answering the number 14 so that it can be concluded that students' understanding is still lacking on this sub-chapter-energy equipartition.

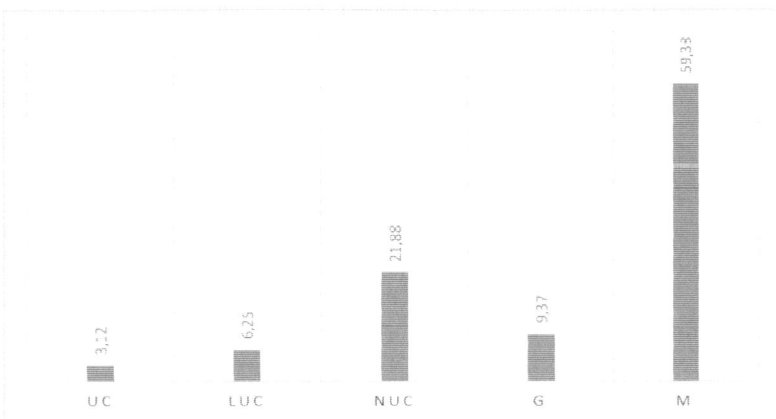

Figure 5. The percentage of Understand the Concept (UC), Less Understand the Concept (LUC), do Not Understand the Concept (NUC), Guessing (G), and Misconceptions (M) for item test number 14.

3.6. The cause of students' misconceptions

Each choice of reason presented in the three-tier diagnostic test instrument is structured on the basis of possible misconceptions that may occur in the student. Based on the causes of misconceptions derived from students according to Suparno [19], in a three-tier diagnostic test instrument used to identify the student's misconception profiles in Driyorejo Senior High School. The misconception causes used in the three-tier diagnostic instrument are (1) preconceptions, (2) associative thinking, (3) reasoning, (4) intuition, and (5) false negative.

The students' misconceptions and the cause of misconception identification for 34 students conducted by analysing the cause of misconception from the student's answer choices for reasoning (second tier). The cause of misconception percentage is shown in the Table 3.

Table 3. The Cause of Misconception Percentage.

Item number	Cause of Misconception (%)					
	Pr	AT	R	HT	I	FN
1	33.33	16.67	25	0	0	25
2	8.333	0	8.333	50	16.67	16.67
3	13.33	0	20	0	46.67	20
4	13.64	4.55	4.55	54.55	13.64	9.09
5	8.33	0	8.33	25	16.67	41.67
6	5.26	36.84	10.53	15.79	15.79	15.79
7	0	0	16.67	66.67	16.67	0
8	38.46	15.38	7.69	7.69	23.08	7.69
9	15	15	50	5	15	0
10	18.75	18.75	6.25	12.5	31.25	12.5
11	11.76	23.53	23.53	0	29.41	11.76
12	27.78	16.67	27.78	0	11.11	16.67
13	29.41	5.882	11.76	23.53	23.53	5.88
14	23.81	19.05	14.29	14.29	19.05	9.52

Pr = Preconception, AT = Associative Thinking, R = Reasoning, HT = Humanistic Thinking, I = Intuitive, FN = False Negative

Based on Table 3, it is known that there are various causes of misconception from the students answer. The cause of misconceptions are: preconception, associative thinking, reasoning, humanistic thinking, intuition, and false negative. Among the cause of student's misconception that existed through this test, humanistic thinking has a highest percentage (66.67%) that respond to the student's misconception in the chapter of gas kinetic theory. The highest percentage of causes of misconception, which is humanistic thinking is on question item number 7 about kinetic energy of gas. In this item, students were trained to analyse the effect of temperature change due to the motion of gas particles in closed container. Students were used humanistic thinking to analyse the condition that given by linking the daily phenomenon which is the gas condition was in open space. The dissonance between given condition and student's humanistic thinking lead to the student's misconception.

For the categories of causes of misconceptions derived from humanistic thought (HT), the largest percentage is found in item 7 about the kinetic energy of the gas. At point 7 of this item, students are trained in conceptual knowledge to analyse the effect of temperature changes on the motion of ideal gas particles in the closed containers. Students experience misconception because students attribute phenomena to problems with events occurring in everyday life so that they do not match the ideal gas conditions presented in the question.

3.7. The students' misconceptions profile on chapter the kinetic theory of gas

The dominant student's misconception profile in the kinetic gas theory material is summarized in Table 4.

Table 4. The list of student's misconceptions on chapter the kinetic theory of gas.

Sub-chapter	Students' misconceptions profile
The Ideal Gas Properties	When the particle strikes the wall, the particle's momentum changes and loses its speed. Collisions between particles produce opposite directions from the direction of the previous motion.
Gas State Equation	The gas in the enclosed space despite suppressed molecular motion remains. The greater temperature of the gas in the enclosed space, the gas undergoes expansion resulting in reduced pressure.
Ideal Gas Laws	Giving pressure from the outside on the opposite side can restore the shape of a dented pimp bulb. The balloon inside the closed syringe can't expand or collapse even though the injecting tube pump is pulled up because the balloon diameter is too small. The wax absorbs the liquid when the candle is covered with glass. When the egg is on the bottle, the air in the bottle is so packed that the bottle can suck the egg. Temperature cannot affect the balloon so that at any temperature change any form of the fixed balloon.
Gas Kinetic Energy	In a closed space, the amount of gas particles is directly proportional to the kinetic energy of the gas. As the kinetic energy increases the gas particles in the closed space also increases. The temperature of the gas in the enclosed space is inversely proportional to the speed of particle motion. When heated, the movement of the gas particles slows down and the gas particles shrink. The kinetic energy of the gas does not depend on the gas pressure
Equipartition of Energy	The energy in a gas is inversely proportional to the temperature of the gas.

4. Conclusion

The student's misconception profile shows that students are experiencing misconceptions throughout the sub-chapter of gas kinetic theory. The highest misconception lies in the sub-chapter of a kinetic energy of the gas caused by humanistic thinking with a percentage of 66.67%.

To overcome misconceptions derived from these students can be tailored to the causes of misconceptions that appear in students. According to Suparno [19], misconceptions derived from preconceptions can be overcome by having students try directly so that students can revise their ideas about the wrong knowledge in students' minds. While misconceptions caused by associative thinking can be overcome by teachers explaining the concept slowly and paying attention to events that make students misrepresent so that students can change their incorrect associative thinking [19]. To overcome misconceptions caused by reasoning, students are confronted with actual events or events so that students can know that the student's thinking is inappropriate. In addition, teachers can also ask students to seek additional information so that students can complete the data they have. Suparno [19] states to overcome misconceptions caused by humanistic thinking, students are invited to see the actual events then the teacher explains in detail so that students do not experience misconception anymore. As for the misconceptions caused by intuition can be overcome by exposing students to events contrary to student intuition through repeated experiments so that student's understanding can be improved [19]. Consequently, conception of physics learning can also be increased [26].

References

[1] Slavin R E 2012 *Educational Psychology: Theory and Practice Tenth Edition* (London: Pearson Education)
[2] Persson A and Musher-Eizenman D R 2003 *Early Child. Res. Q.* **18** 530
[3] Ozcan O and Gercek C 2015 *Proc. Soc. Behav. Sci.* **191** 2195
[4] Aşcı H B, Tan F Z and Altıntaş F 2016 *Proc. Soc. Behav. Sci.* **235** 2
[5] Goodhew J, Pahl S, Goodhew S and Boomsma C 2017 *Energy Res. Soc. Sci.* (in Press)
[6] Kolomuc A, Ozmen H, Metin M and Acisli S 2012 *Proc. Soc. Behav. Sci.* **46** 1761
[7] Chhabra M and Baveja B 2012 *Proc. Soc. Behav. Sci.* **55** 1069
[8] Hussain N H, Latiff L A and Yahaya N 2012 *Proc. Soc. Behav. Sci.* **56** 466
[9] Narjaikaew P 2013 *Proc. Soc. Behav. Sci.* **88** 250
[10] Wenning C J 2008 *J. Phys. Teach. Educ.* **5** 11
[11] Türker F 2005 *Developing a Three-Tier Test to Assess High School Students' Misconceptions Concerning Force and Motion* (Çankaya/Ankara: Middle East Technical University)
[12] Peşman H 2005 *Development of a Three-Tier Test to Assess Ninth Grade Students' Misconceptions About Simple Electric Circuits* (Çankaya/Ankara: Middle East Technical University)
[13] Chan S W and Ismail Z 2013 *Proc. Soc. Behav. Sci.* **93** 1478
[14] Djanette B and Fouad C 2014 *Proc. Soc. Behav. Sci.* **152** 582
[15] Gurel D K, Eryılmaz A and McDermott L C 2015 *Eur. J. Math. Sci. Tech. Educ.* **11** 989
[16] Vamvakoussi X and Vosniadou S 2010 *Cognition Instr.* **28** 181
[17] Nofitasari I and Sihombing Y 2017 *J. Penelit. Fis. Apl.* **07** 44
[18] Azizah R, Yuliati L and Latifah E 2015 *J. Penelit. Fis. Apl.* **5** 44
[19] Suparno P 2013 *Misconception and Concept Change in Physics Education (Miskonsepsi dan Perubahan Konsep Dalam Pendidikan Fisika)* (Jakarta: PT Grasindo)
[20] Gronlund N E 1981 *Measurement and evaluation in teaching* (New York: McMillan Pub. Co. Inc.)
[21] Habibbuulloh M, Jatmiko B and Widodo W 2017 *J. Penelit. Fis. Apl.* **07** 27
[22] Peşman H and Eryılmaz A 2009 *J. Educ. Res.* **103** 208
[23] Kirbulut Z D and Geban O 2014 *Eur. J. Math. Sci. Tech. Educ.* 10 (5) 509-521.
[24] Wiyono F M, Sugiyanto S, and Yulianti E 2016 *J. Penelit. Fis. Apl.* **06** 61
[25] Kutluay Y 2005 *Diagnosis of Eleventh Grade Students' Misconceptions about Geometric Optic by a Three-Tier Test* (Çankaya/Ankara: Middle East Technical University)
[26] Suprapto N, Chang T S and Ku C H 2017 *J. Baltic Sci. Educ.***16** 7

IOP Conf. Series: Journal of Physics: Conf. Series **997** (2018) 012031 doi:10.1088/1742-6596/997/1/012031

Appendix

The sample of test item of GKT three-tier diagnostic test instrument:

Item test 2

An ideal gas-filled tube. The following true statement is...

a. When the gas particles hit the tube wall, the particles lose energy.
b. When the gas particles hit the tube wall, the particles lose their momentum.
c. The gas particles will lose speed (still) after colliding with the wall.
d. Gas particles will experience a change in speed after a collision with the wall.
e. Gas particles experience a change in direction of momentum when collided with a tube wall.

Supportive reasons:

i. Particle energy moves on the tube wall.
ii. When mashing the wall, the particle's momentum is equal to zero so that the particle has no speed.
iii. Particles that pound the dense walls will lose speed because the walls have a larger inert force.
iv. When pounding the particles wall changes the direction of the velocity so that the particle also changes the direction of momentum.
v. Particles of gas suffer a partial collision with walls so that particle speed changes.
vi. When pounding the wall of particle energy out of the particle itself the particles lose energy.
vii. ....

Are you sure about your answer?

a. Sure b. Not sure

Item test 6

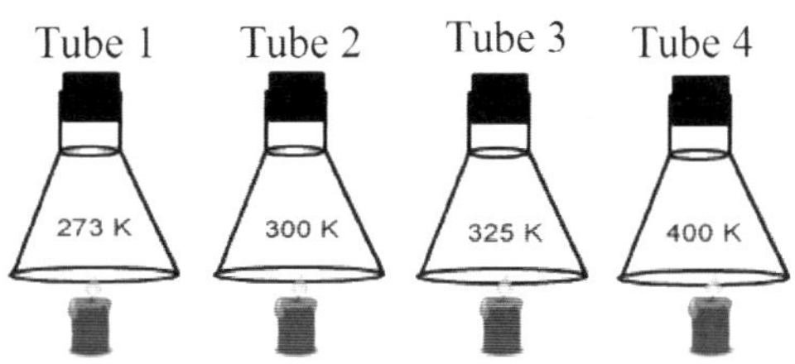

Each tube in the image has the same number of gas molecules. Each tube is heated to reach the temperature as shown. If sorted, the tubes that have the greatest pressure to the smallest are...

a. Tube 4 $<$ Tube 3 $<$ Tube 2 $<$ Tube 1
b. Tube 1 $>$ Tube 2 $>$ Tube 3 $>$ Tube 4
c. Tube 4 $>$ Tube 3 $>$ Tube 2 $>$ Tube 1
d. All tubes have the same pressure
e. Tube 1 $>$ Tube 3 $>$ Tube 2 $>$ Tube 4

Supportive reasons:

i. The greater the temperature, the faster the molecular movement increases the pressure.
ii. As long as the container is covered the gas pressure is the same.
iii. Warming does not affect pressure.
iv. The gas pressure is inversely proportional to the temperature.
v. The greater the gas temperature undergoes expansion resulting in reduced pressure.
vi. The hotter the gas molecule container exits the more so that the pressure decreases.
vii. ....

Are you sure about your answer?

a. Sure b. Not sure

Item test 12

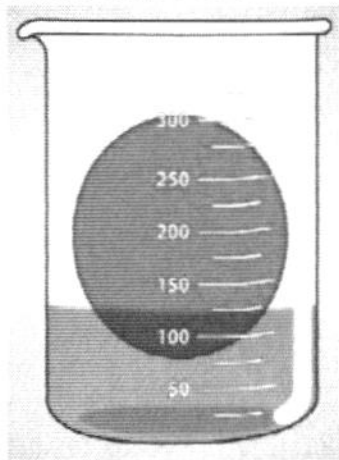

A Tube = 100° C

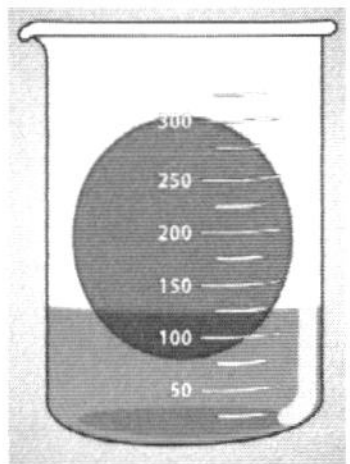

B Tube = 0° C

Look at the picture above! Each container contains different balloons and liquids. If both containers are closed simultaneously, then the corresponding statement is...

a. The balloon on tube A is enlarged, the balloon on tube B remains.
b. The balloon on both tubes did not change.
c. The balloon on tube A is enlarged, the balloon on tube B shrinks.
d. The balloon on tube A is fixed, the balloon on tube B is smaller.
e. The balloon on tube A is smaller, the balloon on tube B enlarges.

Supportive reasons:

i. Low temperatures do not affect the shape of the balloon.
ii. The shape of the balloon is not affected by temperature
iii. In a confined space with the same pressure, the volume is proportional to the temperature.
iv. The pressure on both tubes is the same so that the balloon does not change.
v. Because the balloon is made of an elastic material.
vi. High temperatures can change the shape of a balloon.
vii. ....

Are you sure about your answer?

a. Sure b. Not sure

Item test 13

Here is the correct statement relating to the kinetic energy of the gas in a closed space that is ....

a. The balloon under the hot sun has a greater kinetic energy than the balloon inside the house.
b. If the balloon is placed in the freezer the kinetic energy of the gas in the balloon increases with the reduced balloon temperature.
c. The lower the temperature of a gas, the greater the kinetic energy of the gas.
d. Balloons pumped using helium gas can fly because the kinetic energy of helium gas is greater than the kinetic energy of air in general.
e. Blowing up the balloon using the pump expands faster compared to blowing the balloon by mouth because the kinetic energy of the gas at the pump is greater than the kinetic energy of the gas in the mouth.

Supportive reasons:

i. The scope of the gas in the balloon is narrower so that the particles move faster.
ii. The greater the temperature of the particle movement balloon the faster the oak increases.
iii. Low temperatures can save particle motion energy so that the kinetic energy gets bigger.
iv. By being placed in the particle freezer in a moving stop balloon.
v. Helium gas particles move faster than air in general.
vi. Balloons inside the house get less sunlight.
vii. ....

Are you sure about your answer?

a. Sure b. Not sure

Item test 14

Three tubes A, B, and C contain the same ideal gas.

– Tubes A and B have the same pressure and temperature as the volume on tube A is twice the volume B

– The C tube has twice the temperature of tube B

Here is the correct statement about the energy in the gas owned in accordance with the above circumstances is...

a. The inner energy in tube A is equal to the energy in tube C.
b. The internal energy in tube B is twice as large as the energy in the tube A.
c. The internal energy in the tube C is twice as large as the inner energy of tube B.
d. Each tube has the same inner energy.
e. The inner energy in tube B doubles the inner energy in tube C.

Supportive reasons:

i. The greater the volume the gas molecules move the more freely.
ii. The energy in proportion to the volume of a gas.
iii. If the gas is heated, the gas molecule will move faster and faster, resulting in greater inner energy.
iv. If heated, large-volume gas will have a higher temperature than a gas that has a small volume.
v. The internal energy depends on the type of gas and does not depend on the temperature or volume of the gas.
vi. Tube A and tube B are at the same pressure.
vii. ....

Are you sure about your answer?

a. Sure b. Not sure

The effectiveness of CCDSR learning model to improve skills of creating lesson plan and worksheet science process skill (SPS) for pre-service physics teacher

I Limatahu[1], Suyatno[3], Wasis[2] and B K Prahani[2,a]

[1]Department of Physics, Faculty of Mathematics and Natural Sciences
University of Khairun, Indonesia
[2]Department of Physics, Faculty of Mathematics and Natural Sciences
Universitas Negeri Surabaya, Indonesia
[3]Department of Chemistry, Faculty of Mathematics and Natural Sciences
Universitas Negeri Surabaya, Indonesia

E-mail: [a]binarprahani@gmail.com

Abstract. In the previous research, CCDSR (Condition, Construction, Development, Simulation, and Reflection) learning model has been developed to improve science process skills for pre-service physics teacher. This research is aimed to analyze the effectiveness of CCDSR learning model towards the improvement skills of creating lesson plan and worksheet of Science Process Skill (SPS) for pre-service physics teacher in academic year 2016/2017. This research used one group pre-test and post-test design on 12 pre-service physics teacher at Physics Education, University of Khairun. Data collection was conducted through test and observation. Creating lesson plan and worksheet SPS skills of pre-service physics teacher measurement were conducted through Science Process Skill Evaluation Sheet (SPSES). The data analysis technique was done by Wilcoxon t-test and n-gain. The CCDSR learning model consists of 5 phases, including (1) Condition, (2) Construction, (3) Development, (4) Simulation, and (5) Reflection. The results showed that there was a significant increase in creating lesson plan and worksheet SPS skills of pre-service physics teacher at $\alpha = 5\%$ and n-gain average of moderate category. Thus, the CCDSR learning model is effective for improving skills of creating lesson plan and worksheet SPS for pre-service physics teacher.

1. Introduction

The learning model is a plan or a form which can be used to help students getting information, ideas, skills, values, ways of thinking, and the meaning of their expressions [1]. The role of teachers in the learning process is considered very important [2], because teachers play an important role in planning [3]. Teachers are obliged to plan lessons and implement quality learning process. Lesson planning is very important as the main guide or determinant of activities that teachers will do in the classroom during the learning process [4,5]. Teachers are the determinants of what is taught in the classroom and how to teach it [3].

Findings from previous studies on study literature studies were reinforced by preliminary studies. The results of the preliminary study by [6] in the Physics Education Study Program, FKIP University of Khairun showed that physics learning planning by physics teacher candidate is still low. The results of interviews and observations on some students, teachers and lecturers in the city of Ternate found that (1) Limited time teachers and lecturers develop learning models and tools that emphasize learning planning; (2) Students are not well trained in making learning tools that trains the science process skill indicators including formulating problems, formulating hypotheses, identifying variables, formulating operational definitions of variables, conducting experiments, designing tables, graphs, analyzing data, and formulating conclusions; (3) Physics teachers in the city of Ternate have not yet optimal ability in preparing learning tools; (4) There is no standard guidance on learning tools that will be used by lecturers to teach on pre-service physics teacher, then they will teach to students on the Senior High School. This phenomenon should be handled by a lecturer or lecturer. As the essence of the function of a lecturer are a professional educator and scientist with the main task of transforming, developing, and disseminating science, technology through education, research, and community service. In general, students must gain and be able to learn science process skills that will be useful in real life. The perspective of John Dewey (1916), schools should be the laboratories for solving real-life problems [4].

In the previous research, CCDSR learning model has been developed to improve science process skills for pre-service physics teacher for pre-service physics teacher. The CCDSR model has been specially designed to increase the skills of science process skills for pre-service physics teacher. The CCDSR learning model consists of 5 phases, including (1) Condition, (2) Construction, 3) Development, (4) Simulation, and (5) Reflection. The previous research developed a device of learning physics as an operational form of CCDSR model developed. In this research, CCDSR model designed to improve the skills of creating lesson plan and worksheet SPS for pre-service physics teacher. Actually, the implementation of the CCDSR model that has been developed quality of pre-service physics teacher.

The purpose of this research is to analyse the effectiveness of CCDSR model to improve the skills of creating lesson plan and worksheet SPS for pre-service physics teacher. The focus of the problem in this study included: (1) whether there was a significant increase (statistically) of skills of creating lesson plan SPS for pre-service physics teacher before and after the CCDSR model was applied, (2) whether there was a significant increase (statistically) of skills of creating worksheet SPS for pre-service physics teacher before and after the CCDSR model was applied, (3) how much level of skills of creating lesson plan SPS for pre-service physics teacher increased before and after applied CCDSR model, (4) how much level of skills of creating worksheet SPS for pre-service physics teacher increased before and after applied CCDSR model.

2. Methodology of Research

2.1. *General Background of Research*

This research was conducted at Khairun University (Indonesia). The scope of this research is on pre-service physics teacher in academic year 2016/2017. This research is emphasized on the analysis of the fulfillment of the effectiveness of CCDSR model by analyzing the improvement of skills of creating lesson plan and worksheet SPS for pre-service physics teacher before and after following the CCDSR model. The effectiveness of the CCDSR model was determined based on indicator: a significant increase in scores (statistically) between pre-test and post-test of skills of creating lesson plan and worksheet SPS for pre-service physics teacher, as well as the mean of n-gain determined by criteria: low, medium, and high.

2.2 Sample of Research

The samples in this study were 12 pre-service physics teacher at University of Khairun, Indonesia. Pre-service physics teacher in academic year 2016/2017 take the Study and Field Practice course in physics education.

2.3 Instrument and Procedures
Creating lesson plan and worksheet SPS for pre-service physics teacher measurement were conducted through Science Process Skills Evaluation Sheet (SPSES) [7-14]. SPSES of lesson plan SPS includes: formulation of learning objectives; organizing the material; learning resources/learning media; learning scenarios; assessment; and grammar [7-14]. SPSES of worksheet SPS includes formulate the problem, formulate the hypothesis, identify experiment variables, defines operational definition of experiment variables, designing an experiment, collecting data, create an observation table, data analysis, and formulate conclusions [7-14]. In this study, the indicators of creating lesson plan SPS skills include: formulation of learning objectives; organizing the material; learning resources / learning media; learning scenarios; assessment; and grammar [7-14]. The indicators of creating worksheet SPS skills include: formulate the problem, formulate the hypothesis, identify experiment variables, defines operational definition of experiment variables, designing an experiment, collecting data, create an observation table, data analysis, and formulate conclusions [7-14]. Implementation CCDSR model used Study and Field Practice course used in physics learning. This research uses one group pretest-posttest design, which is O1 X O2 [15]. The learning process begins by giving pre-test (O1). Each pre-service physics teacher is required to complete the SPSES. After the pre-test, the lecturer applies the CCDSR model and learning tool in each group (X). Physics learning is consisting of creating lesson plan and worksheet SPS skills. The process of physics learning ends with post-test (O2). Every pre-service physics teacher is required to post-test of creating lesson plan and worksheet SPS skills by SPSES.

2.4 Data Analysis
Creating lesson plan and worksheet SPS skills of pre-service physics teacher is analyzed based on the assessments obtained by pre-service physics teacher before and after learning using the CCDSR model. The pre-test, post-test, and n-gain data of creating lesson plan and worksheet SPS skills of pre-service physics teacher were further analyzed using inferential statistical tests with the help of SPSS and supported by qualitative descriptive analysis. The n-gain value is determined by the equation: n-gain = (score post-test - score pre-test) / (maximum score - pre-test score) [16]. According to the following criteria: (1) if n-gain $\geq .7$ (high), (2) if $.3 <$n-gain $<.7$ (moderate), and (3) if n-gain $\leq .3$ (low).

3. Result and Discussions of Research
The learning outcomes of all groups related to the creating lesson plan and worksheet SPS skills of pre-service physics teacher are presented in Figures 1 and Table 1. Vertical bar represent the mean of pre-test, Shape bar scores represent the mean post-test scores, and Black bar scores represent the n-gain scores. Figure 1 shows the average post-test scores of creating lesson plan and worksheet SPS skills of pre-service physics teacher is greater than the pre-test score. The average pre-test, post-test, and n-gain scores associated with creating lesson plan and worksheet SPS skills of pre-service physics teacher indicators are presented in detail in Table 1. Figure 1 show the average n-gain value of creating lesson plan and worksheet SPS skills is respectively .49 and .57. The average n-gain value of creating lesson plan and worksheet SPS skills of pre-service physics teacher for all groups is in the moderate category. Table 1 shows that the creating lesson plan SPS skills of pre-service physics teacher scores of each indicator include: formulation of learning objectives, organizing the material, learning resources or learning media, learning scenarios, assessment, and grammar are low and the post-test score is high for all indicators. The n-gain of creating lesson plan SPS skills of pre-service physics teacher score of each indicator includes: formulation of learning objectives; organizing the material; learning resources or learning media; learning scenarios; assessment; and grammar in all groups are moderate or high.

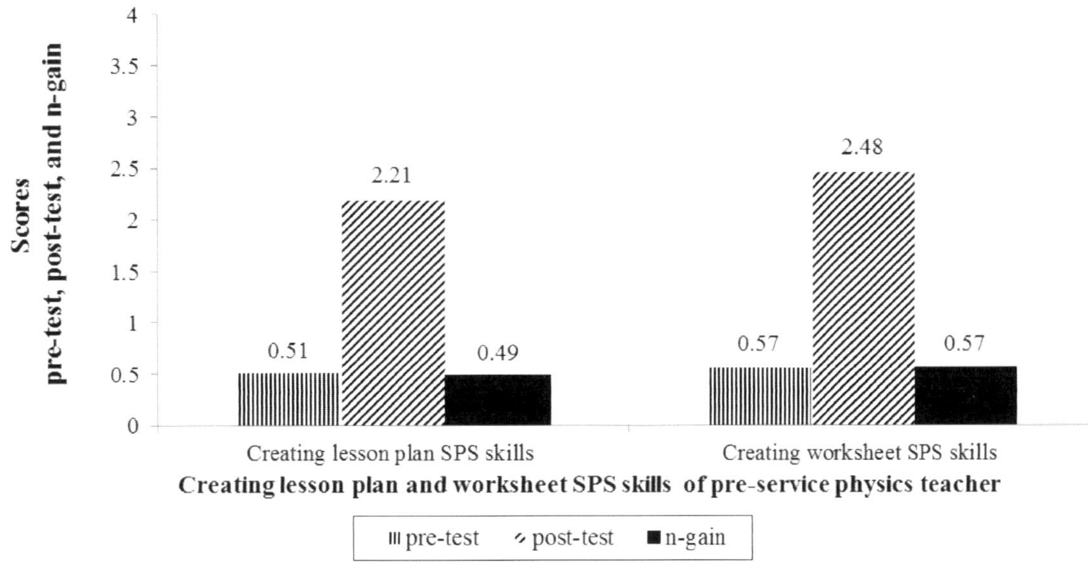

Figure 1. The average pre-test, post-test, and n-gain scores of creating lesson plan and worksheet SPS skills of pre-service physics teacher.

Table 1. The average score of pre-test, post-test and n-gain of creating lesson plan and worksheet SPS skills of pre-service physics teacher.

Indicator of creating lesson plan SPS skills	Pre-test		Post-test		N-gain	
Formulation of learning objectives	.92	Low	3.00	High	.68	Moderate
Organizing the material	1.00	Low	3.08	High	.69	Moderate
Learning resources / learning media	.67	Low	2.67	High	.60	Moderate
Learning scenarios	.33	Low	2.92	High	.70	Moderate
Assessment	.75	Low	2.92	High	.67	Moderate
Grammar	.42	Low	3.08	High	.74	High
Indicator of creating worksheet SPS skills						
Formulate the problem	.75	Low	2.50	Moderate	.54	Moderate
Formulate the hypothesis	.75	Low	2.58	Moderate	.56	Moderate
Identify experiment variables	.58	Low	2.25	Moderate	.49	Moderate
Defines operational definition of experiment variables	.33	Low	2.08	Moderate	.48	Moderate
Designing an experiment	1.00	Low	3.08	High	.69	Moderate
Collecting data	.33	Low	2.67	High	.64	Moderate
Create an observation table	.33	Low	2.33	Moderate	.55	Moderate
Data analysis	.33	Low	2.25	Moderate	.52	Moderate
Formulate conclusions	.50	Low	2.33	Moderate	.52	Moderate

Table 1 shows that the creating worksheet SPS skills of pre-service physics teacher scores of each indicator include: formulate the problem, formulate the hypothesis, identify experiment variables, defines operational definition of experiment variables, designing an experiment, collecting data, create an observation table, data analysis, and formulate conclusions are low and the post-test score is moderate or high for all indicators. The n-gain of creating worksheet SPS skills of pre-service physics teacher score of each indicator includes: formulate the problem, formulate the hypothesis, identify experiment variables, defines operational definition of experiment variables, designing an experiment, collecting data, create an observation table, data analysis, and formulate conclusions are moderate.

Physical learning cannot be separated from the process of systematic scientific investigation. Along with its development, the process contained in scientific inquiry is packed more systematically in the form of skills that must be possessed by a person to conduct a scientific investigation, this skill is referred to as Sciences Process Skills (SPS). Skills of the process of science are procedural, experimental, and investigate systematic science as the basis of scientific science literacy [17-19] so it is important for teachers to have a good understanding of SPS [20]. The effectiveness of the CCDSR model on increasing creating lesson plan and worksheet SPS skills of pre-service physics teacher included (1) an increase in pre-test and post-test scores and (2) n-gain CPLP skills value, as shown in Figure 1 and Table 1. Before the CCDSR model is applied; pre-service physics teacher low mastery the creating lesson plan SPS skills, average score of pre-service physics teacher is under the standard score (minimum score 2.00 in score range 1-4), that is the average score indicator of creating lesson plan SPS skills respectively are .92; 1.00; .67; .33; .75; and .42. All this time, pre-service physics teacher are not used to formulation of learning objectives, organizing the material, learning resources / learning media, learning scenarios, assessment, and grammar. And then, before the CCDSR model is applied; pre-service physics teacher low mastery the creating worksheet SPS skills, average score of pre-service physics teacher is under the standard score (minimum score 2.00 in score range 1-4), that is the average score indicator of creating worksheet SPS skills respectively are .75; .58; .33; 1.00; .33; .33; .33; and .5. All this time, pre-service physics teacher are not used to formulate the problem, formulate the hypothesis, identify experiment variables, defines operational definition of experiment variables, designing an experiment, collecting data, create an observation table, data analysis, and formulate conclusions. These results are reinforced by the results of the preliminary study [6] in the Physics Education Study Program, Khairun University; Ternate shows that the science process skills of physics teacher candidates are still low. The results of interviews and observations on some students and lecturers in the city of Ternate reveal 1) Limited time teachers and lecturers develop learning models and tools that emphasize the skills of the science process; 2) Students are still having difficulty using science process skills in physics learning; 3) Limited time teachers and lecturers develop learning models and tools that emphasize learning planning; 4) Students are not well trained in making learning tools that trains the science process skill indicators including formulating problems, formulating hypotheses, identifying variables, formulating operational definitions of variables, conducting experiments, designing tables, graphs, analyzing data, and formulating conclusions; 5) Physics teachers in the city of Ternate have not yet optimal ability in preparing learning tools; 6) There is no standard guidance on learning tools that will be used by lecturers to teach on pre-service physics teacher, then they will teach to students on the Senior High School.

After the CCDSR model was applied to the pre-service physics teacher, the mastery of creating lesson plan SPS skills rises above average and becomes high; the average score indicator of creating lesson plan SPS skills respectively to 3.08; 2.67; 2.92; 2.92; and 3.08 (well beyond the minimum score of 2.67 in the 1-4 score range). The mastery of creating worksheet SPS skills rises above average and becomes moderate or high; the average score indicator of creating worksheet SPS skills respectively to 2.50; 2.58; 2.25; 2.08; 3.08; 2.67; 2.33; 2.25; 2.33 (well beyond the minimum score of 2.08 in the 1-4 score range). N-gain score creating lesson plan and worksheet SPS skills respectively to .49 and .57 (in the 0-1 score range). The increase in creating lesson plan and worksheet SPS skills is allegedly influenced by design of scenario in phase of the CCDSR model. Each phase emphasizes creating lesson plan and worksheet SPS skills of the CCDSR model.

Table 2. Wilcoxon t-test result of creating lesson plan and worksheet SPS skills of pre-service physics teacher.

Inferential Statistics Test	Data	Asymp. Sig. (p)
Wilcoxon t-test	*Pretest Posttest*	.01

p < .05 (2-tailed)

Table 2 shows the improving in pre-test and post-test of CPLP skills tested using Wilcoxon t-test. The p score gives a value of .01 for creating lesson plan and worksheet SPS skills of pre-service physics teacher. Each score is considered significant, because p < .05. Because Z the result of the calculation is negative so it shows that there is an increase of creating lesson plan and worksheet SPS skills of pre-service physics teacher after applied learning with CCDSR model. Implementation of the CCDSR learning model has an impact on the significant increase of creating lesson plans and worksheet SPS skills as shown in Table 2. The increase of creating lesson plans and worksheets SPS skills cannot be separated from the role of learning tools that have been valid. The process of creating lesson plans and worksheets SPS skills in physics learning by design has been developed in the CCDSR learning model, Phase 3: Developing a Tool oriented science process skills (Development). Students create learning tools to develop learning planning skills on SPS (focus on learning to practice SPS). Students are guided by lecturers to see the skills of creating lesson plan and worksheet SPS skills that students have. Students are proactive in learning activities by contributing to the skills of the science process in their working groups. Lecturers act as mentors, moderators, facilitators, consultants and mediators in the learning process in an effort to improve the skills of the science process. This process is systematically in which trained: (1) the indicators of creating lesson plan SPS skills include: formulation of learning objectives; organizing the material; learning resources or learning media; learning scenarios; assessment; and grammar; (2) the indicators of creating worksheet SPS skills include: formulate the problem, formulate the hypothesis, identify experiment variables, define operational definition of experiment variables, design an experiment, collecting data, create an observation table, data analysis, and formulate conclusions. The SPS skills lesson plan and worksheet activities are relevant to the results of [7-11] research that physics teachers should have and be able to develop a student response data strategy that the majority of students feel the lesson plan and their SPS skills are improved after the implementation of the CCDSR learning model. Reinforced theory Production, the learner needs to convert mental representations created during encoding to motor activity [21-23]. As well as social constructivist theory by Vygotsky, that learners share the perspective of individuals with others to build a common understanding that is impossible to build individually [21-23], SPS improved by physics learning [24,25]. The above description shows that the implemented CCDSR learning model has been proven to be effective in increasing the students' writing lesson plan and worksheet SPS skills with moderate or high criteria.

4. Conclusions

The CCDSR learning model consists of 5 phases, including (1) Condition, (2) Construction, (3) Development, (4) Simulation, and (5) Reflection. The results showed that there was a significant increase in creating lesson plan and worksheet SPS skills of pre-service physics teacher at $\alpha = 5\%$ and n-gain average of moderate category. Thus, the CCDSR learning model is effective for improving skills of creating lesson plan and worksheet SPS for pre-service physics teacher. The implication of this research is that CCDSR model can be used as an alternative to overcome the low of creating lesson plan and worksheet SPS for pre-service physics teacher. To improve the result of this research, it is necessary to do further generalizations in various education levels and countries.

5. Acknowledgements
The author's gratitude goes to the Ministry of Research, Technology and Higher Education who has funded the Research. Likewise, the author's gratitude goes to the Khairun University that has provided research opportunities

References

[1] Joyce B, Calhoun E and Weil M 2003 *Models of Teaching* (New Jersey: Pearson Education)
[2] National Research Council 1996 *Executive Summary of National Science Education Standards*
[3] Abell S K and Lederman N G 2007 *Handbook of Research on Science Education* (NJ: Lawrence Erlbaum)
[4] Arends R I 2012 *Learning to Teach* (New York: McGraw-Hill)
[5] Reigeluth C M and Stein F S 1983 *Instructional Design Theories and Models* (NJ: Erlbaum Associates)
[6] Limatahu I 2016 Pengembangan Model Perangkat Pembalajaran Fisika Menerapkan MPBM untuk Mendukung Program PPL II Mahasiswa FKIP Unkhair Ternate *Pros. Semin. Nas. Universitas Negeri Surabaya* **4th** (Surabaya: Unesa) 545
[7] Abdullah C, Parris J, Lie R, Guzdar A and Tour E 2015 *CBE—Life Sci. Educ.* **14** 1
[8] Dogan I and Kunt H 2016 *J. Eur. Educ.* **6** 32
[9] Karsli F and Şahin Ç 2009 *Asia-Pacific Forum Sci. Learn. Teach.* **10** 1
[10] Etkina E 2010 *Phys. Rev. ST Phys. Educ. Res.* **6** 02011
[11] Loughran J J, Mulhal P and Berry A 2011 *Int. J. Educ.* **30** 1301
[12] Magnusson S, Krajcik J and Borko H 1999 *Examining Pedagogical Content Knowledge* (Dordrecht: Kluwer Academic Publishers)
[13] Ertmer P A and Timothy J N 2013 *Performance Improvement Quarterly* **26** 65
[14] Williams J, Eames C, Hume A and Lockley J 2012 *Res. Sci. Technol. Educ.* **30** 327
[15] Fraenkel J R, Wallen N E and Hyun H H 2012 *How to Design and Evaluate Research in Education* (New York: McGraw-Hill)
[16] Hake R R 1998 *Am. J. Phys.* **66** 64
[17] Dogan I and Kunt H 2016 *J. Eur. Educ.* **6** 32
[18] Zeidan A H and Jayosi M R 2015 *World J. Educ.* **5** 13
[19] Fellenz M R 2004 *Asses. Educ. Higher Educ.* **29** 703
[20] Padilla M J 1990 *Res. Math. Sci. Teach.* **1** 603
[21] Moreno R 2010 *Educational Psycology* (New York: John Wiley & Son)
[22] Slavin E R 2011 *Educational Psychology Theory and Practice* (USA: Pearson)
[23] Santrock J W 2011 *Educational Psychology* (New York: McGraw-Hill)
[24] Habibbullah M and Madlazim 2014 *J. Penelit. Fis. Apl.* **4** 15
[25] Prastowo T 2015 *J. Penelit. Fis. Apl.* **1** 8

The use of quizStar application for online examination in basic physics course

R Kustijono[1] and H Budiningarti[1]

[1]Department of Physics, Universitas Negeri Surabaya

E-mail: rudykustijono@unesa.ac.id

Abstract. The purpose of the study is to produce an online Basic Physics exam system using the QuizStar application. This is a research and development with ADDIE model. The steps are: 1) analysis; 2) design; 3) development; 4) implementation; 5) evaluation. System feasibility is reviewed for its validity, practicality, and effectiveness. The subjects of research are 60 Physics Department students of Universitas Negeri Surabaya. The data analysis used is a descriptive statistic. The validity, practicality, and effectiveness scores are measured using a Likert scale. Criteria feasible if the total score of all aspects obtained is $\geq$ 61%. The results obtained from the online test system by using QuizStar developed are 1) conceptually feasible to use; 2) the system can be implemented in the Basic Physics assessment process, and the existing constraints can be overcome; 3) student's response to system usage is in a good category. The results conclude that QuizStar application is eligible to be used for online Basic Physics exam system.

1. Introduction

The assessment of education is the process of collecting and processing information to determine the achievement of student understanding carried out based on applicable standards, namely standards related to mechanisms, procedures, and instruments of understanding of learners' understanding. Teachers must know the extent to which students have understood the material that has been taught or the extent to which the learning objectives have been achieved, then, the level of achievement of learning objectives that have been achieved is expressed by the score. Students not only learn from the teacher but also get other learning experiences from various learning sources, but the main thing in the assessment is how the material taught by the teacher can be understood by the students [1]. Assessment results are evidence of feedback from teachers, students, and other stakeholders who have access to the learning process [2]. Unfortunately, there are still many teachers who have not mastered the education assessment according to the applicable standards [3]. One of the assessments of learning in the classroom is the measurement through tests or examinations to the students used to measure the extent to which certain subject matter has been understood by the students. There are many types and ways of giving tests to students and currently, many tests are being developed using the computer and implemented online. One of the online exam applications is Quizstar which has several advantages over other applications. The objective of the study is to produce a Basic Physics exam system online using the QuizStar application

Basic Physics is the basic idea that arises from the application of the scientific method which examines the most fundamental ideas about the properties of Physics [4], which deals with the basic concepts and principles of Physics necessary to study further Physics or other sciences. Basic Physics learning objective is to master basic knowledge of mechanics, heat, wave, sound, optics, static electricity, direct current circuit, magnet, electromagnetic induction, and alternating current. and can develop and apply them to study higher knowledge of physics. Physics is a science related to facts, processes, theories, concepts, and generalizations. Physics is not just a science of memorization, but a science that allows students to be able to link one concept with another concept to express the meaningful relationship. In learning meaningful new information is associated with information that already exists in the cognitive structure of students. Therefore, the study of physics must be able to present the phenomenon in everyday life that can encourage and train students to think analytically, critically and creatively. The best way to develop these skills is to give students the opportunity to investigate a Physical phenomenon in everyday life. One effective method that can be used in Physics learning is an inquiry-based learning. The current learning trend is to move away from teacher-centered teaching to a more student-centered approach. Science (including Physics) involves observation, identification, description, experimental investigation, and theoretical explanation of phenomena [5], thus requiring investigative activities. Learning-based learning is the process by which students engage in their learning, formulate questions, investigate extensively and then build new understandings, meanings, and knowledge. Inquiry-based learning can help students become more creative, more independent, and improve achievement [6]. Understanding science and student inquiry skills will increase with an investigative laboratory-based guide [7]. Inquiry-based learning provides us with the opportunity to help students learn concepts by allowing them to explore questions, develop and test hypotheses. Thus it can give students more opportunities to reflect on their own learning, gain a deeper understanding of concepts in an integrated way, and become better critical thinkers. The teaching of science has a specific characteristic, so science teachers avoid memorizing and only the small facts of science alone [8]. Science teachers develop teaching with scientific methods, critical thinking, scientific attitudes, problem-solving approaches, discovery methods, and methods of inquiry. When we start planning lessons, their level of experience will determine the number of structures and modeling that need to be developed [9]. Research on the assessment of the process of applying research-based tasks in science education shows that the most widely used method in science and technology is experimentation [10].

Learning outcomes can be grouped into three domains: the cognitive, the psychomotor, and the affective domains. Each subject always contains the three domains, but the emphasis can be different. Practical subjects focus more on the psychomotor domain, while the conceptual understanding subject emphasizes the cognitive domain, but both contain affective domains. The psychomotor aspect deals with manipulation skills involving muscle and physical activity, the cognitive domain is closely related to the ability to think (memorize, understand, apply, analyze, synthesize and evaluate), while the affective domain includes character and behavior (attitudes, interests, self-concept, values and morals). Assessment of learning outcomes in the classroom is actually aimed to improve the learning objectives [11]. The process of effective learning is goal-oriented, and its objectives are clearly understood. [12] Whatever the emphasis of the given realm, the judgment must be accountable[13]. The purpose of the cognitive aspect is oriented to the ability of thinking that includes simple intellectual ability (remembrance) to the problem-solving skills that require students to connect, incorporating some ideas, methods, and procedures to solve the problem. Thus the cognitive aspect is the sub-taxonomy that reveals the mental activity that often begins from the level of knowledge to the highest level of evaluation. Some aspects to be considered in making judgments are: valid, objective, fair, integrated, open, thorough, systematic, criteria, and accountable. Student assessment is based on data obtained through individual or group assignments, midterm exam, final exam, self-assessment, peer assessment, attitude assessment, and student performance observation through oral or written display. The assessment developed in this study is related to understanding the concept through online exams.

One of many online exam applications is QuizStar which can be used by accessing the site address: http://quizstar.4teacher.org. There are many advantages of using the QuizStar, among others, we can manage problems, classes, attach multimedia files to questions, create questions in various languages, display questions at any given time, accessible from any computer connected to the internet, allow students to complete and review repeated quiz answers, and so forth. QuizStar application provides facilities for two users of User Instructor and User Student, both facilities have different functions. User Instructor is a user who has the right to administer and manage questions, classes, and reports. Manage the questions, give the score, set the timer or when the matter is displayed, determine the type and form of the question whether it is multiple choices, right-wrong, or essay. Managing the class is creating a class, determining which classes can access the tested questions and grouping students in specific classes. The report which is displaying the results obtained from each student for every conducted question, statistics, determining the number of access time for student user, evaluating the level of question difficulties, and so forth. User Student is a user who will access and work on the questions that have been given by User Instructor. In order to access these questions, students must first have an account as User Student. To register an account as User Student does not need an email address as well as on User Instructor registration. Simply enter your first name, last name, username, and password. User Student enrollment can be done by User Instructor or students themselves. The steps of using the QuizStar application are: a) User Instructor registration, b) Class Folder setup, c) User Student registration, d) problem preparation, answer options, and key and e) online exam management [14].

2. Research Methodology

This is a Research and development study with the ADDIE model which is an acronym for the stages of the model: Analysis, Design, Development, Implementation, and Evaluation [15, 16]. This research develops System of Basic Physics Exam online by using QuizStar with the following research procedure: 1) Analysis that is doing analysis include analysis of Basic Physics objectives, Basic Physics material content, Basic Physics problems that can be developed which refers to the rules of making the correct problem, and deepening QuizStar application characteristics; 2) Design is planning the system of Basic Physics Exams online by using QuizStar application based on the results obtained by the analysis phase, which refers to the study of Basic Physics materials, learning evaluation rules, and technical rules of the use of online examination system; 3) Development is developing the system of Basic Physics exam online by using QuizStar application include: a) User Instructor Registration, b) Class Folder Setup, c) User Student Enrollment, d) Preparation of questions, answer options, and key, e) Online exam management, subsequently validated by experts in their field, and revising according to the input of experts; 4) Implementation is to apply an online Basic Physics exam system using a QuizStar application that is valid in both categories on the Basic Physics exam in class; 5) Evaluation is to assess the practicality and effectiveness of QuizStar based on the application stage. Practicality is based on the implementation and constraints of learning evaluation, while the effectiveness is based on the mapping of basic physics understanding and positive responses from students.

The subjects are 60 students of Physics Department, State University of Surabaya consisting of classes A, B, C, and D. Each class consist of 15 students. This research was conducted at Computer Service Unit of Faculty of Mathematics and Natural Sciences of Universitas Negeri Surabaya in the even semester of academic year 2016/2017 and the odd semester of academic year 2017/2018.

The research variables defined and the operational definitions used in this study are: 1) The validity is the level of suitability of the design with the latest study [17]. The study referred to in this study is the study of learning aspects (material and evaluation) and technical aspects. The validity is defined as the suitability of the online Basic Physics exam system by using QuizStar with material review and Basic Physics learning evaluation as well as technical studies. The validity is measured by using a validation sheet instrument assessed by experts; 2) Practicality is the degree to which other users can use it in learning conditions [17]. Practicality is defined as the success of other users using the online Basic Physics exam system by using QuizStar in learning evaluations. Practicality is measured by the instrument of observation sheet of the implementation of the evaluation evaluated by the observer; 3) Effectiveness is the level of conformity of results with the desired goal [17]. Effectiveness is defined as the success of the Basic Physics exam system online by using QuizStar in mapping out basic Physics understanding and positive responses from students. The effectiveness is based on the results of the test of the use of the system in the online Basic Physics exam from the students and the response was obtained from the student response questionnaire instrument.

The instruments used in this study include the validation sheet, the observation sheet for the implementation of the online Basic Physics exam system by using QuizStar, and the student response sheet on the use of the online exam system. The instrument used briefly can be explained as follows: 1) The validation sheet is the instrument used to assess the validity of the online Physics Baseline test system by using QuizStar. The aspects assessed on the validation sheet refer to the Basic Physics material, meet the rules of making the right questions, and the technical suitability; 2) The instrument of practicality is the observation sheet of the implementation and learning constraints. Aspects assessed on the observation sheet of learning constraints and constraints refer to the fulfillment of the principles of evaluation of learning; 3) Student response questionnaire is the instrument used to assess the response of students in the implementation of the Basic Physics exam online. Aspects assessed on this questionnaire refer to the response to the evaluation of Basic Physics and Systems operating techniques. Assessment of validity, practicability, and effectiveness using Likert scale scores (1 = very less, 2 = less, 3 = good, and 4 = very good [18].

Data collection techniques used in this study are questionnaires, observations, and tests. The brief description is as follows: 1) Questionnaires are used to assess the validity of the online Basic Physics exam system using the QuizStar application in terms of their conformity with the Basic Physics aspects, the appropriate rules, and the technical aspects. Questionnaires are also used to assess the effectiveness of the system based on student responses; 2) Observations are used to assess the practicality of an online Basic Physics exam system by using the QuizStar application in terms of the implementation and constraints of learning evaluation. To assess the practicality, observers are given subjectivity observation sheets and learning obstacles that refer to the appropriateness of the exam on the principles of learning and their appropriateness with the technical aspects. 3) The test is used to assess the effectiveness of the online Basic Physics exam system by using the QuizStar application in terms of the system's ability to map the basic physics understanding of the students.

Data analysis techniques used in this study is descriptive statistics to determine the validity, practicality, and effectiveness of the online Basic Physics exam system using the developed QuizStar application. The percentage of validity, practicality, and effectiveness is calculated by summing the average validators/observers score on every aspect of good and very good categories, then divided by maximum scores, then multiplied by 100%. Validity, practicality, and effectiveness are categorized good if the percentage is in the range of 61% - 80% and very good if the percentage is in the range of 81% - 100%. The Basic Physics Examination system online is eligible if the percentage of each validity, validity, and effectiveness is $\geq$ 61%.

3. Research Results

3.1. The validity of Online Physics Exam System by Using QuizStar

The validity of online physics exam system by using the QuizStar application is assessed from learning and technical aspects. Aspects of learning include Basic Physics content and constructions of making a good understanding of the problem. Technical aspects include ease of operation, ease of access, and affordable. The validity was assessed by three experts in the Basic Physics course who understood how to make the matter according to the rules. The results of the validity are shown in Table 1, as follows:

Table 1. The validity of Online Physics Exam System By using QuizStar.

Number.	Aspects	Average score	Explanation
	Compatibility with Basic Physics content		
1	Basic Physics Material	4.0	Very good
2	Concept Accuracy	3.7	Very good
	Conformity with the rules of making the questions		
1	In accordance with the purpose	4.0	Very good
2	The decking works well	3.7	Very good
3	The questions have only one correct answer	4.0	Very good
4	The subject matter is formulated clearly and firmly	4.0	Very good
5	The questions do not give directions to the right answer.	3.7	Very good
6	The questions do not contain double negative statements.	4.0	Very good
7	Choice of answers are homogeneous and logical	3.3	Good
8	The length of the choice formula is relatively the same	3.3	Good
9	The questions do not contain the statement "All are wrong/correct"	4.0	Very good
10	The questions are sorted based on the value of numbers	3.0	Good
11	The images, graphs, tables, diagrams, are clear and functional.	4.0	Very good
12	The questions do not use meaningless words	4.0	Very good
13	The questions do not depend on the answer to the previous question	4.0	Very good
14	The questions use formal language	3.7	Very good
15	The questions do not repeat words that are not part of the understanding.	3.7	Very good
	Technical Conformity		
1	The ease of operation	3.3	Good
2	The ease of access	3.3	Good
3	Affordable	3.7	Very good
Total average score		62.7	
Maximum score		80	

From Table 1, the validity level of the online Basic Physics test system by using QuizStar is 78%. Based on these results can be stated that conceptually the system of Basic Physics exam online is feasible to use. Nevertheless, these results still need to be implemented further to prove its practicality and effectiveness.

3.2. Practicality of Online Physics Exam System By using QuizStar

The level of practicality of the online exam system can be known from the implementation and constraints faced during implementation. The data needed to know the level of practicality of the online Basic Physics exam system by using the QuizStar application is reviewed from the aspect of its ability to meet the rules of good comprehension exam implementation and meet the technical aspects. Practicality is obtained based on the observation of 3 peers who use the online exam system. The result of practicality is shown in Table 2 as follows:

Table 2. Practicality of Online Physics Exam System by Using QuizStar.

Number.	Aspects	Average score	Explanation
	Compliance with test execution rules		
1	The exam reflects the ability to measure	4.0	Very good
2	Clarity of procedures and assessment criteria	4.0	Very good
3	The exam is not affected by the condition of the learners	3.7	Very good
4	Alignment with the learning process	3.3	Good
5	The exam scores are easily accessible to learners	4.0	Very good
6	Continuity of use of the test system	4.0	Very good
7	Planning and systematic exams	3.7	Very good
8	Test results can be used for capability mapping	4.0	Very good
9	Test results can be justified	3.7	Very good
	Technical Conformity		
1	The ease of operation	3.0	Good
2	The ease of access	3.0	Good
3	Affordable	3.7	Very good
Total average score		44.1	
Maximum score		48	

From Table 2, the level of practicality of the online Basic Physics test system by using QuizStar is 92%. Based on these results can be stated that the online exam system can be used in Basic Physics learning. Nevertheless, these results still need to be strengthened by testing their effectiveness by looking at how far the system can map students' basic physics understanding and student responses to the use of the online exam system.

3.3. The effectiveness of the Online Basics Physics System By using QuizStar

The ability of online examination system in mapping the students' basic physics understanding is seen from the report features of the online exam system. The assessment was obtained based on the observation of 3 colleagues who used the online examination system. The results obtained are shown in Table 3 as follows:

Table 3. Online Test System Ability in Reporting Test Results.

Number.	Test results Reporting Feature	Average Score	Explanation
	Individual report		
1	The identity of the participants	4.0	Very good
2	Date	4.0	Very good
3	The time for the test to begin	4.0	Very good
4	The time for the test to end	4.0	Very good
5	Scores obtained by participants	4.0	Very good
	Class report		Very good
1	Scores per examinee	4.0	Very good
2	Average Score	3.3	Good
3	Maximum Score	3.3	Good
4	Minimum Score	3.3	Good
5	Student score chart	3.7	Very good
Total average score		37.6	
Maximum score		40	

From Table 3 can be obtained the calculation of the level of ability of the online exam system in mapping the understanding of Basic Physics students is 94%. Student responses are obtained from a questionnaire given to students by using the QuizStar application. The results obtained are shown in Table 4 as follows:

Table 4. Student Response To the Use of Online Exam System.

Number.	Aspects	Average score	Explanation
1	The test Confidentiality	3.0	Good
2	Reduce the cheating in exams	2.9	Good
3	After completion of the exam, the score is immediately shown	3.5	Very good
4	The correct answer key is shown	3.2	Good
5	Build an exam atmosphere that is not stressful	2.9	Good
6	The ease to operate	3.2	Good
7	The ease of access	3.2	Good
8	Affordability of application operating costs	3.3	Very good
9	Empower smartphone in learning	3.4	Very good
10	Student support for online examination system	3.3	Very good
Total average score		31.9	
Maximum score		40	

Table 4 shows that the effectiveness of the system based on students' response to the online Basic Physics test system by using QuizStar is 79.8%. Based on these results it can be stated that students support the use of the Basic Physics exam system online using the QuizStar application. Since the online examination system has the ability to map the students 'basic physics understanding and get students' responses with the effectiveness level of $\geq$ 61%, then the online Basic Physics exam system using the QuizStar application can be declared effective. Therefore, the online exam system has validity, practicality, and effectiveness in both categories. Based on these results it can be concluded that the online Basic Physics exam system by using QuizStar application is worth using.

4. Discussion

Assessment of student learning outcomes is to test whether the material listed in the learning objectives has been studied and impact students' retention [19]. Of the multiple choice of exam types, multiple choice is the most selected teachers. Multiple-choice exams are used by teachers for the following reasons: 1) fast, easy and economical to achieve goals; 2) can be objectively assessed and thereby provide a fairer and more reliable test view than the description test; 3) score obtained easily converted; 4) reduces the likelihood of students guessing answers compared to true-false issues [20]. Although multiple-choice exams have advantages over other types, it is necessary to take into account the factors that influence the assessment in general. There are several important factors that influence the student's assessment: structure, self-regulation cognition, autonomy, and writing skills of activities and assessment as a process [21]. In making multiple choice questions, in addition to emphasizing the content of Basic Physics, must also meet the criteria of making the multiple choice questions.

Currently, distance learning is beginning to be widely implemented and taking into account the characteristics of distance learning, online assessment becomes a favorable alternative assessment [22]. Alternative assessment is particularly useful for students who are involved in distance learning, as self-assessments can be obtained, and can eliminate time pressures, but are unlikely to be implemented over time [23]. This research develops an online Basic Physics exam system using the QuizStar application and tests its feasibility including validity, practicality, and effectiveness. The instructional aspects (Basic physics matter and criteria of problem-making) are met with a good category, which means the developed online test system has been conceptually feasible. Good execution rules and technical aspects are also met with a good category of online exams using the QuizStar application without any significant constraints. Although there are some students who still feel anxious about the online exam, but with the training of the students before the online exam is done related to the technical operation, and make them not feel depressed, will be able to establish good cooperation between students and instructors, so that the constraints there is resolved [24]. These results indicate that a practical online exam system can be used in the Basic Physics online exam. The developed online exam system can also map exam results and get student responses in either category, which means the developed online test system is effective. Based on the results of the study, the online Basic Physics exam system by using QuizStar is worthy of use in the Basic Physics exam because it has met the aspects of validity, practicality, and effectiveness in both categories.

5. Conclusions

The results obtained are: 1) online test system by using QuizStar is conceptually feasible to use in Physics Elementary examination in terms of learning and technical aspects; 2) the system can be implemented in the Basic Physics assessment process, and the existing constraints can be overcome; 3) students' responses to the use of systems is in good category. This research can conclude that the QuizStar application is worthy of being used for the online Basic Physics exam system.

Suggestions that can be proposed for future researchers are developing the online exam by exploring more multimedia such as video and voice usage which are not yet applied in this research. This is to get the impression that the exam can be fun for the students and not something that is worrying or scary.

6. Acknowledgment

Our gratitude goes to the Dean of the Faculty of Mathematics and Natural Sciences of Universitas Negeri Surabaya who has supported and funded this research. Thank you also to the Chairman of the Physics Department, colleagues, and students who have helped finish this research, because without them this project would not be completed.

References

[1] Wiliam D 2013 *Voices from the Middle* **21** 15
[2] Hanover Research 2014 The Impact of Formative Assessment and Learning Intentions on Student Achievement *District Administration Practice*
[3] Kustijono R, Budiningarti H and Sunarti T 2016 *J. ABDI* **1** 102
[4] Feynman R 2010 *The Feynman Lectures on Physics Basic Physics* Volume 1 Chapter 02.
[5] Bell R L 2008 *Teaching the nature of Science through Process Skills-Activities for Grades 3-8* (Boston: Pearson Education Inc).
[6] Learning A 2004 *Learning and Teaching Resources Branch Focus on inquiry: a teacher's guide to implementing inquiry-based learning* (Alberta Canada)
[7] Brickman P, Gormally C, Armstrong N and Hallar B 2009 *Int. J. Scholarship Teach. Learn.* **3**
[8] Abd-El-Khalick F, Boujaoude S, Duschi R, Lederman N G, Hofstein A, Mamlok-Naama R, Niaz M, Treagust Dand Tuan H 2004 *Inquiry in Science Education: International Perspectives* (Wiley Periodicals Inc)
[9] Lane, Jill L 2007 Schreyer Institute for Teaching Excellence *Inquiry Based Learning* (Penn State University Park)
[10] Akinoglu Q 2008 *Int. J. Instr.* **1** 1
[11] Genesee F and Upshur J 1996 *Classroom-Based Evaluation in Second Language Education* (London: Cambridge University Press)
[12] Jabbarifar T 2009 *Proc. the 2nd Int. Con. Teach. Learn. (ICTL 2009) INTI (*Malaysia: University College)
[13] Scholl K and Olsen H M 2014 *J. Leis. Stud. Recreat. Educ.* **1** 37

[14] Advanced Learning Technologies Center for Research on Learning at The University of Kansas 2008 *Quizstar* (online) Available on http://quizstar.4teachers.org/ Accessed March 23th 2017
[15] Branch R M 2009 *Instructional Design: The ADDIE Approach* (New York: Springer)
[16] Dick W, Carey L and Carey J 2009 *The Systematic Design of Instruction* 7th ed (Boston, MA: Pearson A&B)
[17] Akker J V, Branch R M, Gustafson K, Nieveen N and Plomp T 1999 *Design Approaches and Tools in Education and Training* (Dordrecht: Springer-Science+Business Media)
[18] Fraenkel J R and Wallen N E 2003 *How To Design And Evaluate Research in Education* Fith Edition (New York: McGraw-Hill Higher Education)
[19] Carter S D 2015 *Athens J. Educ.* **2** 179
[20] Bailey K M 1998 *Learning about language assessment: dilemmas, decisions, and directions* (Heinle& Heinle: US)
[21] Vonderwell S and Alderman X L K 2007 *J. Res. Technol. Educ.* **39** 309
[22] Dikli S 2003 *The Turkish Online J. Educ. Technol.* **2** 13
[23] Simonson M, Smaldino S, Albright M and Zvacek S 2000 *Teaching and Learning at a Distance: Foundations of Distance Education Assessment for distance education* (Upper Saddle River, NJ: Prentice-Hall)
[24] Ozden M Y, Ertruk I and Sanli R 2004 *J. Distance Educ.* **19** 77

Implementation of ECIRR model based on virtual simulation media to reduce students' misconception on kinetic theory of gases

A C Prastiwi[1], A Kholiq[1,a] and W Setyarsih[1]

[1]Department of Physics, Faculty of Mathematics and Natural Sciences
State University or Surabaya, Indonesia
Jl. Ketintang, Surabaya 60231, Indonesia

Email: [a]kholiq@unesa.ac.id

Abstract. The purposed of this study are to analyse reduction of students' misconceptions after getting ECIRR with virtual simulation. The design of research is the pre-experimental design with One Group Pretest-Posttest Design. Subjects of this research were 36 students of class XI MIA-5 SMAN 1 Driyorejo Gresik 2015/2016 school year. Students misconceptions was determined by Three-tier Diagnostic Test. The result shows that the average percentage of misconceptions reduced on topics of ideal gas law, equation of ideal gases and kinetic theory of gases respectively are 38%, 34% and 38%.

1. Introduction

Learning activities is designed by teacher in which actual learning occurs and learners actively involved. Teacher acts as a facilitator for learners by encouraging students, explaining concept, fixing misconception and practicing problem solving in learning activities. The main role of the teacher is in explanation of the concept and improvement of the students' misconceptions. Previous **research revealed that one of difficulties in teaching physics is a result of students' misconceptions [1]. Students' misconceptions commonly due to simple phenomena in daily life that they experienced. Students have been already experience various physical phenomena such as free-falling objects, electricity, energy, collisions, gas properties and others.**

Most of the students have had its own understanding about physical phenomena or concepts that they have been experienced which is inconsistence with scientific concepts. When the concept of understanding is incorrectly formed, it will usually be very difficult to repair, because unintentionally the incorrect concept of physics becomes the handle of his life. The existence of this misconception will obviously hamper the process of acceptance and assimilation of new knowledge in the students, so that will hinder the success of students in the process of learning more. This is a major problem in the teaching of physics that can't be tolerated. Along with the growing awareness of this, efforts to tackle the problem of misconception continue to be developed, although the results have not been significant.

The misconception diagnostic instrument is an assessment instrument that can be used to identify misconceptions experienced by students. To find misconceptions in students, there is an instrument known as the three-tier diagnostic test. The result of misconception analysis using the instrument can be identified misconception to know misconception that happened in student. So, it will be easier to help students change their misconception [2].

In conventional learning, misconception can be reduced by a learning model which stages include elicit, confront, and resolve [3]. However, this learning is less effective because it fails to identify student misconceptions, so the teacher cannot know the cause of student misconception. In addition, in this model, the teacher does not provide reinforcement of the correct concept in the student, so students cannot change the wrong initial concept and keep repeating it after the learning is over. Therefore, Wenning in 2006 [8] developed the learning model by adding two stages of identify and reinforce which then called ECIRR learning model. The ECIRR model stands for elicit-confront-identify-resolve-reinforce. The ECIRR model emphasizes the use of cognitive conflicts and constructivist theories that invite students to build their concepts. One of the tips to reduce misconceptions is by the way students are exposed to the correctness of existing concepts, according to the stage of the ECIRR model in the confront stage. At this stage students are exposed to two different phenomena, namely the phenomenon of conception that they already have with a phenomenon that describes the scientific concept. Having been exposed to two different phenomena students will experience cognitive conflicts. Furthermore, at the stage of identifying students will know the error of the concept that they experienced and at the resolve stage students are given activities that will solve student cognitive conflict. In this phase, the role of teacher is needed to direct the students to concept improvement. After the students have improved the initial concept, the teacher reinforces the new concept students receive during the reinforcement phase.

Learning model that will be used for teaching needs to be supported by learning media that can be accepted either by the students, so that the given material can be processed correctly. A virtual simulation that is a means to conduct demonstration activities, and laboratory activities that can be done to meet the needs of the conceptual thinking on students about materials whose concepts are abstract to understand. One of the most commonly used software as a virtual lab is the PhET software. PhET simulation is one of the discovery-based simulation media and is used to clarify physical concepts, thus minimizing misconceptions of students.

Associated with the misconception, has been checked to the students of senior high school at SMAN 1 Driyorejo by using three-tier diagnostic test instrument, obtained the result that the students have misconception on sub material equation gas state equal to 64.58%, gas law 54.89% , and the kinetic theory of gas is 65.74%. Therefore, the researcher tries to do research using ECIRR learning model with virtual simulation media so that the misconception owned by the students can be reduced and the students also get deep understanding about the concept they are learning.

2. Theory

On learning conventional, misconception can be reduce to a learning's model with elicit, confront, and resolve These approaches tend to have in common the requirement that students encounter phenomena that run counter to their existing beliefs. Doing so, they are put in a state of intellectual disequilibrium or cognitive conflict. Becoming aware of the conflict between what they believe to be correct based on prior experiences and know to be correct based on more recent experience helps them to confront and resolve their conflicting perspectives in favor of a proper understanding. Such pedagogical approaches that emphasise conflict and resolution appear to derive from a Piagetian perspective on learning [4].

These and other approaches dealing with alternative concepts typically include three fundamental steps–those identified by the University of Washington Physics Education Group: elicit/ confront/resolve [5]. Firstly, in this model a teacher elicits a response (prediction about what will happen or an indication of agreement or disagreement with a given statement) from students, forcing them to commit to an answer in relation to a specific situation. Secondly, the students confront a situation that challenges their beliefs and answers, typically in an experiment that the students perform. During this second phase, if the students were incorrect in their prediction, they experience cognitive dissonance when confronting the conflict between prediction and experience. Students quickly come to realize the need for a new understanding about the concept under consideration, and are motivated to resolve the conflict with teacher assistance in phase three.

The traditional approach of overcoming misconception consists of eliciting, confronting, and resolving has not always been an effective way for teaching and learning physics as can be inferred from the results certain physics education research. ELICIT-CONFRONT-RESOLVE approach fails to make a substantial lasting difference in the area of misconception because it fails to clearly IDENTIFY the existence of the misconception to students and fails to REINFORCE student learning in the area of the misconception. A better approach to dealing with misconception suggests a more clearly elucidated five-step approach that will be herein referred to as the ECIRR (Elicit-Confront-Identify-Resolve-Reinforce) model.

2.1 How the ECIRR Model Works

ELICIT, the teacher probes for students' misconception through activities that make students' thinking evident such as asking questions, and conducting Socratic dialogues with white boarding [6]. During such practices teachers ask students to predict, explain, and make clarify statements. Of course, this step assumes that the teacher is cognizant that misconception exist and what they are. Previous research has shown that in order for a teacher to effectively address student's misconception, they must be aware of the presence of such ideas [7].

CONFRONT, the teacher uses discrepant events to provide contradictions to students' statements or predictions and place them in a state of cognitive conflict. They confront misconceptions through demonstration, implications, and questions, and encourage discussion [6, 8].

IDENTIFY, After misconception are elicited and confronted, the teacher must clearly and unambiguously identify them as such. Teachers must be careful, however, not to denigrate the value of intuition that often can lead to correct predictions. They must explain the power of misconception to mislead, and state emphatically that students must not be misled and they should divorce themselves from it because the old conception will compete with the new conception [6, 8].

RESOLVE, The teacher should foster the replacement of an alternative conception using any of the following approaches: questions, thought experiments, interactive demonstrations, hypothetical situations, and experiments designed to test hypotheses. They should help reevaluate students' understanding by posing conceptual questions, and eliciting student sources of misconception. To overcome misconception, teachers should place as much attention on students' prior knowledge as possible, but allow students to actively resolve discrepancies by themselves because teaching by telling simply does not work [6, 8].

REINFORCE, When teachers help students develop a new understanding of a phenomenon rooted in an misconception, this does not necessarily extinguish prior learning. As experience shows, there are frequently two competing concepts in students' minds. To address misconception effectively, teachers must reinforce the pathway that leads to the new understanding and extinguish or at least suppress the pathway that leads to the old understanding, or help students to decide in the case of preconceptions. Failure to do so can result in students recalling the misconception preferentially over the desired understanding [6, 8].

This reinforcement should be done repeatedly, over time, and under varying conditions. This is due in part because retrieval pathways are not well established, and effort must be expended on firmly establishing the retrieval mechanism associated with the new understanding. Several important approaches from cognitive psychology can be used to do so.

2.2 Three Tier Diagnostic Test (TTDT)

Therefore, identification of the students' misconceptions is crucial for the planning of effective instruction and remediation of students' difficulties in understanding science concepts. Therefore, identification of the students' misconceptions is crucial for the planning of effective instruction and remediation of students' difficulties in understanding science concepts.

However, multiple-choice tests do not provide reasons for students' holding a particular conception. A student can give a correct answer with a wrong reason or a wrong answer with a correct reason. Because of the above limitations of the aforementioned tools, two-tier multiple-choice instruments have been developed by researchers [5, 9]. The first part of each item includes a conventional multiple choice question and the second part of each item contains a set of possible reasons for the given answer in the first part [10].

Three-tier Diagnostic Test (TTDT) enable researchers to address this limitation by adding an extra tier that require students to state whether or not they are sure about their answers to the first two tiers [11,12]. TTDT are valid tests that can be used efficiently with large samples of students, and help researchers to understand students' reasoning behind their answers without conducting interviews to distinguish misconceptions from lack of knowledge, and to estimate percentages of false positives and negatives [12]. Students could provide correct answers with wrong reasoning as "false positive" and wrong answers with correct reasoning as "false negative".

TTDT instrument have three step to analyze students' misconception:
One-tier scores: The first part of each item includes a conventional multiple choice question. This score was created by using students' answers for only the first tiers of items. Correct answers were coded as 1 and others were coded as 0.
Two-tier scores: The second part of each item contains a set of possible reasons for the given answer in the first part. This variable was based on the first two tiers of items. When a student's answer to both the first and second tiers was correct, it was coded as 1; otherwise, 0.
Three-tier scores: In the form of an affirmation question from a previously selected answer. This score was produced by taking all three tiers into account. When a student's answer to all tiers was correct, it was coded as 1; otherwise, 0. This means that it was coded as 1 if the student answers the first two tiers correctly and she/he selects "I am sure" in the third tier [13].

3. Method

3.1 Sample

In the research, the ECIRR was administered to 36 students of 11th grade at Driyorejo senior high school aged 16-17 years after they were taught about the kinetic theory of gases. Just one class to include of the research this is XI IPA grade.

3.2 Design of research

The research used pre-experimental-design of one group pretest-posttest design without control class. Design of research:

O1 X O2

O1 = pretest score (first observation before treatment)
X = treatment by implementing ECIRR model with virtual simulation media
O2 = value of the second observation (posttest)

3.3 Procedure

Identification of student's conception can be analyzed with TTDT instrument with combination answer were reported (see Table 1).

Table 1. Combine of result Three-Tier Test (TTDT).

Type Diagnostic	Combine of result	Category	Code
Three – tier	True + True + Sure	Understand the concept	SU

Type Diagnostic	Combine of result	Category	Code
	False + False + Not Sure	Lack of knowledge	LK
	False + True + Not Sure True + False + Not Sure	Guessing	G
	True + False + Sure False + T + Sure False + False + Sure	Misconception	M

Adaptation [13].

To calculate percentage of student's misconception can use this formula:

$$(\%)\text{student's misconception} = \frac{N_M}{N} \times 100\% \qquad (1)$$

N_M= quantity of question that resulted student's misconception
N =total of question

The difference between student's misconception during pretest and posttest after the implementation of ECIRR's model with virtual simulation media can be done using t-test[14].

$$t = \frac{M_d}{\sqrt{\frac{\Sigma x_d^2}{N(N-1)}}} \qquad (2)$$

Md = mean from difference between student's misconception during pretest and posttest
Xd = deviation each subject (d-Md)
Σx2d = total of chi square
N = total of subject
d.b = N-1

Hypothesis of research:
H0 : Implementation of ECIRR model with virtual simulation media in theory kinetic of gas not reduce student's misconception on 11th grade Driyorejo Senior high school
Hi : Implementation of ECIRR model with virtual simulation media in theory kinetic of gas reduce student's misconception on 11th grade Driyorejo Senior high school

4. Result

The result of implementation ECIRR's models were reported (see Table 2):

Table 2. Clasification of student's misconception.

Number of students	(%) Student's misconception of Pretest	(%) Student's misconception of Posttest	(%) Reduce of Student's Misconception	Category
3	79	14	-64	High Reduce of Student's Misconception
1	93	36	-57	
6	71	14	-57	
11	93	36	-57	
15	86	29	-57	
12	93	43	-50	Middle Reduce of Student's Misconception
14	64	14	-50	
10	57	7	-50	
33	57	7	-50	
18	86	36	-50	
4	50	7	-43	
20	71	29	-43	
21	64	21	-43	
25	64	21	-43	
27	64	21	-43	
28	64	21	-43	
23	79	36	-43	
29	86	43	-43	
36	64	29	-36	
2	36	0	-36	
5	50	14	-36	
24	79	43	-36	
26	50	14	-36	
8	57	21	-36	
16	43	7	-36	
17	57	21	-36	
19	50	21	-29	
31	79	50	-29	
35	50	29	-21	Low Reduce of Student's Misconception
7	29	7	-21	
13	29	7	-21	
34	21	0	-21	
9	21	7	-14	
22	57	43	-14	
30	57	43	-14	
32	43	36	-7	

Table 2 describes that the category of middle student's misconception more with the other categories. It's show that this sample of research matches with the normality graph, the data that used obtained comes from a normally distributed sample. If the Table 2 change to graph (see figure 1).

IOP Conf. Series: Journal of Physics: Conf. Series **997** (2018) 012034 doi:10.1088/1742-6596/997/1/012034

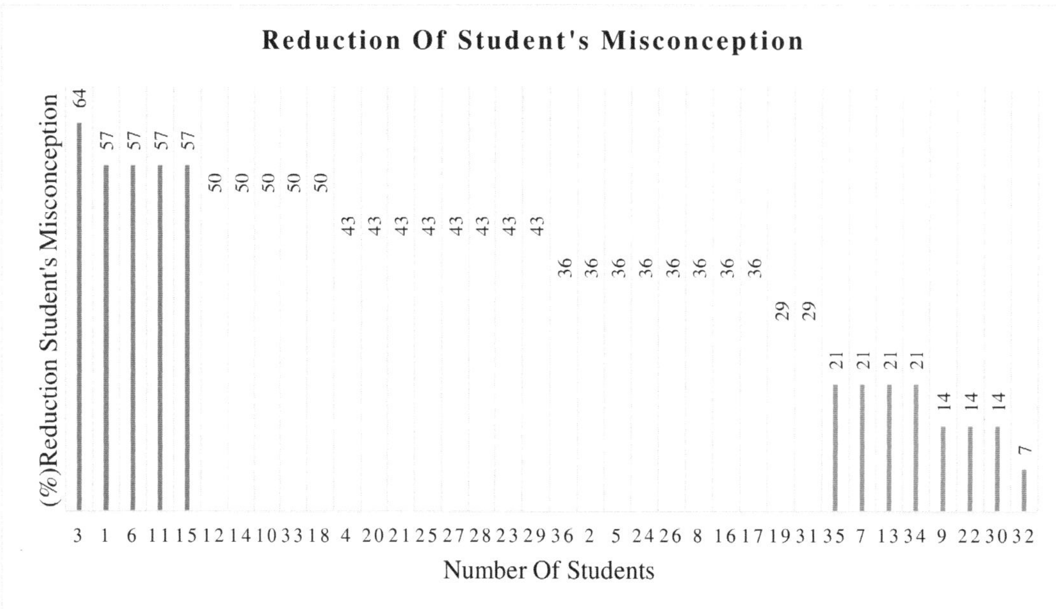

Figure 1. Graph reduction of student's misconception.

There are reduce student's misconception because the student get a treatment ECIRR's models with virtual simulation. Phase of ECIRR's model that can be used to change students conception is confront. On the confront phase, the teacher uses discrepant events to provide contradictions to student's statements or predictions and place them in a state of cognitive conflict. They confront misconceptions through demonstration, questions, and encourage discussion. There are cognitive conflict on confront phase, it make a students know that his conception wrong. Cognitive conflict suggest that when a new concept is learned it weakens or destroys an existing memory. So, cognitive conflict have an important phase to change incorrect conception to correct conception.

4.1 Discussion

The following will be discussed about the misconceptions experienced by some students.

Student 1

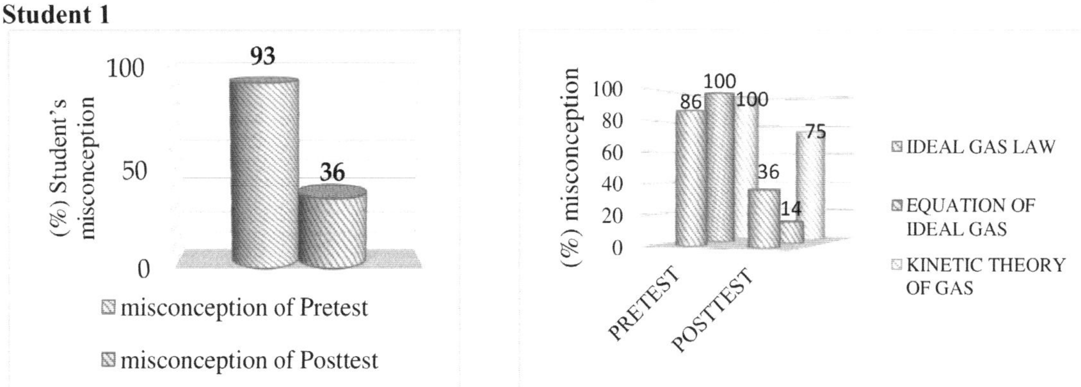

Figure 2 student's misconception diagram.

Based on the Figure 2 can be discussed as follows:

1) Student 1 has misconception 93% for pretest. The misconceptions occurred in several sub-materials, among others, the sub-material on ideal gas law about the numbers 1, 4, 5, 9, 10 and 13, because the number of questions for ideal gas law is 7 so students 1 get misconceptions of 86% in this sub

material. For sub-material equation of ideal gases, student 1 get misconceptions of the numbers 6, 7, 8 and 11, in this sub material total number is 4, so students 1 get misconceptions 100% in this sub material. On sub-material kinetic theory of gases about numbers 2,3, and 12, because the number of questions for kinetic theory of gases is 3 so students 1 get misconceptions of 100% in this sub material .

2) The result of identification misconception during posttest shows the change of misconception level experienced by Student 1. The misconception experienced by Student 1 when posttest becomes 36%, that is in sub-material ideal gas law in question number 13, so students 1 get misconceptions of 14%. For sub-material equation of ideal gases in question number 6, 7, and 11, so she/he get misconception 75%. As well as sub- material kinetic theory of gases get problem or number 12, so students 1 get reduce misconceptions become 33%.
3) The existence of students misconception reduction on some number, because students have been given treatment of ECIRR learning model with virtual simulation media. In the ECIRR learning with virtual simulation media, the teacher tries to create a learning that raises the cognitive conflict in the students through the confront phase by confronting the students with a reality different from the students' initial thinking. When students experience it, students feel uncomfortable and will be more open to new concept changes [15].

Student 32

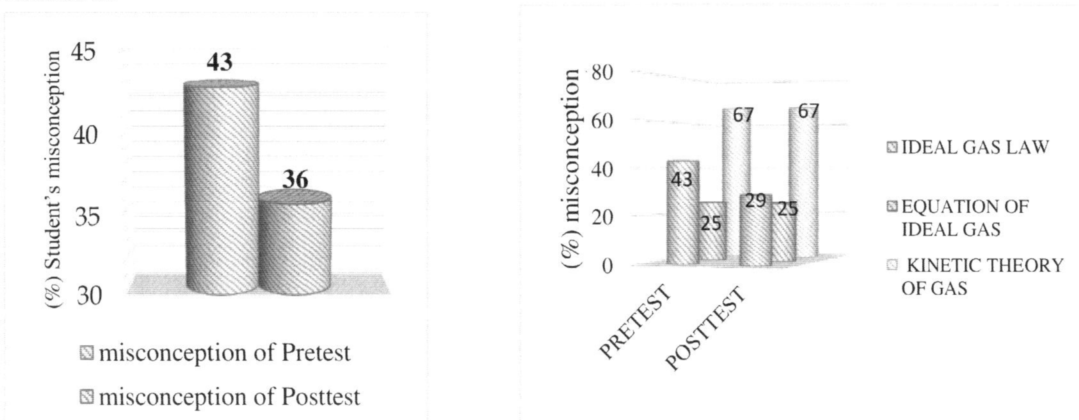

Figure 3 student's misconception diagram

Based on the Figure 3 can be discussed as follows:

1) Misconception of students number 32 during pretest can viewed from picture 2 is 43%. The misconception in all sub materials, on the sub material ideal gas law of misconceptions identified in number 9, 10, and 13, for the sub-matter equation of ideal gases in question 11, whereas in the sub-material of kinetic theory of gas at question number 3 and 12. If this data change into percent form to each submitter is 43% for ideal gas law, 25% for equation of ideal gases, and 67% for kinetic theory of gases.
2) Based on the posttest result shows the reduction of misconception to become 36%. The still-identified misconception lies in the matter of ideal gas law in questions 9 and 10, on the sub-matter of kinetic theory of gas still in questions 3 and 12, as well as on the sub-matter of the equation of ideal gases still in question number 11. So, student 32 just get reduce misconception on the matter of ideal gas law about 29%.
3) The percentage reduction of student misconception 32 is quite low at only 7%. This is probably because the percentage of misconceptions of students based on pretest and posttest is not much different, so the difference between the two results in a low percentage value. In the 32 students, there is not much change of misconception, only one number is reduced number 13. This can happen

because the learning treatment given to the students is interpreted slowly, so there is only a small change from the concept that students have [12]. It can also be from affection, social, and context factors. Elements of affection such as motivation, values, student interests can influence concept changes. Students who do not have the motivation and interest in physics learning usually do not quickly experience a change of concept [15].

5. Conclusion

Based on the results of research and discussion, the conclusion of this study are:

1) ECIRR learning model with virtual simulation media can significantly reduce misconception of students of class XI MIA-5 in SMAN 1 Driyorejo on the material of kinetic theory of gas. This is evidenced by the t test calculation, which shows that the t count is greater than the t table with < t count > = 15.60 and < t table > = 2.04. The result showed that the percentage of misconception of each student decreased after learning using ECIRR model with virtual simulation media. The mean percentage reduction in misconceptions for gas law submissions is 38%, for submission the gas state equation is 34%, and for sub chapter the kinetic theory of gas is 38%.
2) Effectively addressing misconception requires more than just eliciting, confronting, and resolving a false notion. Forming memories that are easily and accurately retrieved requires more than a desire to remember. Efforts must also include identifying the presence of misconception and reinforcing new learning. Forgetting takes work, and it is important to include activities that weaken memories and enhance recall of preferred understandings.
3) Explanations and reminders reinforce learning and are important to the habit-breaking process. Study and practice are required if students are to develop a long-lasting change in understanding and the ability to recall that knowledge accurately under a variety of new conditions. While the EICRR model for dealing more effectively with misconception is conjectural, findings from both craft wisdom and cognitive psychology would seem to suggest that it is also important to identify misconception and reinforce student learning in this area. This ECIRR conjecture could well be a fruitful area of work by physics education researchers.

References

[1] Berg V D 1991 *Miskonsepsi Fisika dan Remidiasi* (Salatiga: Universitas Kristen Satya Wacana)
[2] Fia M W, Sugiyanto and Erni Y 2016 *J. Penelit. Fis. Apl.* **6** 61
[3] McDermott L C 1991 *Am. J. Phys.* **59** 301
[4] Scott P H, Asoko H M and Driver R H 1998 International Commission on Physics Education Teaching for conceptual change: A review of strategies, in Connecting Research in Physics Education with Teacher Education, Andrée Tiberghien, E. Leonard Jossem, Jorge Barojas (Eds.)
[5] Peterson R, Treagust D F and Garnett P 1986 *Res. Sci.Edu.* **16** 40
[6] Wenning C J, Holbrook T W and Stankevitz J 2006 *J.Phys.Tchr.Educ.* **4** 10
[7] Da Silva C, Mellado V, Ruiz C and Porlán R 2007 *Sci. Educ.* **91** 461
[8] Wenning C J 2005 *J.Phys.Tchr.Educ.***3**3-10
[9] Tan K C D, Goh N K, Chia L S & Treagust D F 2002 *J. Res. Sci. Teach.* **39** 283
[10] Kirbulut Z D 2014 *Eurasia J. Math. Sci. Technol. Educ.***10**509
[11] Caleon and Subramaniam R 2010 *Int. J. Sci.Edu.* **32** 939
[12] Pesman H and Eryilmaz A 2010 *J. Educ. Res.* **103** 208
[13] Yuniarti S H, Ida K and Johar M 2014 *Proc. Math. Sci. Forum* 205
[14] Arikunto S 2006 *Prosedur Penelitian Suatu Pendekatan Praktik* (Jakarta: PT Rineka Cipta)
[15] Suparno P 2013 *Miskonsepsi Dan Perubahan Konsep Dalam Pendidikan Fisika* (Jakarta: Grasindo)

Identification student's misconception of heat and temperature using three-tier diagnostic test

Suliyanah[1], **H N P A Putri**[1] **and L Rohmawati**[1]

[1]Department of Physics, Faculty of Mathematics and Natural Science
Universitas Negeri Surabaya, Indonesia

E-mail: suliyanah@unesa.ac.id

Abstract. The objective of this research is to develop a Three-Tier Diagnostic Test (TTDT) to identify the student's misconception of heat and temperature. Stages of development include: analysis, planning, design, development, evaluation and revise. The results of this study show that (1) the quality of the three-tier type diagnostic test instrument developed has been expressed well with the following details: (a) Internal validity of 88.19% belonging to the valid category. (b) External validity of empirical construct validity test using Pearson Product Moment obtained 0.43 is classified and result of empirical construct validity test obtained false positives 6.1% and false negatives 5.9% then the instrument was valid. (c) Test reliability by using Cronbach's Alpha of 0.98 which means acceptable. (d) The 80% difficulty level test is quite difficult. (2) Student misconceptions on the temperature of heat and displacement materials based on the II test the highest (84%), the lowest (21%), and the non-misconceptions (7%). (3) The highest cause of misconception among students is associative thinking (22%) and the lowest is caused by incomplete or incomplete reasoning (11%). Three-Tier Diagnostic Test (TTDT) could identify the student's misconception of heat and temperature.

1. Introduction

Learning is a teacher's effort to help students do learning activities to gain knowledge [1-8]. Physical learning is the acquisition of knowledge developed by students in accordance with the cognitive preparedness of each [9-14]. However, the reality found many students studying physics is not interested and has no understanding after learning physics [15-22]. The results of previous research literature studies have found that there are still many students who experience misconceptions [19-26]. Misconception is a conception that is inconsistent with the conception advanced by the experts, so that students have difficulty learning physics. Misconception is a problem that needs to be reduced because if misconception is left it will be a barrier to students in learning the next material. Misconceptions occur not only among students, but occur throughout education, including students and teachers. Misconceptions are difficult to change in the self because each student builds his knowledge with the experience they have.

Misconceptions in physics include: (1) Preconception occurs when students have not followed formal learning, the students have already had an initial concept based on their experience and the phenomenon of everyday life. (2) Associative thinking has occurred when misinterpreting the terms scientific concepts or the relationship between concepts and other concepts. (3) Humanistic thinking, students think that the nature of the object or the behavior of analogous things such as human nature. (4)

Incomplete reasoning is caused by incomplete information so that the students draw wrong conclusions other than it happens because of wrong logic in taking conclusions or in generalizing. (5) Wrong intuition, usually derived from continuous observation, resulting in a spontaneous feeling when faced with the problem. (6) The stage of cognitive development of students has influenced misconception because students have not been able to abstract a problem to be solved. (7) Ability of students, students who are less talented physics or less able to study physics so that difficulty in capturing the correct concept in the learning process. (8) Student interest in the form of low talent or level of intelligence that exist in students as a basis for understanding a scientific concept [19-26]. Specifically more misconceptions have been found in heat and temperature materials [23-26] including (1) misconceptions of heat and temperature topics of respondents difficult to identify temperature and heat differences. (2) misconceptions for heat and temperature material at undergraduate level ie 5% sub temperature chapter and 48% sub heat at master student 4% sub chapter temperature and 49% sub heat and for doctoral program 7% sub chapter temperature and 40% sub chapter heat. (3) Supported by the results of preliminary tests on January 22, 2016 from 36 Unesa students class of 2014 it was found that 59% misconceptions at subheading temperature, 40% in the heat subspace as change of form, 58% on heat and temperature change and 50% on sub-chapter of heat transfer by assuming better student ability than students. From the results of these studies stated that the material temperature, heat and heat transfer of almost all students have an understanding of the concept itself and many still experience misconception.

The results are reinforced by the results of preliminary study at State Senior High School of 1 Sooko Mojokerto. (1) The results of interviews in March 2016 with one of the physics teacher at State Senior High School of 1 Sooko Mojokerto, the results of daily physics test that reached 75% PG and the value of national examination subjects physics in 2015 at State Senior High School of 1 Sooko Mojokerto averages 78.78 highest to three after Indonesian and English. (2) The average student in State Senior High School of 1 Sooko Mojokerto has achieved the learning objectives made by the teacher but, in order to achieve the basic competency the students are still experiencing learning difficulties. (3) Daily repeats given mostly only in the realm of application (C3) that is applied. (4) There is no comprehension of the concept of students as a whole. (5) The teacher has not maximally detected the location of student learning difficulties. Based on these facts there is a problem that must be solved that is necessary early detection to detect misconceptions of high school students and as a basic idea to develop learning that can reduce misconceptions of high school students.

Diagnostic tests are tests used to identify student weaknesses so that proper handling is possible [27-31]. Benefits diagnostic test can be used by teachers to know the level of mastery of student concepts, lack of knowledge, misconception, and difficulties learning [27-31]. Diagnostic tests such as multiple choice and two-tier tests are less efficient to identify misconceptions widely, test results are not sufficient enough to describe misconceptions of lack of knowledge. The alternative is a three-tier test to diagnose student misconceptions on physical materials. The three-tier type diagnostic test is an activity of constructing multiple-choice tests [32-39]. At the first level includes a regular multiple choice test, at the second level covers the reason of the selected answer from the first level and at the third level is the confidence level of the selected answer. Students answer on each item considering the truth. For example, students answer wrongly on the first and second level and have high confidence at the third level then the student misconception. The three-tier test on physics material is very accurate for measuring student misconceptions from lack of conceptual understanding, misconceptions to not understanding the concept [40-41]. Based on literature study and preliminary study there is a problem that must be solved that is need to detect and diagnose misconception of high school students in physics material using Three-tier test. The results of this study will be used as a basic idea and empirical evidence to develop learning that can reduce misconceptions of high school students on the material physics. The research problem in this study as follows:

a. How are the qualities of Three-Tier Diagnostic Test (TTDT) developed?
b. What is the profile of student misconception on the temperature, heat, and displacement material through a Three-Tier Diagnostic Test (TTDT)?

c. What is the result of identifying the causes of student misconceptions on the temperature, heat, and displacement materials through a Three-Tier Diagnostic Test (TTDT)?

2. Methodology of Research

2.1. Sample
This research was conducted at State Senior High School of 1 Sooko Mojokerto with research subjects of three classes of X MIA, ie X MIA 1, X MIA 2, and X MIA 3. The research was conducted in State Senior High School of 1 Sooko Mojokerto in the even semester of academic year 2015-2016. Initial trials were conducted in January 2016, I Trial and II trials in April 2016 in 30 students.

2.2. Procedure
This type of research is a developmental study with the development method proposed by [2]. In this study developed a Three-Tier Test Diagnostic Test (TTDT) instrument on the temperature, heat, and displacement materials. Design instruction [2] there are five stages: analysis, planning, design, development, evaluation and revision are described as follows.

(1) The analysis stage includes: (a) Analysis of problems based on literature; (b) Analysis of the temperature, heat, and displacement material; (c) Identify the student's initial concept.
(2) The planning stage includes: (a) Place of study; (b) Choosing the form of questions; (c) Selects diagnostic test instruments related to the temperature, heat and displacement materials.
(3) Design stage, Preparation of three-tier type diagnostic tests as follows: (a) Prepare the diagnostic test gratings; (b) Prepare a diagnostic test; (c) Preliminary trials to determine the choice of answers to the causes of misconceptions on the second level with first-degree answer options for divers; (d) Evaluate the preliminary test results and revise the diagnostic test instrument.
(4) The development stage includes: (a) Developing a three-tier diagnostic test on the basis of closed answers; (b) The preparation of a three-tier diagnostic test begins with the preparation of reasons at the second level. Preparation of second-level reasons based on preliminary trial results to categorize the causes of misconceptions so that a closed three-tier diagnostic test is obtained. Selection of three-tier diagnostic test is closed to make it easier to analyse the causes of student misconceptions; (c) Design validation, design validation is used to generate quality questions, at this stage validated content, language, and conformity with basic competence by the experts ie two physics lecturers then revise the diagnostic test instruments based on the input of the validator and the developed instrument is declared valid; (d) Test I, Test I conducted on students of State Senior High School of 1 Sooko Mojokerto as many as 50 students, to determine the validity, reliability, level of difficulty, and different instruments that have been made of 20 items; (d) Problem quality analysis, this quality analysis is used to determine the validity, reliability, differentiation, and degree of instrument difficulty as well as determine the number of good quality questions used to identify student misconceptions. Selected 19 questions from 20 questions that meet the quality of the good question; (e) II Trial, II Trial to students of State Senior High School of 1 Sooko Mojokerto X class MIA 3 as many as 30 students.
(5) Analysis phase includes: (a) Conducting data analysis from result of II trial to identify student misconception and its misconceptions cause 30 students of class X MIA 3; (b) Make a report of research results.

2.3. Instrument
The TTDT is a 19-item consisting of three tier items for assessing students' understanding of state of mater concepts. The first tier consists of ordinary multiple choice questions with five choices. The second tier includes one correct reason and 4 alternative reasons. The alternative reasons are the misconceptions identified from open-ended questions and then make cause of misconceptions as humanistic thinking, associative thinking, preconception, intuition, and partial reasoning. The third tier requires students to state how confident they are about their answers for the first two tiers. The TTDT

examines the conceptual area of temperature and thermal equilibrium (5 items), concept heat 8 item (thermal capacity, changed phase functions, thermal changed of heat) and heat flow (6 item). For known about student's conceptions, table 1 is combining of result three-tier diagnostic test.

Table 1. Combine of result Three-Tier Test Diagnostic Test (TTDT) [32-39].

Type diagnostic	Combine of result	Category	Code
Three –tier	T + T + Agree	Sound understanding	SU
	T + F + Disagree	Partial understanding	PU
	F + F + Disagree	Lack knowledge	LK
	F + T + Disagree	Riddle	R
	T + F + Disagree		
	T + F +Agree	Misconception	M
	F + T + Agree		
	F + F + Agree		

2.4. Data Analysis

2.4.1. Analysis of Three-Tier Test Diagnostic Test (TTDT) quality. The instrument to be used should be a good instrument of quality by calculating the validity, reliability, difficult level, and different power [42-43].

2.4.2. Analysis of misconceptions by percentage. Categorize the results of all students' answers by using equation (1). This percentage result illustrates the conception of students from concepts, lack of conceptualism, not understanding concepts, misconception, and guessing, then every misconception is analysed.

$$PKJ = \frac{n_i}{N_s}.100\% \tag{1}$$

Explanations:
PKJ = result percentage
ni = quantity students conceptions (SU, PU,LK, R, M)
Ns = all of sample

2.4.3. Analysis of the causes of misconceptions. The result of the calculation of the students entering the misconception group is why it is grouped based on the causes of misconception which includes preconception, associative thinking, humanistic thinking, incomplete (wrong) reason, and false intuition; these results are then analysed and can be calculated by equation (2).

$$PPM = \frac{n_i}{N_m}.100\% \tag{2}$$

Explanations:
PPM = percentage cause of misconceptions
ni = quantity cause of misconceptions either one
Nm = all off students misconception.

3. Result of Research

3.1 The quality of Three-Tier Test Diagnostic Test (TTDT)
Instrument has been developed

3.1.1 Internal validity of Three-Tier Test Diagnostic Test (TTDT) Internal validity validated by two expert lecturers of 88.19% belong to valid category. After doing the assessment of internal validity then conducted experiment I with a sample of 50 students of State Senior High School of 1 Sooko Mojokerto to know the external validity of the instruments that have been developed.

3.1.2 External validity of Three-Tier Test Diagnostic Test (TTDT). External validity of empirical validity test constructs using Pearson Product Moment obtained = 0.43 are classified and empirical construct validity test results obtained false positives 6.1% and false negatives 5.9% then the instrument is said to be valid.

3.1.3 Reliability of Three-Tier Test Diagnostic Test (TTDT). Test reliability of three-tier diagnostic test by using Alpha Cronbach's of 0.98 which means it can be accepted.

*3.1.4 Difficult Level of Three-Tier Test Diagnostic Test (TTDT).*The 80% difficulty level test is 20% moderate and 80% categorized as difficult.

*3.1.5 Different power of Three-Tier Test Diagnostic Test (TTDT).*The diagnostic test instrument developed 45% in sufficient category and 55% in the bad category, so 55% of the developed instruments cannot distinguish clever and clever students, but they can still be used to diagnose student misconception profiles. Recapitulation of result of instrument quality test based on validity, level of problem difficulty, different power with reliability 0.98.

3.1.6. Student misconceptions profiles on temperature, heat, and displacement materials through a three-tier diagnostic test

The II trial results of the Three-Tier Test Diagnostic Test (TTDT) type of temperature, heat and displacement were given to the students of grade X MIA 3 State Senior High School of 1 Sooko Mojokerto who had obtained the temperature, heat, and displacement materials. Three-Tier Test Diagnostic Test (TTDT) consists of 19 questions that have met the quality of the instrument a good question. The problem consists of three multiple-choice levels. The first level is the answer to the question that has five answers for the cognitive domain of C4 (analyse) and C5 (evaluate) whereas for C6 has six answers and the sixth answer is an empty answer option because, C6 (create) is the cognitive domain of students' freedom to be creative. The second level is the choice of reason consisting of 5 reasons for the deception (indicating the cause of misconception), one true reason and the choice of the last reason is an empty excuse. This empty excuse is used for students who disagree with the choice of reasons already provided, so that students express in their own language. The third level is a belief of both levels that have been answered. Misconceptions of the students of grade X MIA 3 State Senior High School of 1 Sooko Mojokerto have temperature material of heat and displacement based on the highest test of 84%, 21% low and non-misconception 7%.

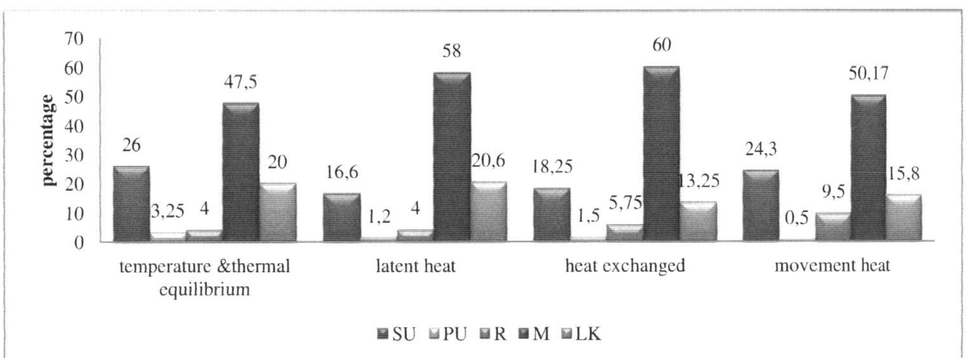

Figure 1. Graph students' conception in sub the subject.

Based on Figure 1 it is seen that students as a whole have misconceptions on the temperature, heat and displacement materials. Students experiencing the highest misconception on the material sub-heat can change the temperature of objects, heat capacity, and heat type of 60% and 18.25% of students understand the concept. 80% of students consider the heat capacity to be proportional to temperature change and cannot distinguish between the concept of heat transfer (conduction) and the parallel event i.e. heat can change the temperature of the object and the heat is used to change its shape. In sub heat material can change the form of objects 58% of students experiencing misconception and 16.6% of students understand the concept. In this sub-material 77% of students assume that water whose particles bond more closely than gas has higher energy and considers that adding heat to an object always leads to an increase in the temperature of the object. In the matter of heat transfer 51.5% of students who experienced misconception, while students who understand 24.3%. 67% of students experiencing misconceptions do not even understand the concept of material conductivity. In the sub temperature and thermal equilibrium matter the percentage of students who understood the highest concepts than the other sub-material is 26% and the student percentage of students with the lowest misconception than the other sub material is 47.5%. In this sub-material 60% of students assume that the plastic container is temperature higher than the liquid in the container and the student has not understood the concept of thermal equilibrium.

3.1.7. *Result of identification of the cause of student misconception on temperature, heat, and displacement material through three-tier diagnostic test*

The causes of misconceptions that come from students can only be analyzed in five categories of eight categories namely preconceptions (P), wrong intuition (I), wrong reasoning (R), humanistic thinking (PH), and associative thinking (PA). For the other three categories of cognitive developmental stages, students' abilities and student learning interests cannot be detected using these developed instruments. The student's cognitive development has been mapped out by Piaget by age so that it can be assumed that the cognitive development of the class X students is the same. Figure 2 shows that the highest cause of misconception in the high misconception category is associative thinking so that it can be said that the students in the high category experience misconception after obtaining the temperature, heat and displacement material and this student wrongly concepts one concept with the other. The highest cause of student misconception in the medium misconception category is faulty intuition so students spontaneously answer the phenomenon in the question. In the low misconception category the highest misconception is preconception, so before getting formal material in school students have gotten the wrong concept from their experience. The cause of student misconception on the temperature sub-matter and the highest thermal equilibrium is faulty intuition, students respond spontaneously about the given phenomenon.

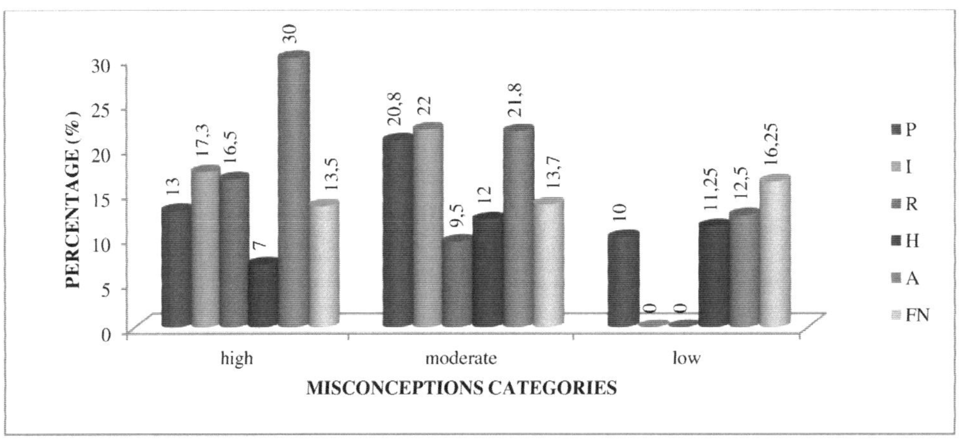

Figure 2. Percentages cause of students misconception.

In sub heat matter can change the form of objects that cause the highest misconception is associative thinking. It can be said students learn on this sub-material mix with other concepts. The causes of misconceptions experienced by students in the material sub-heating and temperature changes of objects, heating capacity, and the highest body heat are due to a preconception which means students have learned before attending formal schooling based on their experience. The cause of the misconception on the highest substantial heat transfer material experienced by students is the faulty intuition that students spontaneously respond to the phenomenon in question.

4. Discussions

The quality of the Three-Tier Test Diagnostic Test (TTDT) instrument that has been developed is stated well with the following details: (a) internal validity validated by two expert lecturers of 88.19% belong to valid category. (b) External validity of empirical construct validity test using Pearson product moment obtained = 0.43 is classified and result of empirical construct validity test obtained false positives 6.1% and false negatives 5,9% then the instrument is said valid. (c) Test reliability by using Alpha Cronbach's of 0.98 which means acceptable. (d) The 80% difficulty level test is classified and the 55% different test is classified as bad, although the differentiation and degree of difficulty is difficult and difficult to use because the diagnostic test function is to know students' learning difficulties not to distinguish interpersonal skills [31]. The results of this study received empirical support from previous studies that a qualified three-tier diagnoses test can be used to diagnose student misconceptions [27-31]. Diagnostic tests can be used to identify students' weaknesses so that appropriate handling can be done [27-31]. The three-tier diagnoses test on physics material is very accurate for measuring student misconceptions from lack of conceptual understanding, misconceptions to not understanding the concept [40-41].

Based on Figure 1, the trial results of the Three-Tier Test Diagnostic Test (TTDT) type of temperature, heat and displacement are given to students of class X MIA 3 State Senior High School of 1 Sooko Mojokerto who have obtained the material temperature, heat, and displacement. Three-Tier Test Diagnostic Test (TTDT) consists of 19 questions that have met the quality of the instrument a good question. The problem consists of three multiple-choice levels. The misconceptions that occur in the sub-material temperature and thermal equilibrium as much as 47% of the 30 students, namely on the concept: (a) Temperature, According to the students' temperature is not the result of thermal equilibrium between the system and the environment. (b) Thermal equilibrium The factor of occurrence of thermal equilibrium according to the student is influenced by: (1) the thickness of the object, the thinner the faster the thermal equilibrium; (2) The surface area the larger the surface area the faster the thermal equilibrium; (3) The material of the object, as well as the thermal equilibrium, never occurs on two objects having different materials although they are contacted in the same room. 58% of the 30 students of misconception on the sub-matter of heat can alter the existence of objects, that is, in concept (a) when the change in the form of ice becomes water, the water is never at 0 ° C and the ice never keeps the heat. (b) When the change in the form of water becomes a gas, the vapour mass is lighter than the water mass. (c) The heat used to change the form is affected by the heat, mass and surface area and shape of the object.

Misconceptions on sub heat material can change the temperature of the object at most 60% of the 30 students that is on the concept: (a) The heat capacity of the students considers the heat capacity is proportional to the temperature change of the object, so the greater the heat capacity the faster the object undergoes temperature changes. (b) The heat used for temperature change is affected by the surface area and is considered a conduction process (c) Lack of skill to relate magnitudes in thermodynamics i.e. type heat and latent heat in graphical form. In the sub-matter of heat transfer from 30 students experiencing misconceptions of 50% i.e. on the concept. (a) Conduction, understanding students about the concept of conduction is a cold flow event and conduction is proportional to the capacity of heat and latent heat. (b) Emissivity, Students understand the concept of emissivity is the ability to absorb it only for grey matter (real body). (c) Reflectivity, Reflexivity is the ability to radiate heat for a body of grey (real body). The results of this study in accordance with previous findings that still lack of understanding of students there is the concept of physics [15-22].

Figure 2 shows that the causes of misconceptions derived from students can only be analyzed in five categories from eight categories namely preconceptions (P), wrong intuition (I), wrong reasoning (R), humanistic thought (PH), and associative thinking (PA). The analysis causes of student misconceptions. Based on the results of misconception analysis and the causes of the students of class X MIA 3 State Senior High School of 1 Sooko Mojokerto can be mapped according to sub-material to be as follows. The cause of student misconception on the temperature sub-matter and the highest thermal equilibrium is faulty intuition, students respond spontaneously about the given phenomenon. Suppose the students spontaneously that the temperature of the plastic bottle is higher than the canned drink even though both are placed in the same place. In sub heat material can change the form of objects that cause the highest misconception is associative thinking that is equal to 24%. It can be said that students learn on this sub-material mix with other concepts. As well as on the phenomenon of comparing the energy that gives the worst impact of burns between water and water vapor. Students associate with the concept of heat capacity, the students believe that the steam heat capacity of water is higher than that of water so that it has greater energy and gives more burns. The concept of heat capacity is already great but it becomes a misconception because in that case the impact of burns is the bonding of the free gas particles making it easier to penetrate the skin.

The causes of misconceptions experienced by students in the material sub-heating and temperature changes of objects, heating capacity, and highest body heat are due to preconceptions which mean students have learned before attending formal schooling. One of them is the students have learned based on his experience one of them is the observation about the ice for melting faster than the ice chunk, so more quickly lower the temperature. The cause of the misconception on the ultimate heating material substance that students experience is false intuition. One example is the students think that the conduction process that flows is cold instead of heat. Based on the above exposures the highest cause of misconception on the temperature sub is the faulty intuition of 15%, in the heat sub material is a preconception of 23% and in the sub-matter of heat transfer is an incorrect intuition of 18%. The results of this study are well supported by previous research that there are still many misconceptions in the matter of heat and temperature [23-26]. The above description shows that the Three-Tier Test Diagnostic Test (TTDT) developed in this study has been shown to detect and diagnose the causes of student misconceptions on heat and temperature materials.

5. Conclusion

The results of this study show that (1) The quality of the three-tier type diagnostic test instrument developed has been expressed well with the following details: (a) Internal validity validated by two expert lecturers of 88.19% belonging to the valid category. (b) External validity of empirical construct validity test using Pearson Product Moment obtained 0.43 is classified and result of empirical construct validity test obtained false positives 6.1% and false negatives 5,9% then the instrument was valid. (c) Test reliability by using Cronbach's Alpha of 0.98 which means acceptable. (d) The 80% difficulty level test is quite difficult. (2) Student misconceptions on the temperature of heat and displacement materials based on the II test the highest (84%), the lowest (21%), and the non-misconceptions (7%). (3) The highest cause of misconception among students is associative thinking (22%) and the lowest is caused by incomplete or incomplete reasoning (11%). Three-Tier Diagnostic Test (TTDT) could identify the student's misconception of heat and temperature. Implications and further research include: (1) the need for physics learning that develops conceptually before solving the physics and theoretical problems in order to reduce misconceptions. (2) Research findings can be used as a reference for teachers using a three-tier test to detect misconceptions. (3) For further research it is necessary to observe the learning process in the classroom in order to explain the cause of the complete misconception.

References

[1] Barthelemy S R, Van Dusen B and Henderson C 2015 *Phys. Rev. ST Phys. Educ. Res.* **11** 020107
[2] Fenrich P 2004 *Proceeding of The 2004 Informing Science and IT Education Joint Conference*
[3] Griffin P and Care E 2015 *Assesment and teaching of 21st century skills: Methods and approach* (New York: Springer)
[4] Honebein P C and Honebein C H 2015 *Educ. Tech. Res. Dev.* **63** 937
[5] Batdi V 2014 *Educ. Res. Rev.* **9** 272
[6] Jenkins J S, Bugeja A D and Barber L K 2014 *Col. Teach.* **62** 129
[7] Rudolph L A, Lamine B, Joyce M, Vignolles H and Consiglio D 2014 *Phys. Rev. ST Phys. Educ. Res.* **10** 010103
[8] Shubert W C and Meredith C D 2015 *Phys. Rev. ST Phys. Educ. Res.* **11** 020138
[9] Ding L, Reay N, Lee A and Bao L 2011 *Phys. Rev. ST Phys. Educ. Res.* **7** 020109
[10] Lin Y S, Henderson C, Mamudi W, Singh C and Yerushalmi E 2013 *Phys. Rev. ST Phys. Educ. Res.* **9** 010120
[11] Lin 2015 *Phys. Rev. ST Phys. Educ. Res.* **11** 020105
[12] Noroozi O, Teasley S D, Biemans H J A, Weinberger A and Mulder M 2013 *Inter. J. Comp. Supp. Coll. Learn.* **8** 189
[13] Zhang P, Ding L and Mazur E 2017 *Phys. Rev. ST Phys. Educ. Res.* **13** 1
[14] Zwiki M B, Hu D, Finkelstein N and Lewandowski J H 2015 *Phys. Rev. ST Phys. Educ. Res.* **11** 020113
[15] Goodhew L M and Robertson A D 2017 *Phys. Rev. ST Educ. Res.***13** 1
[16] Hull M M, Lindsey A B, Archambault M, Davey K and Liu Y A 2016 *Phys. Rev. ST Phys. Educ. Res.* **12** 010101
[17] Prahani B K, Nur M and Yuanita L 2015 *Semin. Nas. Fis.*
[18] Prahani B K, Yuanita L and Soegimin W W 2015 *J. Pembel. Pend. Sains* **4** 503
[19] Snyder J L 2000 *Int. J. Sci. Educ.* **22** 979
[20] Hançer A H and Durkan N 2008 *World App. Sci. J.* **3** 45
[21] Bilal E and Erol M 2012 *J. Baltic Sci. Educ.* **11** 236
[22] Palmer 2001 *Int. J. Sci. Educ.* **23** 691
[23] Alwan A A 2010 *Misconception of Heat and Temperature Among Physic Students.*
[24] Georgio H 2010 *A Report on Preliminary Diagnostic for Indentifying Thermal Physics Conception of Tertiary Student.*
[25] Gönen S and Kocakaya S 2010 *Eur. J. Phys. Chem. Educ.* **2** 1
[26] Suparno P 2013 *Miskonsepsi perubahan konsep dalam pendidikan fisika* (Jakarta: Grasindo).
[27] Gurel D K, Ali E and Lilian C M 2015 *Eur. J. Math. Sci. Tech. Educ.* **11** 989
[28] Hastenes D and Ibrahim H 1995 *Phys. Teach.* **33** 502
[29] Lemma A 2012 *African J. Chem. Educ.* **2** 2.
[30] Lemma A 2013 *African J. Chem. Educ.* **3** 39.
[31] Suwarto 2013 *Pengembangan tes diagnostik dalam pembelajaran* (Yogyakarta: Pustaka Belajar)
[32] Wiyono F M, Sugianto and Yulianti E 2016 *J. Penelit. Fis. Apl.* **6** 61
[33] Pesman H and Eryilmaz A 2009 *J. Educ. Res.* **10** 208
[34] Dindar A C and Geban O 2011 *Proc. Soc. Behav. Sci.* **15** 600
[35] Türker F 2005 *Developing a three-tier test to assess high school students' misconceptions concerning force and motion* (Tesis: Middle East Technical University)
[36] Pesman H 2005 *Development of a three-tier test to asses ninth grade students' misconceptions about simple electric circuits* (Unpublished Master's Thesis, Middle East Technical University, Ankara, Turkey)
[37] Kutluay Y 2005 *Diagnosis of eleventh grade students' misconceptions about geometrikoptic by a three-tier test* (Unpublished Master's Thesis, Middle East Technical University, Ankara, Turkey).
[38] Kirbulut D, Zubeyde and Geban O 2014 *Eur. J. Math. Sci. Tech. Educ.* **10** 509

[39] Hidayatullah M, Humairoh F, Wachidah U, Iswati D A and Suliyanah 2015 *J. Penelit. Fis. Apl.* **5** 28
[40] Susantini D, Waskito S E and Surantoro 2014 *J. Pend. Fis.* **2** 20
[41] Arslan H O, Cigdemoglu C and Moseley C 2012 *Int. J. Sci. Educ.* **34** 1667
[42] Fraenkel J R, Wallen N E and Hyun H H 2012 *How to design and evaluate research in education* (8th ed.) (New York: McGraw-Hill)
[43] Suharsimi A 2013 *Dasar – dasar evaluasi pendidikan edisi 2* (Jakarta: Bumi Askara)

Characterization of magnetite (Fe_3O_4) minerals from natural iron sand of Bonto Kanang Village Takalar for ink powder (toner) application

M R Fahlepy[1], V A Tiwow[1] and Subaer[1]

[1]Department of Physics, Faculty of Mathematics and Natural Sciences, State University of Makassar, Jl. Daeng Tata Raya, Makassar 90224, Indonesia

E-mail: muhrizalfahlepy@gmail.com

Abstract. This research is about magnetite's characterization (Fe_3O_4) from natural iron sands of Bonto Kanang Village, District of Takalar for ink powder (toner) application. This study aims to determine the process parameters to obtain magnetite of high purity degree and to observe its physical characteristics as a supporting toner material which synthesized through co-precipitation method. The iron sand was first separated by the magnetic technique and dissolved into HCl solution before conducting the precipitation process. Precipitation was done by dripping ammonium hydroxide (NH_4OH). The precipitated powder was dried at 100°C, and then calcined at 400°C. The purity degree and magnetite mineral grain size were analyzed by XRD and SEM-EDS. The EDS elemental test before and after precipitation shown an increase of iron oxide composition from 66.70% to 87.76%. Diffractogram of XRD before and after precipitation showed Fe_3O_4 compounds with magnetite phase of 59% and 98%, respectively. The crystal structure iron sand powder structure before and after precipitation is cubic with each lattice parameters a = b = c = 8.384971 Å, V = 589.528423 $Å^3$ and a = b = c = 8.386829 Å and V = 589.920291 $Å^3$ when angle $\alpha = \beta = \gamma = 90°$. SEM images (using SE and HV 20kV) showed inhomogeneous magnetite morphology. The magnetite phase percentage that obtained based on the XRD analysis gives information that magnetite precipitation has been successfully performed with high degree of purity. The material obtained can be applied as a support toner material.

1. Introduction

South Sulawesi province has the abundance of iron sands as in the District of Takalar [1]. Iron sand deposits contain magnetic minerals such as maghemite (γ-Fe_2O_3) [2], hematite (α-Fe_2O_3) and magnetite (Fe_3O_4) [3]. The content of minerals in the iron sand is very potential as industrial materials. Mineral magnetite can be used as a biomedical application [4], diagnose including MRI (Magnetic Resonance Imaging) [5, 6], cell separation [7], drug delivery system [8-10] antibacterial activity [11], protein separation [12], hyperthermia [13-15], as magnetic catalyst on a magnesium based hydrogen storage material [16], and manufacture of ink powder (toner) [17].

Toner is a composite powder that contains a polymer, pigment, magnetite (Fe_3O_4), and additives which are used for electrophotography printing and photocopying processes. One of the most important ingredients of toner supporters is magnetite minerals because this it acts as a provider the tribocharging property for toner particles [17]. Therefore, the presence of magnetite minerals (Fe_3O_4) is needed as a toner support material.

Various methods have been reported to synthesize Fe_3O_4 among others is chemical reduction [18], sol-gel [19], electrochemical [20], flow injection [21], solvothermal [22], hydrothermal [23], microwave-assisted [24], polyol method [25], co-precipitation [26] etc. In this research, the method that used is co-precipitation method, where the parameters to be optimized are the duration of heating, grinding duration, extraction and calcination process. The use of co-precipitation methods is currently the most widespread concern because the process is simple, easy and cheap [27].

In this research the characterization of micro structure of natural iron sand in Bonto Kanang Village Takalar is done in Microstructure Laboratory of Physics Department of FMIPA UNM by using SEM-EDS (Scanning Electron Microscopy-Energy Dispersive Spectroscopy) to know morphology and composition of mineral compound and to know crystal structure, phase of Fe_3O_4 compounds and crystal size through XRD (X-Ray Diffraction) measurement data by using Debye Scherrer analysis.

2. Experimental Method

The main material used in this research is the iron sand of Bonto Kanang Village, Takalar District taken at some point randomly. In the first stages, iron sand was separated by the magnetic technique using a permanent magnet. Subsequently, the sample was crushed for 6 hours then tested using XRD and SEM-EDS methods to examine its the crystal structure, phases, morphological conditions and composition of the compound in the iron sand.

The separation result was weighed by a ratio of 25 grams of iron sand and 150 millilitres of HCl while stirred and heated to a temperature of 100°C using a magnetic stirrer at 350 rpm in 1 hour. Precipitation was done by dripping ammonium hydroxide (NH_4OH), dissolved to pH 6 and forming the precipitated reaction:

$$2FeCl_3 + 3H_2O + 6NH_4OH \longrightarrow 2Fe(OH)_3 + 6NH_4Cl + 3H_2O$$

The precipitated powder washed with H_2O and dried in the memmert oven at 100°C for 19 hours, the reaction:

$$2Fe(OH)_3 + 6NH_4OH + 3H_2O \longrightarrow 2FeOOH + 6NH_4Cl + 5H_2O$$

Furthermore, precipitated powder was calcined at 400°C with heated time for 1 hour.

Characterization was used XRD (X-Ray Diffractometer) Rigaku MiniFlex II method with PDXL2 software to find out the crystal structure, compound phase and crystal size used Deybe Scherrer analysis and SEM-EDS (Scanning Electron Microscopy- Energy Dispersive Spectroscopy) Tescan Vega 3SB to know the morphology and composition of the compound.

3. Results and Discussion

The result of EDS spectrum test of iron sand can be seen in figure 1; these spectrums show the compound composition in the iron sand is an iron oxide especially ferrous oxide (FeO).

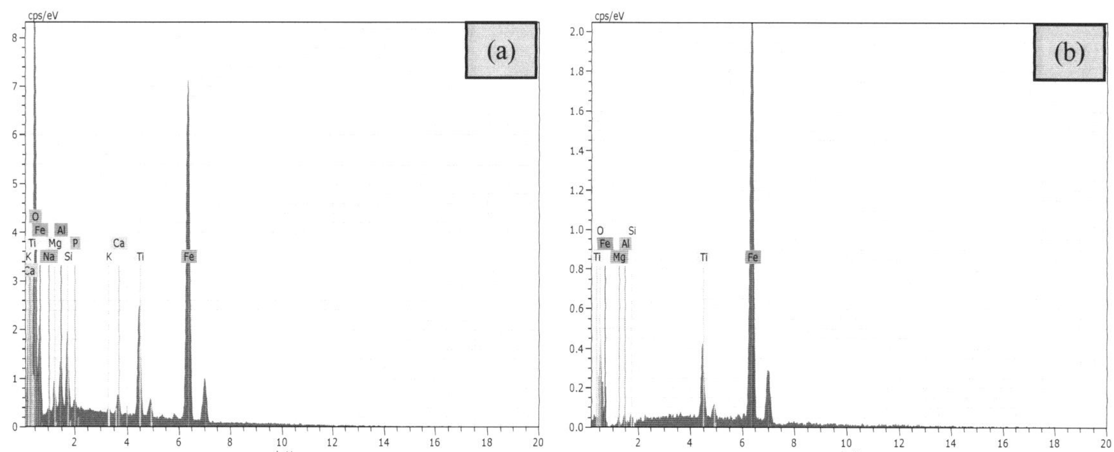

Figure 1. The EDS result of iron sand spectrum test, (a) before precipitation and (b) after precipitation.

Table 1. Result of EDS spectrum before synthesis.

Element	Unn.C	Norm.C	Atom.C	Compound norm.C	Comp.C	Error (3 sigma)
	[wt.%]	[wt.%]	[at.%]		[wt.%]	[wt.%]
Sodium	1.02	1.19	1.57	Na_2O	1.61	0.37
Magnesium	1.85	2.17	2.70	MgO	3.60	0.46
Aluminium	2.38	2.79	3.13	Al_2O_3	5.27	0.48
Silicon	2.40	2.81	3.02	SiO_2	6.01	0.43
Phosphorus	0.31	0.37	0.36	P_2O_5	0.85	0.14
Potassium	0.14	0.16	0.12	K_2O	0.19	0.11
Calcium	0.83	0.97	0.73	CaO	1.36	0.19
Titanium	7.38	8.64	5.46	TiO_2	14.42	0.77
Iron	44.25	51.85	28.05	FeO	66.70	3.73
Oxygen	24.79	29.05	54.87		0.00	9.98
Total	85.34	100.00	100.00			

Table 2. Result of EDS spectrum after synthesis.

Element	Unn.C	Norm.C	Atom.C	Compound norm.C	Comp.C	Error (3 sigma)
	[wt.%]	[wt.%]	[at.%]		[wt.%]	[wt.%]
Titanium	6.20	5.97	4.25	TiO_2	9.95	0.90
Iron	70.92	68.21	41.69	FeO	87.76	6.44
Oxygen	25.56	24.58	52.44		0.00	19.63
Magnesium	0.49	0.47	0.66	MgO	0.78	0.33
Aluminium	0.52	0.50	0.64	Al_2O_3	0.95	0.30
Silicon	0.27	0.26	0.32	SiO_2	0.56	0.20
Total	103.97	100.00	100.00			

The result of EDS analysis above shows the iron oxide compound (FeO) is major compound present in the sample. After synthesis, FeO composition content was higher than before synthesis 66.60% increased to 87.76 wt.%. This result had been accordance with previous studies when iron sand contains the dominant FeO compounds [18]. The impurity compounds present decreased composition content that is TiO2 14.42 wt.% to 9.95 wt.%, SiO2 6.01 wt.% to 0.56 wt.%, Al2O3 5.27 wt.% to 0.95 wt.% .

The presence of iron oxide or ferrous oxide (FeO) compounds in iron sand samples is according to the theory. Because basically iron sand deposits contain iron oxide compounds is magnetic. Also, the formed of Fe_3O_4 compounds expected in this research, that is reaction result between FeO (reduced) and ferric oxide Fe_2O_3 (oxidized) or $FeO + Fe_2O_3 \rightarrow Fe_3O_4$, so the presence FeO is required as the dominant compound in the sample.

Diffractogram of XRD iron sand to present is shown in figure 2. It can be seen that material have dominant composition compound is Fe_3O_4 with magnetite phase. Figure 2 (a) shows the pattern of iron sand diffraction before sample passes through the precipitation stage. XRD analyzed result showed that magnetite phase emerged with composition level 59%, then followed by impurities that Al_2O_3 and CrO_3 respectively 11% and 30%. The crystal structure is cubic with lattice parameters a = b = c = 8.384971 Å where V = 589.528423 Å^3 with angle $\alpha = \beta = \gamma = 90^o$. The crystal size was analyzed using Deybe Scherrer equation [28] by taking the highest peak in figure 2 (a). Obtained Fe_3O_4 crystals size before synthesis is 6.39244803 nm.

Figure 2 (b) shows the XRD diffractogram after passing through the precipitation process. From the analyzed results obtained Fe_3O_4 compounds with magnetite phase increased purity from 59% to 98% and ε-Fe_2O_3 impurities 1.7%. The results of the Fe_3O_4 purity study were reported previously by

Zulkarnain et al. 2014 [36], with purity levels obtained 85.80% using mechanical milling method. That means the co-precipitation method performed in this research is far superior compared the mechanical milling method if reviewed based on the purity level. The identification of Fe_3O_4 with high purity levels had been previously reported [18, 25, 31] and the results obtained in this research is close to perfect purity.

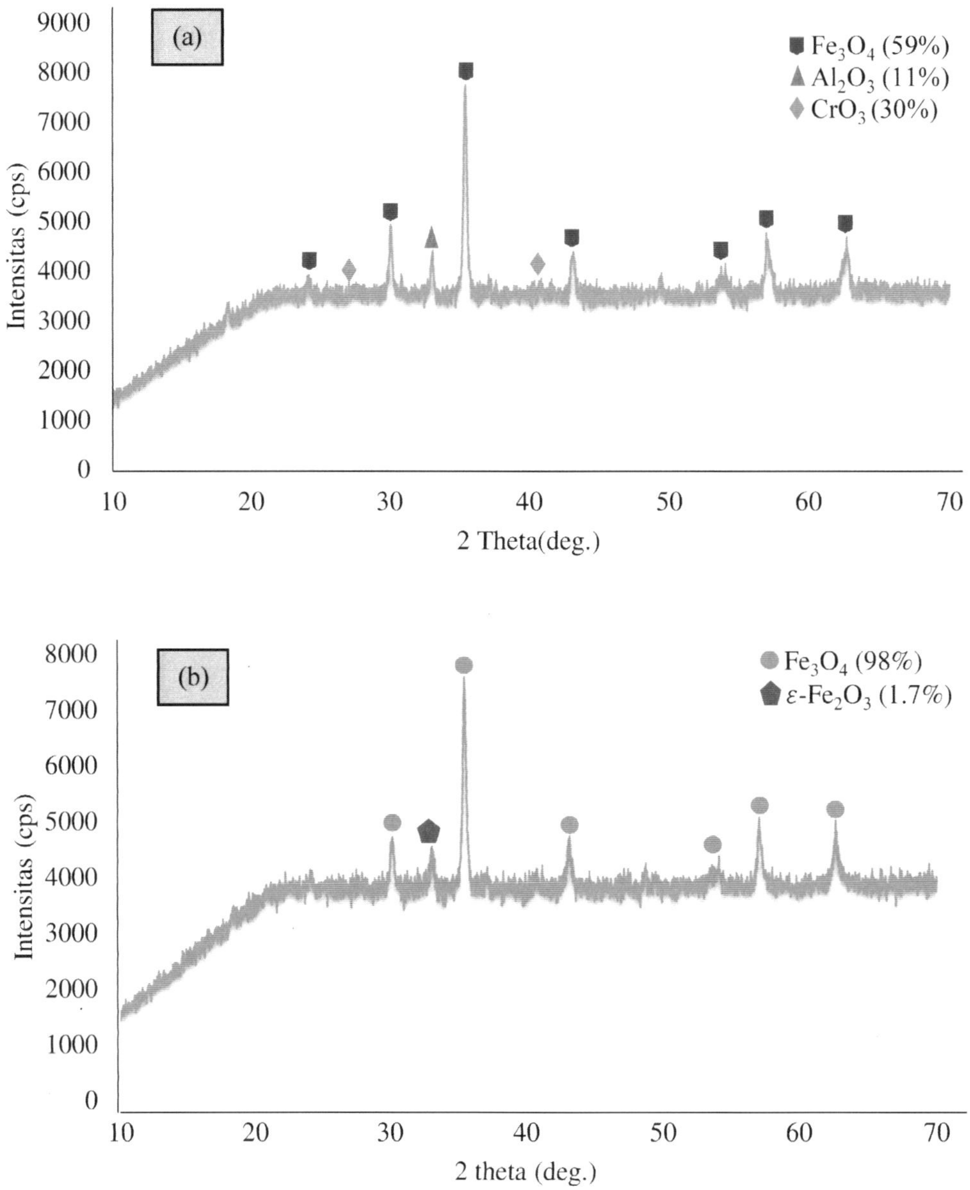

Figure 2. Diffractogram XRD from mineral Fe_3O_4 (a) before precipitation and (b) after precipitation.

The crystal structure sample after precipitation is cubic with lattice parameters a = b = c = 8.386829 Å and V = 589.920291 Å^3 with angle $\alpha = \beta = \gamma = 90^o$. The parameters obtained were similar with research result by Chaki et al, 2015, Setiadi et al, 2016, Puspitaningrum et al, 2017 [18, 29, 30]. The maximum intensity value was obtained is I (cps) = 7120 at bragg angle $2\theta = 35.497°$, which is reflected at the maximum peak (311) [10, 31-35] and corresponds to (JCPDS card No. 19-629) [37, 38].

The crystal size Fe_3O_4 was obtained based on Debye Scherrer analysis [39]. The results of the analysis obtained in this study was appropriate with previous research [18] where the crystal size obtained was 7.20 nm. In this research, the magnetite crystal size of natural iron sand of Bonto Kanang Village Takalar was found 5.57994062 nm.

SEM investigations were conducted to identify the morphological conditions and homogeneity of the iron sand sample. SEM images of the Fe_3O_4 mineral is shows in figure 3.

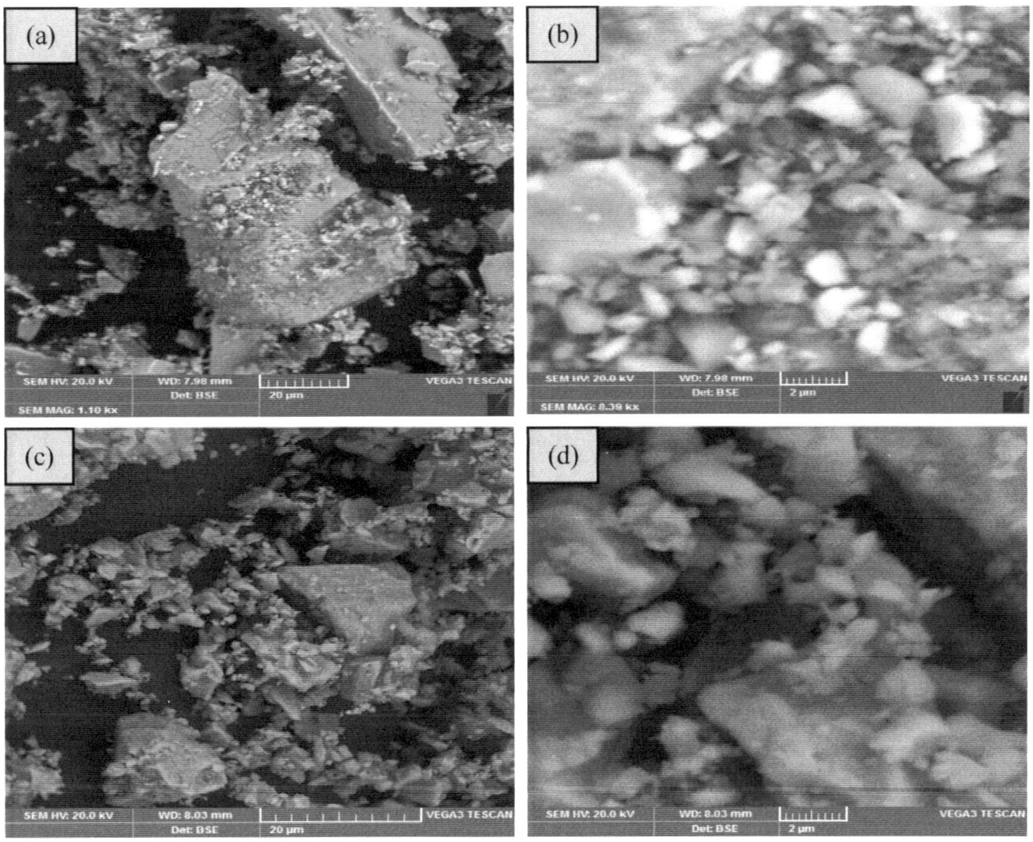

Figure 3. Morphology of iron sand (a, b) before precipitation and (c, d) after precipitation.

Figure 3 (a), (b) and (c), (d) each shows the morphological condition of iron powder before and after precipitation at 20 μm and 2 μm respectively. Based on SEM image using SE and HV detector 20 kV, it is seen that morphological conditions of iron sand appeared inhomogenous because of clumps. These visible clump is an agglomeration of tiny particles formed by scouring and calcination processes.

4. Conclusions

Characterization of mineral magnetite of natural iron sand of Bonto Kanang Village Takalar was successfully by co-precipitation method. The results of this study indicated that based on EDS spectrum analysis the composition content of FeO compound increased from 66.70 wt.% to 87.76 wt.%. While the result of XRD diffractogram before and after precipitation showed that iron sand successfully obtained high purity level from 59% to 98%. The purity level obtained is greatly influenced by the process parameters which is conducted in the research that is optimization during the precipitation process, drying the precipitation result at 100°C for 19 hours, and the optimum calcination time is 1 hour where the calcination temperature is constant 400°C. The results obtained from this research has been accordance with the results of research previously reported and can be applied as a support toner.

Acknowledgements

Acknowledgments for the financial support from Ministry of Research, Technology, and Higher Education (Kemristekdikti) the Republic of Indonesia through Student Creativity Program (PKM) 2017 and the operator of Microstructural Laboratory of Physics Department FMIPA UNM on behalf Syamsidar D and Kharisma Noor Afifah so this research can be done properly.

References

[1] Aly S H and Takdir T 2011 *J. Transportasi* **11** 123
[2] Jarzebski M, Kościński M and Białopiotrowicz T 2017 *Journal of Physics: Conf. Series* **885**-012007.
[3] Kurniawan C, Eko A S, Ayu Y S, Sihite P T A, Ginting M, Simamora P and Sebayang P 2017 *Mater. Sci. Eng: Conf. Series* **202** 012051
[4] Mishra A 2014 *Int. J. Mech. Eng. Robot.* **3** 410
[5] Phillips M A, Gran M L and Peppas N A 2010 Targeted nano delivery of drugs and diagnostics, nano today, tom 5, p. 143 – 159
[6] Lee J H, Jun Y W, Yeon S I, Shin J S and Cheon J 2006 *Angew. Chem. Int. Edn* **45** 8160-2
[7] Jing Y et al 2007 *Biotechnol. Bioeng.* **96** 1139–54
[8] Neuberger T, Schöpf B, Hofmann H, Hofmann M and Von Rechenberg B 2005 *J.Magn. Magn. Mater.* **293** 483–96
[9] Bucak S, Yavuztürk B and Sezer A D 2012 *Recent Advances in Novel Drug Carrier Systems* ed A D Sezer (Rijeka: InTech)
[10] Xuan N P, Tan P, Tuyet N P, Thi T N T and Thi V T T 2016 *Adv. Nat. Sci.: Nanosci. Nanotechnol.* **7** 045010
[11] Inbaraj B S, Tsai T-Y and Chen B-H 2012 *Sci. Technol. Adv. Mater.* **13** 015002
[12] Gu H, Xu K, Xu C and Xu B 2006 *Chem. Commun.* **9** 941
[13] Ito A et al. 2003 *Cancer Sci.* **94** 308
[14] Spiers K, Cashion J and Gross K 2003 *Key Eng. Mater. Bioceram.* **213** 254
[15] Múzquiz-Ramos E M, Guerrero-Chávez V, Macías-Martínez B I, López-Badillo C M and García-Cerda L A 2015 *Ceram. Int.* **41** 397
[16] Huang Z G, Guo Z P, Calka A, Wexler D, Lukey C dan Liu H K 2006 *J. Alloy. Compd.* **422**
[17] Ataeefard M, Ghasemi E and Ebadi M 2014 *Sci. World J.* **2014** 706367
[18] Chaki S H, Tasmira J M, Chaudhary M D, Tailor J P and Deshpande M P 2015 *Adv. Nat. Sci: Nanotechnol.* **6** 035009
[19] Xu J, Yang H, Fu W, Du K, Sui Y, Chen J, Zeng Y, Li M and Zou G 2007 *J. Magn. Magn. Mater.* **309** 307
[20] Cabrera L, Gutierrez S, Menendez N, Morales P and Herrasti P 2008 *Electrochim. Acta.* **53** 3436
[21] Alvarez S, Muhammed M and Zagorodni A 2006 *Chem. Eng. Sci.* **61** 4625
[22] Liu M and Kim K 2009 *Mater. Lett.* **63** 428
[23] Zhang J, Chen Y, Wang N and Shi W 2008 *J. Cryst. Growth* **310** 5453
[24] Hong Y, Pan T and Li Z 2006 *J. Magn. Magn. Mater.* **303** 60
[25] Vega-Chacón J, Picasso G, Avilés-Félix L and Jafelicci J M 2016 *Adv. Nat.: Nanotechnol.* **7** 015014
[26] Venkatesan M, Nawka S, Pillai S and Coey D 2003 *J. Appl. Phys.* **93** 8023
[27] Shen L, Qiao Y, Guo Y, Tan J 2012 Tianjin University **39** 737
[28] Mai T T T, Le P T H, Pham H N, Do H M and Phuc X 2014 *Adv. Nat. Sci: Nanosci. Nanotehnol.* **5** 045007
[29] Setiadi E A, Sebayang P, Ginting M, Sari A Y, Kurniawan C, Saragih C S and Simamora P 2016 *J. Phys: Conf. Series* **776** 012020

[30] Puspitaningrum A, Taufiq A, Hidayat A, Sunaryono, Hidayat N and Samian 2017 *Materials Science and Engineering: Conf. Series* **202** 012054
[31] Cuenca J A, Bugler K, Taylor S, Morgan D, Williams P, Bauer J and Porch A 2016 *J. Phys: Condens. Matter* **28** 106002
[32] Fahmiati, Nuryono and Suyanta 2017 *Materials Science and Engineering: Conf. Series* **172** 012020
[33] Guan B, Ding D, LeFan W J and Xiong R 2017 *Mater. Res. Express* **4** 056103
[34] Kurniawan C, Eko A S, Ayu Y S, Sihite P T A, Ginting M, Simamora P and Sebayang P 2017 *Materials Science and Engineering: Conf. Series* **202** 012051
[35] Yusmaniar, Adi W A, Taryana Y, Muzaki R 2017 *Materials Science and Engineering: Conf. Series* **196** 012033
[36] Jalil Z, Sari E N, Ismail A B and Handoko E 2014 *Indones. J. Appl. Phys.* **4** 1
[37] Rahmawati R, Melati A, Taufiq A, Sunaryono, Diantoro M, Yuliarto B, Suyatman S, Nugraha N and Kurniadi D 2017 *Materials Science and Engineering: Conf. Series* **202** 012013
[38] Khashan, Dagher S, Omari S A and Tit N *Mater. Res. Express* **4** 055701
[39] Chaki S, Deshpande M, Chaudhary M and Mahato K 2013 *Adv. Sci. Eng. Med.* **5** 285

IOP Conf. Series: Journal of Physics: Conf. Series **997** (2018) 012037 doi:10.1088/1742-6596/997/1/012037

The effectiveness of flipped classroom learning model in secondary physics classroom setting

B D Prasetyo[1], N Suprapto[1,a] and R N Pudyastomo[2]

[1] Physics Department, Faculty of Mathematics and Natural Science, Universitas Negeri Surabaya, Surabaya 60231, Indonesia
[2] Head of Physics Department of Singapore School – Kelapa Gading, North Jakarta 14250, Indonesia

E-mail: [a]nadisuprapto@unesa.ac.id

Abstract. The research aimed to describe the effectiveness of flipped classroom learning model on secondary physics classroom setting during Fall semester of 2017. The research object was Secondary 3 Physics group of Singapore School Kelapa Gading. This research was initiated by giving a pre-test, followed by treatment setting of the flipped classroom learning model. By the end of the learning process, the pupils were given a post-test and questionnaire to figure out pupils' response to the flipped classroom learning model. Based on the data analysis, 89% of pupils had passed the minimum criteria of standardization. The increment level in the students' mark was analysed by normalized n-gain formula, obtaining a normalized n-gain score of 0.4 which fulfil medium category range. Obtains from the questionnaire distributed to the students that 93% of students become more motivated to study physics and 89% of students were very happy to carry on hands-on activity based on the flipped classroom learning model. Those three aspects were used to generate a conclusion that applying flipped classroom learning model in Secondary Physics Classroom setting is effectively applicable.

1. Introduction

Physics as a very basic of various science makes it not limited to one particular topic. Physics as a science that studies various natural phenomena along with all the interactions that occur on it make it can't be separated from everyday life. Those various phenomena are led to various innovations in multi sector of life.

In this recent digital era, information and communication technology (ICT) develops rapidly. There are a lot of new innovations produced every day, where some of the innovations are related to education field of study. The innovations are presented with the aim to create a generation that has the skills needed in today's digital era that has the character of a thinker who has the skills to think critically, innovate, create, communicate, solve problems, and use technology and have a responsibility [1].

One of the new innovations in the education field is related to the learning model conducted in the classroom. Currently, the flipped classroom learning model is favoured by many teachers. As a new learning model, it makes many teachers want to try to apply the flipped classroom learning model.

The idea of a flipped classroom learning model arises from the problems faced by students who have to leave the class frequently for some reason. When students have to leave the class, automatically the student will miss various explanations given by the teacher during the learning process in class. On the other hand, teachers with various time constraints are very difficult to provide additional lessons outside the class to the students. From these conditions, ideas arise to make learning videos easily accessible to students [2].

2. Literature Review

2.1. Definition of flipped classroom

At the heart of the flipped classroom is moving the "delivery" of material outside of formal class time (through the use of extensive notes, video-recorded lectures and other appropriate means) and using the formal class time for students to undertake collaborative and interactive activities relevant to that material [3].

Flipped classroom learning model is an approach to education that emphasizes active learning. The flipped classroom learning model was developed by Jonathan Bergmann and Aaron Sams in 2007 [4]. Simply put, the concept of a flipped classroom is to replace activities that are usually done at home for work in school and vice versa [2]. The flipped classroom is a learning model consisting of two parts: interactive learning in groups in the classroom, and direct computer-based individual instruction outside the classroom [6].

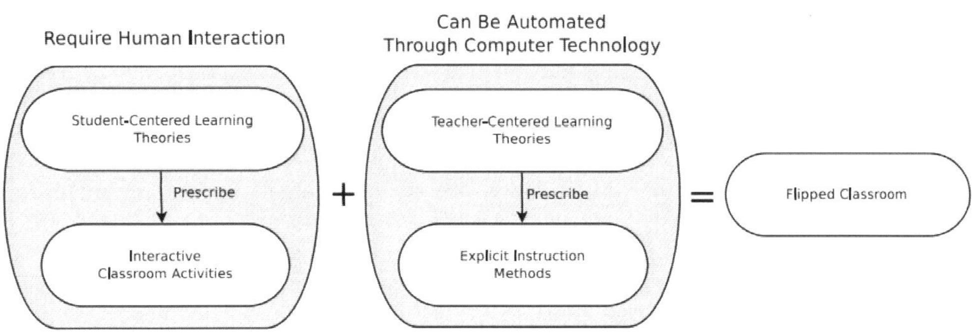

Figure 1. Diagram of flipped classroom [5].

2.2. Learning strategy of flipped classroom

The flipped classroom learning model is characterized by the materials provided to the students as homework before starting school activities. The material given can be learning videos, hand outs, or other learning materials. Students are assigned to study the materials that will be discussed in schools through various media that teachers have provided. With this, students are expected to already have enough knowledge to do activities in the classroom.

One of the advantages of using video learning media is that students can easily repeat or pause learning videos. When students feel left behind by the explanations given by teachers in the traditional classroom, students are unable to repeat the teacher's explanation at a moment's notice. However, by using video learning the students can repeat the learning video to feel understood about the material discussed in the video [2]

Classroom learning activities begin with motivations aimed at improving students' interest in learning. Then, followed by question and answer relating to the material that has been studied by the students through the medium of learning that has been given by the teacher. Here the teacher can know how much the level of understanding of students after watching the learning video given or studying other learning materials.

After the question and answer activities are completed, the activities are continued with investigation activities that can be in the form of problem-solving, lab work, or other activities aimed at improving students' understanding as well as the application of the material already learned. Here the teacher's function is changed dramatically, where the teacher is no longer a source of student information, but the teacher acts as a student facilitator in conducting activities. Where in the traditional classroom teachers should explain the material first before starting to conduct investigation activities either in the form of problem-solving or laboratory activities? [6]

Table 1. Comparison of class time in traditional versus flipped classroom [2, 4].

Traditional Classroom Activity	Time	Flipped Classroom Activity	Time
Warm-up activity	5 min.	Warm-up activity	5 min.
Go over previous night's homework	20 min.	Q&A time on video	10 min.
Lecture new content	30 – 45 min.	Guided and independent practice/or lab activity	75 min.
Guided and independent practice and/or lab activity	20 – 35 min		

2.3. Students' view of flipped classroom
Active learning was described as requiring students "to do meaningful learning activities and think about what they are doing" [7]. Baeten et al. [8] find that student-centred learning approaches are more likely (among other factors) to lead to a deep approach to learning [9] by students.

A recent survey of research into the flipped classroom learning model with results being relatively consistent in that general student opinion of the flipped classroom tends to be positive with a significant minority being opposed [5].

A research find that 80% of survey respondents in a first-year physics course preference the flipped structure to a traditional approach [10]. Where in the business course, a largely positive response from students in a flipped introductory business course was found, although 32% of survey respondents advocate a return to a traditional lecture format [11].

In the other hand, students in a flipped classroom introductory statistics class were less satisfied with classroom structure than those in the traditional class, but that they become more open to co-operation and innovation as the semester progressed [12].

Another research find that student response to the flipped classroom structure of Actuarial Techniques was largely positive, indicating it to be an approach worth pursuing in future years [3]. While around a 50/50 split of respondents thought the proposed structure would be beneficial at the start of the semester, by the end of the semester over 75% of total respondents viewed the flipped classroom as being beneficial to their learning experience compared to a didactic lecture structure. The 25% of students who viewed the flipped classroom as not being beneficial to learning is relatively consistent with the 32% in Schullery et al. [11] and the 20% in Bates and Galloway [13].

Based on the literature, students' view of the flipped classroom is very varied. Mostly the students give a good response to the flipped classroom learning model, but on the other hand, there are some students that less satisfied with the result of flipped classroom learning model. It means that the flipped classroom learning model is still debatable in students' perspective.

3. Method
The research model used is the quasi-experiment. The quasi-experiment is an experimental study that controls the research situation using non-random methods. The research design used is one group pre-test and post-test design. The subject used in this research will get treatment in the form of application

of flipped classroom learning model. Subject used in this study is students of Secondary 3 SIS Kelapa Gading who choose Physics.

O_1	X	O_2

O_1 = Pretest
X = Treatment
O_2 = Posttest

Figure 2. One group pretest-posttest design.

Before doing research, it begins by doing an observation of learning activities undertaken. It aims to know the characteristics of students who are used as research subjects. After that, preparing the learning tools that required. The learning media are designed to be suitable for the characteristics of the students themselves. This is important because, in the flipped classroom learning model, students will learn more independently through the media that have been prepared. Use of appropriate media is expected to increase student motivation to learn.

The research begins by giving a pre-test that aims to determine the initial ability of students. The pre-test is only given once at the first meeting. The pre-test was given in type of multiple choice related to the topic to be studied. Then proceed with the treatment that is in the form of application of flipped classroom learning model.

Learning activities in the class begin with a question and answer session related to the materials that have been learned by students through the given media by the teacher. Students are given the opportunity to clarify the knowledge gained by asking questions or providing answers to questions raised by other students. In this session, the teacher acts as a facilitator while ensuring the concepts built by the students are appropriate.

After the question and answer session, learning activities continued with exploration activities. Exploration activities conducted in the form of experiments (either directly or virtual computer-based) and solving a problem/phenomenon (based on the worksheet). The teacher actively guides students in conducting exploration activities. At the end of the exploration activities, the teacher provides a review of the learning activities that have been done. Before ending the learning activities, teachers always remind students to learn the next topic through the media provided (Video, PDF, PPT).

At the last meeting, post-test is given in aims to know the results of learning activities undertaken. The post-test was given in the form of multiple-choice questions and open-ended question. After that, students fill out the student response questionnaire sheet of the application of the flipped classroom learning model.

Data analysis of research result includes analysis of lesson plan implementation, analysis of pre-test and post-test result and student response questionnaire analysis.

Analysis of the implementation of the lesson plan using the following formula:

$$P(\%) = \frac{Score}{Total\ Score} \times 100\% \tag{1}$$

with the criterion of assessment score interpretation as follows:

Table 2. Interpretation Criteria.

Percentage	Criteria
25% - 43.7%	Bad
43.8% - 62.5%	Enough
62.6% - 81.25%	Good
81.26% - 100%	Very God

Analysis of pre-test and post-test data involves determining the completeness and the improvement of students' comprehension skills by calculating normalized n-gain values.

Based on the post-test result, it can be determined the percentage of students who have met the minimum criteria set. The minimum criterion specified is 50% of the total post-test value. The percentage can be calculated by the following formula:

$$P(\%) = \frac{Score}{Total\ Score} \times 100\% \tag{2}$$

Based on the pre-test and post-test results, a normalized n-gain score calculated aims to define the improvement of knowledge after being given treatment to the students. The normalized n-gain score is calculated by the following formula:

$$< g >= \frac{Posttest\ score - Pretest\ Score}{100 - Pretest\ Score} \tag{3}$$

with the following score categories :

Table 3. Gain score interpretation.

Gain Score	Category
$(< g >) \leq 0.3$	Low
$0.3 < (< g >) < 0.7$	Moderate
$(< g >) \leq 0.7$	High

In addition, based on the results of the questionnaires that distributed to students, it will be analysed in order to determine the students' responses to the application of the flipped classroom learning model. Analysis of student response questionnaire data used the following formula:

$$P(\%) = \frac{Score}{Total\ Score} \times 100\% \tag{4}$$

with the following criteria :

Table 4. Interpretation criteria.

Percentage	Criteria
25% - 43.7%	Bad
43.8% - 62.5%	Enough
62.6% - 81.25%	Good
81.26% - 100%	Very God

4. Analysis

4.1. Analysis of lesson plan implementation

Based on the observation data of lesson plan implementation which conducted by an observer obtained data as follows :

Table 5. Learning implementation aspects.

Aspects	Percentage	Interpretation
Activities Before Learning and Teaching Process	80.4 %	Good
Level of Understanding of Learning Materials	76 %	Good
Implementation of Teaching Strategy	82.4 %	Good
Use of Learning Resources	83.4 %	Very Good
Learning Instrument	81.8 %	Good
Class Management	80.4 %	Good
Implementation of Student Assessment and Evaluation	85 %	Very Good
Interrelationship with students	83.4 %	Very Good

Based on all of the observed aspects, Use of Learning Resource, Implementation of Students Assessment and Evaluation, and Interrelationship with Students included to Very Good criteria while the other aspects are included in Good criteria. It shows that the learning activities conducted by using the flipped classroom learning model on the Secondary 3 students performed well.

4.2. Analysis of pre-test and post-test result

Based on the data of pre-test and post-test results obtained as follows:

Table 6. Pre-test and post-test results.

Pre-test		Post-test	
Minimum Score	28	Minimum Score	36
Maximum Score	78	Maximum Score	92
Average Score	54	Average Score	72

Based on posttest result obtained that 89% of students had passed the minimum criteria of standardization. Furthermore, to find out the increment of students understanding of the concept by calculating the normalized n-gain score. The normalized n-gain score that obtained is 0.4. That normalized n-gain score is fulfilled in the moderate category.

4.3. Analysis of questionnaire
Based on the analysis of questionnaire that distributed to the students, obtained that 93% of students become more motivated to study physics and 89% of students were very happy to carry on hands-on activity based on the flipped classroom learning model.

The use of appropriate media makes students more motivated to learn physics. This is shown by 93% of students who become increasingly motivated to study physics by using the flipped classroom learning model. In addition, the appropriate media allows students to understand the concept of physics well so that makes students more easily during the exploration activities.

5. Conclusion
Based on the data analysis, 89% of students had passed the minimum criteria of standardization. The increment level in the students' mark analysed by normalized n-gain formula, obtaining a normalized n-gain score of 0.4 which fulfil medium category range. Obtains from the questionnaire distributed to the students that 93% of students become more motivated to study physics and 89% of students were the to carry on hands-on activity based on the flipped classroom learning model. Those three aspects were used to generate a conclusion that applying flipped classroom learning model in Secondary Physics Classroom setting is effectively applicable.

References
[1] Trilling B and Fadel C 2009 *21st Century Skills: Learning for Life in Our Times* (California: Jossey-Bass)
[2] Bergmann J and SamS A 2012 *Flip Your Classroom: Reach Every Student in Every Class Every Day* (USA: Library of Congress Cataloging-in-Publication Data)
[3] Butt A 2014 *Bus. Ed. Acred.* **6** 33.
[4] Bergmann J and SamS A 2009 *Learn. Leading With Technol.* **36** 22
[5] Bishop J L and Verleger M A 2013 *Proc. 120th American Soc. of Engr. Edu. Ann. Conf. Expo.* (United States: Atlanta/ASEE) 6219
[6] Jihyun L Cheolil L and Hyeonsu K 2017 *Educ. Tech. Research Dev* **65** 427
[7] Prince M 2004 *J. Eng. Educ.* **93** 159
[8] Baeten M, Kyndt E, Struyven K and Dochy F 2010 *Educ. Res. Rev.* **5** 243.
[9] Marton F and Säljö R 1976 *Br. J. Educ. Psychol.* **46** 4
[10] Guoqing Z and Xuefeng J 2014 *Appl. Mechanics and Matls* 543-547 4312
[11] Schuerly N M, Reck R F and Schuerly S E 2011 *Int. J. Bus. Humn. Tech.* **1** 1
[12] Strayer J F 2012 *Learn. Environ. Res.* **15** 171
[13] Bates S and Galloway R 2012 *The Inverted Classroom in A Large Enrolment Introductory Physics Course: A Case Study* The Higher Education Academy.

Characterization of natural puya sand extract of Central Kalimantan by using X-Ray Diffraction

K G Suastika[1], K Karelius[2] and I N Sudyana[2]

[1]Program Studi Fisika, FMIPA Universitas Palangka Raya, Indonesia, 73112
[2]Program Studi Kimia, FMIPA Universitas Palangka Raya, Indonesia, 73112

E-mail : suastika@mipa.upr.ac.id

Abstract. Start Zircon sand extraction in this study use natural sand material from Kereng Pangi village of Central Kalimantan, also known as Puya sand. There are only three ways to extract the Puya sand. The first is magnetic separation, the second is immersion in HCl, and the third is reaction with NaOH. In addition, sample of each extraction step is analyzed with X-Ray Diffraction (XRD). Then based on the quantitative analysis using X'Pert Highscore Plus software, the samples are identified mostly as zircon (ZrSiO4) and silica (SiO2). Moreover, after the immersion process with HCl, the silica compound goes down and the zircon compound climbs to 74%. In the reaction process with NaOH zircon compound content further to increase to 88%.

1. Introduction

Puya sand or commonly known as zircon sand is by-product from the processing of heavy mineral sand that commonly found in river and shore. It can also be acquired from the processing of rare earth minerals [1-2]. Zircon is the common accessory mineral in the majority of igneous and metamorphic rocks with Zr as an essential structural constituent [3]. Zircon minerals are mostly associated with some valuable minerals such as monasite, senotim, and ilmenite [4]. Pure Puya sand has density of 4.68 g/cm3 with varied colors like yellowish, pink, reddish, brownish, colorless, and sometimes green, blue or black [5-6].

Nowadays Zircon mineral in Indonesia can be discovered in many areas of Sumatra, Kalimantan, Sulawesi and Papua [4]. Its existence has been known for a long time as an alluvial sediment with tin sand and other mineral by-products at the waters of Bangka Belitung. In addition, Zircon is also present along the rivers in rural Kalimantan following the spread of gold alluvial sediment and swamp. In the Borneo island, the sediment containing Zircon are identified from the results of alluvium recycling to obtain gold and diamond [5].

The natural resources that become the object of this research is the Puya sand which is mined in Kereng Pangi area, as by-product of gold mining. Puya sand can be processed into zirconia as one of the ceramic material with toughness and strength as its advantages [7]. In general, zircon compound in nature is discovered zirconium silicate compound ($ZrSiO_4$) commonly referred as zircon hence it has to be purified first in order to gain zirconia.

Extraction is one of purification methods with magnetic separation, immersion with acid, and reacting with a strong base as the steps [8-9]. In this research, X-Ray Diffraction (XRD) is used to determine the crystallographic character and composition of Puya sand before and after the extraction process such as magnetic separation, immersion with acid and reaction with NaOH. Furthermore, it is also used to know other details like the arrangement of various types of atoms in the crystal, the presence of defects, orientation, and crystal defects [7].

2. Research Method

The initial phase of this research is started with the preparation of the Puya sand samples from Kereng Pangi, Central Kalimantan. The Puya sand is watered with aquades to remove the initial visible impurities before drying. Prior to the treatment, Puya sand samples are analyzed first using XRD. The extraction process begins with a magnetic separation using a permanent magnet to separate the Fe element [8]. The sand is smoothed using the planetary milling ball for 2 hours to reduce the particle size. The powder is soaked with 2M HCl for 12 hours. The precipitate is watered with aquades until it reaches pH 7 and then is dried. Dry powder is reacted with 7M NaOH at a temperature of 300 °C until the sample is shoveled, then settled to room temperature [9]. Moreover, aquades is added to the sample and stirred for 1 hour using magnetic bar and stirrer. Then it is filtered to obtain sediment and Na_2SiO_3 solution. The result of the precipitate is watered again with aquades until it reaches pH 7 and dried to produce Zircon powder subsequently characterized using XRD. XRD test aims to obtain the characteristics of crystallography and the composition of Puya sand. The XRD was setup at voltage of 40 kV with 35 mA in current. Angle shots start from 15-90 deg, a step size of 0.02 deg, time/step 0.5 s, and a scan speed of 0.04 deg/s. XRD data is analyzed using Highscore Plus software which is the commercial powder diffraction analysis software from PAN analytical [10].

3. Results and Discussion

3.1. Puya extraction process

Magnetic separation on the extraction process aims to reduce magnetic element that contained in Puya [8]. This process was done because Zircon ($ZrSiO_4$) is a non-magnetic mineral [11]. The following images of Puya sand before and after treatment are presented: XRD analysis of the composition of Puya

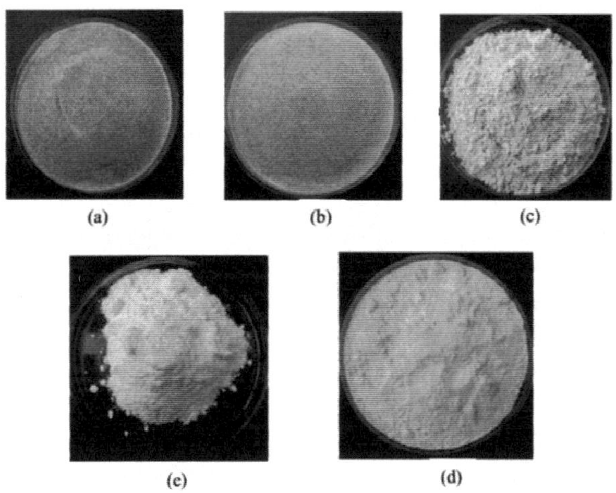

Figure 1. Samples of Puya sand; (a) Puya sand (b) after magnetic separation (c) after milling process (d) after immersion in HCl (e) after reaction with NaOH.

Figure 1(a) shows the samples of Puya before treatment. While Figure 1(b) shows the samples of Puya after magnetic separation treatment. Based on Figure 1(a) and Figure 1(b) can be seen the difference of color between before and after the treatments. The color of sample before the magnetic separation is darker than after the magnetic separation. Elsner said that dark color is caused by the presence of impurities particles, especially compounds of magnetic element [6]. Figure 1(c) shows the samples after milling process. While Figure 1(d) and Figure 1(d) show the samples after the immersion treatment with

HCl and after thermal immersion treatment with NaOH, both of these samples resulted a white powder that was estimated as zircon

3.2. XRD analysis of the Composition of Natural Puya

One of important functionalities in Highscore Plus software is quantification of phases with partially know or unknown crystal structure. In this software, empirical structure factors are extracted via a Pawley or a LeBail fit and are used for HKL file fit (together with a pseudo-formula mass derived from the known concentration in the binary mixture) for further quantifications following the Rietveld formalism [10].

Figure 2 shows the results of XRD data that has been analyzed by using software Highscore Plus. The result shows that the Puya contains the composition of $ZrSiO_4$ 58% and SiO_2 23% and $CaSi_2$ 19%. From the result can be seen that there is exist two main elements of Zr and Si. Elsner said that the sand which contains element of Zr and Si can be used as the basic material for zirconia production, with the main target being zircon ($ZrSiO_4$) [6] so that need extraction process to get Zr and Si with higher purity.

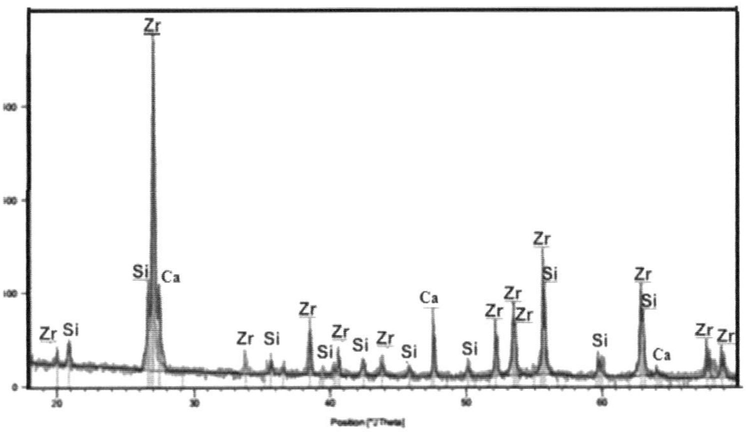

Figure 2. XRD Result of Natural Puya.

3.3. XRD Analysis of Puya Sand Composition after Magnetic

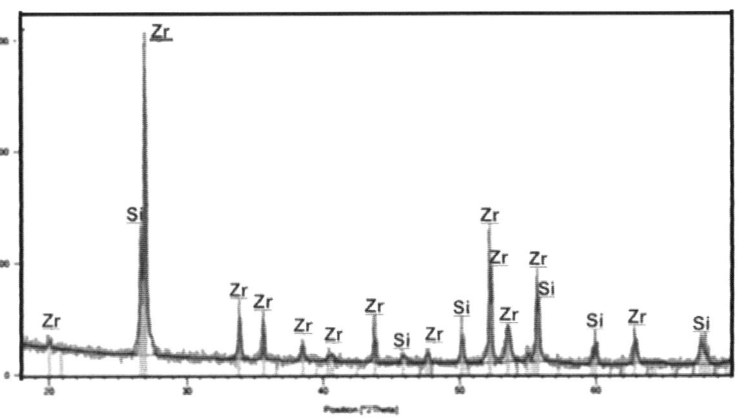

Figure 3. XRD data Result of Puya Sand after Magnetic Separation.

Magnetic Separation of the extraction process aims to to separate the magnetic element from sample, especially Fe element. This process is done because zircon ($ZrSiO_4$) is a non-magnetic mineral so that can be separated with Fe element by utilizing magnetic field.

Figure 3 shows the results of XRD data Result of Puya Sand after Magnetic Separation that has been analyzed by using software Highscore Plus. The result shows that the sample contains the composition $ZrSiO_4$ 65% and SiO_2 35%. The increase of zircon and silica content occurs due to the loss of impurities particles, especially the Fe containing compounds so that obtain zircon and silica sample with a better purity level.

3.4. XRD analysis of the Puya Sand composition after HCl Immersion

Before the HCl immersion process, the milling process was done to increase the particle's reactivity because with the small particle size the contact surface area of the reaction becomes larger making it easier in the dissolution process.

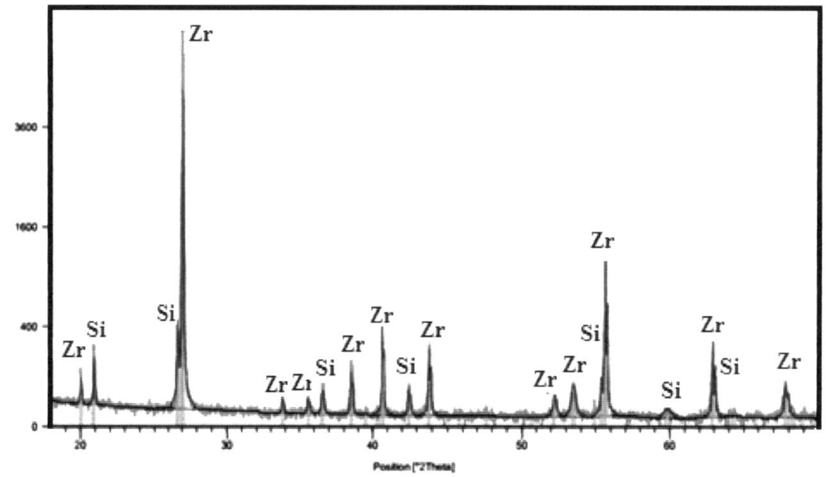

Figure 4. XRD data Result of Puya Sand after HCl Immersion.

Figure 4 shows XRD data results after HCl immersion process that has been analyzed by using software X'Pert Highscore Plus. The result shows that there was occured the increase of zircon content from the previous treatment from 65% to 74% and silica decreased from 35% to 26%. The immersion process was using high concentration of HCl that aims to dissolve the impurities that can not be separated by magnetic separation, so that resulted the increase the purity of zircon. While the decrease of silica percentage caused by some non-framework of silica also dissolved due to HCl immersion process.

3.5. XRD analysis of the composition of the Puya Sand after NaOH Reaction

Figure 5 shows XRD data result after NaOH reaction that has been analyzed by using software Highscore Plus. The result shows that there was an increase of zircon content from the previous treatment from 74% to 88% and silica decreased from 26% to 12%. The decrease of silica percentage occurs because when the sampel was dissolved in NaOH solution there was occurs demolition of chemical bond of SiO_2 due to reaction with NaOH solution so that it was formed the dissolved of sodium silicate (Na_2SiO_3) precursor, and then there was occurs the decrease of SiO_2 percentage in sample. Based on the extraction results, the purity of zircon after reaction with NaOH was quite high. So the sample can be synthesized into zirconia (ZrO_2).

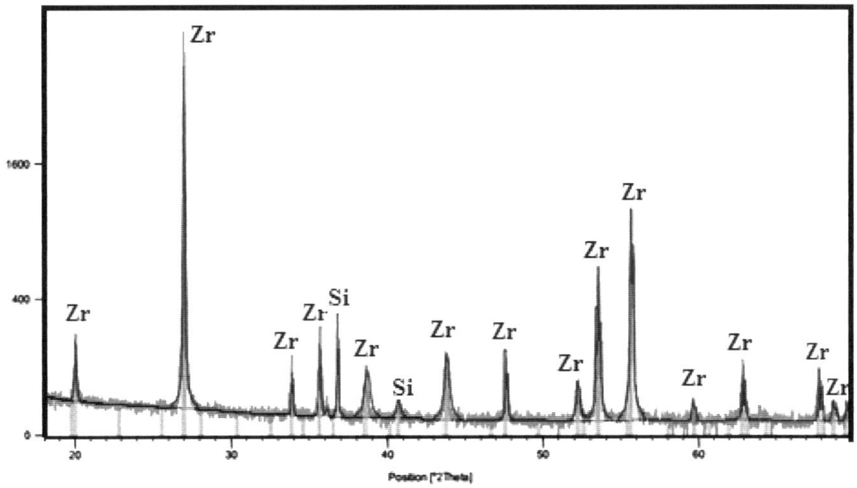

Figure 5. XRD Result of Puya Sand after NaOH Reaction.

Less purity of zircon obtained as a result of extraction because treatment with HCl and NaOH still leaves the Hf element whose forming phase that was not identified XRD. The presence of Hf elements in zircon powder was difficult to be identified because all of Zr compound contain Hafnium about 1-3%. Chemically, Zr and Hf have very similar properties that can not be separated by the extraction process in general [2,4,5].

3.6. Characteristic Crystallography of Puya Sand

Further analysis of the samples was performed with XRD to determine the crystallographic characteristics of the Puya sand before and after the extraction process. Based on the analysis of XRD data, most of the identified peaks were zircon ($ZrSiO_4$) with crystal parameters a = b = 6.607 Å and c = 5.982 Å and $\alpha = \beta = \gamma = 90°$, with density 4.66 g/cm^3 and cell volume 261.13 µm^3 with crystal system is tetragonal, and silica having quartz structure (SiO_2) with crystal parameter a = b = 4.913 Å and c = 5.404 Å and $\alpha = \beta = 90°$ and $\gamma = 120°$, with density 4.66 g/cm^3 and cell volume 112.96 µm^3 with crystal system is hexagonal.

4. Conclusion

The characterization result of Puya using X-Ray Diffraction after extraction process, we can describe as follows: (1) After the final extraction process with NaOH, the content of zircon compound in the Puya sand is 88%, so that the sand extract can be processed into zirconia. (2) Most of the identified peaks were zircon compounds ($ZrSiO_4$) with crystal parameters $a = b \neq c$ and $\alpha = \beta = \gamma$ with crystal system is tetragonal, and silica having quartz structure (SiO_2) with crystal parameters $a = b \neq c$ and $\alpha = \beta \neq \gamma$ with crystal system is hexagonal.

References

[1] Biswas R K, Habib M A, Karmakar A K and Islam M R 2010 *Hydrometallurgy* **103** 124
[2] Skidmore C 2005 *Zirconium and Hafnium* (Stoke/UK: MinChem Ltd.)
[3] Hoskin P W O and Schaltegger U 2003 *Rev. Mineral. Geochem.* **53** 27
[4] Poernomo H, Dwi Biyantoro and Maria V P 2016 *Eksplorium* **2** 73
[5] Poernomo H 2012 *Informasi Umum Zirkonia* (Yogyakarta: Badan Tenaga Nuklir Nasional Pusat Teknologi Akselerator dan Proses Bahan)
[6] Elsner H 2013 *DERA Rohstoffinformationen Zircon – Insufficient Supply in the Future* (Berlin: DERA)
[7] Priyono S and Febrianto E Y 2012 *J. Ilmu Pengetah. Teknol.* **30**

[8] Mun S P, Cai Z and Zhang J 2013 *Mater. Lett.* **96** 5
[9] Amaral J C B S and de Morais C A 2016 *World J. Eng. Technol.* **4** 138
[10] Degen T, Mustapha D, Egbert B, Uwe K and Gwilherm N 2014 *Powder Diffr.* **29** S13
[11] Abubakar F 2009 *Pengelolaan Zirkon di PT Timah Tbk.* (Yogyakarta: Workshop Keselamatan dan Keamanan Pertambangan Zirkon Bagi Pekerja, Masyarakat dan Lingkungan)

IOP Conf. Series: Journal of Physics: Conf. Series **997** (2018) 012039 doi:10.1088/1742-6596/997/1/012039

The performance assessment of undergraduate students in physics laboratory by using guided inquiry

H Mubarok[1], A Lutfiyah[1], A Kholiq[1], N Suprapto[1] and N P Putri[1,a]

[1]Department of Physics, Faculty of Mathematics and Natural Science, Universitas Negeri Surabaya, Surabaya 60231, Indonesia

E-mail: [a]nugrahaniprimary@unesa.ac.id

Abstract. The performance assessment of basic physics experiment among undergraduate physics students which includes three stages: pre-laboratory, conducting experiment and final report was explored in this study. The research used a descriptive quantitative approach by utilizing guidebook of basic physics experiment. The findings showed that (1) the performance of pre-laboratory rate among undergraduate physics students in good category (average score = 77.55), which includes the ability of undergraduate physics students' theory before they were doing the experiment. (2) The performance of conducting experiment was in good category (average score = 78.33). (3) While the performance of final report was in moderate category (average score = 73.73), with the biggest weakness at how to analyse and to discuss the data and writing the abstract.

1. Introduction

The laboratory becomes an important part of physics learning [1], in addition to learning how to perform experiments perfectly, the undergraduate physics students are also trained to have good scientific performance [2] and can communicate the results of experiments both orally and in writing [3].

The creativity of undergraduate physics students is needed in scientific performance and communicates the results of the experiment. At the cognitive domain level, the undergraduate physics students are already above the ability to understand (C2) [4]. Experiments can increase the level of critical thinking and effectiveness on learning [4-5], it shows that the cognitive ability of undergraduate physics' students has reached the evaluation stage (C5) [4].

Worksheets or experiment guidebook are needed for designing experiment [7]. In this research used guidebook based guided inquiry. Inquiry becomes a trending topic to improve the performance of undergraduate physics students in conducting experiments [8] to train scientific performance and to improve the critical thinking [9].

The undergraduate physics students' performance are obtained from assessment result of pre-laboratory activity, laboratory activity and experiment report [6][9–11]. Students' preparation improvements for actual activity during laboratory are known from pre-laboratory result [13]. Besides that, students' performance assessment is authentic assessment to assess the students' ability of concept to apply in real situations and train students in having a positive mental attitude [14]. The guidebook based on inquiry using is expected to improve the ability of critical thinking and excellent performance.

2. Method

Participants of the research included 116 undergraduate physics students at State University of Surabaya, Indonesia, which are distributed from four classes, with four topics of basic physics experiment: density, pulley system, mathematical pendulum and spring constant. The research design used a descriptive quantitative approach by using the guidebook of basic physics experiment [7][15]. This is grouped three assessment stages: pre-experiment, conducting experiment and final report. Data was obtained from the score of all stages. It will be analysed in descriptive of the undergraduate physics students' performance as the average score at three stages and then identified their performance capability.

Based on the guidebook of basic physics experiment [7], the scoring system used two types. First, pre-laboratory score and conducting experiment score by giving the check list (√) on the assessment sheet (Table 1 and Table 2) [7] and then converted into numeric form (Table 3) [7]. While the taking score of the final report by giving the score directly using the number (Table 4) [7]. Second, changing the score to the student's performance (Table 5) [7].

Table 1. Pre-laboratory assessment aspects.

No	Aspects	Good (G)	Moderate (M)	Low (L)
1	Understand of the experiment goals			
2	Understand of the concepts that related to experiment			
3	Identify the variables that must be measured and calculated			
4	Understand of the tool's specification and how to use it			
5	Able to determine the data to be obtained, how to obtain and how to analyse			
	Total			

Table 2. Experiment assessment aspects.

No	Aspects	Good (G)	Moderate (M)	Low (L)
1	The ability to set tools correctly			
2	The ability to use and read the scale of the measuring instrument properly			
3	Cooperation among group members			
	Total			

Table 3. The conversation score.

Term	Symbol	Score
Good	G	80
Moderate	M	70
Low	L	65

The three stages of physics performance have different assessment aspects, at the pre-laboratory have five aspects assessed (Table 1) [7], whereas at the conducting experiment have three aspects (table 3) [7], and for the final report have six aspects assessed (Table 4) [7].

Table 4. The final report assessment aspects.

No	Aspects	Score
1	The ability to write abstract	
2	The ability to write basic theories that support the experiment	
3	The ability to formulate experimental methods	
4	The ability to analyse the data with uncertainty measurement and answer the questions	
5	The ability to discuss the results of the data	
6	The ability to formulate the conclusions	
Total		

The final score is obtained by each student based on three stages, with different weights. The pre-laboratory has a weight of 2, the conducting experiment has a weight of 3 and the final report has a weight of 5, which all points are divided by 10 (equation 1). The students' performance category can be grouped into 5 levels (table 5) [7].

$$The\ final\ score = \frac{(\mathbf{2}\ \boldsymbol{x}\ \boldsymbol{Pre-Laboratory})+(\mathbf{3}\ \boldsymbol{x}\ \boldsymbol{Conducting\ Experiment})+(\mathbf{5}\ \boldsymbol{x}\ \boldsymbol{Final\ Report})}{\mathbf{10}} \quad (1)$$

Table 5. The range of performance category.

Range	Students' performance
80.1 – 85	Excellent
75.1 – 80	Good
70.1 – 75	Moderate
65.1 – 70	Low
60 – 65	Very low

3. Result and Discussions

3.1. Pre-Laboratory

Pre-laboratory becomes the first assessment stage process of the basic physics experiment. This activity can improve students' performance skills before experiment [11][13]. Based on the five pre-laboratory aspects assessed from four experiment topics obtain results that are not much different. The students' ability to understand the experiment purpose is in a good category, with the score obtained from the four topics, which are 78 from the average of all participants. On the understanding concepts that related to practical topics, the most difficult topic is the spring constant, this is due to the ability of physics students in describe the formula of total constants from various arrangements of spring is not perfect. Based on the score obtained the ability to understand the concepts related to experiment on the topics of spring is in a moderate category. While on the other three topics is in a good category. For the third aspect, the ability of students is in good category, that is identifying the variables that must be measured and calculated.

The fourth aspect of the pre-laboratory stage is to understand of the tool's specifications and how to use it, from this aspect the ability of students is in good category, this is because before doing the experiment, students have been briefed on the tools and experiment tools to be used for basic physics experiment. The Topics of the pulley system get the lowest score compared to the other three topics, the problem is students' performance in a set of pulleys system not perfect. However, from the average score obtained, the category of this topic is good with score 77 (table 6), while the score of the other topics is 79 (table 6).

The ability to determine the data to be obtained, how to obtain, and how to analyse it is the fifth aspect of the pre-laboratory stage. This aspect becomes the basic substance for reporting the results of the experiment in the final report. In this aspect the topic of the pulley system also gets the lowest score, the score is 74 (table 6) that into the moderate category. The weakness of student performance in this aspect is how to analyse the data, they are confused in managing the data obtained. Both in analyse the calculations and graphs. While on the other topics is in a good category. Overall, based on the average score obtained from each experiment topic in the pre-laboratory stage of student performance is in a good category.

Table 6. Assessment of basic physics experiment.

Stage	Aspect	Score			
		Density	Pulley system	Mathematical pendulum	Spring constant
Pre-laboratory	Understand of the experiment goals	78	78	78	78
	Understand of the concepts that related to experiment	76	77	76	75
	Identify the variables that must be measured and calculated	77	77	79	80
	Understand of the tool's specification and how to use it	79	77	79	79
	Able to determine the data to be obtained, how to obtain and how to analyse	78	74	78	77
	Average	77	77	78	78
Conducting experiment	The ability to set tools correctly	79	77	78	78
	The ability to use and read the scale of the measuring instrument properly	75	79	78	77
	Cooperation among group members	80	80	79	80
	Average	78	79	78	78
Final report	The ability to write abstract	70	73	76	73
	The ability to write basic theories that support the experiment	76	68	77	76
	The ability to formulate experimental methods	75	79	78	78
	The ability to analyse the data with uncertainty measurement and answer the questions	72	76	79	77
	The ability to discuss the results of the data	65	67	74	75
	The ability to formulate the conclusions	71	72	73	71
	Average	**71**	**73**	**76**	**75**
	Total score	**75**	**75**	**77**	**76**

3.2. *Conducting experiment*

The second stage of performance assessment of the basic physics experiment is conducting an experiment. In this activity there are three aspects assessed. The first is the ability to set tools correctly. The student's performance in this aspect is in a good category. The second aspect of this stage is the ability to use and read the scale of the measuring instrument properly, on the topic density get the lowest score compared to other topics. this is because the students' performance in reading the measurement tool of length (vernier callipers) is not precise, especially when reading the nonius scales on the vernier callipers. Besides that, how to measure the height of the liquid in the U pipe is not good. One case is when measuring the height of the liquid, they don't read the scale with parallax position. The effects of concave and convex meniscus are not concerned, so the data obtained is not accurate. Based on the value obtain in the second aspect is in a good category.

The third aspect of the conducting experiment stage is cooperation among group members. Theirs average score in third stage is in a good category. It because the task is divided equally and each member do it well.

Overall, the conducting experiment stage is in good category, with the average score is 78 and 79 from four topics. These results show that there is a relationship between the pre-laboratory stage and conducting the experiment stage. This is consistent with the study of Maria et.al that pre-laboratories can improve the preparation skills of practitioners before conducting experiment [13].

3.3. *Final report*

The last stage of basic physics performance assessment is the final report in writing systematics, list in the guidebook of basic physics experiment [7][12]. This stage has six aspects of performance assessment. The first is the ability to write abstract. The results of this obtained: less (70), moderate (73) and good (76). The problem of this aspect is students' ability to choose words and to write important sections in abstract. On the topic of density, students' performance is in a low category, they reveal too many unimportant things in abstract, beside that there are students who write abstracts very short so that the important thing that should be in the abstract they lose. While on the topic system pulley and the spring constant is in moderate category, and for the topic of mathematical pendulum student performance in writing abstract is in good category.

The ability to write basic theories that support the experiment is the second aspect of the final report. The pulley system topic has the lowest score. Most students write the final formula without any explanation, whereas the pulley system has different mechanical advantages based on the pully position. Based on table 6, the score in this aspect is 68, in low category but in the second aspect of the pre-laboratory stage, understand of the concepts that related to experiment is 77, in good category. These different results were caused different standard from the different assessor in that stages. Therefore, there should be standardization for each point of rubric (aspects and stages) (appendix 1) [16]. In addition, the common mistakes of this aspect are no information about the images, tables and equations.

The third aspect at the final report stage is the ability to formulate experimental methods. The topics of pulley system, mathematical pendulum and spring constant are included in a good category, whereas on the density topic is in moderate category. Students can't determine the density of two different methods. The first method for measuring the density of the solid state uses volume calculation by using the measurement tool of length (vernier callipers and micrometer screw) while the second method uses the volume of spilled water, as for the first liquid method of measuring the volume of the liquid in the measuring cup and the second method using pupa U. With different methods between solids and liquids make students still wrong in formulating experimental methods.

The ability to analyse the data with uncertainty measurement and answer the questions correctly on the fourth aspect has a relationship with the fifth aspect of the ability to discuss the results of data. The students' performance in analysing and discussing the data of the experiment result is less. The greatest hope of performance students on the final report are analysing and discussing of experiment data [7][10]. The data of experiment from the four topics can be made a graph to know the relationship between variables and also get the point from the equation graph. As well as on the topic of mathematical pendulum, not only using accurate calculations the acceleration of gravity [17][18] but also through graphical analysis. In the fourth aspect, indicate students' critical thinking skill that is in a good category. For the fifth aspect, the problem is the ability of students in connecting the results of experiments with the theory. The lowest score of ability to discuss at the fifth aspect at the topic of density and pulley systems, the score is in low category.

The last aspect of the final report stage is the ability to formulate conclusions. In formulating the conclusions refers to the formulation of the problem and the purpose of the experiment [7]. The students' performance in formulating the conclusion is in moderate category. The problems of this aspect are the low ability of writing conclusion, where they answer the formulation of the problem incorrectly. Based on the average score in the final report stage, only the topics of mathematical pendulum that have good category. Meanwhile, the other three topics are in moderate categories. If the average scores of four topics is in moderate level, so the overall of this aspect at the final report stage is in moderate category. The average score gives a good estimate of the score variety [20]. Thus, all the three stages in laboratory activities, including pre-laboratory, conducting experiment, and final report may increase conception of physics learning [19].

3.4. Final score

Based on the guidebook of basic physics experiment [7], the determination of weights at each stage is clear [equation 1]. From the average data obtained, the physics student in the basic physics experiment can be categorized moderate on the topic of density and the pulley system with a score of 75, while on the topic of mathematical pendulum and spring constant are in good category, the scores are 77 and 76. Overall, the results of physics students' performance are in good category and critical thinking skill is in good category. This is in accordance with the study of Emiliannur et.al, which states that the implementation of a performance assessment model in physics subjects is increasing students' critical thinking disposition [21].

4. Conclusion

The performance assessment of the students in the physics laboratory using guided inquiry show that in the pre-laboratory stage is in a good category (average score = 78) and for the conducting experiment stage is in a good category (average score = 78) while in the final report stage is in a moderate category (average score = 74). The performance of physics students in the basic physics experiment is in moderate category on the topic of density and pulley system, whereas on the topic of mathematical pendulum and spring constant are in good category. Overall, the results of physics students' performance and critical thinking skills are in good category.

References

[1] Admoko S and Supriyono 2016 *J. Penelit. Fis. Apl.* **6** 34
[2] Samsudin A, Suhendi E, Efendi R and Suhandi A 2012 *Indones. J. Phys. Educ.* **8** 15
[3] Syafi'i A, Handayani L and Khanafiyah S 2014 *Unnes Phys. Educ. J.* **3** 10
[4] Tutkun O F, Guzel D and İlhan H 2012 *Online J. Couns. Educ.* **1** 23
[5] Riyadi U 2008 *Model Pembelajaran Inkuiri dengan Kegiatan Laboratorium untuk Meningkatkan Keterampilan Berfikir Kritis Siswa Pokok Bahasan Fluida Statis* [Inquiry Learning Model With Laboratory Activities to Improve Student Critical Thinking Skills at the Points Of Static Fluid] (semarang: Unnes University Press) p 1
[6] Gunhaart A and Srisawasdi N 2012 *Proc. Soc. Behav. Sci.* **46** 5750
[7] Laboratory Team 2017 *Buku Panduan Praktikum Fisika Dasar 1* [Guidebook of Basics Physics Experiment] (Surabaya: Unesa University Press) p 1
[8] Ariesta R and Supartono 2011 *J. Pendidik. Fis. Indones.* **7** 62
[9] Brewe E, Kramer L and Sawtelle V 2012 *Phys. Rev. ST Phys. Educ. Res.* **8** 010101

[10] Ganiel U and Hofstein A 1982 *Sci. Educ.* **66** 581
[11] Timothy F, Slater J and Ryan M 1993 *Phys. Teach.* **31** 306
[12] Allie S, Buffler A, Kaunda L and Inglis M 1997 *Phys. Teach.* **35** 399
[13] Limniou M, Papadopoulos N and Whitehead C 2009 *Comput. Educ.* **52** 45
[14] Septiana A and Rustaman N Y 2017 *J. Phys.: Conf. Ser.* **812** 012052
[15] Suprapto N, Syahrul D A, Agustihana S, Pertiwi C A and Ku C-H 2016 *Chemistry* **25** 718
[16] Peeters M J, Sahloff E G and Stone G E 2010 *Am. J. Pharm. Educ.* **74** 1
[17] Nelson R A and Olsson M G 1986 *Am. J. Phys.* **54** 112
[18] Smith H J T and Blackburn J A 1992 *Am. J. Phys.* **60** 909
[19] Suprapto N, Chang T-S and Ku C-H 2017 *J. Baltic Sci. Educ.* **16** 7
[20] Bell S 1999 *A Beginner's Gide to Uncertainty of Measurement* (United Kingdom: Crown) p 1
[21] Emiliannur E, Hamidah I, Zainul A and Wulan A R 2017 *J. Phys.: Conf. Ser.* **895** 012143

IOP Conf. Series: Journal of Physics: Conf. Series **997** (2018) 012039 doi:10.1088/1742-6596/997/1/012039

Appendix

1. Rubric of pre-laboratory assessment

No	Aspect	Performances' indicator		
		Low	Moderate	Good
1	**Understand of the experiment goals**	Experiment goal don't based on relation between independent variable and dependent variable	Experiment goal based on relation between independent variable and dependent variable, but with help	Experiment goal based on relation between independent variable and dependent variable
2	**Understand of the concepts that related to experiment**	Student can understand basic concepts, can't derive the equation and misconceptions	Student can understand basic concepts, derive the equation and misconceptions	Student can understand basic concepts, derive the equation and aren't misconceptions
3	**Identify the variables that must be measured and calculated**	Student can't identify independent, dependent and control variable	Student can identify independent, dependent and control variable, but don't based on experiment goal	Student can identify independent, dependent and control variable based on experiment goal
4	**Understand of tool's specification and how to use it**	Student can't choose the tools and specification, but according to the experiment	Student can choose the tools but the specification don't according to the experiment	Student can choose the tools and specification according to the experiment
5	**Able to determine the data to be obtained, how to obtain and how to analyse**	Student can't determine the data, how to obtain it and analyse	Student can determine the data and analyse but how to obtain it with help	Student can determine the data, how to obtain it and analyse

2. Rubric of conducting experiment assessment

No	Aspect	Performances' indicator		
		Low	Moderate	Good
1	**The ability to set tools correctly**	Student can't string up the tool and modify the circuit	Student can string up the tool and modify the circuit incorrectly	Student can string up the tool and modify the circuit correctly
2	**The ability to use and read the scale of the measuring instrument properly**	Isn't doing calibration and don't read the scale with parallax position	Doing calibration but don't read the scale with parallax position	Doing calibration and read the scale with parallax position
3	**Cooperation among group members**	The task isn't divided equally and it is done by certain dominant student	The task is divided equally but each group member doesn't do it	The task is divided equally and each member do it

3. Rubric of final result assessment

No	Aspect	Performances' indicator				
		60	65	70	80	85
1	**The ability to write abstract**	The abstract consists of experimental goals and methods, the conclusions are incorrect	The abstract consists of experimental goals and methods, but the analysis and conclusions are incorrect	The abstract consists of experimental goals, methods, and conclusions the analysis result are incorrect	The abstract consists of experimental goals, methods, results analysis and conclusions are incorrect	The abstract consists of experimental goals, methods, results analysis and conclusions
2	**The ability to write basic theories that support the experiment**	Basic theory is too general, no derive equation and don't give information of the equation, table and figure	Basic theory is specific but no derive equation and don't give information of the equation, table and figure	Basic theory is specific and give information of the equation, table and figure, but there is no derive equation	Basic theory is specific and derive equation and don't give information of the equation, table and figure	Basic theory is specific and derive equation and give information of the equation, table and figure
3	**The ability to formulate experimental methods**	Experiment method isn't based on goals and not written in a row	Experiment method based on goals, not written in a row and point form	Experiment method based on goals, not written in a row and paragraph form	Experiment method based on goals, written in a row and point form	Experiment method based on goals, written in a row and paragraph form
4	**The ability to analyse the data with uncertainty measurement and answer the questions**	Student can't analyse data, calculate uncertainty and answer question incorrectly	Student can't analyse data and calculate uncertainty, but answer question	Student can analyse data and calculate uncertainty, but not answer question	Student can analyse data, calculate uncertainty and answer question incorrectly	Student can analyse data, calculate uncertainty and answer question correctly
5	**The ability to discuss the results of the data**	Students can't connect the result of data with theory and don't give explanation	Students can't connect the result of data with theory, but give explanation	Students can connect the result of data with the scientist result and not give explanation it	Students can connect the result of data with the scientist result and explanation it incorrectly	Students can connect the result of data with the scientist result and explanation it correctly

No	Aspect	Performances' indicator				
		60	**65**	**70**	**80**	**85**
6	**The ability to formulate the conclusions**	All of the conclusion aren't answer the formulation of problems and incorrectly	some of the conclusion aren't answer the formulation of problems but incorrectly	some of the conclusion aren't answer the formulation of problems but correctly	some of the conclusion are answer the formulation of problems correctly	All of the conclusion are answer the formulation of problems correctly

The use of physics practicum to train science process skills and its effect on scientific attitude of vocational high school students

E Wiwin[1] and R Kustijono[2]

[1]SMK Negeri 12 Surabaya, Surabaya, Indonesia
[2]Department of Physics, Universitas Negeri Surabaya, Indonesia

E-mail: buelok@gmail.com

Abstract. The purpose of the study is to describe the use of Physics practicum to train the science process skills and its effect on the scientific attitudes of the vocational high school students. The components of science process skills are: observing, classifying, inferring, predicting, and communicating. The established scientific attitudes are: curiosity, honesty, collaboration, responsibility, and open-mindedness. This is an experimental research with the one-shot case study design. The subjects are 30 Multimedia Program students of SMK Negeri 12 Surabaya. The data collection techniques used are observation and performance tests. The score of science process skills and scientific attitudes are taken from observational and performance instruments. Data analysis used are descriptive statistics and correlation. The results show that: 1) the physics practicum can train the science process skills and scientific attitudes in good category, 2) the relationship between the science process skills and the students' scientific attitude is good category 3) Student responses to the learning process using the practicum in the good category, The results of the research conclude that the physics practicum can train the science process skill and have a significant effect on the scientific attitude of the vocational highschool students.

1. Introduction

Nowadays, the world of education is required to be able to equip students with 21st-century skills such as: flexible, adaptive, initiative, self-control, social, productive, leadership, and responsibility [1]. These abilities must in line with the view of life, life attitudes, and life skills from students, and they can thrive in school, out of school, and family. In schools, these abilities can be trained through scientific literacy and science process skills [2]. One subject that supports the field of expertise in vocational high schools is Physics, which is the science related to facts, processes, theories, concepts and generalizations, which are not merely about memorization, but also processes that enable students to link one concept with another to express a meaningful relationship. Therefore, the study of physics must be able to present the phenomenon in everyday life that can encourage and train students to think analytically, critically and creatively. The best way to develop that ability is to give students the opportunity to investigate a physical phenomenon in everyday life. So physics learning requires investigative activities as part of scientific work involving process skills and also involving scientific attitudes. One effective method that can be used in physics learning is practicum, because it can be used to trill the three domains of intelligence (cognitive, psychomotor, and affective) simultaneously.

Science process skills are empowering the basic skills of students by thinking and acting to discover, develop, and apply natural science, involving both intellectual and psychomotor skills [3]. Therefore, Physics practicum basically also applies science process skills. The elements of science process skills include: observing, classifying, measuring, communicating, inferring, predicting, collecting-recording and interpreting data, identifying and controlling variables, defining operationally, making a hypothesis, experimenting, making and using models [4]. In addition, Physics practicum can also integrate many attitudes such as: objectivity, accuracy, precision, honesty, collaboration, discipline, responsibility, open-mindedness, courage, humility, decision-making, integrity, diligence, persistence, curiosity, etc. Concurrent development of many attitudes in a learning process is certainly impossible to do, but integrating some selected attitudes as scientific attitudes in Physics learning is a real step. Practicum is a part of teaching that aims to get students to test and execute in real terms what is obtained in theory. Three functions of practicum are training, feedback, and improving motivation, practicum requires adequate information-processing skills (intellectual) and psychomotor skills that are conducted through a process that demands scientific attitude (curiosity, honesty, collaboration, responsibility, and open-mindedness). The purpose of this study is to describe the use of physics practicum to train the science process skills and its influence on the scientific attitudes of vocational students.

Physics is a branch of science that studies the nature of the phenomenon of things in nature that provides life lessons aligned according to the laws of nature, and used as a vehicle to develop the ability to think to solve problems in everyday life. Physics taught in secondary schools includes basic physics that is the basic idea that arises from the application of scientific methods that examine the most fundamental ideas about the physical properties that include the basic concepts and principles of physics necessary to study further physics or other sciences [5]. Physical learning is ideally implemented in an inquiry to foster the ability to think, work, and be scientific. Therefore, physics learning traces the scientific skills, which are skills related to scientific products, scientific processes, and scientific attitudes, which are used to express the most important procedures, processes, and methods scientists use when they construct science and solve experimental problems [6]. Vocational secondary school (SMK) is one form of formal education units that provides vocational education at secondary education as the next step from a secondary junior high school or other equivalent or advanced form of recognized learning outcomes. In vocational schools, there are many skill programs, but not all of them require physics. In the meantime, education in SMK demands the development of human resources quality who are also creative, and have competitiveness in facing global competition. An attempt to meet the needs of human resources at the secondary level is the fostering of vocational secondary schools quality. The role of Physics is to help improve the quality of program learning skills that require physics with practicum activities to trace the science process skills that are expected to increase scientific attitudes.

There are many science teachers use practicum in learning activities that develop scientific methods, critical thinking, scientific attitudes, problem-solving approaches, discovery methods, and inquiry methods [7]. Hands-on activities and workouts in the laboratory are essential so that students can conduct an investigation as scientists [8]. The use of inquiry-based laboratory instruction enhances the science literacy and research skills of students [9], in addition, those activities also improve student's achievement [10]. The most widely used method in science and technology is experimentation [11]. Although investigation activities have a positive impact on student research skills, however, teachers should not be excessive in providing students with experience [12]. The learning model needed to anticipate the 21st century is what makes possible the scientific thinking skill, the development of a sense of inquiry and the creative thinking ability of students [13]. The required learning model is capable of generating the ability of how the knowledge, skills, and attitudes are derived by students [14]. The learning used should integrate the science process skills into the material presentation system [15]. In addition, learning should help students learn to learn, and help students acquire knowledge by finding it themselves [16]. The science-based learning process skills emphasize the development of high order thinking skills [17]. Scientific experiences that need to be given and developed to students are science process skills, thinking skills, and scientific reasoning skills [4]. Science process skills need to be applied in physics learning because students need not only knowledge, but also need to be trained to discover, develop, and apply that knowledge [2].

Physical learning cannot be separated by the development of character, the nature of a person in responding to the situation in a moral manner is shown by the action through a good behavior. The counterproductive natures that attack today's students such as low ethics and morality, requires teachers and schools to take them seriously. The teacher is a vital and fundamental component in the education process, which emphasizes the process of mental maturation, mindset and the formation and development of student character to realize the whole person. The character of the student will be seen from his or her behavior and can be observed when the learning process takes place. Therefore, Physics learning cannot be separated from the scientific attitudes in addition to science process skills. Learning-based learning produces self-confidence in scientific abilities [9]. It is generally found that laboratory investigation activities can significantly improve student attitudes [19]. Scientific skills in physics practicum are the ability to perform a scientific process that requires adequate intellectual skills and psychomotor skills and is conducted through a process that demands the scientific attitudes of the students. Scientific skills can be maximized if the science process skills and the scientific attitudes in a Physical practicum can be trained in an integrated manner [20].

Science process skills and scientific attitudes in the practicum will be interrelated, for example, a student will be able to assemble the experimental tools correctly (science process skills) if the student is doing it thoroughly and scientifically (scientific attitudes). Another example, scientifically written experimental data (scientific attitudes), although not in accordance with the theory will encourage students to develop the ability to analyze the results lead to problem-solving (science process skills). From various observations so far there are some students who have a bad attitude, although not yet known exactly how many, but should be attempted to be minimized. Teachers as the key actors in learning need to play a maximum role in the formation of student characters through the formation of a good attitude in learning. The formation of student attitudes in Physics learning can be fully obtained when students perform the process of practicum activities that focus on the integrity of training in science process skills and scientific attitudes.

2. Research Methodology

The type of research is an experimental research with the one-shot case study design, ie. giving treatment to one group then the dependent variable is measured with the aim to know the effect of the treatment [21]. Variable manipulation in this research is learning using practicum, while the dependent variables are science process skills and scientific attitudes. The science process skills defined in the study are: observing, classifying, inferring, predicting, and communicating. Scientific attitudes established in this study are: curiosity, honesty, collaboration, responsibility, and open-mindedness.

The procedure implemented in this research is applying Physics learning using practicum, then, at the execution of the learning science process skills and scientific attitudes are measured through observation using attitude scale. Science process skills and scientific attitudes scores obtained are then described to know the performance of students when the learning is in progress. To find out how far the relationship of science process skills with scientific skills, the correlation coefficient is calculated based on the average score of students. To complete the results, students are then given a questionnaire to get their response to Physics learning using the practicum. The Practicum topic applied in this research is Simple Pendulum. The subjects are 30 students from the Vocational High School (SMK) 12 Surabaya.

Data collection techniques use observations and performance tests. Science process skills and scientific attitudes score are derived from the score of performance tests using performance appraisal instruments with Likert scale (very less = 1, less = 2, enough = 3, good = 4, and very good = 5). The questionnaire used to know the student's response to Physics learning using the practicum also use the Likert scale.

IOP Conf. Series: Journal of Physics: Conf. Series **997** (2018) 012040 doi:10.1088/1742-6596/997/1/012040

The data analysis uses the average score description of each element of science process skills and scientific attitudes, while the relationship between the two uses correlation. The criteria used for the average score are: very less (1≤ x <1.5), less (1.5≤x <2.5), enough (2.5≤x <3.5), good (3, 5≤ x <4.5), very good (4.5≤ x ≤5). The correlation criterion (r) used is: less (0≤r <3), enough (3≤ r <7), and good (7≤r ≤ 1). Criteria of student response is obtained by summing all the average score given by students divided maximum score multiplied by 100%, good criterion if score obtained ≥ 61%. Physics learning using practicum is declared to be successful in scaling up the science process skills and having good effect on the scientific attitudes if the average score of student performance is in either good or very good category, the correlation coefficient obtained is enough or good, and the student's response to Physics learning using the practicum is also in good category.

3. Results of The Research

3.1. The Effect of Practicum on Science Process Skills and Scientific Attitudes
The average score of students science process skills in the process of learning physics using practicum work is shown in Table 1, while the average score of students scientific attitudes are shown in Table 2 as follows:

Table 1. The average score of Science Process Skills.

Aspects	Score	Category
Observing	4.4	Good
Classifying	3.7	Good
Predicting	3.7	Good
Inferring	3.5	Good
Communication	3.6	Good

Table 2. Average score of Scientific Attitudes.

Aspects	Score	Category
Curiosity	4.4	Good
Honesty	3.9	Good
Collaboration	3.8	Good
Responsibility	3.9	Good
Open-mindedness	4.1	Good

Based on Table 1 and Table 2, it can be seen that the scores of all elements of science process skills and scientific attitudes are in good categories in which the range of values is 3.5 ≤ x <4.5. These results indicate that the Physics learning process by using practicum has a good impact on science process skills and scientific attitudes.

3.2. The relationship between Science Process Skills and Scientific Attitudes in Practicum
The relationship between science process skills and scientific attitudes can be seen from the correlation score (r) as presented in Table 3 as follows:

Table 3. The relationship between Science Process Skills and Scientific Attitudes.

r	Category
0.82861	Good

Based on Table 3, it can be seen that the relationship between science process skills and scientific attitudes shows the correlation coefficient of 7 ≤ r ≤ which means the correlation is in good category.

It shows that in the process of Physics learning using practicum, there is a significant relationship between science process skills and scientific attitudes.

3.3. Student's Response to Physics Learning Using Practicum

To complete the results of the science process skills and scientific attitudes in the Physics learning process by using the practicum, students are given questionnaires to assess the effectiveness of the practicum. The results obtained are shown in Table 4 as follows:

Table 4. The Average Assessment of Students on the Practicum.

Aspects	Score	Category
Provide an opportunity to observe	4.1	Good
Provide an opportunity to classify	4.0	Good
Provide an opportunity to predict	4.1	Good
Chance to infer	3.8	Good
The opportunity to communicate	3.9	Good
Encourage curiosity	4.1	Good
Encourage honesty	4.1	Good
Encourage work together	4.3	Good
Encourage responsibility	4.0	Good
Encourage openness	4.1	Good
Total average score	40.5	
Maximum average score	50	

Based on Table 4, it can be seen that all the elements asked about the practicum is considered good by students. The elements related to the science process skills (observing, classifying, inferring, predicting, and communicating) and scientific attitudes (curiosity, honesty, collaboration, responsibility, and open-mindedness) all received good judgment from students, and overall practicum effectiveness based on students' responses have a good category (81%).

4. Discussion

Science process skills such as observing, classifying, measuring, communicating, inferring, predicting, etc [4] are the abilities of a scientist which can be trained to students. In this study, the ability trained is limited to: observing, classifying, inferring, predicting and communication. Observing is determining the nature of an object or event by using the senses. Each practicum will always start from sensory observation, and the results of these observations will be the basis of the next step, because if there is a mistake on the observation it will affect the final result. Therefore, observation becomes important to the students so that they are familiar with Physical Phenomena carefully. The student's ability to observe can be seen from how far the observation of the object is done in accordance with the actual situation. Classifying is the ability to group objects or events by their nature. In practicum activities, each group formed is given the opportunity to observe the object under investigation, then they are given the opportunity to group objects or events according to their nature which is aimed to: 1) the ability to group the various information obtained, 2) the ability to classify objects (tools and materials) used in the practicum. The ability to classify students that can be seen from the extent to which students are able to group information based on its nature (facts, concepts, and principles), and able to sort and choose the right object.

Inferring is an explanation of a particular event based on observations and data. The ability of the inference needs to be trained to the students so that they are used to express a Physics event clearly without incurring other perceptions that deviate from the actual intent.The students inferring can be seen from the following indicators: the students can write the results of observations well, describe systematically, and write with the standardized Indonesian. Those abilities require the skill to think with good accuracy, and honesty when observing the events of an investigation conducted in order to get a series of words that describe the facts which are easy to understand and systematic. Predicting is to anticipate the consequences of a new situation or change past experience and previous observations. The ability to predict needs to be trained to students so that they are used to anticipate the consequences that will occur due to deeds or behavior performed. The example in Physical practicum is: the heat travels on the metal (metal is the heat conductor). If using a metal as a hot liquid stirrer for example (boiling water), then the tip of the metal must be coated with a heat resistant insulator (e.g. wood). Communication is the use of written and oral words, graphics, tables, diagrams, and other information presentations, including technology-based ones. Ability is considered necessary to be trained to students, because it is necessary for everyday life. In Physics practicum, students are required to write a report of results in writing and present it to other friends. With this experience, students familiarize themselves with an idea systematically and easily understood by utilizing graphs, tables, and diagrams, and trying to utilize the features of information technology, especially the use of multimedia. The use of multimedia is expected to motivate students to carry out a better Physics practicum.

Scientific attitudes are humanist performances which are personal and interpersonal behaviors in the scientific context, which describe the person's nature, determination, control and way of thinking in conveying subjects related to science [20]. A scientist always has the scientific attitudes that can be trained to students. The study is limited to: curiosity, honesty, collaboration, responsibility, open-mindedness. Curiosity is a person's curiosity on a symptom or event. This nature is always present in a person and needs to be developed by the student, for his curiosity is accustomed to events or symptoms related to the development of science. The emphasis of curiosity is directed to the benefit of deepening the previously taught knowledge. Student's honesty in the practicum can be seen from the following indicators: they do not manipulate data, and dispose of data that is considered bad. In fact, they give information about the data that is considered bad. Cooperation is an active role of students in carrying out the task and success of the group. Feeling the same fate among peers in a group is a positive factor. Cooperation is a condition where students can synergize and support each other to achieve the achievement or goal set in practicum.

When students carry out practicum, they can share the task and role in completing the investigation and each student can play a role according to his or her task, synergize and integrate the results into group successes. So cooperation does not mean every member of the group should do the same job, but rather on how each group member can support the joint success. Student co-ops are seen with indicators of all group members performing the practicum in groups, with a clear division of tasks, and each member completing the task. Responsibility is a consequence that must be borne as a result of the actions that have been done. When students take a conclusion, they must be prepared to face the consequences of the decision that has been chosen. At that time students have actually set a decision that has the consequence that the decision may be accepted or not by others, and when others can not accept the conclusion, the students are ready to give an explanation. Open-mindedness is the attitude of not imposing ideas/opinions themselves and can appreciate the ideas/opinions of others, and it is done when students carry out reporting activities through the expression of ideas with the presentation. Conveying ideas requires good presentation techniques in order to demonstrate the superiority of ideas delivered, and an open attitude in receiving responses to the ideas. Open attitude impacts effective communication because it produces a win-win solution, which raises the basic concept of effective communication that no one is to blame, so it needs an open attitude from all parties.

These results indicate that the science process skills are related to scientific attitudes and co-exist with other studies such as self-confidence in scientific abilities [9] and laboratory investigation activities improve student attitudes [9]. Practicum is an inquiry-based and science process skill that emphasizes the process of how knowledge, skills, and attitudes are acquired by students [14], helps students acquire knowledge by finding it themselves [16], and provides an opportunity for the development of high order thinking skills [17]. The science process skills applied in practicum provide an opportunity for students to discover, develop, and apply knowledge [2]. The result of this research shows that physics practicum can be able to train science process skills well and have a significant influence on students' scientific attitudes.

5. Conclusions

The results of the research show that: 1) learning process by using practicum can train science process skills and scientific attitudes in good category 2) the relationship between science process skills and scientific attitudes of students is in good category 3) Students' response to Physics learning process by using practicum is in good category. The conclusion is that the process of learning physics by using practicum can train science process skills and have a significant effect on the scientific attitudes of Vocational High School students.

Suggestions that can be submitted is that scientific attitudes should be planned and trained as part of the Physics learning process objectives by using the practicum and not just as a side impact alone. To produce the maximum ability of students to the learning process, scientific attitude should be integrated into the science process skills.

6. Acknowledgments

Our gratitude goes to the Headmaster of SMK Negeri 12 Surabaya who has supported this research. Thank you also to colleagues, and students who have helped finish this research, because without it this project would not be completed.

References

[1] Trilling B and Fadel C 2009 *21st Century Skills: Learning for Life in Our Times* (California: Jossey-Bass)
[2] Turiman P, Omar J, Daud A D and Osman K 2012 *Procedia Soc. Behav. Sci.* **59** 110
[3] Sheeba M N 2013 *Educ. Confab* **2** 108
[4] Valentino and Catherine 2000 *Developing Science Skills* (Houghton Mifflin Company)
[5] Feynman R 2010 *Basic Physics* The Feynman Lectures on Physics Volume 1 Chapter 02
[6] Etkina E, Heuvelen A V, White-Brahmia S, Brookes D T, Gentile M, Murthy S, Rosengrant D and Warren A 2006 *Scientific abilities and their assessment* Physical Review Special Topics-Physics Education Research 2, 020103 (2006)
[7] Abd-El-Khalick F, Boujaoude S, Duschi R, Lederman N G, Hofstein A, Mamlok-Naama R, Niaz M, Treagust D and Tuan H 2004 *Inquiry in Science: International Perspectives* (New York: Wiley Periodicals, Inc)
[8] Bell R L 2008 *Teaching the nature of Science through Process Skills-Activities for Grades 3-8* (Boston: Pearson, Education, Inc)
[9] Brickman P, Gormally C and Armstrong N 2009 *Int. J. Scholarsh. Teach. Learn.* **3** 2
[10] Learning A 2004 *Learning and Teaching Resources Branch* (Focus on inquiry: a teacher's guide to implementing inquiry based learning Alberta Canada)
[11] Akinoglu O 2008 *Int. J. Instr.* **1** 1
[12] Lane J L 2007 *Inquiry-Based Learning. Schreyer Institute for Teaching Excellence* (Penn State University Park)
[13] De Vito A 1989 *Creative Wellsprings for Science Teaching West Lafayette* (Indiana: Creative Venture)
[14] Joice B and Weil M 1996 *Model of Teaching* (Boston: Allyn and Bacon).
[15] Beyer B K 1991 *Teaching Thinking Skill: A Handbook for Elementary School Teachers* (New York, USA: Allyn and Bacon)
[16] Carin A A and Robert B S 1989 *Teaching Science Through Discovery* (Columbus, Ohio: Merril Publishing Company)
[17] Houston W R 1988 *Touch the Future Teach* (St. Paul, MN: West Publishing Company)
[18] UNESCO 2013 *Handbook for Science Teachers*
[19] Ural E 2016 *J. Educ. Train. Stud.* **4** 217

[20] Kustijono R, Jatmiko B and Ibrahim M 2017 *The use of physics practicum to train science Process Skills and Its Effect on Scientific Attitude of Vocational Highschool Students* The 2nd Int. Joint Conf. on Science and Technology September 27-28 2017 Bali-Indonesia
[21] Fraenkel J R and Wallen N E 2003 *How to design and evaluate research in education (Fith Edition)* (New York: McGraw-Hill Higher Education)

Designing laboratory activities in elementary school oriented to scientific approach for teachers SD-Kreatif Bojonegoro

Dwikoranto[1], W A Surasmi[2], A Suparto[2], S Tresnaningsih[2], D Sambada[2], T Setyowati[2], A Faqih[2] and R Setiani[3]

[1]Department of Physics, Faculty of Mathematics and Natural Science, Universitas Negeri Surabaya, Surabaya, Indonesia
[2]UPBJJ Universitas Terbuka Surabaya, Surabaya, Indonesia
[3]STKIP PGRI Tulungagung, Tulungagung, Indonesia

E-mail: dwikoranto@unesa.ac.id

Abstract. Important science lessons are introduced to elementary school students through inquiry. This training is important to do because one key determinant of succesful laboratory activities is teachers. This course aims to enable teachers to design an inquiry-based Laboratory Activity and be able to apply it in the classroom. The training was conducted at SD-Kreatif Bojonegoro by Modeling, Design Laboratory activities and Implementing. The results of Laboratory Activities designed to trace the seven aspects that can support the development of inquiry skills in either category. The teacher's response in this activity is positive. The conclusion of this training can improve the ability of teachers in designing and implementing laboratory activities of Science and then expected to positively affect the frequency of science laboratory activities. Usually teachers use learning by using this Laboratory Activity, it will be affected on the pattern of inquiry behavior to the students as well so that will achieve the expected goals. Teachers are expected to continue for other topics, even for other similarly characterized subjects. This habitation is important so that the teacher's skill in making Laboratory Activity continues to be well honed and useful for the students.

1. Introduction

Teachers' learning tends to teach like when teachers are taught lecturers during college. If the teacher is taught using the conventional method, then when the teacher also prefers to teach by the way of lecture method, although the learning is not suitable or less suitable for the students. Distinguishing the learned science is difficult for teachers, thus teaching it to the students [1]. The teacher should be given the opportunity to examine the nature of the subject matter, to understand not only what is known, but how facts, from which knowledge is acquired rationally. Understanding the basic conceptual sains and the ability to argue scientifically will provide a strong foundation for effective teaching rather then learning about the high material but superficial understanding [1].

Science learning in some elementary schools (SD), teachers generally perform the lecture method by the teacher directly, asking students to summarize the taught topics, or to ask students to read certain material, to compose papers in groups, then to present it in front of the class. Then the discussion, question and answer session about the contents of the paper. In discussion, question and answer session, students ask questions about things that considered unclear to the presenter group. The renderer group takes turns answering these questions. If the presenter group can not provide an answer, than the moderator, one of the presenter groups, throws the question to the teacher. Furthermore, the teacher gives a direct answer to the students. During the lesson, laboratory activities or field activities are lacking. This kind of learning has weaknesses, among others: (a) the concept, principle or theoretical knowledge obtained by the students from the reading of a book or information from the teacher, is not the result of the discovery of the science process done by the students. As a result, the knowledge gained is limited to knowing, not understanding. Information obtained in this way will be more quickly forgotten; (b) students are more likely to study the topics that are part of their presentation tasks, while other topics are studied only. According to students, the way of learning is like they experience does not make them better understand the material.

Science is a collection of knowledge about the objects or symptoms of nature that have been tested the truth [2]. Science encompasses two aspects of science as a process, known as the scientific method and science as a product known as the body of knowledge [3]. Science also has scientific values or values of science inherent in scientific knowledge [3, 4]. Science as a process begins with observations of natural phenomena by way of work as is done by scientists [5]. Therefore, science learning should start from observation of natural phenomena. Through the scientific process can be developed students' scientific attitude. Such scientific attitudes include curiosity, respect for evidence, critical thinking, creative thinking, speaking based on concrete evidence or data, and care about the environment. Through the process of science can be developed skills to observe, explain, think, solve problems, and make decisions [6]. This is consistent with the purpose of learning science that is to educate students to be able to develop observation and experimentation and obedient principle thinking through: observing, understanding, and utilizing natural phenomena involving substances and energy [7].

It is suspected that the ability of teachers in designing inquiry laboratory activities is still low. This prediction is supported by the result of research [8] the science process of students and elementary school teachers are generally low (4.08% and 65.79%), and in elementary school of science process skill generally developed and integrated with deductive learning, this contradicts the scientific process skills that are taught through scientific inquiry learning to foster thinking, work and scientific ability and communicate it as an important aspect of life skills by emphasizing the provision of direct learning experiences through the use and development of process skills and scientific attitudes [7].

Based on the news uploaded by Tribunnews.Com Bojonegoro [9] stated that a total of 1,177 units of 4,397 classrooms of Bojonegoro District Primary School in damaged condition, in the data of the Education Office, the damage was categorized as moderate, mild to severe. Meanwhile, data from the recapitulation on the use of primary school Funds 2015 issued by the Directorate of Elementary School Development [10]. Bojonegoro Regency received rehabilitation of classrooms (12 SD, 42 classes) Rp 4,009,360.00, library room (14 SD, 14 Space) Rp 2,082,360.00 and ICT facilities (2 SD, 2 rooms) Rp 104,000,000.00, based on the two data sources mentioned above, it can be predicted that the future of Bojonegoro Regency Government in the development of education field will focus on the improvement of infrastructure in the form of rehabilitation of the classroom that will require substantial funds. Seeing the condition of infrastructure and rehabilitation done above it can be ascertained for infrastructure facilities of science laboratory activities in many elementary schools in Bojonegoro regency still can not be fulfilled.

How teacher will conduct inquiry laboratory activities, if the teacher does not have the means of infrastructure that support the activity and the teacher has not mastered the skills needed. Conversely, if teachers have the ability to design laboratory activities then teachers who have a laboratory will be motivated to carry out experiments, while teachers who do not have laboratory equipment can carry out experimental activities using the laboratory by utilizing the tools and materials available in the environment or use the help of Virtual Laboratory to be able operationalizing and having ability.

There are two main factors that can influence teachers to develop inquiry laboratory activities, namely ability and willingness factor. Therefore, in order to improve the quality of science education in Primary School is not necessary two things as follows. First, the improvement of the quality of teachers, especially in matters relating to the improvement of the ability to design and organize laboratory activities based on inquiry. Second, the creation of conditions that can improve and keep teachers motivated to always seek learning science in elementary school quality.

Based on these two things: ability and willingness, this training is focused on increasing the quality of science teacher teachers in Creative Elementary School members of Bojonegoro District Teachers Meeting, with the aim to develop teachers' ability in designing inquiry laboratory activities in the form of Student Activity Sheets (LKS) for science subjects. This inquiry-based LKS at Creative Schools needs to be developed because of teacher's characteristics and students support for it.

This effort is important to do, because one of the factors determining the success of laboratory activities is the teacher [11], and the results of the training are expected to improve the ability of teachers in designing and implementing science laboratory activities, and then expected to positively affect the frequency of science laboratory activities in Primary school. With the use of teachers using the learning by using this inquiry LKS, it will be affected to the pattern of inquiry behavior to the students as well, so that will achieve expected goals [12, 13].

After the training activity in the newly SD-Kreatif is implemented, the teacher of the trainees is expected to have the ability: (1) Designing the laboratory activities of science in the form of Creative LKS based on Inquiry, (2) Implement science laboratory activity using Creative LKS based on Inquiry.

After the training activities at the Creative School are conducted, the participants are expected to benefit: (1) The result of the activities to be carried out is expected to be a pioneering activity of LKS development in elementary school for other subjects, (2) LKS generated can be made as a learning resource (real teaching) for the world of education in order to realize research-based education, (3) Motivate Primary School Teachers to pioneer create various Creative LKS to facilitate in studying certain material to students.

2. Method Implementation

The objective of providing skills training to make this Student Worksheet is the teachers of Primary School Creative and Primary School Mulyoagung members of MGMP Bojonegoro. The selection of the objectives of this activity is taken with the consideration of those direct actors and can provide information about the learning conditions in the real class. This LKS skill training course will work together between the Unesa Community and Universitas Terbuka's Community Service team with the elementary school teachers of the Bojonegoro MGMP. Implementation will be coordinated with Bojonegoro District Supervisor and Education Office.

Training activities through community service are also aimed at tracing aspects that can support the development of the capability of designing the laboratory activities, namely (1) determining the purpose of laboratory activities, (2) determining the type of experiment in accordance with the objectives, (3) determining the appropriate tools and laboratory materials with the required specifications, (4) determining the experimental circuit and describing the diagram, (5) plotting the experimental procedure and implementing it, (6) preparing an inquiry-based student worksheet, and (7) designing an evaluation of laboratory activities.

This program is considered successful if the trainees are able to design the laboratory activities in the form of Student Worksheet and the team of trainers successfully carry out the training of Designing the Laboratory Activities Oriented on Scientific Approach for the Primary School teachers of MGMP members in Bojonegoro Regency well.

Training methods and techniques for improving these skills are developed through three stages of training:

1. Modeling. Implementation of an example or modeling of science laboratory activities based on inquiry by a team of Community Service from Universitas Terbuka and Universitas Negeri Surabaya.
2. Duplication. The design of the inquiry laboratory activities by the trainees are the teachers of SD Kreatif and SDN Mulyoagung Bojonegoro Regency.

3. Implementation. Implementation of LKS draft results in the simulation of learning in each elementary school class of trainees.

In the implementation of activities carried out monitoring by both parties concerned and at the end of the activity carried out an evaluation of the implementation of activities and training results.

As described in the introduction that there is a lack of laboratory equipment which continues to be pursued by the government through the Bojonegoro education office. With the many problems of school facilities that are not necessarily insurmountable, but activities in the laboratory must keep running, then the teacher must be creative. These limitations should be a motivation to improve things.

Problems have been found and alternative troubleshooting has been offered then makes a plan to solve problems in the field. Elementary teachers will be awakened and will be easy to receive if counseling by way of direct practice or training activities that go directly to the target.

3. Results and Discussion

Training Activities conducted at SD Kreatif Jalan Mangga, campus complex of Universitas Terbuka in Bojonegoro Learning Group, which starts from July to November 2017 from three planned training stages has been successfully done with the following results.

3.1 Modeling

The execution of an example or modeling of Inquiry-based science laboratory activities by the team is designed to tackle seven aspects that can support the development of inquiry skills, namely: (1) Determining the purpose of laboratory activities. (2) Determine the type of experiment that fits the purpose. (3) Determine laboratory equipment and materials in accordance with the required specifications. (4) Determine the series of experiments and illustrate the diagrams. (5) Plan your own trial procedure and implement it. (6) Prepare an inquiry based on student workbook. (7) Design evaluation of laboratory activities. The ability of the instructor in modeling when this activity takes place can be seen in Table 1 below.

Table 1. Instructor Capabilities in Modeling.

No	Aspect Skill	Achievement			Fluency	
		high	medium	less	fluent	less
1	Determining the purpose of laboratory activities	80			80	
2	Determine the type of experiment that fits the purpose		72		75	
3	Determining laboratory equipment and materials in accordance with the required specifications		75		78	
4	Determine the series of experiments and illustrate the diagrams	85			78	
5	Plan your own trial procedure and implement it.	88				65
6	Prepare an inquiry based student workbook		72		80	
7	Designing evaluation of laboratory activities	85			82	

Modeling activities there are few constraints in terms of providing variations of tools and materials for laboratory activities in the form of lamps in the series and parallel circuits, the number of lights of a kind. This has little impact on the activities of planning your own trial procedures and carrying them out.

3.2. Duplication

The design of Laboratory Activities Inquiry in the form of LKS by the elementary school teachers of the training results can be seen in Table 2 below.

Table 2. Master's Ability in LKS Design.

No	Aspect Skill	Thruth			Activity	
		before	After	gain	Active	Passive
1	Determining the purpose of laboratory activities	40%	80%	40%	0.75	0.25
2	Determine the type of experiment that fits the purpose	30%	75%	45%	0.80	0.20
3	Determining laboratory equipment and materials in accordance with the required specifications	50%	80%	30%	0.85	0.15
4	Determine the series of experiments and illustrate the diagrams	25%	75%	50%	0.80	0.20
5	Plan your own trial procedure and implement it.	20%	65%	45%	0.70	0.30
6	Prepare an inquiry based student workbook	30%	60%	30%	0.60	0.40
7	Designing evaluation of laboratory activities	40%	80%	40%	0.75	0.25

Prior to the truth modeling in determining the skill aspect that is trained at highest 50%. In the components determine the laboratory tools and materials and the lowest on the component plan the experimental procedure itself (20%) and implement it. While the activity of participants in the training above 60% (proportion 0.60).

3.3. Implementation

The next training activity undertaken by the team is the implementation of the design result in the simulation of the learning in the elementary school, class of each participant. Real-time simulation was carried out in October 2017. In the implementation of monitoring activities by both parties, namely Team and Principal Elementary School and at the end of the evaluation, the implementation of activities and training results.

Student's response after applying LKS, result of what teacher designed in each class indicates that most of student feels new with LKS, feels guided and motivated with existence of laboratory equipment tangible like tools around them, available on market and cheap.

As in theb teacher's design in finding light relationship with the number of batteries that are used, it was easily observed and did because it uses large battery, battery container, big lamps, and tights that are commonly found in market, easy to find, and cheap.

4. Conclusion

Training through modeling, duplication, and implications can improve 7 aspects of inquiry skills by at least 30%. This training is important to do because the one of the determinants of the success of primary laboratory activities is teachers and training results can improve the ability of teachers in designing and carrying out science laboratory activities and then expected to positively affect the frequency of science laboratory activities in elementary school. Usually teachers use learning by using this inquiry LKS, it will be affected to the pattern of inquiry behavior to the students as well so that will achieve expected goals.

After the training activities at the Creative School are conducted, the trainees are expected to continue to make LKS-based inquiry for other science material topics, even for other similarly characterized subjects. This habitation is important so that teachers' skill in making LKS continues to be well honed and beneficial to their students.

Acknowledgments
Thanks to Universitas Terbuka (UT) as through excellent research with contract research Number: 042.01.2.400981/2017 Date, 05 April 2017 has been supporting the funding of this research.

References

[1] Dermott L C, Shafer P S and Constantinou C P 2000 *Phys. Educ.* **35** 411
[2] Hungeford 1990 *Science-Technology-Society: Investigating and evaluating STS issues and solution (*Illinois: STIPES Publishing)
[3] Trowbridge, Leslie W and Bybee R W 2010 *Becoming a Secondary School Science Teacher Edition 5* (The University of Virginia: Merril Publishing Company)
[4] NSTA 2013 *Standart for Science Teacher Preparation: Revised 2003* (online) available on http://nsta.org/pdfs/NSTA Accessed September 2nd 2017
[5] Rutherford F J and Ahlgren A 1990 *Science for All Americans* (Oxford University: New York)
[6] Yager R E 1996 *S-T-S: As Reform in Science Education* (Al Bary State University: New York Press)
[7] Depdiknas 2007 *Kurikulum Tingkat Satuan Pendidikan* (Jakarta: Depdiknas)
[8] Haryono A, Sadiman A S, Raharjo R and Rahardjito 2008 *Media Pendidikan: Pemahaman, Pengembangan dan Pemanfaatan* (Jakarta: PT Raja Grafindo Persada)
[9] Tribune.com. Bojonegoro 2015 *Improvement of School Facilities. Decompose the needs of school buildings* (Bojonegoro: Tribune News)
[10] Kemendikbud 2016 *Petunjuk Teknis Penyaluran Bantuan Pemerintah kepada Sekretariat Jenderal Badan Keuangan Kementerian Pendidikan dan Kebudayaan* (Jakarta: Kemendikbud)
[11] Lazarowitz R and Tamir P 1994 *Research on Using Laboratory Instruction in Science: Handbook of Research on Science Teaching and Learning* (New York: Macmillan Publishing Company)
[12] Atwood R K and Atwood V A 2006 *J. Res. Sci. Teach.* **33** 553
[13] Nur M 2004 *Application of Innovative Ideas of MIPA Education in Research Setting*. Paper presented at the Nat. Semin. MIPA Education held by FMIPA Unnes on February 28, 2004

IOP Conf. Series: Journal of Physics: Conf. Series **997** (2018) 012042 doi:10.1088/1742-6596/997/1/012042

The use of a four-tier wave diagnostic instrument to measure the scientific literacy among students in SMA Negeri 2 Karanganyar

A Krisdiana[1], N S Aminah[1] and F Nurosyid[2]

[1]Physics Education, Postgraduate Program, Universitas Sebelas Maret Surakarta, 57126, Indonesia
[2]Physics Department, FMIPA, Universitas Sebelas Maret Surakarta, 57126, Indonesia

E-mail: krisdianaanita@gmail.com.

Abstract. This study aims to investigate the scientific literacy among 12th grade science students in SMA Negeri 2 Karanganyar. The instrument used is a four-tier wave diagnostic instrument. This instrument was originally used to diagnose students' conceptions about nature and propagation of waves. This study using quantitative descriptive method. The diagnostic results based on dominant students' answers show the lack of knowledge percentage of 14.3%-77.1%, alternative conceptions percentage 0%-60%, scientific conceptions percentage 0%-65.7%. Lack of knowledge indicated when there is doubt about at least one tier of the student's answer. The results of the research shows that the students' dominant scientific literacy is in the nominal literacy category with the percentage of 22.9% - 91.4%, the functional literacy with the percentage 2.86% - 28.6%, and the conceptual/procedural literacy category with the percentage 0% - 65.7%. Description level of nominal literacy in context of the current study is student have alternative conceptions and lack of knowledge. Student recognize the scientific terms, but is not capable to justify this term.

1. Introduction

1.1. Identification Students' Alternative Conceptions

Students often answer questions about physics using intuition or naive ideas based on experience from the world around them. The students' intuition or ideas are often wrong if viewed in a scientific perspective [1]. Researchers use several terms to describe students' mistakes in understanding scientific facts. Terms such as alternative conceptions, misconceptions, preconceptions, child's scientific intuitions, children's sciences, general knowledge concepts, and spontaneous knowledge are used by researchers to describe the errors of students in relating scientific facts [2]. The present study will use the term alternative conception. Alternative conceptions can be interpreted also as an initial conception that is different from scientifically accepted concepts. Investigations of students' alternative conceptions can provide information about the diversity they have in content knowledge, explanatory knowledge, and the power of understanding or incomprehension about a particular concept [3]. Investigations about understanding scientific facts should be made before starting a particular subject matter. The results of study are use to select appropriate teaching activities for students [4].

It's important to identify students' alternative conception correctly using a valid and reliable instrument. Diagnostic tests of alternative conceptions are assessment tools focusing on alternative conceptions that have been held or are being possessed by students and not handled resulting in learning difficulties. From the results of questionnaires given to physics teachers in SMA Negeri 2 Karanganyar obtained information that teachers consider the need to identify alternative conceptions of students. Unfortunately, teachers never give diagnostic tests to diagnose students' alternative conceptions.

One of diagnostic tests is a four-tier wave diagnostic test to diagnose alternative conceptions on nature and propagation of waves. Each item of four tier wave diagnostic test consists answer and reason tiers with the additional tiers of confidence rating specification for selected options at the answer and reason tier separately. Including a confidence rating at answer and reason tiers separately make the determination of conception categories more precise. The use of four tier wave diagnostic test is capable of diagnosing the nature and strength of alternative conceptions with error-free and less detectable knowledge in an easier way, especially for nature and propagation of waves [5]. Five categories are based on the students' respons on each tier of four tier wave diagnostic test. The five categories consist of scientific conceptions, false positives, false negative , alternative conceptions and lack of knowledge. The division of the four tier test categorization will make the proportion of less knowledge not lessened and the proportion of alternative conceptions not exaggerated.

1.2. Identification Students' Scientific Literacy Level

The main goals of science education curriculum are to provide an opportunity for students to be able to use scientific understanding in public debates, resulting in a balance of information and decisions about sociosaintific issues that affect their lives [6]. The term used to indicate coverage and affirming the goals of science education is the scientific literacy. Paul De Hart Hurd [7] in an article entitled "Science Literacy It's Meaning for American Schools," describes scientific literacy as an understanding of science and its application to social experience.

The achievement of scientific literacy becomes the main goal of learning. The two most comprehensive survey programs aimed at measuring science literacy are: The Program for International Student Assessment (PISA) and Trends in Mathematics and Science Studies (TIMMS). The ability of scientific literacy of Indonesian students can be seen based on the average score of PISA (Program for International Student Assessment). In the achievement, Indonesia is still at a low level with a score below the average score. In 2000, the subject of science placed Indonesia at number 38 out of 41 participating countries. In 2003, Indonesia was ranked 38 out of 40 participants. In 2006, the number of participants increased, Indonesia ranked 50th out of 57 countries, and in 2009 Indonesia ranked 60th out of 65 participating countries. In 2012 Indonesia ranked 64 out of 65 participants, and the latest in 2015 Indonesia ranked 62 out of 70 participants.

Bybee [8] suggests a comprehensive theoretical scale that is more appropriate for the measurement of scientific literacy at the 8] suggests a comprehensive theoretical scale that is more appropriate for the measurement of scientific literacy at the school level because it is easily transferred into learning objectives. The scale created suggests scientific literacy levels as a nominal, functional, conceptual or procedural, and multidimensional level of literacy. This level of scientific literacy was originally developed with reference to science programs and teaching in schools, but the level of scientific literacy can also be identified when measuring student achievement in science learning. For example, Sumarni, et al., [9] uses an evaluation instrument of achievement of generic science skills to measure students' scientific literacy in the integrated chemistry learning of ethnosciences and Ranikmae [10] found that instruments for measuring problem-solving and decision-making ability can also be used to measure the attainment of the scientific literacy aspect.

Assessment of students' scientific literacy is not used to determine the final students' literacy level but to know the indication of the existence of the seeds of literacy in the minds of students. Measurement of science literacy of students only aims to measure the effectiveness of science learning in providing attitudes, values, basic skills, knowledge, and understanding of science.

One of the challenges in science education is to create a link between theoretical framework and taxonomy with the practical assessment of learning [11]. The present study will use a conception diagnostic test instrument to measure students' scientific literacy levels according to Bybee's theoretical framework on nature and propagation of waves. Knowledge of the concepts and ideas of science or called content knowledge is one of the aspects of scientific knowledge. By using one instrument that is the four-tier wave diagnostic test, can be obtained the diagnosis of alternative conception and also information students' competence of specific scientific literacy.

The objectives of this study are using four-tier wave diagnostic test instrument to identify the students' alternative conceptions profile and scientific literacy levels on nature and propagation of waves. The results obtained can be used as an information and reference materials about of students alternative conceptions and scientific literacy levels, as a discourse to determine learning strategies that are able to remedy/reduce alternative conceptions, provide input to determine and develop conceptualized learning approach.

Table 1. Determining Students' Conceptions Categories.

Tier 1 (Answer)	Tier 2 (Rating)	Tier 3 (Reason)	Tier 4 (Rating)	Categories
1[a]	High[f]	1	High[h]	SC[c]
1	High	1	Low[i]	LK[g]
1	Low[g]	1	High	LK
1	Low	1	Low	LK
1	High	0[b]	High	FP[d] rarely AC
1	High	0	Low	LK
1	Low	0	High	LK
1	Low	0	Low	LK
0	High	1	High[h]	FN[e]
0	High	1	Low[i]	LK
0	Low	1	High	LK
0	Low	1	Low	LK
0	High	0	High[h]	AC[f] rarely Mistakes
0	High	0	Low[i]	LK
0	Low	0	High	LK
0	Low	0	Low	LK

[a]True response score '1'
[b]False response score '0'
[c]SC = Scientific Conceptions
[e]FN = False Negative
[g]LK = Lack of Knowledge
[h]High confidence rating ≥ 4
[i]Low confidence rating < 4
[d]FP = False Positive
[f]AC = Alternative Conceptions

2. Research Methods

The study was conducted in SMA Negeri 2 Karanganyar. This study uses a four-tier wave diagnostic test with a total of 12 items of questions. The sample of this research are 35 students of grade 12th class of science 1 in SMA Negeri 2 Karanganyar, Central Java Indonesia. Students are asked to do the test independently by ensuring that this test is only to diagnose their conception not to measure their learning achievement.

Students response by selecting A, B, C, D or E for the answer and reason tiers, and option 1, 2, 3, 4, 5 or 6 for each confidence ratings. In the four-tier wave diagnostic test, each correct response was scored '1' and the wrong response was scored '0'. The confidence rating has a range of 1 to 6. Number 1 for 'guess only', 2 for 'very unsure', 3 for 'unsure', 4 for 'confident', 5 for ' very confident ', and 6 for ' be completely sure'. The guidance of Table 1 uses to determine students' conceptions.

Table 2. Rubric of Students' Conceptions Categories.

Categories	Score
Scientific Conceptions	5
False Positive	4
False Negative	3
Alternative Conceptions	2
Lack of Knowledge	1

The diagnostic results of the students' conceptions then inserted into the scientific literacy levels with the following comparisons.

Table 3. Comparison of Students' Conceptions Categories and Scientific Literacy Levels.

Conceptions Categories	Scientific Literacy Levels
Lack of Knowledge Alternative Conceptions	Nominal
False Negative False Positive	Functional
Scientific Conceptions	Conceptual or procedural

Table 4. Description of Students' Competencies on Each Levels of Scientific Literacy.

Scientific Literacy Level Measured	Specific Competency
Nominal	student have alternative conceptions and lack of knowledge. Student recognizes the scientific terms but is not capable to justify this term
Functional	Students are still experiencing both false positive and negative. Students are able to use scientific vocabulary, as well as remember and write down basic information. But students are not able to determine the exact relationship of the information because of the limitations of understanding.
Conceptual or procedural	Students have a scientific conception. Students understand the principles and theories in science. Studentss understand the relationship of concepts from various disciplines. Studentss learn about the process of inquiry, how to determine the choice based on the correct information from the text, graphics or tables.

IOP Conf. Series: Journal of Physics: Conf. Series **997** (2018) 012042 doi:10.1088/1742-6596/997/1/012042

Statistical analysis using the QUEST program is performed after the student's scoring process is completed. Four-tier wave diagnostic test results were analyzed using classical test theory elements. The `program QUEST will be measured descriptive statistics, reliability coefficients (Cronbach α), the degree of difficulty, and discrimination.

2.1. Test Statistics
Table 5 summarizes the mean, realibility, degree of dificulty, and discrimination of four-tier wave diagnostic test results.

Table 5. Test Statistics.

N Students	N items	Mean ± St. Dev.	Reliability	Mean Degree of Dificulty	Mean Discrimination index
35	12	13.54±5.60	0.47	0.12	0.17

Reliability is obtained based on Internal Consistency value using Cronbach alpha (α) statistic. The α value for the four-tier wave diagnostic test is 0.47. A value below 0.7 indicates a weak correlation between items so it raises the question of whether the data is considered reliable or not. The value of α is an inappropriate indicator for conception test questions. Low α values indicate when student construct the knowledge they have gaps and errors, and negligence in learning based on the prior experience [12].

The degree of difficulty is a statement about how easy or difficult a test item is for the students concerned. Problem items with difficulty levels below 0.3 are categorized as difficult and above 0.8 are categorized as easy. The difficulty of obtaining a score of 5 (scientific conception) on a four-tier wave diagnostic test was in the range of 0-0.65 with an average of 0.12. From this data it can be seen that students found it difficult to provide response answers and reason according to the correct concepts on properties and propagation of mechanical waves.

The discrimination index (D) on the four-tier wave diagnostic test results is in the range of 0.17 to 0.45 with an average of 0.17. Based on the result, it can be concluded that the instrument is capable enough to distinguish high testee ability and low testee ability.

3. Results and Discussions
The four-tier wave diagnostic test in previous research by Caleon and Subramaniam has successfully diagnosed the nature and strength of alternative conceptions of 598 students from 9th and 10th grade in Singapore. The use of the four-tier wave diagnostic test on wave material has been able to identify 19 alternative conceptions, with 9 of which are original alternative conceptions (due to lack of understanding and false reason). The use of four-tier wave diagnostic tests can be used as the window of cognitive process information, to measure and detect variations in content knowledge, explanatory knowledge, and the power of understanding or misunderstanding certain concepts.

The same instrument, the four-tier wave diagnostic test, is used to identify the students' conceptions in Indonesia. Problems given to students in Indonesia are rewritten using Bahasa. The students' conceptions can be studied based on the response to each item of the four-tier wave diagnostic test. The four-tier wave diagnostic test indicators are understand the general characteristics of waves and waveform representations, understand about frequency, source, and medium, and understand the velocity of waves with mediums having constant properties.

In Figure 1. we can see that the dominant response of students indicates a lack of knowledge, followed by alternative conceptions, scientific conceptions, and a small portion of both negative and positive false. The highest percentage of the less knowledge category is found in 8 problem numbers. Less knowledge category has a percentage of 14.3% -77.1%. Lack of knowledge is shown when students provided low confidence ratings (<4) for each selected response. The response that shows lack knowledge is the result of guessing because students did not have the confidence to choose the answer. The number of questions with the dominance of lack knowledge response are 2, 3, 4, 5, 6, 7, 8, and 10.

Alternative conceptions are shown when students gave a high confidence rating (> 4) but their answers are wrong. The category of alternative conception has a percentage of 0%-60%. The question numbers with the dominance of alternative conceptions are 1 and 11. The dominance of alternative conceptions is indicated when students gave respond to question number 1 about the interpretation of wave properties shown in the time-displacement graph (y-t). In accordance with the findings of Maurines [13], that students tend to ignore the labels on the axes of the graph and assume that the graph is a representation of waves traveling on the medium. The dominant response for question number 1 given by 25% of students assumes that the maximum displacement between two possible particle points in the time-displacement graph (y-t) is the wavelength. The dominance of alternative conceptions is also shown by students when responding to the question number 11. Questions number 11 contains the determination of the frequency of waves propagating at the connection of two ropes with different masses. Students were not able to understand the relationship between wave sources and frequency. The dominant responses to number 11 are given by 57% of students assuming a heavier rope provides greater resistance to vibration thereby reducing the frequency of waves propagating on the rope. Students are unable to provide the correct response that the fact is that waves travel through two different mediums so that the wavelengths remain the same as the waves are generated by one common source and the two mediums are connected. The teacher must know the alternative conceptions and difficulties that the students experience and fix them for the purpose of effective learning [14].

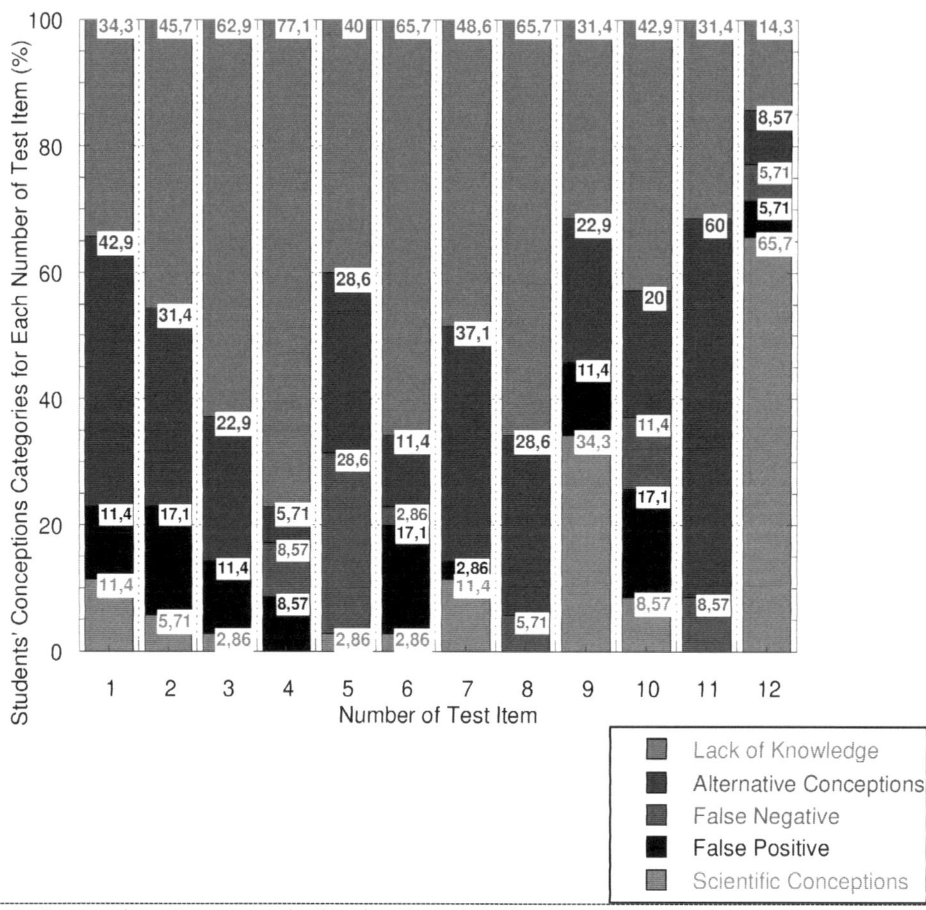

Figure 1. ConceptionsCategory Distribution Based on Students Respons for Each Number of Test Item.

The highest percentage for the category of scientific conception is found in questions 9 and 12. The category of scientific conception has a percentage of 0%-65.7%. Scientific conception is shown when students are able to choose the right choice and give a high confidence rating (> 4). The dominance of scientific conception is demonstrated when students respond to the number 9 of the effect on the magnitude of the wave as the energy carried by the waves decreases. The dominance of scientific conception is also shown when students respond to the number 12 predicted changes in wave velocity as it travels from a light-medium to a heavier medium.

The diagnostic results of students' conceptions are inserted into scientific literacy levels according to the comparison in Table 3. distribution of students' scientific literacy levels is shown in Figure 2.

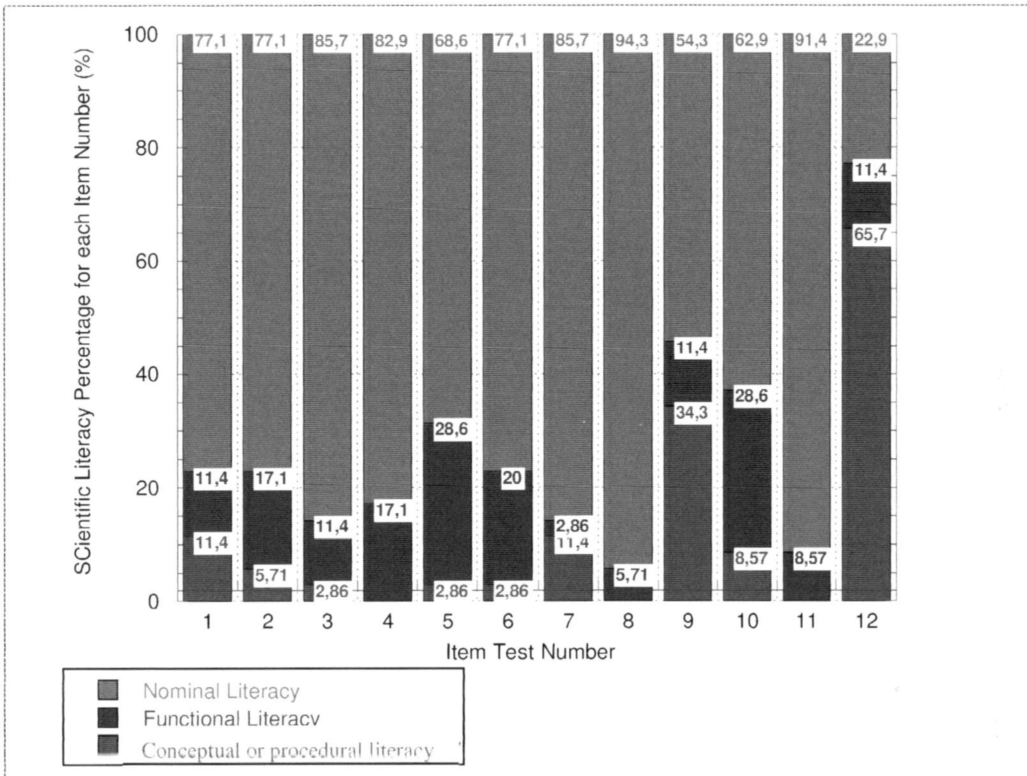

Figure 2. Scientific Literacy Distribution Based on Students Respons for Each Number of Test Item.

In Figure 2, can be seen the results of four-tier wave diagnostic test in students of grade XII IPA 1 in SMA Negeri 2 Karanganyar. Students' scientific literacy is dominant at the level of nominal literacy percentage 22.9% - 91.4%, functional literacy level 2.86% -28.6%, and level of conceptual or procedural literacy 0% -65,7% . These results mean student have alternative conceptions and lack of knowledge. The student recognizes the scientific terms but is not capable to justify this term. Students have the ability of science literacy on the category able to give opinions agree and disagree with an idea but unable to explain the reasons behind it [15]. The results of this study are supported by the results of some research on the identification of students' scientific literacy level in Indonesia and found that the majority of students are at the level of nominal literacy and the percentage of literacy's achievement is below 50% [16-17].

4. Conclusion

Based on the results of the four-tier wave diagnostic test, showed that students of grade XII IPA 1 in SMAN 2 Karanganyar majority have less knowledge and level of nominal literacy on materials nature and properties of waves. The results of this study are also consistent with Caleon and Subramaniam's statements in 2010 that the use of four-tier wave diagnostic test can be used as a window of cognitive process information experienced by students, providing it can be used to measure and detect variations in content knowledge, explanatory knowledge, and the strength understanding or misunderstanding of certain concepts [3]. The category of conception of students can be associated with the level of science literacy. According to the PISA 2015 terminology, knowledge of facts, concepts, and explanatory theories become one of the knowledge that construct science literacy.

References

[1] Wood A K, Galloway R K and Hardy J 2016 *Am. Phys. Soc.* **12** 023101(1)
[2] Ozkan G and Selcuk G S 2015 *J. Baltic Sci. Educ.* **14** 753
[3] Caleon I S and Subramaniam R 2010 *Res. Sci. Educ.* **40** 313
[4] Ozkan G and Selcuk G S 2015b *Univers. J. Educ. Res.* **3** 981
[5] Gurel D K, Eryilmaz A and McDermott L C 2015 *Eur. J. Math. Sci. Technol. Educ.* **11** 989
[6] OECD 2013 *PISA 2015 Draft Science Frame Work*
[7] Hurd P D 1958 *Educ. Leadersh.* **16** 13
[8] Bybee R W 1997 *Scientific Lit. Int. Symp.* 37
[9] Shwartz Y, Ben-Zvi R and Hofstein A 2006 *Chem. Educ. Res. Practice* **7** 203
[10] Sumarni W, Sudarmin, Wiyanto and Supartono 2016 *Int. J. Evaluation Res. Educ.* **5** 221
[11] Soobard R and Rannikmäe M 2011 *Sci. Educ. Int.* **26** 263
[12] Luxford C J and Bretz S L 2014 *J. Chem. Educ.* **91** 312
[13] Maurines L 1992 *Int. J. Sci. Educ.* **14** 279
[14] Maries A and Singh C 2016 *Phys. Rev. ST Phys. Educ. Res.* **12** 010131(1)
[15] Odja A H and Payu C S 2014 *Pros. Semin. Nas. Kimia Univ. Neg. Surabaya*
[16] Fakhriyah F, Masfuah S, Rosya M, Rusilowati A and Rahayu E S 2017 *J. Pendidik. IPA Indonesia* **6** 81
[17] Rusilowati A, Kurniawati L, Nugroho S E and Widyatmoko A 2016 *Int. J. Environ. Sci. Educ.* **11** 5718

Potential of iron sand from Betaf beach, Sarmi regency and river sand from Doyo, Jayapura regency, Papua as basic materials of mortar as nuclear radiation shielding

E Haryati[1] and K Dahlan[1]

[1]Department of Physics, Faculty of Mathematics and Natural Science, Cenderawasih University, Jayapura 99351, Indonesia

E-mail: [a]endanghfis@gmail.com

Abstract. According to the SNI, the type of concrete for use as nuclear radiation shielding is concrete or mortar that contains hematite, ilmenite, magnetite, barite, or ferrophosphorus synthesis. This study is focused on the characterization of iron sand from and river sand from Papua. The purpose of this research was to determine the specific content of gravity and minerals in iron sand and river sand from Papua. The specific gravities of the sands were measured by manual experiment in a laboratory, while their mineral content were calculated using XRF and SEM - EDS method. The result showed that the specific gravities of iron sand and river sand were 2.66 and 2.50, respectively. The XRF method revealed that the iron sand contained 41.68% Silica, 33.84% Iron, and 24.48% other minerals. The river sand on the other hand contained 58.98% Silica (Si), 26.87% Iron (Fe), and 14.15% other minerals. The SEM -EDS method showed that the iron sand was composed of 39.99 % SiO_2, 21.67 % FeO, and, and 38.34 % others; while the river sand contained 39.28% SiO_2, 17.45 % FeO, and 43.27 % others. The result showed that both sands have the potentials to be isolated from hematite minerals or magnetite.

1. Introduction

Sand is an abundant natural substance in Indonesia. It generally contains silica (Si), Iron (Fe), Aluminium (Al), Magnesium (Mg), and other minerals. The river sand from Bramaputra river of Bangladesh contains iron oxide (21.604%), titanium oxide (2.368%), silicon oxide (39.132%) and other minerals [1], while the heavy fractions of sands along the Brahmaputra River are primarily the source of iron titanium and iron oxide minerals magnetite and ilmenite [2]. The mineral composition of sand in Southern Ontario comprises hornblende, garnet, micas, magnetite, pyroxenes and sphene [3]. Amazon river sand is composed of floodpain sedimen, while Brazilian sand is composed of Silica, Alumunium, Iron, Natrium, Calsium, Kalium, Magnesium and Titanium [4]. Iron sand from the Southern coast of Yogyakarta is composed of magnetite, maghemite, ilmenite and titanomagnetite; whereas its non-magnetic residue mainly consists of silica, alumina and calc [5]. Minerals of sand and rock from Tuban and Sumenep, Tulungagung, Bawean-Gresik Island contain 98.23% CaCO3 in rock samples and 65.9 -76.8% SiO_2 with the highest impurities CaO and Fe2O3 (20- 28%) in sand samples (Tuban and Sumenep) [6]. The quantity of mineral content on every type of sand varies depending on the condition of rocks in the area and sedimentation process.

This research is focused on the analysis of mineral composition in sand substance from two different areas in Papua, Indonesia: Betaf beach in Sarmi regency and Doyo sub-district in Jayapura regency. Sarmi is one of the regencies in Papua that have unlimited amount of natural resources, particularly black sand, also known as iron sand, in its river and coastal areas. In a similar subject, Doyo is a sub-district in Jayapura Regency that also has iron sand resources, although it spreads only on its river zones and the local people mainly use it for buildings constructions.

The widespread use of nuclear technology has put the security factor in its application into serious attention. The technology possesses a certain side effect of radiation that could be completely dangerous for its users. Today, the development of nuclear technology is rapid, especially for medical industry. Despite its benefits, the risk in using nuclear technology must be addressed properly. One of which is through the application of a radiation shield with better quality and reliability in minimizing radiation hazard. The use of nuclear radiation shield is therefore imperative in order to avoid radiation during nuclear technology applications.

According to the SNI 03-2494-2002, the aggregate material for ionizing radiation retaining concrete is a natural material that is accumulated for heavy-weight content that contains hematite minerals (Fe_2O_3), ilmenite ($FeTiO_3$), magnetite (Fe_2O_4), barite ($BaSO_4$), or heavy ferrophosphorus synthetic aggregates (FenP), which is a mixture of iron phosphide [7]. Therefore, in this study we investigated specific gravities and mineral contents of iron sand and river sand as fine aggregate for mortar composition in a nuclear radiation shield [8].

2. Materials and Methods

The iron sand used in this study was obtained from Betaf beach in Sarmi, while the river sand was extracted from a river in Doyo. Both areas are located in Papua, Indonesia. Before the experiment, the iron sand was dried and then purified by using a permanent magnet. The iron sand was put on the magnet to determine the specific gravity value and its mineral content. As for the river sand, it was not treated before the examination.

2.1. Specific Gravity Test

The experiment to determine the specific gravity value was conducted at the Laboratory of Mining and Energy in Papua, by manual experiment method using a pycnometer [9].

2.2. Minerals Content of Iron Sand and River Sand

The examination of mineral contents in both iron sand and river sand was performed by X-ray Fluorescence (XRF) method in the Science Laboratory of FMIPA in the Hassanudin University in Makassar. After that, the samples were sent to the Geological Laboratory of ITB Bandung, for further examination by Scanning Electron Microscopy with Dispersive Spectrometer (SEM - EDS) method.

3. Result and Discussions

3.1. Specific Gravity Value

The specific gravity values for the iron sand from Sarmi and the river sand that was obtained from Doyo, in Jayapura are presented in Table 1.

Table 1. Specific gravity values of iron sand from Sarmi and river sand from Doyo Jayapura.

No	Material	Specific Gravity
1	River sand	2.50
2	Iron sand	2.66

Table 1 shows that the specific gravity value of iron sand is greater than the river sand. This result revealed that the iron sand has a smaller volume than river sand. The Indonesian National Standard (SNI 03-1970-1990) on specific gravity values assumes that the specific aggregate gravity rate ranges from 2.5 to 2.7 [10]. Therefore, from this result we can see that the two aggregates have met the minimum established standard and they can be applied as materials in mortar composition.

3.2. Mineral Contents of iron sand and river sand

3.2.1. XRF Method

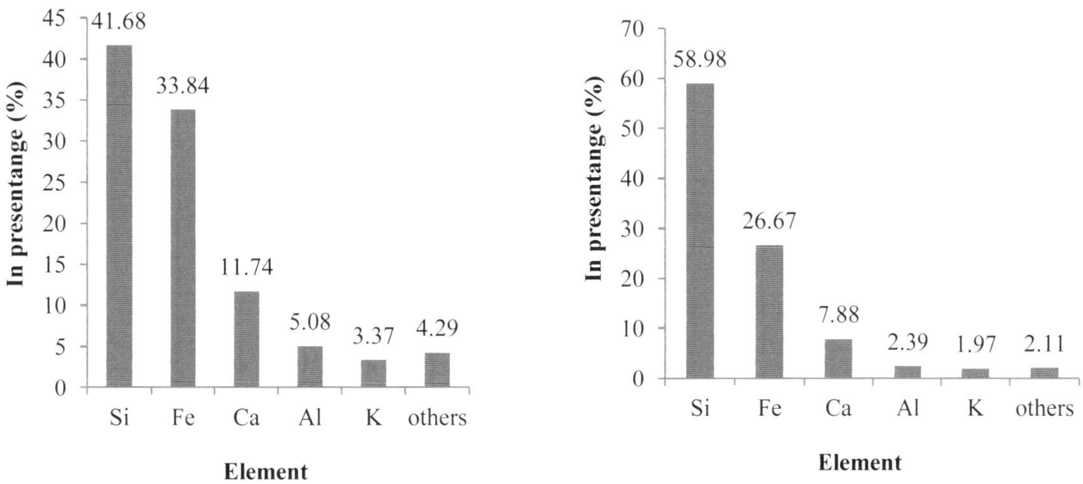

Figure 1. Mineral contents of (a) iron sand and (b) river sand.

Figures 1 displays that mineral contents in both sands are dominated by Silica (Si) and Iron (Fe). The iron sand contains 41.68% Si and 33.84% Fe, in contrast to the river sand that contains 58.98% Si and 26.67% Fe. The iron sand contains more iron than the river sand. However, the silica content of the river sand is much higher than the iron sand. The mineral compositions of both samples were influenced by the sand formation process and the site where the samples were obtained.

3.2.2. SEM-EDS Method

(a) (b)

Figure 2. SEM image of (a) River Sand and (b) Iron sand.

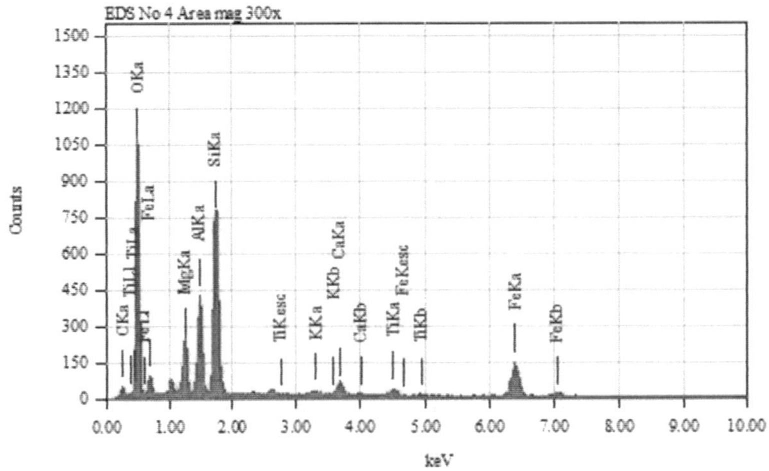

Figure 3. Grafik EDS river sand.

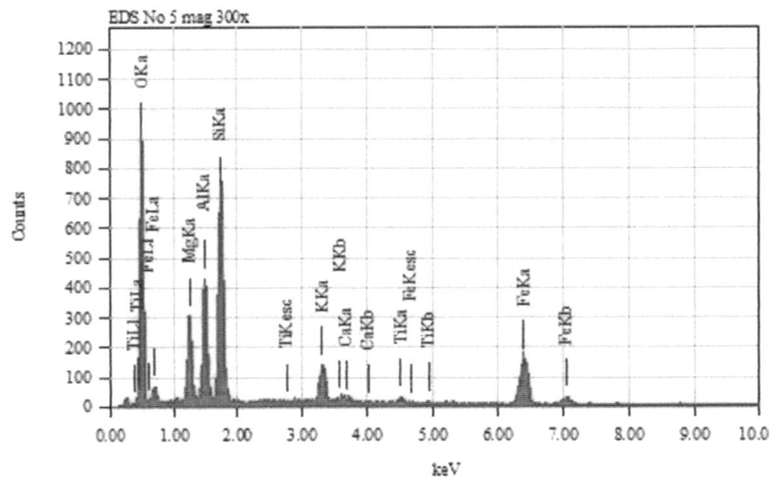

Figure 4. EDS spectra of iron sand.

From Figure 3 and 4, we can see that the highest peak for the iron sand and the river sand is the maximum value of Silica (Si), followed by the apex of iron (Fe), while the remaining numbers represent other minerals contained in the samples.

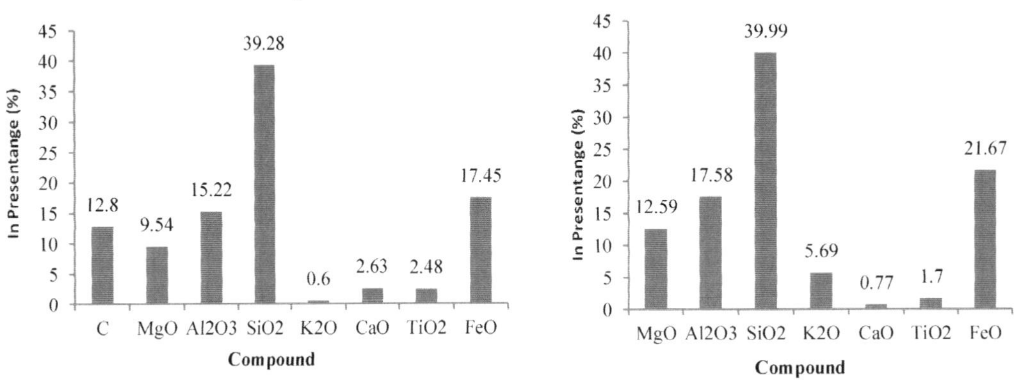

Figure 5. Mineral compounds of iron sand (a) and river sand (b).

Figures 5 show that SiO2 and FeO are the main contents in both samples. However, in the iron sand, the FeO content is greater compared to the river sand. From the results of XRF and SEM-EDS testings, we have revealed that both sands contain large quantities of iron (Fe).

4. Conclusions

The Iron sand from Betaf beach in Sarmi Regency and the river sand from Doyo, Jayapura Regency have been characterized by XRF and SEM-EDS method. Both of sands contain large quantities of Iron (Fe), it is therefore possible to isolate hematite minerals (Fe_2O_3), limeite ($FeTiO_3$), or magnetite (Fe_2O4) from iron sand and river sand [11]. Thus we can conclude that these materials can be used in a nuclear radiation shield, when it is applied into composition for concrete and mortar.

Acknowledgment

This research is funded by BOPTN Cenderawasih University, Jayapura Indonesia 2017.

References

[1] Rahman M A, Biswasb P K, Zamanb M N, Miaha M Y, Hossaina T and Huqa S M I 2012 *Bangladesh J. Sci. Ind. Res.* **4** 7167
[2] Khalil Md I, Khan Md N I, Kabir Md Z, Majumder R K, Ali Md I, Paul D and Islam S M A 2016 *Int. J. Geosci.* **7** 47
[3] Dell C I 1959 *Canadian J. Soil Sci.* **39** 185
[4] Martinelli L A, Victoria R L, Dematte J L I, Richey J E and Devol A H 1992 *App. Geochemistry* **8** 391
[5] Nugraha P A, Sari S P, Hidayati W N, Dewi C R and Kusuma D Y 2016 The Origin And Composition Of Iron Sand Deposit In The Southhern Coast Of Yogyakarta *AIP conferences Proceeding* 1746 020028
[6] Munasir, Triwikantoro, Zainuri M and Darminto 2012 *J. Penelit. Fis. Apl.* **2** 20
[7] SNI 03-2494-2002 Badan Standar Nasional (*Spesifikasi Agregat Beton penahan Radiasi)*
[8] Ouda A S, Hamdy A and Gawwad A 2017 *HBRC J.* **13** 255
[9] Braja M D 1993 *Mekanika Tanah Jilid I* (Erlangga: Jakarta)
[10] SNI 03-1970-1990 Badan Standar Nasional (*Pengujian Berat Jenis dan Penyerapan Agregat Halus)*
[11] Rusianto T, Wildan M W, Abraha K and Kusmono K 2015 *Int. J. Techno.* **6** 1017

IOP Conf. Series: Journal of Physics: Conf. Series **997** (2018) 012044 doi:10.1088/1742-6596/997/1/012044

Developing physics learning media using 3D cartoon

M Wati[1], S Hartini[1,a], N Hikmah[1] and S Mahtari[1]

[1]Physics Education Program, Universitas Lambung Mangkurat, Banjarmasin, Indonesia

E-mail: [a]srihartini_pfis@unlam.ac.id

Abstract. This study focuses on developing physics learning media using 3D cartoon on the static fluid topic. The purpose of this study is to describe: (1) the validity of the learning media, (2) the practicality of the learning media, and (3) the effectiveness of the learning media. This study is a research and development using ADDIE model. The subject of the implementation of media used class XI Science of SMAN 1 Pulau Laut Timur. The data were obtained from the validation sheet of the learning media, questionnaire, and the test of learning outcomes. The results showed that: (1) the validity of the media category is valid, (2) the practicality of the media category is practice, and (3) the effectiveness of the media category is effective. It is concluded that the learning using 3D cartoon on the static fluid topic is eligible to use in learning.

1. Introduction

The Act No. 19 of 2005 Section 19 Article (2) states the educational process in the educational unit is organized in an interactive, inspiring, fun, challenging, and motivates learners to actively participate and provide enough space for innovation, creativity, and independence according to the talents, interests, physical, and psychological development of learners [1]. To stimulate the thinking ability of learners, teachers are required to be more innovative in every teaching and learning activity. Learning media as a means of teaching aids teachers can facilitate the delivery of information from the teacher to the learners with the stimulus interesting, so students will be easy in processing the received information. It could be packaged in a media that support a learning process in understanding the concept without limited space and time [2].

Learning media is a tool that serves to explain some of the overall learning programs that is difficult to explain verbally [3]. Media in the learning process tends to be interpreted as graphics tools, photographic, or electronically to capture, process, and reconstitute the visual or verbal information [4]. The utilization of learning media makes learners not to be bored; thus, they are more enthusiastic about learning [5].

The low level of preference for school ultimately became would have an impact on the students outcome to study identified the investing public caused by media elections to resolve months of learning which less appropriately. In this case, the teachers may not utilizing learn in an optimum manner. In the learning process, teacher often neglect of using the media, despite the fact that media serves to increase the preference for school students and in turn will improve the quality of education students.

Based on the observations in SMAN 1 Pulau Laut Timur, especially in physics learning, the acquired learning media is merely whiteboard. Hence, students appear to be bored and lack of motivated to keep learning. The whiteboard is only able to deliver a visual symbol. Based on Edgar Dale Cone experience blackboard is still abstract [6]. Most students had difficulty learn physics because many charge material that should be learned in a semester. This resulted in the interest of learning to decrease, consequently study results obtained by students unsatisfactory. Meanwhile, physics is a lesson that contains many abstract concepts that it requires a learning tool that can make abstract concepts to be more concrete direction so that the material is more easily understood.

Various types of learning media need to be considered in the utilization and development. Film cartoon is a very capable media to attract the attention of students so that they can develop their interests and motivation and clarify abstract concepts. Cartoons can develop students' interest and motivation as well as clarify abstract concepts [6]. According to Sadiman, et al [7], the great cartoon is able to attract attention, affect attitude, and behavior. The Interest of a person to cartoons compared to the media another is also due to symbols certain in cartoons that cause jocularity, in addition, the contents of cartoons tell various phenomena in daily life. Cartoon film media has undergone many developments [7]. One of them is a three-dimensional cartoon (3D). One software that can be used to create a 3D cartoon is Muvizu. Creating a 3D cartoon with Muvizu is easy and intuitive, and it is equipped with the tools that help predetermined characters to be adjusted [8]. Media learning to learn physics still very limited to the print of textbook and learning monotonous, students become saturated and learning less attention. In this case, it takes some examples of the concrete object that draws so that it can reduce saturation and boredom in learning. To forced need a media interesting, in this case, the media animated cartoons.

Based on the above-mentioned consideration, the research and development using 3D cartoon learning media are conducted on the static fluid topic for senior high school students. This study aims to produce an eligible instructional media using 3D cartoon on the static fluid topic for senior high school students based on the validity, practicality, and effectiveness of the developed learning media.

2. Method

2.1. Type of Research

It is a research and development (R &D). R & D is an official procedure in order to develop a new product or perfecting existing products accountable. The steps of this study were carried out using ADDIE model. The ADDIE model steps are analysis, design, development, implementation, and evaluation. The research subjects in this study is a 3D cartoon movie instructional media and the subjects of the implementation were the students of the XI class of SMA Negeri 1 Pulau Laut Timur Academic Year 2015/2016 as many as 30 students. The object of this study is a feasibility media of this 3D cartoon movie learning media on the static fluid topic.

2. 2. Data Analysis Technique

2.2.1 Validity. The media validity was measured through the validation sheet. Media validation was validated by three validators. Table 1 shows the validity criteria for the instructional media [9].

Table 1. Criteria for the validity learning material.

Interval	**Category**
$x > 4.21$	very valid
$3.40 < x \leq 4.21$	valid
$2.60 < x \leq 3.\ 40$	fairly valid
$1.79 < x \leq 2.60$	less valid

2.2.2 Practicality. Student questionnaire responses are the instrument used to measure the practicality of learning material. The practicality learning criteria are shown in Table 2 [9].

Table 2. Criteria for practicality learning material.

Interval	Category
$x > 4.21$	Very practical
$3.40 < x \leq 4.21$	Practical
$2.60 < x \leq 3.40$	Fairly practical
$1.79 < x \leq 2.60$	Less practical

2.2.3 Effectiveness. The effectiveness of learning material was measured from learning outcomes test through pre-test and post-test. The testing and inferred to know the effectiveness of the learning material. Analysis of the students learning result was done by using a normalized gain equation as follows [10].

$$\langle g \rangle = \frac{\% \langle s_f \rangle - \% \langle s_i \rangle}{100 - \% \langle s_i \rangle} \qquad (1)$$

where are $\langle s_f \rangle$ = Score from the post-test and $\langle s_i \rangle$ = Scores from the pre-test. Table 3 shows the effectiveness criteria for pre-test and post-test.

Table 3. Effectiveness criteria for the learning material.

Value	Category
$(\langle g \rangle) \geq 0.7$	very effective
$0.7 > (\langle g \rangle) \geq 0.3$	Effective
$(\langle g \rangle) < 0.3$	less effective

Development media animated cartoons made in the research efforts to improve the liveliness learn through interesting, fun, and reduce. For that media saturation developed have to valid, practical, and effective.

3. Result and Discussion

Development media animated cartoons made in the research is efforts to improve liveliness learns through interesting, fun, and reduce saturation. The developed media have valid, practical, and effective categories.

Table 4. Validity result of the learning media.

Aspect	Value	Criteria
Media Design	3.3	Valid
Content	3.3	Valid
Media Appearance	3	Valid
Technical Quality	3.56	Valid
Reliability	0, 8	High

The Validity of the learning media covers four aspects namely design of the media, content, media appearance, and technical quality. The aspect of the media design is categorized as valid. It shows that the media is developed from the maintenance, ease of use, and ability to operate in several movie player software programs are good. The content aspect category is valid. It shows that the material arrangement is in accordance with the syllabus, the presented material is appropriate, up to date, and systematically arranged. The appearance media categorizes the valid category. It shows that the developed media is interesting to the students. The technical quality is categorized valid. It shows that the media developed for the figures selection and characters are appropriate, the language used in the presentation of the material is well and communicative, the dubber is appropriate to the character, the animation presentation is appropriate to the story, the media can clarify the material clearly, the setting makes the material clearer, and the questions are appropriate to the topic. In addition, the voice can be clearly heard and the animation appearance is clearly seen. The reliability of the developed media is 0.82. This suggests that the high category and the assessment of the validators are consistent. The criteria selection of the media must meet the criteria of appropriateness, ease, attractiveness, and usefulness [11]. It also meets the criteria for selection of the other media by [4], which is in accordance with the objectives to be achieved, the precision of supporting defenders distance and meet the technical quality. This media for receiving a recommendation of the people to be able to try out on students.

Table 5. Practicality result from the learning media.

Aspect	Value	Criteria
Ease	3.96	Good
Benefit	3.90	Good
Reliability	0.84	High

The practicality of the teaching media consists of two aspects, namely the ease and usefulness. The ease aspect is categorized as good. This proves that the learning media is easy to use, can be operated in various player applications, as well as easy to maintain. The benefit aspect is categorized as good. This shows that the media can facilitate learners to master the material, help them independently, make the learners remember the material better, help students to focus, make students interested in the material, and make students learn more quickly than usual according to the needs of students. This is in line with the opinion of Nieven in [12] that the learning media is practical if it is easy and useful to use for students. In addition, this result fulfilled the requirement or criteria of the media selection based on the ease aspect by [11] that is easy to use and operational. It is also in line with [6] opinion that the media can be used to accelerate the learning material. This media also fulfilled the media function according to Levied and Lentz [4] which includes the function of attention that attracts and directs students' attention to concentrate, cognitive function to facilitate students understanding and recalling information. This is in line with the opinions presented by Arsyad [4] pertaining to an election media focusing on several components, that is the purpose of instructional what will be reached, material learning ran will discuss the degree to which the depth to which must be achieved, characteristic of students quantitatively and qualitatively, and consideration for the media to be developed.

Table 6. The effectiveness of the learning media.

N-gain	Category
0.76	High

The effectiveness of the learning media was analyzed using N-gain equation. Based on the results of pre your test and post the test then can be determined the more so or to an increase in the understanding of the related students. N-gain value based on the analysis is 0.74. Based on these results so, is likely to increase their level of understanding of students who indicated by the value of the gain. This shows that the media is selected to be used in learning. These results are consistent with what was said by [6] that media can improve the quality of learning in this case in terms of students learning outcomes. According to Ploom [13], the effectiveness is seen from the achievement of the students learning outcomes improvement.

As for weakness media animated cartoons in research is students need more controlled when learning held because sometimes they are fun see the cartoon picture without regard to the material that was delivered so as to cause a lack of concentration to material substance. In addition, have weaknesses whether the contents of material and technology, so that expected can be used as a reference in subsequent development

4. Conclusion

Based on the implementation of the developed product, it is seen that the media validity is categorized as valid, the media practicality is good, and the media effectiveness is effective. In conclusion, this learning media is feasible to use.

5. Acknowledgment

The authors thank Faculty for Teacher Training and Education, Universitas Lambung Mangkurat due to the funding of DIPA-2016 for this research.

References

[1] Undang-undang RI No 20 2003 *Sistem Pendidikan Nasional* (Jakarta: Republik Indonesia)
[2] Anggraeni R D and Kustijono R 2013 *J. Penelit. Fis. Apl.* **3** 11
[3] Hadibin M M, Purnama B E and Kristianto G 2013 *Indonesian J. Comput. Science-Speed.* **1** 1
[4] Arsyad A 2014 *Media Pembelajaran* (Jakarta: Raja Grafindo Persada)
[5] Syafrudin C and Pujiyono W 2013 *J. Sarj. Teknik Informatika* **1** 2
[6] Musfiqon 2012 *Pengembangan Media & Sumber Pembelajaran* (Jakarta: Prestasi Pustaka Karya)
[7] Sadiman A S, Raharjo and Haryono 2014 *Seri Pustaka Teknologi Pendidikan Nomor 6-Media Pendidikan-Pengertian, Perkembangan, dan Pemanfaatannya* (Jakarta: Raja Grafindo Persada)
[8] Kainz O, Jakab F and Kardoš S 2013 *Proc.11th IEEE Int. Conf. on Emerging eLearning Technologies and Applications* (Slovakia: IEEE). p 203
[9] Widoyoko EP 2016 *Evaluasi Program Pembelajaran* (Yogyakarta: Pustaka Pelajar)
[10] Hake R 1999 *Analyzing Change/Gain Score* (Online) Available on www.physics.indiana.edu/-sdi/AnalyzingChange-gain.pdf. Accessed September 28th 2016
[11] Mulyanta St and Marlon L 2009 *Media Pembelajaran* (Yogyakarta: Universitas Atma Jaya)
[12] Rochmad 2012 *J. Kreano* **3** 70
[13] Hafiz M 2013 *Ta'dib* **16** 32

The Stark Effect on the Wave Function of Tritium in Relativistic Condition

B Supriadi[1], S H B Prastowo[1], S Bahri[1], Z R Ridlo[1] and T Prihandono[1]

[1]Physics Education Departement, University of Jember, Jember, Indonesia

E-mail: bambangsscmsc@gmail.com

Abstract. Tritium Atom is one of the isotopes of Hydrogen that has two Neutrons in the nucleus and an electron that surrounds the nucleus. The Stark Effect is an effect of a shift or polarization of the atomic spectrum caused by the external electrostatic field. The interaction between the electrons and the external electric field can be reviewed using an approximation method of perturbation theory. The perturbation theory used is a time Independent non-degenerate perturbation and reviewed to second order to obtain correction of Tritium Atomic wave function. The condition that used in the system is a relativistic condition by reviewing the movement of electrons within the Atom. The effects of relativity also affect the correction of the wave function of Atom Tritium in the ground state. Tritium is radioactive material that is still relatively safe, and one of the applications of Tritium Atom is on the battery of betavoltaics (Nano Tritium Battery).

1. Introduction

Quantum physics has replaced the theory of classical physics for some cases and has grown faster science over the last 100 years. The theory of quantum has changed many components in the physics such as Quantum Mechanics, Quantum Electrodynamics, Quantum Thermodynamics and Astrophysics. All of the microscopic phenomena like an atomic case can be explained by using the theory of quantum mechanics and using the principle of wave-particle dualism. Calculations of the spectrum energy and wave functions of Hydrogen atoms use a second-order equation that known as the Schrodinger equation. Schrodinger equation is able to give a definition of the wave function of a particle [1]. This equation is formed by using the law of conservation of energy, obedient to the hypothesis of de Broglie so as to produce a complex analytical solution in the form of wave functions of a single value, continuous and finite. The Schrodinger equation can explain the problem of microscopic like atom until the macroscopic problems like Stars or universe [2]. The wave function must be normalized by using mathematical calculation. Based on the general characteristics of the wave function, the Schrodinger equation can be divided into two part, the first is time-dependent Schrodinger equations and time-independent Schrodinger equations or often referred to as steady state. The simplest form of Schrodinger equation is the time-independent schrodinger equation, where the wave function is only influenced by the potential (V) and position (r).

In this research will be studied about the effect of external electrostatic field (Stark Effect) on Tritium atom wave function in the ground state ($1s$). The approximation method is time-independent non-degenerate perturbation theory and reviewed until second order correction of the wave function in the ground state of Tritium. The motion of an electron in Tritium has a high velocity (2.1×10^6 m/s) and will be faster when the mass number (Z) is used is also large. So to get the real condition of the hydrogenic atom should to calculate the relativity effect of the movement of electrons. The relativistic correction used in this study is only in first order because for second order correction it gives very small effect in calculations.

Tritium is one of the isotopes of hydrogen that contains two neutrons in the nucleus and one of an electron that surrounds the nucleus. Tritium is a radioactive atom that have a half-life of 12.32 years[3]. Tritium is a radioactive material that is still relatively safe. The application of Tritium Atom is on the battery of betavoltaics (Nano Tritium Battery). Beta-voltaic battery is one type of radioactive isotope battery that has advantages such as a long service life, high energy density and free of maintenance [4].

The power of Tritium Beta-voltaic battery is about 0.5 – 1.5 μW and voltages between 500 – 1500 V [5]. In this research will be discussed the characteristics of Tritium due to the influence of external electric field on energy and wave function of Tritium. Then, It will examine the effect of the external electric field to the power of Tritium battery theoretically.

2. Tritium Atoms in Ground State

Tritium atom has nuclei containing a proton and two neutrons, and an electron that moving around the nucleus. The mass used in this research is the reduction mass (μ_T) that occurs due to the interaction between the mass of electrons (m_e) and the mass of the nucleus (m_N) and the equation is written as:

$$\frac{1}{\mu_T} = \frac{1}{m_e} + \frac{1}{m_N} \quad (1)$$

The general formula of schrodinger equation which describes the hamiltonian system in time-independent for the mass particles μ_T and moves in one dimension in the potential field $V_r(r)$ shown in equation (2). We used the schrodinger equation because this equation can describe the characteristic of wave in a particle on microscopics phenomena.

$$-\hbar^2\nabla^2(r)\psi + V_r(r)\psi = \xi\psi \quad (2)$$

the spherical coordinate given by,

$$\nabla^2 = \frac{1}{r^2}\frac{\partial}{\partial r}\left(r^2\frac{\partial}{\partial r}\right) + \frac{1}{r^2\sin\varphi}\frac{\partial}{\partial\varphi}\left(\sin\varphi\frac{\partial}{\partial\varphi}\right) + \frac{1}{r^2\sin^2\phi}\frac{\partial^2}{\partial^2\phi} \quad (3)$$

By solving this schrodinger equation, we obtain a solution of radial wave function $(\Re_{nl}(r))$ and a spherical harmonic function symbolized by $\Theta_{lm}(\varphi,\phi)$, with n, l and m are a quantum number. The atomic energy of tritium, the solution of the radial function equation and the solution of the spherical harmonic equations are given by equations (4), (5) and (6),

$$\xi_n = -\frac{\mu_T e^4}{2(4\pi\varepsilon)^2\hbar^2}\frac{1}{n^2} = -\frac{21.79477\times10^{-19}}{n^2}\,joule \quad (4)$$

$$\Re_{nl}(r) = -\left(\frac{2}{na_1}\right)^{3/2}\sqrt{\frac{(n-l-1)!}{2n[(n+l)!]^3}}\left(\frac{2r}{na_1}\right)^l e^{-r/na_1}L_{n-l-1}^{l+1}\left(\frac{2r}{na_1}\right) \quad (5)$$

$$\Theta_{lm}(\varphi,\phi) = (-1)^{(m+|m|)/2}\sqrt{\frac{2l+1}{4\pi}\frac{(l-|m|)!}{(l+|m|)!}}\Omega_l^m(\cos\varphi)e^{im\phi} \quad (6)$$

where $L_{n-l-1}^{l+1}\left(\frac{2r}{na_1}\right)$ is the Associated Laguere Polynomial equation, a_1 is the first radius of Tritium atomic state and $\Omega_l^m(\cos\varphi)$ is an associated Legendre polynomial [6]. Using the equations (4), (5) and (6), the magnitude of the energy and the function of the Tritium wave in the ground state is,

$$\xi_1 = -21.79477\times10^{-19}\,joule = -13.6032255eV \quad (7)$$

$$\psi_{1s} = \Re_{10}\Omega_{00} = \frac{1}{\sqrt{\pi a_1^3}}e^{-r/a_1} \quad (8)$$

IOP Conf. Series: Journal of Physics: Conf. Series **997** (2018) 012045 doi:10.1088/1742-6596/997/1/012045

3. Time-Independent non Degenerate Perturbation theory

A Hamiltonian initial condition without any perturbation show in equation (9),

$$H\psi_n = E_n^{(0)}\psi_n \tag{9}$$

In perturbation theory, Hamiltonian system will divide by two part, part one is unperturbed Hamiltonian system (H_0) and the second part is Hamiltonian with small perturbances (*W*). In equation (10), α means the expansion parameter for the perturbation correction order[7].

$$H = H_0 + \alpha W \tag{10}$$

Because the perturbation *W* assumed to be small, it should be possible to expand ψ_n and E_n as a power series in *W*. The expansion of energy and wave function are,

$$E = E^{(0)} + \alpha E^{(1)} + \alpha^2 E^{(2)} + ... \tag{11}$$

$$\psi = \psi^{(0)} + \alpha\psi^{(1)} + \alpha^2\psi^{(2)} + ... \tag{12}$$

and then, subtitute eq. (10), eq. (11) and eq. (12) to eq. (9).

$$\begin{aligned}(H_0 + \alpha W)(\psi^{(0)} + \alpha\psi^{(1)} + \alpha^2\psi^{(2)} + ...) = (E^{(0)} + \alpha E^{(1)} + \alpha^2 E^{(2)} + ...)\\ (\psi^{(0)} + \alpha\psi^{(1)} + \alpha^2\psi^{(2)} + ...)\end{aligned} \tag{13}$$

By doing operation between two sides, and the coefficients of successive power of α on both sides of this equation must be equal. We obtain the results are,

$$(H_0 - E^{(0)})\psi^{(0)} = 0 \tag{14}$$

$$(H_0 - E^{(0)})\psi^{(1)} = (E^{(1)} - W)\psi^{(0)} \tag{15}$$

$$(H_0 - E^{(0)})\psi^{(2)} = (E^{(1)} - W)\psi^{(1)} + E^{(2)}\psi^{(0)} \tag{16}$$

Equation (14) is the zero-order correction solution, equation (15) is the first –order correction solution and equation (16) is the second-order correction solution.

From Equation (15), we can find the results of the first order correction of energy and wave function of Tritium,

$$E_n^{(1)} = \langle\psi_n|W|\psi_n\rangle = \int_V \psi_n W\psi_n dV \tag{17}$$

$$\psi_n^{(1)} = \sum_{n\neq k}\frac{\langle\psi_k|W|\psi_n\rangle}{E_n^{(0)} - E_k^{(0)}}\psi_k^{(0)} \tag{18}$$

From equation (16), we can get the result of second order correction for energy and wave function of Tritium,

$$E_n^{(2)} = \sum_{n\neq k}\frac{|\langle\psi_k|W|\psi_n\rangle|^2}{E_n^{(0)} - E_k^{(0)}} \tag{19}$$

$$\begin{aligned}\psi_n^{(2)} = \sum_{r\neq n}\left[\sum_{m\neq n}\frac{\langle\psi_r|W|\psi_m\rangle\langle\psi_m|W|\psi_n\rangle}{\left(E_n^{(0)} - E_r^{(0)}\right)\left(E_n^{(0)} - E_m^{(0)}\right)} - \frac{\langle\psi_n|W|\psi_n\rangle\langle\psi_r|W|\psi_n\rangle}{\left(E_r^{(0)} - E_n^{(0)}\right)^2}\right]\psi_r^{(0)}\\ -\sum_{r\neq n}\frac{1}{2}\frac{|\langle\psi_r|W|\psi_n\rangle|^2}{\left(E_r^{(0)} - E_n^{(0)}\right)^2}\psi_n^{(0)}\end{aligned} \tag{20}$$

Where ψ_n mean the wave function of Tritium in the ground state that consist of ψ_{1s}. ψ_m is the wave function of Tritium in the first excited state, that consists of ψ_{2s}, ψ_{2pz}, ψ_{2px} and ψ_{2py}. And ψ_r is the wave function of Tritium in the second excited state, that consists of ψ_{3s}, ψ_{3pz}, ψ_{3pz}, ψ_{3pz}, ψ_{3dzz}, ψ_{3dyz}, ψ_{3dxz}, ψ_{3dxy}, and $\psi_{3dx^2-y^2}$.

4. Relativistic Correction

In principle, the relativistic effect for Tritium atoms will affect the magnitude of Energy for various circumstances. Relativistic kinetic energy of electrons

$$K = \sqrt{p^2c^2 + \mu_T^2c^4} - \mu_T c^2 \tag{21}$$

With using Taylor expansions to get the simple solution of eq. (21), and the result is,

$$\sqrt{p^2c^2 + \mu_T^2c^4} - \mu_T c^2 \cong \frac{p^2}{2\mu_T} - \frac{p^4}{8\mu_T^3c^2} + ... \tag{22}$$

The next step is subtitute eq. (10) to total Hamiltonian system,

$$H = \left(\frac{p^2}{2\mu_T} - \frac{p^4}{8\mu_T^3c^2} + ...\right) - \frac{e^2}{4\pi\varepsilon}\frac{1}{r} = H_0 + W \tag{23}$$

$$W = -\frac{p^4}{8\mu_T^3c^2} \tag{24}$$

Equation (24) is perturbed Hamiltonian dan H_0 is unperturbed Hamiltonian of Atom. To find the Tritium first-order energy correction due to the effect of relativity is to use the theory of perturbation with using equation (17).

$$\xi R_{nl}^{(1)} = \left\langle \psi_{nlm} \middle| \frac{p^4}{8\mu_T^3c^2} \middle| \psi_{nlm} \right\rangle \tag{25}$$

By using schrodinger equation, we can get the solution of quadratic momentum of electron to solve the equation (25). Quadratic momentum that derive from schrodinger equation given by eq. (26).

$$p^2 = 2\mu_T(\xi_n - V(r)) \tag{26}$$

Then substitute equation (26) into equation (25) to obtain the first order energy correction of relativity effect .

$$\xi R_{nl}^{(1)} = \frac{1}{8\mu_T c^2}\left\langle \psi_{nlm} \middle| \xi_n^2 - \frac{2\xi_n e^2}{4\pi\varepsilon}\frac{1}{r} + \frac{e^4}{(4\pi\varepsilon)^2}\frac{1}{r^2} \middle| \psi_{nlm} \right\rangle$$

$$\xi R_{nl}^{(1)} = \frac{1}{8\mu_T c^2}\left(\xi_n^2 - \frac{2\xi_n e^2}{4\pi\varepsilon}\left\langle\frac{1}{r}\right\rangle_{nl} + \frac{e^4}{(4\pi\varepsilon)^2}\left\langle\frac{1}{r^2}\right\rangle_{nl}\right) \tag{27}$$

To got the solution of $\left\langle\frac{1}{r}\right\rangle_{nl}$, we can use recursion method. From the radial equation of Tritium atom,

$$\left[-\frac{\hbar^2}{2\mu_T r^2}\frac{d}{dr}\left(r^2\frac{d}{dr}\right) + \frac{l(l+1)\hbar^2}{2\mu_T r^2} + V(r)\right]U_{nl}(r) = EU_{nl}(r)$$

$$\left[-\frac{\hbar^2}{2\mu_T r^2}\frac{d}{dr}\left(r^2\frac{d}{dr}\right) + \frac{l(l+1)\hbar^2}{2\mu_T r^2} - \frac{e^2}{4\pi\varepsilon_0}\frac{1}{r}\right]U_{nl}(r) = \left(-\frac{\mu_T e^4}{2\hbar^2(4\pi\varepsilon_0)^2}\frac{1}{n^2}\right)U_{nl}(r) \tag{28}$$

Where $U_{nl}(r)$ is the radial wave function that depended of *r*. And then do differential on equation (28) to e^2, and we could got the result in equation (29):

$$\left[\frac{1}{r}\right]U_{nl}(r)=\left[\frac{2\mu_T e^2}{2\hbar^2(4\pi\varepsilon_0)}\frac{1}{n^2}\right]U_{nl}(r) \tag{29}$$

The next step is to multiply in the both sides of equation (29) with the conjugate radial wave function and integrating to *r*.

$$\int_0^\infty U_{nl}^*\left[\frac{1}{r}\right]U_{nl}(r)dr=\int_0^\infty U_{nl}^*\left[\frac{2\mu_T e^2}{2\hbar^2(4\pi\varepsilon_0)}\frac{1}{n^2}\right]U_{nl}(r)dr$$

$$\left\langle\frac{1}{r}\right\rangle_{nl}=\left[\frac{2\mu_T e^2}{2\hbar^2(4\pi\varepsilon_0)}\frac{1}{n^2}\right]$$

With the radius of Tritium atom in ground state is $a_1=\frac{4\pi\varepsilon_0\hbar^2}{\mu_T e^2}$, so the result is:

$$\left\langle\frac{1}{r}\right\rangle_{nl}=\frac{1}{a_1 n^2} \tag{30}$$

Whereas to find the value of $\left\langle\frac{1}{r^2}\right\rangle_{nl}$, we would to differentize the equation (28) to the quantum number *l* on each segment.

$$\frac{d}{dl}\left[-\frac{\hbar^2}{2\mu_T r^2}\frac{d}{dr}\left(r^2\frac{d}{dr}\right)+\frac{l(l+1)\hbar^2}{2\mu_T r^2}-\frac{e^2}{4\pi\varepsilon_0}\frac{1}{r}\right]U_{nl}(r)=\frac{d}{dl}\left(-\frac{\mu_T e^4}{2\hbar^2(4\pi\varepsilon_0)^2}\frac{1}{n^2}\right)U_{nl}(r) \tag{31}$$

Because the value of $n=N+l+1$, so $dn/dl=1$ or $dn=dl$, the eq. (31) become

$$\frac{d}{dl}\left[-\frac{\hbar^2}{2\mu_T r^2}\frac{d}{dr}\left(r^2\frac{d}{dr}\right)+\frac{l(l+1)\hbar^2}{2\mu_T r^2}-\frac{e^2}{4\pi\varepsilon_0}\frac{1}{r}\right]U_{nl}(r)=\frac{d}{dn}\left(-\frac{\mu_T e^4}{2\hbar^2(4\pi\varepsilon_0)^2}\frac{1}{n^2}\right)U_{nl}(r)$$

$$\left[\frac{(2l+1)\hbar^2}{2\mu_T r^2}\right]U_{nl}(r)=\left(\frac{\mu_T e^4}{\hbar^2(4\pi\varepsilon_0)^2}\frac{1}{n^3}\right)U_{nl}(r)$$

$$\left[\frac{1}{r^2}\right]U_{nl}(r)=\left(\frac{2\mu_T^2 e^4}{\hbar^4(2l+1)(4\pi\varepsilon_0)^2}\frac{1}{n^3}\right)U_{nl}(r) \tag{32}$$

The next step is to multiply in the both sides of equation (32) with the conjugate radial wave function and integrating to *r*.

$$\int_0^\infty U_{nl}^*\left[\frac{1}{r^2}\right]U_{nl}(r)dr=\int_0^\infty U_{nl}^*\left[\frac{2m^2 e^4}{\hbar^4(2l+1)(4\pi\varepsilon_0)^2}\frac{1}{n^3}\right]U_{nl}(r)dr$$

$$\left\langle\frac{1}{r^2}\right\rangle_{nl}=\frac{1}{a_1^2 n^3} \tag{33}$$

Then substitute equation (30) and equation (33) to equation (27), so that a first-order tritium energy correction solution will be obtained,

$$\xi R_{nl}^{(1)}=-\frac{\beta^2\xi_n}{4n^4}\left(\frac{8n}{2l+1}-3\right) \tag{34}$$

Where β is a fine structure constant and $\xi R_{nl}^{(1)}$ is the first-order relativistic correction of Energy Tritium in various state.

5. Result and Discussion

5.1. Result of Relativistic Correction in the Various of State

Relativistic correction of Tritium depend of the primary quantum number (n) and the orbital quantum number (l). The total momentum angular of Tritium given by $j = l \pm \frac{1}{2}$, so by using the equation (30) we could to find the correction of energy Tritium atom for the ground state , the first excitted state and the second excited state such as $1s, 2s, 2p_z, 2p_x, 2p_y, 3s, 3p_z, 3p_x, 3p_y, 3d_{zz}, 3d_{xz}, 3d_{yz}, 3d_{xy}, 3d_{x^2-y^2}$ and the result of correction shown in table 1.

Table 1. The result of relativistic correction until state $n \le 3$

State	Initial Energy (Joule)	Correction Energy (Joule)
$1s$	$-21.7947692\times10^{-19}$	1.45075116×10^{-22}
$2s$	-5.4486923×10^{-19}	$2.357470986\times10^{-23}$
$2p_z, 2p_x, 2p_y$	-5.4486923×10^{-19}	$4.231357811\times10^{-24}$
$3s$	$-2.42164102\times10^{-19}$	7.52241373×10^{-24}
$3p_z, 3p_x, 3p_y, 3d_{xz}, 3d_{yz}$	$-2.42164102\times10^{-19}$	1.79105103×10^{-24}
$3d_{zz}, 3d_{xy}, 3d_{x^2-y^2}$	$-2.42164102\times10^{-19}$	6.44778389×10^{-25}

From table 1, we can know that in every state have a different correction of relativistic. It's because of relationalistic correction depend of electron's motion.We can compare the relativistic correction in the ground state ($n=1$), the first excited state ($n=2$) and in the second excited state ($n=3$). If the principal quantum number (n) became larger, the correction relativistic in the first order will get smaller. The correction relativistic energy between $2s$ ($l=0$) and $2p$ ($l=1$) have a different, It prove that orbital quantum number (l) affect to the relativistic energy of atom. If the orbital quantum number get higher in each state, the relativistic correction energy will get smaller. The relativistic correction of Tritium atoms leads to a reduction in the value of classical energy proposed by Bohr models[8]. In the classical atomic theory, the energy of an atom is only influenced by the principal quantum number, whereas in relativistic atomic theory the energy level of the atom is also influenced by the orbital quantum number. In Relativistic condition, It will give more accurate calculations about energy and wave functions of atom[9].

5.2. Result of Stark Effect on Wave Function of Tritium in Relativistic Condition

The solution from the problem of stark effects on Tritium in a relativistic condition is to combine these two corrections between external electric field and relativistic correction. By using equation (17) we could to find the first-order correction of energy from the influence of the external electric field and will give a zero value. while in second order correction we use equation (19) to found the solution and it gives the effect on the energy shift for the ground state [10,11]. The effect of electric field will influenced to the energy atoms or crystals[12]. The solution of the resulting equation is shown as follows,

$$\xi_{1s} = \xi R_{1s}^{(0)} - \frac{(e\mathrm{E})^2}{\left(\xi R_{1s} - \xi R_{2pz}\right)} \left(\frac{256a_1}{243\sqrt{2}}\right)^2 \qquad (35)$$

In equation (35), the result of the difference of energy is negative value, so that the disturbance given by the external electric field will increase the binding energy of electron and nucleus. If the binding energy get higher, the radiation of beta will get smaller and the power of Tritium Battery will get smaller too.

By using the equations (18) and (20), we could find the correction of wave function Tritium until second-order correction. And the result of wave function of Tritium in the ground state with the correction in the first order is,

$$\psi_{1s}^{(1)} = \frac{\frac{256a_1 e\text{E}}{243\sqrt{2}}}{\xi R_{1s} - \xi R_{2pz}} \psi_{2pz}^{(0)} \tag{36}$$

And the second order correction of the wave function Tritium is,

$$\begin{aligned}\psi_{1s}^{(2)} = & \frac{0.7443554a_1^2 e^2 E^2}{(-16.3446682\times 10^{-19})}\left(\frac{25.36243}{(-19.37169511\times 10^{-19})}\psi_{3pz}^{(0)} + \frac{0.54178767}{(-19.37169511\times 10^{-19})}\psi_{3s}^{(0)}\right. \\ & + \frac{2.45185216}{(-19.37169511\times 10^{-19})}\psi_{3dzz}^{(0)} + \frac{1.06168313}{(-19.37169511\times 10^{-19})}\psi_{3dxy}^{(0)} + \frac{1.06168313}{(-19.37169511\times 10^{-19})}\psi_{3dx^2-y^2}^{(0)} \\ & \left. + \frac{2.12336625}{(-19.37169511\times 10^{-19})}\psi_{3dyz}^{(0)} + \frac{2.12336625}{(-19.37169511\times 10^{-19})}\psi_{3dxz}^{(0)}\right) \\ & - \frac{0.0889892553a_1^2 e^2 E^2}{2(3.752625714\times 10^{-36})}\psi_{1s}^{(0)}\end{aligned} \tag{37}$$

The wave function of Tritium atom with correction until second-order perturbation in relativistic condisition is,

$$\psi_{1s} = \psi_{1s}^{(0)} + \psi_{1s}^{(1)} + \psi_{1s}^{(2)} \tag{38}$$

6. Conclusion

The effect of the external electrostatic field on Tritium atom with the relativistic condition will cause a shift in energy and wave function at the ground state. The magnitude of the shift from energy correction and wave function is influenced by the principal quantum number (n), the orbital quantum number (l) and the magnitude of the given external electric field. The stark effect on the ground state of Tritium will increase the binding energy of electron and nucleus. It will decrease the power of Tritium battery because the radiation beta decay got smaller when the binding energy got higher.

References

[1] Schleich W P, Greenberger D M, Kobe D H and Scully M O 2015 *Phys. Scr.* **90** 108009
[2] Nakashima H and Nakatsuji H 2010 *Astrophys. J.* **725** 528
[3] Pourimani R and Aghamohammadi M 2014 *World J. Nucl. Sci. Technology* **4** 170
[4] Lei Y, Yang Y, Liu Y, Li H, Wang G, Hu R, Xiong X and Luo S 2014 *Appl. Radiat. Isot* **90** 165
[5] Braun J, Fermvik L and Stenback A 1973 *J. Phys. E: Sci. Instrum.* **6** 727
[6] Campos P, Martins M G R and Vianna J D M 2017 *Phys. Lett. A* **381** 1129
[7] Rajasekar S and Velusamy R 2015 *Quantum Mechanics I: The Fundamentals* (Tiruchirapali: CRC Press) p.332
[8] Sakho I 2012 *J. At. Mol. Sci.* **3** 23
[9] Grant I P 2010 *J. Phys. B: At. Mol. Opt. Phys.* **43** 074033
[10] Goswami A 1997 *Quantu Mechanics Second Edition* (United States of America: Waveland Press, Inc) p.384
[11] Silitonga A T C and Men L K 2016 *AIP Conf. Proc.* **1712** 040004
[12] Yurov V M, Guchenko S A and Gyngazova M S 2015 IOP Conf. *Ser. Mater. Sci. Eng* **110** 012019

IOP Conf. Series: Journal of Physics: Conf. Series **997** (2018) 012046 doi:10.1088/1742-6596/997/1/012046

A Low Cost Mobile Robot Based on Proportional Integral Derivative (PID) Control System and Odometer for Education

R Haq[1], H Prayitno[1], Dzulkiflih[1], I Sucahyo[1], E Rahmawati[1,a]

[1]Departement of Physics, Fakulty of Mathematics and Natural Sciences, Universitas Negeri Surabaya, Surabaya 60231, Indonesia

[a]E-mail: endahrahmawati@unesa.ac.id

Abstract. In this article, the development of a low cost mobile robot based on PID controller and odometer for education is presented. PID controller and odometer is applied for controlling mobile robot position. Two-dimensional position vector in cartesian coordinate system have been inserted to robot controller as an initial and final position. Mobile robot has been made based on differential drive and *sensor magnetic rotary encoder* which measured robot position from a number of wheel rotation. Odometry methode use data from actuator movements for predicting change of position over time. The mobile robot is examined to get final position with three different heading angle 30°, 45° and 60° by applying various value of KP, KD and KI constant.

1. Introduction

Over the past decades mobile robots with various functions, designs and sizes have been developed for fulfilling the needs of education and research as well. Learning and developing robots involve multidisciplinary field of study including mechanical design, electronics, programming and control system, etc. Motivation in this research is to accommodate the needs of mobile robot in a wide area of scientific learning by experiments.

The mobile robot has been implemented as a learning media in control system topic which is a course of physics instrumentation in Physics Department of Universitas Negeri Surabaya. Based on previous research [1-4], in order to support theoretical concepts on control system that were explained by lecturers in classroom, experiment based works can be done by students must be provided. By conducting laboratory works, it is believed that the students will learn the theoretical concepts easily and grasp them more. Rickel [1] mentioned that students retain 25% of what they hear, 45% of what they hear and see, and 70% if they do what they hear and see or known as "learning by doing" method.

Currently, the evolution of technology achieved a significant reduction in the cost of prototype mobile robot that introduce new and more sophisticated capabilities. In the beginning, mobile robots used to line follower sensor for navigation by following a line and then develop rapidly by implementing various navigation system such as odometer for navigation system. The mobile robot of this research meets a low manucaturing cost about 800000Rp or about 60$.

Odometry is one of the most common relative position estimation methods. In most mobile robots, the offset from known starting position is computed by data of encoders which are mounted on wheels to monitor the wheel rotation steering or heading angle of robot wheels. Implementation of odometry can be independently without external information such as additional sensor or system such as camera. Therefore, odometry is simple, inexpensive and easy to be apllied in real time [5-9].

However, odometry is strongly influenced by internal systemic factors and non-systemic factors which cause the accumulation of error. Systemic error has nothing to do with external environment which almost constant for a periode of time. In order to reduce systemic error, PID control method have been implemented to model the systemic odometry errors and implemented practically on actual robots for validity. Calibration of error for for accurate odometry is necessary. In this article, the mobile robot is calibrated by straight line path and a turning on the spot experiments which only needs smaller experiment site [8].

2. Research Method

2.1. Determining position

One of the commonly used mobile robot drive controllers is the differential controller. This drive system consists of two wheels mounted in a single axis and each wheel can be independently controlled to rotate either clockwise or counter-clockwise.

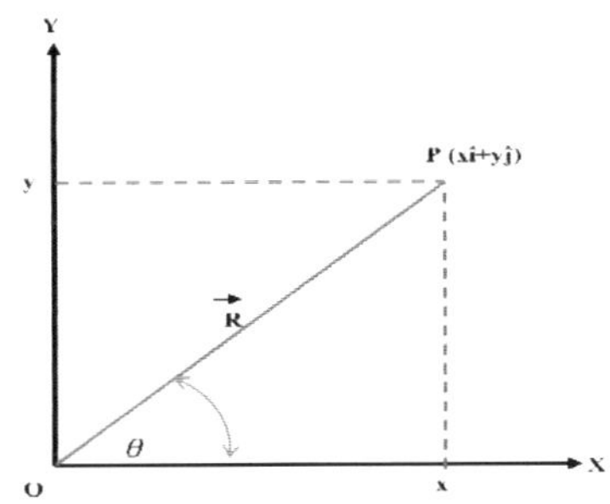

Figure 1. Position vector in cartesian coordinate

The position vector is a quantity having a value and direction indicating the position of a point to the reference point of a plane. In this research using position vector on 2-dimensional plane which has 2 axes x and y. The position of an object P to point O is written (x, y) or $x\hat{\imath} + y\hat{\jmath}$ and R is the length of the vector.

$$\bar{R} = \sqrt{x^2 + y^2} \quad (1)$$

$$\theta = \tan^{-1}\frac{y}{x} \quad (2)$$

To estimate the relative position of the robot, a constant variable required to convert the counter count generated by the rotary encoder sensor is the length of the path the robot takes. This constant is obtained by using the following equation:

$$C = \frac{K_{wheel}}{resolution\ of\ encoder} \quad (3)$$

Where C is the variable constant (mm / counter). K_{wheel} is the circumference of the wheel used (mm) and the encoder resolution is the number of counters in one turn. The distance of the right wheel and the left wheel are respectively d_{right} and d_{left} and the distance between the two wheels is given by l, the distance distance and the orientation angle θ can be formulated by the following equation:

$$d_{right} = count_{right}\ x\ C \quad (4)$$

$$d_{left} = count_{left}\ x\ C \quad (5)$$

$$distance = \frac{(d_{right} + d_{left})}{2} \quad (6)$$

$$\theta_{rad} = \frac{(d_{right} + d_{left})}{l} \quad (7)$$

IOP Conf. Series: Journal of Physics: Conf. Series **997** (2018) 012046 doi:10.1088/1742-6596/997/1/012046

Furthermore, θ in equation (7) is in radians form and should be converted to degree by equation:

$$\theta_{heading} = \theta \; x \frac{180}{\pi} \tag{8}$$

By knowing $distance$ and orientation of angle $\theta_{heading}$ hence can be determined coordinate of position of robot by using concept of position vector in real time.

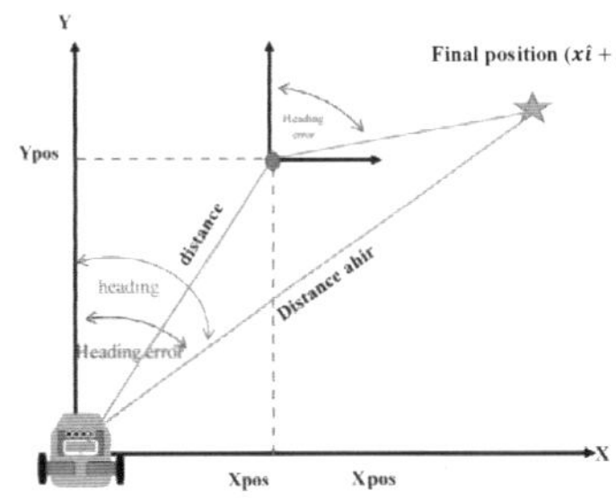

Figure 2. Determining mobile robot position

$$x_{pos} = distance\,.\sin\theta_{heading} \tag{9}$$
$$y_{pos} = distance\,.\cos\theta_{heading} \tag{10}$$

2.2 Hardware and control system

This mobile robot uses two DC motors as its motion device. This DC motor which connected to the rencoder were installed across the rear of the robot. The speed comparison of these two DC motors will determine the direction of robot motion. When the right motor and left motor rotate at the same speed then the robot will run straight. When the right motor speed is greater than the left motor speed then the robot will tend to turn left. When the right motor speed is smaller than the left motor speed then the robot will tend to turn right.

Figure 3. Hardware of mobile robot

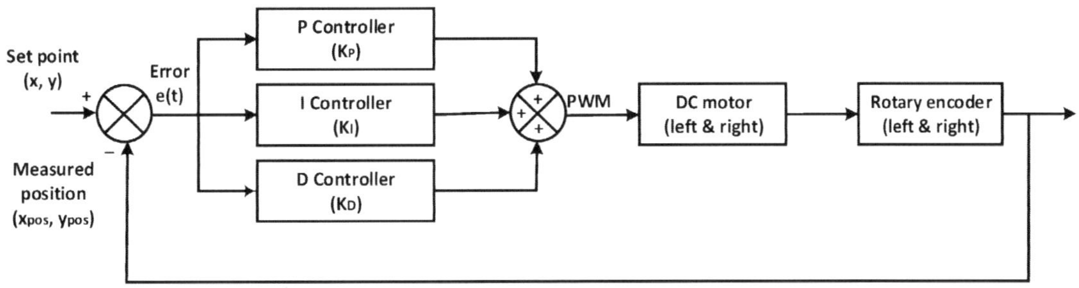

Figure 4. Block diagram of control system

Target in the form of coordinate $x\hat{\imath} + y\hat{\jmath}$ is inserted on the microcontroller and will be stored on EEPROM. From the data (x, y) the microcontroller will calculate the angle (θ) the destination which is then regarded as a set of points, then drive the drive system connected to the rotary encoder sensor. The data from the rotary ecoder will be fed back to the microcontroller and converted into position data $(x_{pos}, y_{pos}, \theta_{pos})$. The data will be compared with the set points so that the error is obtained. Error will be entered into the PID equation on the microcontroller which will then affect the increase or decrease the speed of the drive system.

In this research manipulation variable is the value of the constant PID is K_P, K_I, and K_D. The control variable used is the diameter of the wheel and the distance between the two wheels mobile robot. Respons variable is the value to be obtained in a study. Variable respons in this research is the position of ahir robot $x\hat{\imath} + y\hat{\jmath}$ which has been passed by mobile robot.

2.3. Calibration

The data retrieval process begins with rotary encoder sensor calibration using equation (1). By inserting x and y position, then the mobile robot will move towards the coordinates automatically. When the mobile robot has reached the intended coordinates then the mobile robot will stop and display will inform final coordinates (x', y').

Rotary encoder sensor testing is done by moving the mobile robot with a distance of 10 cm to 100 cm and compare the value of the distance read by the sensor with the actual distance measured by the ruler and then repeated as much as 3 times the measurement for each mobile robot mileage. This measurement aims to test the accuracy of rotary encoder sensor precision. Figure 5 shows that the right and the left wheel data are fit together and there is almost no difference.

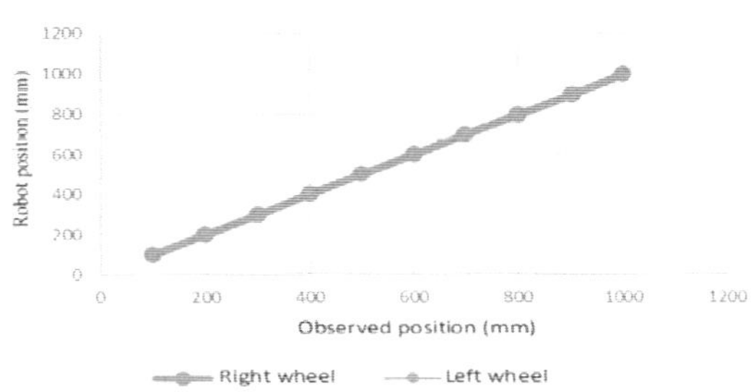

Figure 5. Calibration of right and left wheel

3. Result

Testing of mobile robot control system is done on a flat field that has been given a pattern of diesi kartesius to make it easier to observe the position of mobile robot. At first mobile robot is placed at the center of Cartesian diagram with the front of the robot parallel to the Y axis, then the researcher gives data input in the position of $x\hat{\imath} + y\hat{\jmath}$ which becomes the end of mobile robot. From this data the controller will calculate the angle of heading robot and the distance that will be taken mobile robot, then mobile robot will move towards that position automatically. The test is repeated five times for each manipulation

3.1. Heading error

Final position is the position when the robot has stopped. Error is percentage of mobile robot position difference to set point. Different tuning of PID constant induce different respons of robot to attain the set point.

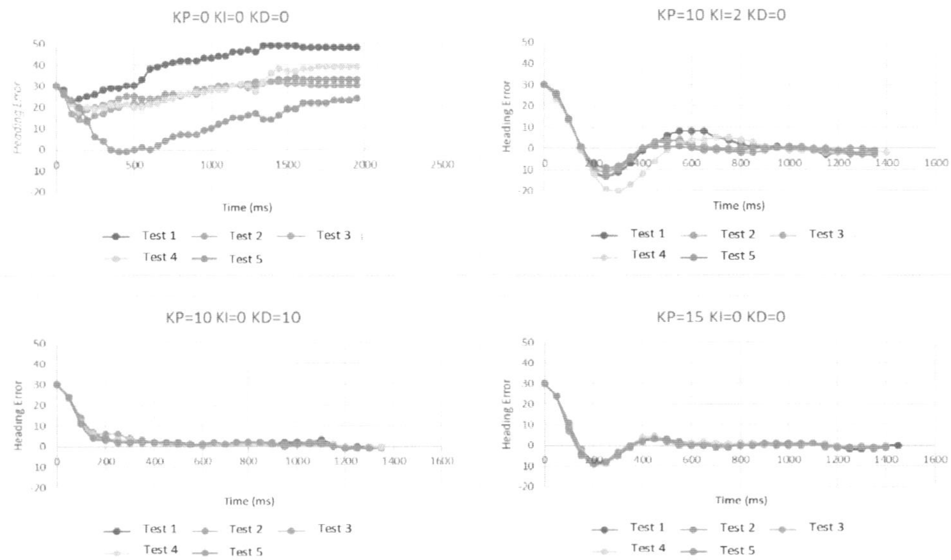

Figure 6. Error of heading for 30°

Heading error is the difference between the robot heading angle in any time with the heading of the robot as set point. Figure 6, 7 and 8 show the change in heading error value every 50 ms when mobile robot starts to move until stop. When $K_P = 0, K_I = 0, K_D = 0$ (without PID controller) the heading error value is randomly changed and can not reach zero. This happens because mobile robots try to reach set point without feedback for correcting its position. While the PID controller is applied to the robot, initially the error heading is at a maximum value of 30° and continues to fall to close to or around zero. This result showed that the control system will try to maintain the value of the heading robot according to the set points and the heading error is zero.

Figure 6 shows four graph of heading error for heading angle of 30° with different PID constant include of; $K_P = 0, K_I = 0, K_D = 0$ (without controller); $K_P = 10, K_I = 2, K_D = 10$; $K_P = 10, K_I = 0, K_D = 10$ and $K_P = 15, K_I = 0, K_D = 0$. PID constant of $K_P = 10, K_I = 0, K_D = 10$ reach the set point 100 ms faster then the other constants and has the smallest error of heading angle that is 0,44%. Data from five times running test for each tuning of PID constant shows a good repeatability.

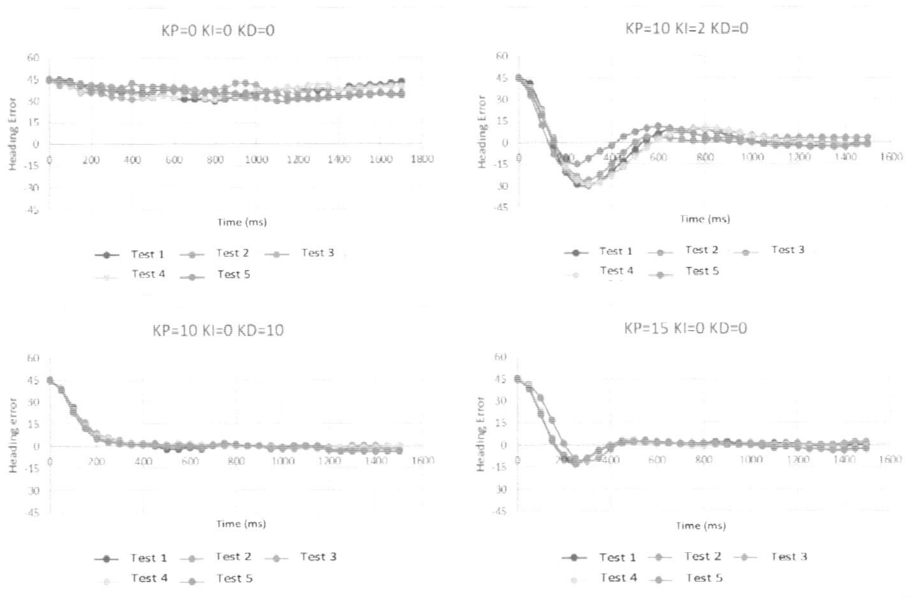

Figure 7. Error of heading for 45°

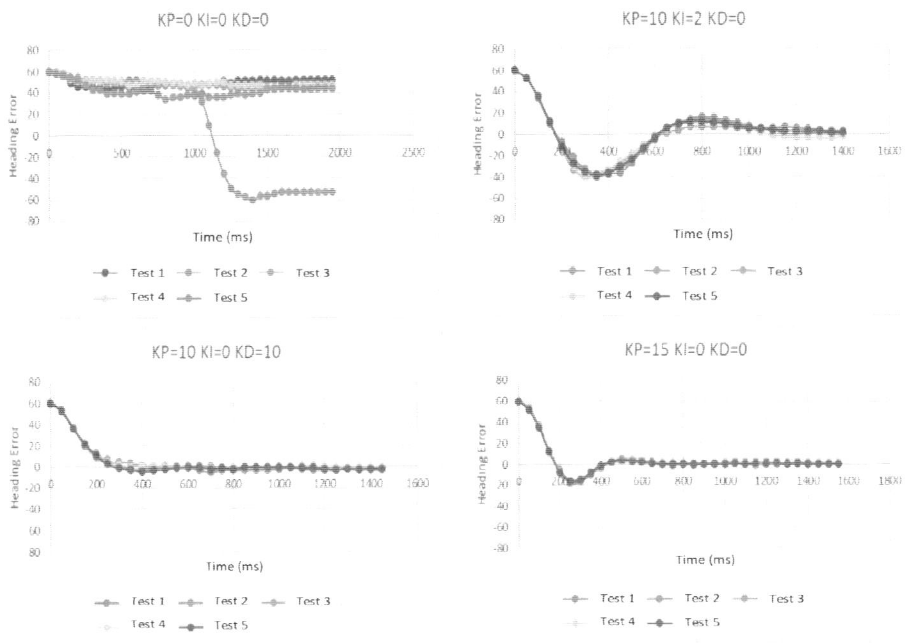

Gambar 8. change of error for heading 60°

Figure 7 shows four graph of heading error for heading angle of 45° with different PID constant tuning include of; $K_P = 0, K_I = 0, K_D = 0$ (without controller); $K_P = 10, K_I = 2, K_D = 0$; $K_P = 10, K_I = 0, K_D = 10$ and $K_P = 15, K_I = 0, K_D = 0$. PID constant of $K_P = 10, K_I = 0, K_D = 10$ and of $K_P = 10, K_I = 0, K_D = 0$ cause the same error of heading of 0,42%. Heading error of PID contanst $K_P = 10, K_I = 0, K_D = 10$ try to reach set point without oscillation, the other experienced it before get

IOP Conf. Series: Journal of Physics: Conf. Series **997** (2018) 012046 doi:10.1088/1742-6596/997/1/012046

the set point.

By manipulating the destination position of 1200î + 700ĵ, we get the heading angle of robot about 60° and the different respons by four different PID constant tuning was depicted in figure 8. Maximum and minimum heading error respectively are 10% (6°) and 0,06% (less then 1°)

3.2. PID respons

In general, the respons of a control system shows a respon characteristic such as rising time, settling time and error. Rising time is the time it takes the system to pass through 90% set point value, settling time is the time it takes the system to reach the smallest error value.

Figure shows the comparison of robots heading respon to set point at any time for each PID constant. Initially the heading value is 0° and will rise towards the set of points that have been determined. This illustrates the mobile robot that originally faces parallel to the y −axis and then gradually toward the set of points. It appears that the system works fairly quickly with rising time less than 200 ms for all PID constants values. Rising time is influenced by the magnitude of KP, the greater the value of KP then rising time will be faster.

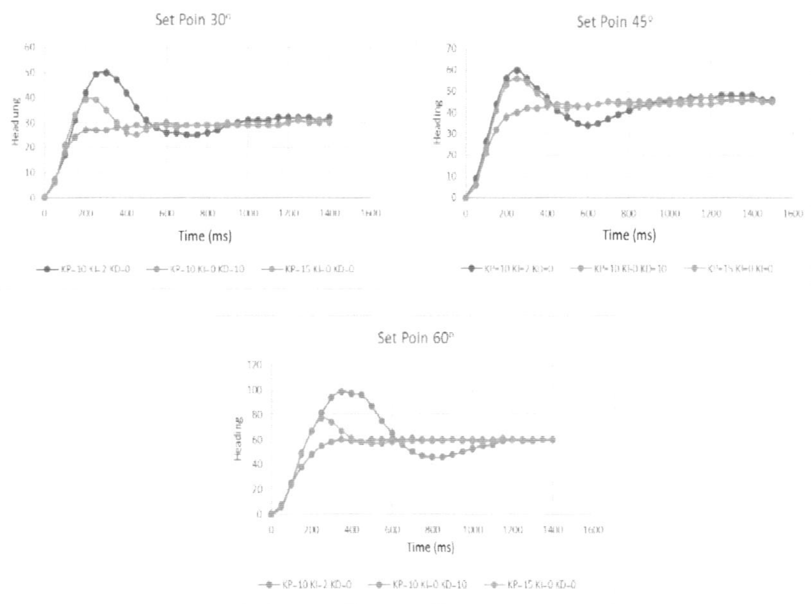

Figure 9. PID respon for heading 30°, 45° and 60°

The rising time indicated by curve with $K_P = 10$ and $K_P = 15$ is almost similar about 600 ms. After reaching the heading corresponding to the set of points, the system will experience overshoot. The magnitude of the overshoot value is influenced by the value of K_P, but at the same K_P value the value of K_I and K_D also affect. Seen in the figure 9 the largest overshoot is achieved when $K_P = 10$, $K_I = 2$, $K_D = 0$.

The time required to achieve the smallest error varies for each PID constant. Settling time is influenced by K_I and K_D. The correct tuning of K_I and K_D will result in a stable system respons. At set point 30° with $K_I = 2$, $K_D = 0$ obtained settling time 900 ms, when $K_I = 0$, $K_D = 10$ obtained a fast settling time of 400 ms, whereas when $K_I = 0$, $K_D = 0$ obtained 600 ms settling time.

3.3. Implementation in Education

The last a year the mobile robot were used for the laboratory exercises of an undergraduate robotics learning at Physisc Department of Universitas Negeri Surabaya. Typical undergraduate exercise consist of electrical and programming control system for navigation. The robot also help the students to have a research for a competion. The robot proved user friendly and helped the undergraduate students to quickly learn how use them and adapt them to another projects. The hardware and software design of the robot is a supporting tool for students to comprehend and practice about robotics.

4. Conclusion

In this article, the development of a prototype mobile robot for education is presented. The low cost of mobile robot has been fulfilled the needs of navigation system of robot in laboratory scale. This robot already proved to be reliable for experiments and research for period a year both for laboratory excercising and research for competition preparation. The use of odometry system and PID controller have been implemented for mobile navigation. The implementation provides the position and orientation of mobile. Through PID controller, the error of heading and position can be corrected in real time. Maksimum and minimum error for several PID tuning respectively are less then 10% (3°) and 0,42%(0,5°). So, the approach proposed in this article is viable to reduce the error of navigation or localization.

References

[1] Rickel J W 1989 *IEEE Trans. Syst. Man, Cybern* **19** 40
[2] Rahmawati E, Ekawita R, Budiman M, Abdullah M and Khairurrijal 2012 *Telkomnika* **10** 733
[3] Rahmawati E, Sucahyo I and Khairurrijal 2014 *ICAET Proc.* **1** 155
[4] Khairurrijal, Abdullah M and Budiman M 2011 *Comp. Appl. Eng. Educ.* **19** 10
[5] Ardilla F, Marta B S and Besari A R A 2011 *IES Proc.* **13** 1
[6] Dudek G and Jenkin M 2010 *Computational Principles of Mobile Robotics second edition* (Cambridge University Perss: New York)
[7] Ang K H, Chong G C Y and Li Y 2005 *IEEE Trans. Control Syst. Technol.* ***13*** 14 559
[8] Chen W and Zhang T 2017 *Int. J. Adv. Robot Syst.* **13**
[9] Martinelli A 2002 *IEEE Trans. on Robot. Autom.* **18** 1

IOP Conf. Series: Journal of Physics: Conf. Series **997** (2018) 012047 doi:10.1088/1742-6596/997/1/012047

Socialization of Solar Energy Utilization in Ponpes Al Hidayah, Arjasa, Kangean Island, Sumenep

Y Cahyono[1], Y Setyaningrum[1], A Sarasechan[1], R G Nafsi[1], Setiyono[1], M D Salamah[1], N A Triyuliana[1], L Silvia[1], B A Subagyo[1], M Zainuri[1], Triwikantoro[1], M A Baqiya[1], Endarko[1], M Z Asrori[1], S Pratapa[1], Suasmoro[1] and Darminto[1]

[1]Department of Physics, Faculty of Sciences, Institut Teknologi Sepuluh Nopember (ITS), Sukolilo ITS Campus, Surabaya 60111, Indonesia

E-mail : yoyok@physics.its.ac.id and yoyok201@gmail.com

Abstract. Electricity problem of most small islands in Indonesia has become a serious problem and need to be immediately resolved. In this present paper, Kangean Islands, Sumenep district of Madura, Indonesia, is one of the most suitable islands for an example. In this island, the existing electricity supply is mainly generated by diesel generators. Even though there are also electricity supplies from the government and private companies, it is very limited capacities just a few families. It is clear that the daily electricity requirements in the Kangean Islands are not adequately met. There is no self-supporting from the local residents to meet their daily energy needs. The community service activity helps to improve the understanding and the self-supporting of the Kangean Island community, especially for the young generation, in the field of electrical energy by utilizing renewable energy sources, especially solar cell system technology. Thus, it is expected that natural resources in Kangean Island can be utilized properly and able to increase the productivity. Finally, in this paper, the light intensity and surface temperature effects on the performance of a monocrystal solar cell are discussed.

1. Introduction

The problem of electricity availability on the islands of Kangean has become a protracted problem. Electricity limitations will certainly make the activities of the community to be limited [1]. Power supplies in the island are still generated by diesel generators, and some solar power plants are also still concentrated only for a few families. It is still very limited, so it can be said that it is not sufficient to meet the electricity daily needs for every family in the Kangean Islands. Solar power plants in the hamlet of Batu Putih and Pabitta is only limited to several families. Moreover, it is limited to 3 lamps for every family.

Geographical conditions in Indonesia consist of millions of islands, and the Kangean Island is a good example as a place for a renewable energy. It is known that Indonesia is very abundant for solar energy [2]. This energy can be utilized by converting the energy into electricity using solar cell panels [3]. Therefore, it should be utilized to overcome the lack of electricity in this kind of region [4]. The main purpose of this community service activity is to improve the understanding of the community in the field of electrical energy by utilizing renewable energy sources, especially solar cell science and technology. Finally, it may increase the productivity and independence of the community.

The ability of solar cell to produce electric energy is related to the photovoltaic effect and its ability to absorb electromagnetic waves. The properties of a solar cell for converting light energy into electric energy depend on environmental factors; those are weather condition [5], the intensity of light [6], panel orientation [7] and position of solar panel to the Sun [8,9], wind direction [10], and the temperature of the cell [11]. Photovoltaic parameters that determine the performance of a solar cell are open-circuit voltage, short circuit current, maximum output power, fill factor, and efficiency [12]. Generally the efficiency of the solar cells is used as parameters to compare the performance of a se solar against other solar cells with similar treatment. In this paper, we also study the effect of sunlight intensity on the solar cell performance, in the case of maximum output power, fill factor, and its efficiency.

2. Methods of Activities

The socialization of implementing solar cell technology was held at one school in Kangean Island, namely Al-Hidayah Senior High School, Ponpes Al-Hidayah, Arjasa, Kangean Island, Sumenep District, Indonesia. The audience involved in this activity is especially for the young generation in Al-Hidayah ponpes, which are students of Upper Secondary School, where they are originally from the Kangean Island and the surrounding islands.

This socialization combined with training to the audience was conducted in the form of lectures, discussion, demonstration, and exercises to the young generations of Kangean Island to utilize solar cell panel as renewable energy. After a class lecture on the short-term theory of renewable energy (and photovoltaic effects), operation, and maintenance, the activity was continued by installing a solar power system as seen in Figure 1. Figure 2 shows the installation setup for the experimental purpose in order to analyze the efficiencies of Silicon monocrystalline solar cell panel with area of 0.27 m^2. The measurements were performed in the sunny and cloudy condition. From the experiment, the fill factor (FF) and the efficiency were obtained using Equation 1 and Equation 2. From these experimental results, it was considered as the achievement of socialization activities.

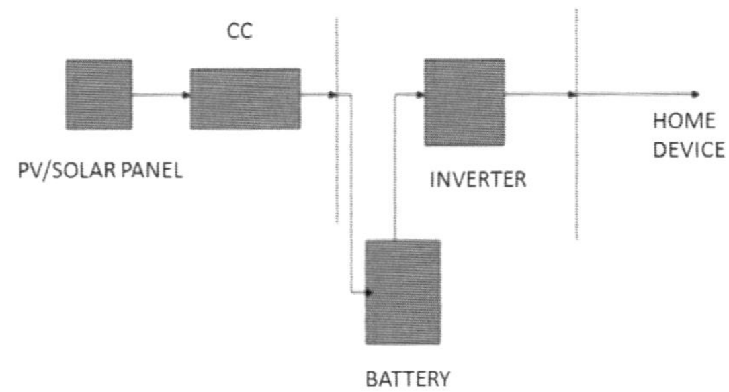

Figure 1. Installation setup for Solar Cell Panel.

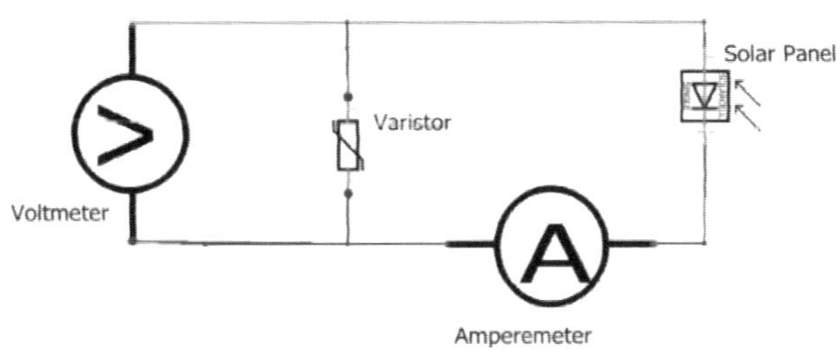

Figure 2. Experimental Setup for Photovoltaic measurements.

$$FF = \frac{I_m \times V_m}{I_{SC} \times V_{OC}} \tag{1}$$

$$\eta = \frac{FF \times I_{SC} \times V_{OC}}{P_{input}} = \frac{V_m \times I_m}{Intensity \times A} \tag{2}$$

3. Results and Discussion

The main discussion of this socialization is the utilization of renewable energy, with a primary focus on the utilization of solar cell power plants and studies on its performance. It needs to be explained to the participants that the use of new and renewable energy sources needs to be developed. It is also explained how solar cell power systems are designed. A solar cell power system consists of a solar module, a regulator circuit, an electrical energy store (batteries), an inverter, an installation cable, and other mechanical equipment. The type of solar module used is monocrystal silicon. The main consideration for using this type of solar module is because it is the most efficient type of solar module available, with efficiency reaching 16-25% [13].

The important thing to be noted in the installation is that it should be located in an open area that is not blocked by trees or tall buildings. The battery should be placed in a humid place and out of reach of children. While the other solar panel accessories located outside the room and it should be resistant to sunlight. The regulator's position should be easily accessible for easy checking and maintenance. The next session is observing and performing experiments on the phenomenon of photovoltaic effects.

Figure 3(a) and 3(b) show I-V characteristics of the solar cell during sunny and cloudy condition, respectively. In the measurements, the values of Voc (voltage in the circuit without resistance) and Isc (electric current in the circuit without resistance) were 21.56 V and 2.68 A, respectively. The I-V measurements were conducted with the variation of resistance in the range of 1 to 1 kΩ. In the case of minimum resistance, the electric current will have maximum value, whereas the voltage will have low value.

It can be seen from Figure 3 that the maximum values of both electric current and voltage can be obtained. They are important to calculate the value of Fill Factor (FF). The FF value corresponds to how much intensity of the sunlight can be absorbed by the solar cell. In the case of sunny condition, the FF value is 0.62 or 62%, meaning that the ability of the solar cell to absorb sunlight during the measurement is about 62% with the intensity average of about 1.44 W/m^2. In fact, the FF value is proportional to the area of the solar cell. The wider area, the higher FF value would be obtained. In Figure 3(a), it can be obtained that the maximum values of electrical current, voltage, and output power are 2.24 A, 16.08 V, and 36.07 W, respectively. By calculation using Equation 2, the efficiency of the solar cell panel in sunny condition is approximately 13.5%.

It can be seen from Figure 3(b), the values of the maximum output of electrical current, voltage, and power are 0.67 A, 16.94 V, and 11.42 W. The calculation results of FF value and its efficiency are 65% and 18.9%, respectively. The increase of FF value may result in the increase of efficiency of the solar cell. From both Figure 3(a) and 3(b), it can be analyzed that the weather (sunny or cloudy) may cause the change of both solar cell efficiency and light absorption characteristics. The light illumination of solar cell in the sunny condition provides higher electrical current with almost the same voltage compared with in the cloudy condition.

In order to study more detail on the relationship between the light intensity and the solar cell performance, for instance output power and efficiency, we have performed the intensity measurements collected in both sunny and cloudy condition. Figure 4(a) and 4(b) show the sunlight intensity dependence on the maximum output power of the solar cell in sunny and cloudy condition, respectively. Both Figures show a nonlinear behaviour because of the effect of surrounding temperature on the surface of the panel. As the temperature close to room temperature, the solar cell produces decreasing output voltage and increasing electrical current. Otherwise, as the temperature on the solar cell surface is high enough, the output voltage decreases, however, the electrical current is relatively constant. Hence, it produces a lower output power.

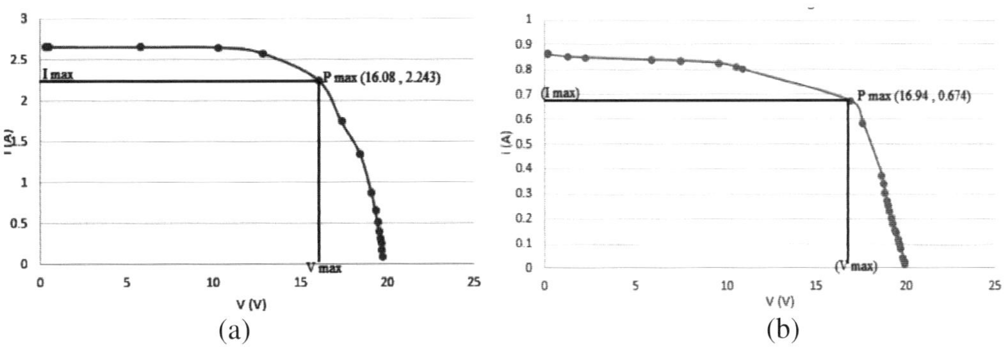

Figure 3. I-V characteristics of the solar cell panel in (a) sunny and (b) cloudy condition.

In a short summary, the output power of solar cell may change significantly due to the change of temperature on the surface of the panel. In Figure 4(b), it can be seen that the output power increases as the sunlight intensity increases. This can be understood that the surface temperature remains constant at around room temperature, while the output voltage decreases and the electrical current increases with increasing the intensity.

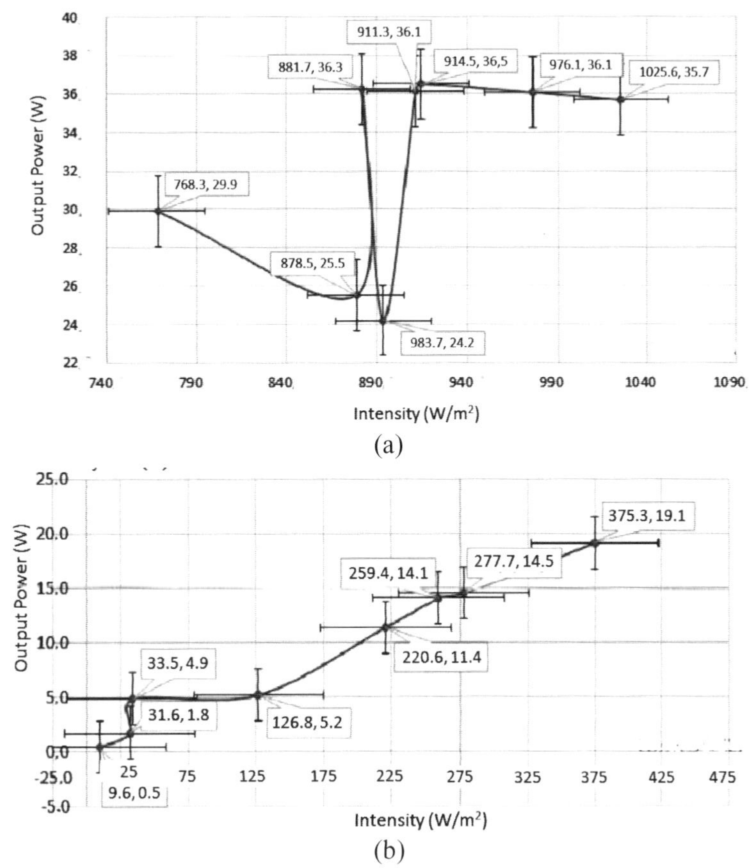

Figure 4. Sunlight intensity dependence of maximum output power of the solar panel in (a) sunny and (b) cloudy condition.

The sunlight intensity dependence on efficiencies of the solar cell in sunny and cloudy condition are illustrated in Figure 5(a) and 5(b), respectively. It can be seen from both Figures that it has irregular pattern meaning that there is a contribution from the surface temperature of the solar cell panel affecting the output power and hence change its efficiency significantly. For example, in Figure 5(a), the solar cell has efficiency of about 9.9% at the sunlight intensity of 893.7 W/m^2, even in this case, the intensity is higher than before. This is because the measurement at a relatively high temperature results in the decrease of output voltage and relatively constant electric current, and hence, the efficiency decreases. In the cloudy condition, as seen in Figure 5(b), the surface temperature is relatively constant. Then, this results in the decrease of output voltage and the increase of electric current with increasing sunlight intensity. At a high intensity and low output power, the efficiency of the solar cell becomes low. Otherwise, the efficiency becomes high when the intensity decreases and the output power is high. Therefore, it can be summarized that the change in sunlight intensity and also the temperature on the surface influence the efficiency of the solar cell.

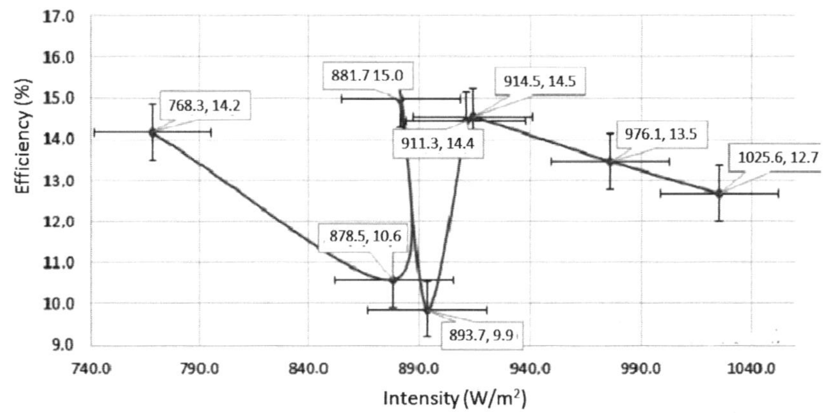

(a)

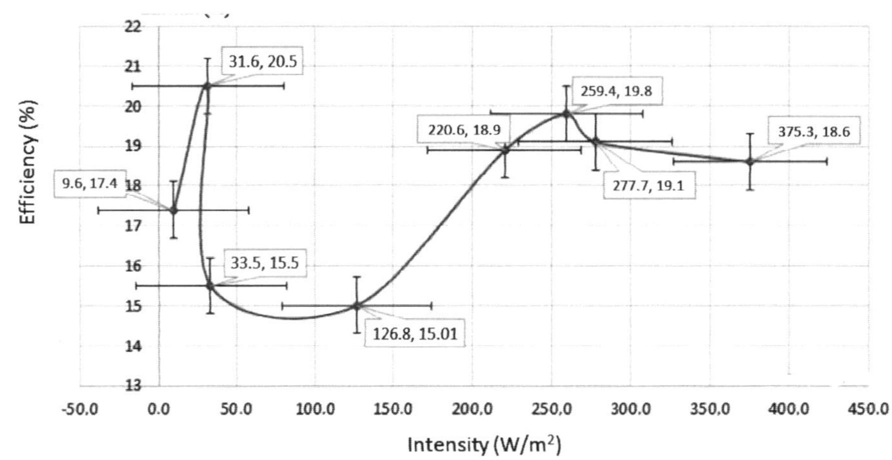

(b)

Figure 5. Sunlight intensity dependence of efficiency of the solar panel in (a) sunny and (b) cloudy condition.

4. Conclusions

From the activities of the community service that has been done in Kangean Island, Indonesia, it can be concluded two important points. First, the activity has been successfully implemented. The participants have the ability to understand about the utilization of renewable energy, especially the utilization of solar energy using solar cells. Second, from the experiment using installed solar cell, it can be summarized that the maximum values of the output power and efficiency of the solar cell are strongly influenced by the surrounding weather and the surface temperature of the panel. For the Silicon monocrystal solar cells, the minimum efficiency during cloudy weather is 10.5%, while the maximum efficiency during sunny weather is 20.5%.

References

[1] Khan J and Arsalan M H 2016 *Renew. Sust. Energ. Rev.* **55** 414
[2] Fathoni A M, Utama N A and Kristianto N A 2014 *Procedia Environ. Sci.* **20** 89
[3] Lee N A, Gilligan G E and Rochford J 2018 *Green Chemistry* (Amsterdam: Elsevier) 881
[4] Kabir E, Kumar P, Kumar S, Adelodun A A and Kim K H 2018 *Renew. Sust. Energ. Rev.* **82** 894
[5] Vasisht M S, Srinivasan J and Ramasesha S K 2016 *Sol. Energ.* **131** 39
[6] Agbo S N, Merdzhanova T, Rau U and Astakhov O 2017 *Sol. Energ. Mat. Sol. Cells.* **159** 427
[7] Miloudi L, Acheli D and Chaib A 2013 *Energy Procedia* **42** 103
[8] Conrado L S 2018 *Renew. Sust. Energ. Rev.* 82 2128
[9] Gulin M, Vašak M and Perić N 2013 *Appl. Energ.* **108** 429
[10] Vasel A and Iakovidis F 2017 *Energ. Convers. Manage.* **153** 455
[11] Merabet S and Birouk B 2017 *Int. J. Hydrog. Energ.* **42** 29026
[12] Fouad M M, Shihata L A and Morgan E I 2017 *Renew. Sust. Energ. Rev.* **80** 1499
[13] Bruton T M 1995 *Renew. Energ.* **6** 299

IOP Conf. Series: Journal of Physics: Conf. Series **997** (2018) 012048 doi:10.1088/1742-6596/997/1/012048

Microstructure of Sinter Deposit Formed at Hot Springs in West Sumatera

A Putra[1], D Y Inanda[1], F Buspa[1] and A F Salim[1]

[1]Laboratory of Earth Physics, Departement of Physics, Faculty of Mathematics and Natural Sciences, Andalas University, Kampus Unand Limau Manih, 25163, Padang

email: ardhee@fmipa.unand.ac.id

Abstract. Sinter deposit emerged and spread at several hot springs in West Sumatera is divided into three types, they are full silica, half silica-carbonate and full carbonate. This work intends to investigate the characteristic of each type by its crystalline structure and morphology and its correlation to surface temperature. The research is focused on Sapan Maluluang hot spring (full silica), Garara hot spring (half silica-carbonate) and Bawah Kubang hot spring (full carbonate). Crystalline structure is analyzed by X-Ray Diffraction (XRD) methods, it showed that deposit from Sapan Maluluang has opal-A structure, Garara has opal-CT structure and Bawah Kubang has crystalline structure. The Scanning Electron Microscopy (SEM) methods is applied to describe its morphology surface, in which spherical, almost rounded and irregular textured was formed at each deposit, respectively. Surface temperature of hot spring also has given impact on deposit texture.

1. Introduction

One of the mineral deposits present around hot springs is silica and carbonate sinter, material that is dominated by silica minerals (SiO_2) and carbonate ($CaCO_3$). The formation of silica sinter in hot springs indicates that geothermal reservoirs have high temperatures. Therefore, the presence of silica sinter is one of the important manifestations for the existence of geothermal reservoir. Silica is classified into three categories based on their atomic structure, ie opal-A, opal-CT, and opal-C [1]. At early formation, silica has noncrystalline (Opal-A) or amorphous properties. The change of noncrystalline mineral structure into Opal-CT and C-opal occurs after 10,000 years, this process is called diagenesis. Opal-CT has α-cristobalite and α-tridymite irregularities, and C opal has a neatly arranged α-cristobalite [2].

The sinter is hypothesized as silica based on the color taken from the sediment around hot springs in Solok Selatan and Solok regencies, in Sapan Malulung hot springs, as well as hot springs Garara, and Bawah Kubang. Based on literature study, the district has hot springs with high silica content. The characteristics of silica sinter can be expressed by describing its structure and morphology. Each structure gives further explanation of the geothermal system. Therefore, X-Ray Diffraction (XRD) and Scanning Electron Microscopy (SEM) are used for this research.

2. Method

The material used in this research is sinter found in three hot springs, namely in Sapan Malulung, Garara, and Bawah Kubang. Samples were taken using a hammer and crushed using mortar and pestle until smooth. After forming a fine sample, it was sieved with a 100 mesh sieve. Further characterization is done using XRD for identifying structure of samples and SEM for describing its morphology. For additional data, temperature measurements of the hot springs surface used a thermometer.

3. Results

3.1. Sinter Structure

XRD results from sinter sediment samples from Sapan Malulung, Garara and Bawah Kubang hot springs are shown in Figures 1 (a), (b) and (c), respectively. Figure 1 (a) shows that the sinter extracted from the Sapan Malulung hot springs is dominated by non-crystalline silica opal-A or commonly called amorphous, indicating a large peak (hump) produced between 2Θ 18° to 30° with d-spacing between 2.97 Å - 4.92 Å. A silica having A-opal phase has a relatively young sinter life. Opal-A silica in Steamboat Spring, Nevada is described for 6500 years old, and opal-A silica at Mound, Utah is estimated for 1600 years old [3,4]. The structural change of noncrystalline (opal-A) to opal-CT and C-opal occurs after a period of 10,000 years [5]. Based on these data it can be estimated that the silica sinter found in Sapan Malulung hot springs is <10,000 years old.

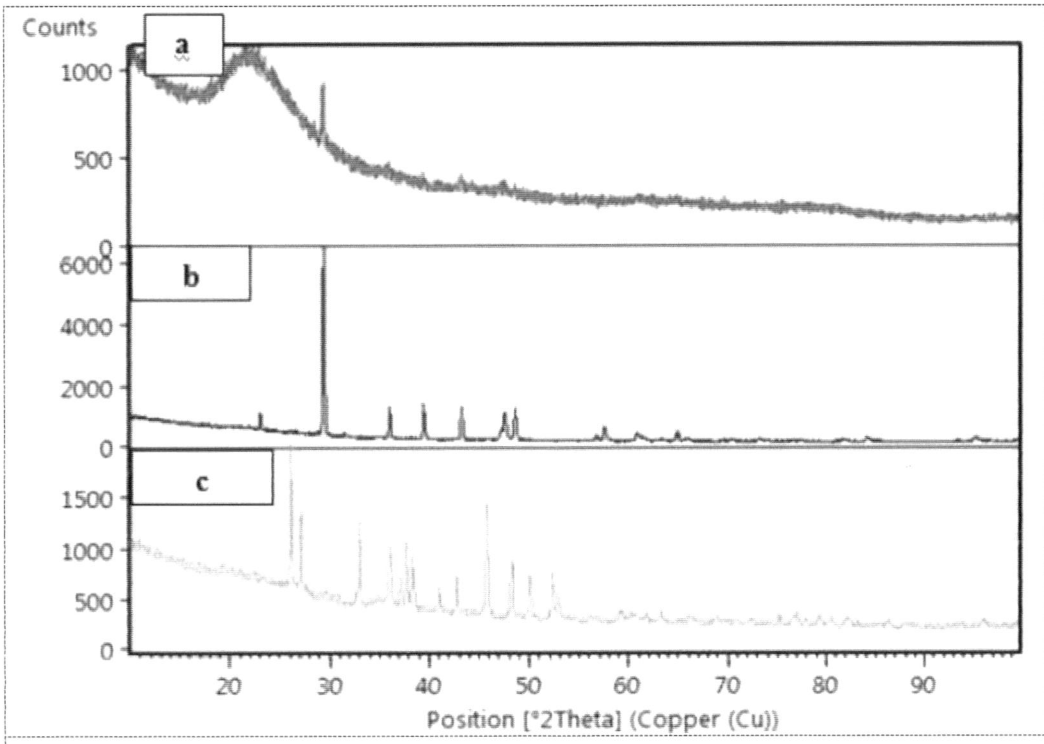

Figure 1. The XRD pattern for each sample

Deposit from Garara hot springs shows different XRD patterns with Sapan Malulung hot springs or can be seen in Figure 1 (b). Garara hot springs have a sharper diffraction peak and higher intensity compared to Sapan Malulung hot springs. The resulting crystalline pattern is opal-CT, where there is a regular layered crystal structure that is tridymite and cristobalite. Based on the data ICCD 00-003-0271 tridymite and cristobalite are the phases of silica minerals (SiO_2). This pattern is a typical pattern of opal-CT.

The hot springs that produce the opal-CT structural silica, which has been formed in Atiamuri and Ohakuri have a silica age ranging from 50 to 40 000 years [6]. The opal-CT produced by silica sinter in Years, TVZ has a 14 500-year-old, and based on petrology data, the Geology Department of

University of Auckland, the CT opal produced from silica sinter at Umukuri has an age of 40 000 - 120 000 years [7]. Based on this data, it is estimated that silica sinter deposits in the Garara hot springs have an older age compared to Sapan Malulung hot springs.

Figure 1 (c) is the XRD result of sinter present in hot springs Bawah Kubang. Based on the figure, the peak of detected diffraction is aragonite, this is known based on ICCD data with the code 01-075-9984. Aragonite is a phase of mineral calcium carbonate ($CaCO_3$). Based on this data, it can be concluded that the sample of hot spring Bawah Kubang classified sinter carbonate. The common phrase used in geothermal systems is the emergence of silica sinter deposits on the surface, indicating that the hot springs have high reservoir temperatures.

3.2. Sinter Morphology

SEM results can be seen in Figures 2, 3, and 4. The magnification used is 5000x and 20.000x. Figure 2 is a SEM sample of sinter samples found in Sapan Malulung hot springs with the magnification of 5000x (left) and 20000x (right). Based on the image, it has a rounded texture neatly arranged together like a long chain, and has a particle diameter between 0.641 μm - 0.670 μm as indicated by the arrows. The diameter of this particle is much smaller than that of Herdianita (2000), that sinter having a crystalline phase has a round texture with a particle size of 4-8 μm and the smallest size of opal-A is 1-2 μm .

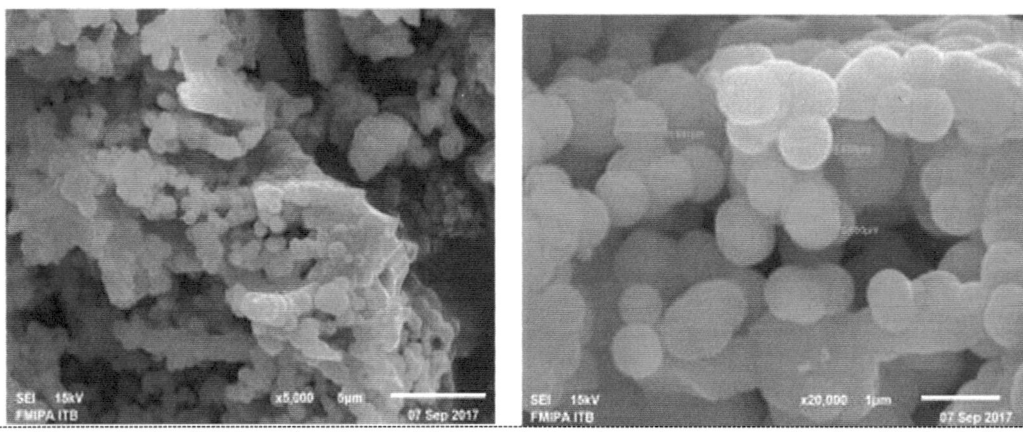

Figure 2. The SEM result of Sapan Malulung hot springs sinter, magnification 5000x (left), magnification 20000x (right)

Figure 3 left and right shows the sinter SEM results found in the Garara hot spring. This sinter has a slightly rounded texture, the slab, and conical shape. And has a particle size between 0.626 μm to 0.783 μm. This size is larger than the sinter found in Sapan Malulung hot springs. Based on research done by Herdianiata (2000) that older sinters usually have an opal-CT phase with a slightly rounded texture (resembling opal-A), and a 20-30 μm particle size plate. Based on field data hot springs has a surface temperature of 58 °C. The conical, cone-like springs have a moderate surface temperature (35-59 °C) [8].

IOP Conf. Series: Journal of Physics: Conf. Series **997** (2018) 012048 doi:10.1088/1742-6596/997/1/012048

Figure 3. The SEM result of Garara hot spring sinter, magnification 5000x (left), magnification 20000x (right)

Based on Figure 4 left and right, sinter at Bawah Kubang hot spring has a texture like bubbles with a smooth and dense aggregate and has a particle diameter of 0.385 μm - 0.424 μm. The particle size is smaller than the sinter found at Sapan Malulung and Garara hot spring.

Figure 4. SEM result of Bawah Kubang hot spring sinter, magnification 5000x (left), magnification 20000x (right)

4. Conclusion

Deposit at Sapan Malulung and Garara hot springs contain silica sinter, but at Bawah Kubang hot spring contains carbonate sinter. Sapan Malulung sinter deposits have opal-A, rounded and connected structures that form like long chains. The sinter deposit at Garara has an opal-CT phase and is a slightly rounded texture and is in the shape of plates and cones. The identification of silica sinter in Sapan Malulung and Garara can be a reference for further research on geothermal research.

References

[1] Jones B and Segnit E R 1971 *J. Geol. Soc. Australia* **18** 57
[2] Valles M G, Turiel J L, Torrente D G, Alonso J S and Manent S M 2008 *Am. Mineral.* **93** 1373
[3] Lynne B Y, Campbell K A, Moore J and Browne P R L 2008 *Sediment. Geol.* **210** 111
[4] Lynne B Y, Campbell K A, Moore J and Browne P R L 2005 *Sediment. Geol.* **179** 249-278
[5] Herdianita N R, Brown P R L, Rodges K A and Campbell K A 2000 *Mineral. Deposita* **35** 1

[6] Smith D K, Tomiano G P and Wright A C 1998 On the paracrystallinity. Krauikopf, K.B. (1956) Dissolution and precipitation of silica at low temperatures of opal *Proc. XVII Conf. Appl. Crystallogr.*, *Geochimica et Cosmochimca Acta* **10** 1
[7] Campbell K A, Sannazzaro K, Rodgers K A, Herdianita N R and Browne P R L 2004 *J. Sediment. Res.* **71** 727
[8] Lynne B Y 2013 Siliceous Sinter An Early Exploration Tool and Direct Link toa Geothermal Reservoir *Proc.Standford Univ.* 11-13

Effectiveness of guided inquiry learning model to improve students' critical thinking skills at senior high school

E K Nisa[1], T Koestiari[2], M Habibbulloh[3] and Budi Jatmiko[4,a]

[1]Physics Teacher of Public Senior High School-1 Tarik, Sidoarjo, Indonesia
[2]Department of Chemistry, Universitas Negeri Surabaya, Surabaya, Indonesia
[3]Physics Teacher of Vocational High School IKIP Surabaya, Indonesia
[4]Department of Physics, Universitas Negeri Surabaya, Surabaya, Indonesia

[a]E-mail: budijatmiko@unesa.ac.id

Abstract. This research aimed to describe the effectiveness of guided inquiry learning model to improve students' critical thinking skills. Subjects in the research were 90 students at three groups of senior high school grade X on Tarik (Indonesia), which follows a physics lesson on static fluid material in academic year 2016/2017. The research was used one group pre-test and post-test design. Before and after being given physics learning with guided discovery learning model, students in the three groups were given the same test (pre-test and post-test). The results of this research showed: 1) there is an increased score of students' critical thinking skills in each group on $\alpha = 5\%$; 2) average N-gain of students' critical thinking skills of each group is a high category; and 3) average N-gain of the three groups did not differ. The conclusion of this research is that learning model of guided inquiry effective to improve students' critical thinking skills.

1. Introduction

The era of globalization and the development of science and technology require a Human Resources (HR) quality. Education is essentially a conscious effort made to improve the skills of human resources in the 21st century challenges. Skills in the 21st century challenges by Trilling and Hood [1] includes: life skills and career skills, skills in technology, media and information as well as critical thinking skills and to innovate. These skills can be acquired through a learning process that is conducted both formal and non-formal. The focus of the 21st century learning process is not only aimed at the provision of material and concepts, but also on the development of students' skills in science learning [2].

According to Krulik and Rudnick [3] critical thinking is a way of thinking that test, connect, and evaluate all aspects of a problem situation, including the ability to gather information, memorize, analyze the situation, to read and understand and identify things required. Ennis [4] states there are six basic elements that need to be considered in critical thinking, namely: focus, rationale, conclusions, situations, clarity and overall examination. Facione and Facione [5] defines critical thinking as a make reflective decisions and resolve what is believed and done wisely. Halpern [6] argues that critical thinking is "cognitive skills and strategies that may enhance the desired result of thinking, i.e., purposeful, reasoned, and the actual destination - Miscellaneous think includes resolve the problem, formulate a conclusion, taking into account the possibility, and make a decision".

Critical thinking skills have to be applied in physics learning both in learning procedural knowledge so that the facts contained in each stage is easy to understand, as well as conceptual knowledge is more in need of the quality of higher thinking in order to connect these facts so that the concept was intended to be well [7]. Science education in the modern era focusing on the ability of students to be active in the learning process through the exploration process. Specifically, the current education try to help students learn to organize and construct opinions, formulate problems, develop hypotheses, and look for evidence himself. Such processes can help students formulate their thinking abilities and manage their ability to solve problems that facilitate the learning of science concepts [8]

Preliminary study results show that the critical thinking skills of physics on the topic of Hooke's Law in Public Senior High School (SMAN)-1 Tarik (Indonesia) is still relatively low. 36.36% of students have not been able to formulate the problem correctly, 72.73% of students have not been able to provide arguments, 84.85% of students have not been able to do the induction, 66.67% of students have not been able to do evaluation, and 90.91% of the students have not been able to decide an action correctly.

Based on the results of the preliminary study, the problem arises how to improve students' critical thinking skills physics? One of the alternative solution according to Eggen [9] is to carry out learning inquiry model. Eggen believes that inquiry learning model can provide an opportunity and assist students in gaining an understanding of the scientific method to develop critical thinking skills, self-regulation, and an understanding of specific topics. This is supported by research results Jatmiko, Widodo, Martini, Budiyanto, Wicaksono & Pandiangan [10] which showed that the learning experience, experimentation, discussion sharing, and solving the problem is effective in increasing mastery of concepts, problem solving skills and decision-making. In addition, it is also supported by the results of research L.M. Sartorelli and R. Swartz [11] which concluded that by improving the analysis and develop the skills of observation / observed, the students' critical thinking skills can be improved. Likewise, the Nurs' research [12] concluded that guided inquiry learning can train scientific thinking skills of students of grade X Natural Sciences (IPA)-4 SMAN-2 Malang, Indonesia on the topic of glasses and magnifying glass. Bilgin [13] states that guided inquiry as a student-centered approach has an influence on a student's academic success and develop scientific process skills of students. While Tyler [14] found that learning gives students the opportunity to acquire skills in problem solving will improve critical thinking skills.

Learning inquiry model is a learning model that based on constructivism. Constructivism is a point of view in learning that considered student must be active to build their own knowledge in order to understand theory and gain knowledge. Teacher isn't take the role as person who transfer information but facilitation in learning which helped student to build their own knowledge. Madden [15] and Khulthau [16] said that the learning model of guided inquiry is a learning model that fulfill many curriculum requirements through engagement, motivation, and learning challenging in line with the purpose for the 21st century for educational institutions to guide students to think and learn through inquiry. Student obtain knowledge through exploration with their senses, including observe, listen, grope, feel and smell. Learning inquiry model invite student to explore their understanding through inquiry. Teachers' role in learning inquiry model are determine kind of research that student doing and give guidance actively to student on collect the data, analysis, and make conclusion [17].

2. Method

This research is a pre-experimental design using one group pre-test and post-test, O_1 X O_2. O_1 is the pre-test scores, X represents the physics lesson on the topic of static fluid guided inquiry model, and O_2 is the post-test scores. Before the guided inquiry learning models performed, three groups of students grade X, namely: X IPA-1, X IPA-2, and X IPA-3 SMAN-1 Tarik (Indonesia) on Odd Semester Academic Year 2016/2017 given the critical thinking skills test (pre-test) on the subjects of physics on the topic of static fluid and after learning the three groups were also given the same test (post-test). The test of critical thinking skills consisted of 10 problems in essay (narrative form), with indicators, including: (1) the skills to formulate the problem, (2) give the argument, (3) induction, (4) evaluating, and (5) to decide a course of action [4]. Criteria scoring is done by using a scoring guidelines were adapted from Washington State University Critical Thinking Project [18].

Data in the form of students' critical thinking skills that have been collected and analyzed using paired t-test, the calculation of the gain is normalized (N-gain), and analysis of variance (ANOVA). Paired t-test was used to analyze whether there is a difference between the scores of pre-test and post-test scores significantly on $\alpha = 5\%$; N-gain is used to determine the level increase in critical thinking skills after learning static fluid model guided inquiry; and ANOVA were used to analyze the consistency of average N-gain. Critical thinking skills score is calculated using the formula:

$$Critical\ thinking\ skills = \frac{number\ of\ scores\ obtained}{number\ of\ total\ scores}\ x\ 100\% \quad (1)$$

Paired t-test and ANOVA was conducted after the assumption of normality and homogeneity of the sample met, while the N-gain is calculated using the formula Hake [19]:

$$N - gain = \frac{S_{post} - S_{pre}}{S_{max} - S_{pre}} \quad (2)$$

with :
N-gain = The normalized gain scores
S_{post} = Post-test scores
S_{pre} = Pre-test scores
S_{max} = Maximum score

Furthermore, from the calculation of the N-gain is then converted to the criteria in Table 1.

Table 1. Criteria Normalized Gain Scores, N-Gain (modification from Hake [90])

N-Gain	*Normalized Gain Criteria*
N-Gain > *0.70*	High
0.30 ≤ *N-Gain* ≤ 0.70	Moderate
N-Gain < 0.30	Low

3. Result and Discussion

3.1. Result

Data from the pre-test and post-test students' critical thinking skills for three groups of grade X: IPA-1 IPA-2, and IPA-3 are shown in Figure 1. While the average N-gain for the three groups of students grade X are shown in Table 2.

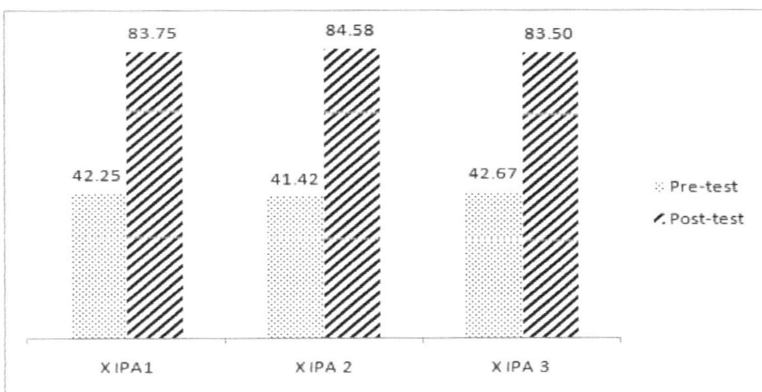

Figure 1. Pre-test and Post-test Students Critical Thinking Skills

Furthermore, based on the data score pre-test and post-test students' critical thinking skills and the fulfillment of the assumptions of normality and homogeneity of the sample, and then carried paired t-test and the results are shown in Table 3, Table 4 and Table 5.

Table 2. The Average N-gain Three Groups of Students of Grade X

No.	Grade	*Average N-gain*	Category
1.	X IPA 1	0.72	High
2.	X IPA 2	0.74	High
3.	X IPA 3	0.71	High
Average		0.72	High

Table 3. Paired t-Test Scores Pre-test and Post-test Students Grade X IPA-1

Paired Differences									
		Mean	Std. Deviation	Std. Error Mean	95% Confidence Interval of the Difference		t	df	Sig (2-tailed)
					Lower	Uper			
Pair 1	Pretest-posttest	-41.500	6.632	1.211	-43.976	-39.024	-34.274	29	.000

Table 4. Paired t-Test Scores Pre-test and Post-test Students Grade X IPA-2

Paired Differences									
		Mean	Std. Deviation	Std. Error Mean	95% Confidence Interval of the Difference		t	df	Sig (2-tailed)
					Lower	Uper			
Pair 1	Pretest-posttest	-42.867	7.229	1.320	-45.566	-40.167	-32.479	29	.000

Table 5. Paired t-Test Scores Pre-test and Post-test Students Grade X IPA-3

Paired Differences									
		Mean	Std. Deviation	Std. Error Mean	95% Confidence Interval of the Difference		T	df	Sig (2-tailed)
					Lower	Uper			
Pair 1	Pretest-posttest	-40.867	7.138	1.303	-43.532	-38.201	-31.360	29	.000

Table 6. Result of Uji One-Way ANOVA test for N-gain critical thinking skills of the students Grade X

	Sum of Squares	df	Mean Square	F	Sig.
Between Groups	.006	2	.003	.515	.599
Within Groups	.474	87	.005		
Total	.480	89			

Based on data from the value of average N-gain critical thinking skills in three groups of students grade X: IPA-1 IPA-2, and IPA-3 and the fulfillment of the assumptions of normality and homogeneity of the sample, then performed statistical tests One-Way ANOVA. Statistical Results of One-Way ANOVA test for average N-gain critical thinking skills of the students are shown in Table 6.

3.2. Discussion

Based on Figure 1, before learning the fluid static guided inquiry model, the average score of critical thinking skills of students in each group in grades X: IPA-1 IPA-2, and IPA-3 shows a low value, less than 50 of scale range 0-100; These things happen perhaps because students have not been trained in critical thinking skills by using guided inquiry learning model. Meanwhile, after learning the static fluid guided inquiry model, the average score of student thinking skills in three groups becomes high, the average score of each group of more than 80. This is possible because the students have gained a guided inquiry learning models. Table 2 shows that the average level of increase in student scores (N-gain) in the three groups each of more than 0.7 on the scale range 0-1, or a high category. This is probably because the students become accustomed to working critically and smoothly in: formulating the problem; arguing; induction; an evaluation; and decide a course of action. Table 3, Table 4, and Table 5 show that average of students' critical thinking skills entire group grade X, including: IPA-1, IPA-2, and IPA-3, increasing significantly at $\alpha = 5\%$. This shows that the improvement of students' critical thinking skills scores in the three groups as do static fluid learning with guided inquiry model. Meanwhile Table 6 shows that the average level of increase in scores before and after the learning on three groups of students were not different (statistically), or it can be said that learning the fluid static model of guided inquiry gives the effect of increased scores critical thinking skills of students in the three groups with the level of improvement consistent.

An increase students' critical thinking skills score after the static fluid learning with guided inquiry model; with an average level of increase in high category and is not different for the three groups in accordance with the findings of Trundle [20], Brickman [21], Zawadski [22] and Minderhout & Loertscher [23], namely that students in early learning is not enjoying the guided inquiry-based learning as much activity they should do themselves although in conjunction with this, the skills students grow and be able to build his own knowledge. The application of guided inquiry learning not only improves students' ability to understand the material, but also can improve the skills of process and scientific work [24]; [25]; [26]; [27]. The stages in the inquiry learning model from stage to stage to formulate the problem deduce proven to assist students in making the concept of accommodation and assimilation as the theory of Piaget [28].

Social interaction within groups and among groups in inquiry learning model, according to Vygotsky can increase the level of potential development of students so that increased its closest development zone [29]. Active student involvement in inquiry-based learning can increase the storage of information in long-term memory, thus increasing mastery of concepts [28]. Results were also consistent with studies Kiumars, Ebrahim Mostafa Mohammad and Maboud [30] that there was a significant effect of guided inquiry method of learning to critical thinking skills and tentative conclusions. Nor was there any significant difference between men and women in the total value of critical thinking. Results of learning through inquiry model can improve the understanding of science, improve the learning objectives, the use of critical thinking [31], and improve the skills of predicting [32]. In addition, the results of this research also was supported by the results of research conducted by Prasetyowati and [33] which concluded that the inquiry model can improve the mastery of concepts and critical thinking skills of students, and there is a strong relationship and significant correlation between critical thinking skills to mastery of concepts students. Likewise, the results of research supported Jatmiko, Widodo, Martini, Budiyanto, Abrams, and Pandiangan [10] that the learning experience, the experiment is effective in increasing mastery of concepts, capabilities problems solving, and decision-making. This is consistent with the thinking Johnson and Siegel [34] that the thinking skills can help students to construct knowledge in the mastery of the whole concept.

4. Conclusion

Based on the research and the discussion concluded that the static fluid learning with guided inquiry model is effective in improving critical thinking skills of high school students grades X, as indicated by: (1) an increase in the average score of students' critical thinking skills before and after the learning significantly at $\alpha = 5\%$; (2) the increase in the average level of critical thinking skills, average N-gain high category; and (3) average N-gain is not different or consistent in the three groups of students.

Acknowledgments

Our gratitude goes to the Head of SMAN-1 Tarik that has given permission for research in SMAN-1 Tarik, Sidoarjo, Indonesia.

References

[1] Trilling B and Hood P 1999 *Educ. Technol.* **39** 5
[2] Semiawan CR 1992 *Pendekatan Keterampilan Proses* Jakarta: Gramedia
[3] Krulik S and Rudnick J A 1999 *Developing Mathematical Reasoning in Grades K-12. 1999 Yearbook* Ed. Stiff LV and Curcio FR (Reston, Virginia: NCTM) Innovative Tasks to Improve Critical and Creative Thinking Skills
[4] Ennis R H 1995 *Critical Thinking* (New York: Prentice Hall)
[5] Facione P A and Facione N C 2007 *Talking Critical Thinking Change* **39** 39
[6] Halpern D 2003 *Thought and Knowledge 4th Ed. Mahwah* (New Jersey: Erlbaum)
[7] Arends R 2012 *Learning to Teach, 9th Edition* (New York: Mc-Graw Hill)
[8] Tabak A, Sigri U and Turkoz T 2009 *Natl. Manag. Organ. Proc.* 303
[9] Eggen P D 2012 *Strategies for Teachers Teaching and Thinking Skills* (Boston: Allyn and Bacon)
[10] Jatmiko B, Widodo W, Martini, Budiyanto, Wicaksono I and Pandiangan P 2016 *J. Baltic Sci. Educ.* **15** 441
[11] Hassoubah Z I 2004 *Mengasah Pikiran Kreatif dan Kritis* (Jakarta: Nuansa)
[12] Nur M 2000 *Pengajaran Berpusat Pada Siswa dan Pendekatan Konstruktivis dalam Pengajaran* (Surabaya: Unipress)
[13] Bilgin I 2009 *Sci. Res. Essay* **4** 1038
[14] Redhana I W 2003 *J. Pendidik. Pengajaran* **XXXVI** 11
[15] Madden K 2011 *The use of inquiry-based instruction to increase motivation and academic success in a high school biology classroom*. Master Thesis (Bozeman: Montana State University)
[16] Khulthau C C and Todd R J 2008 *Guided Inquiry* (Online) Available on www.icwc.wikispaces.com/file/view/Guided+Inquiry.doc Accessed December 13th 2016
[17] Sund R B and Trowbridge L W 1973 *Teaching Science by Inquiry in the Secondary School Second Edition* (Columbus: Charles E. Merill Publishing Company)
[18] Riley D K, Brown G, Condon B and Law R 2006 *Washington State University Critical Thinking Project* (Online) Available on http://wsuctproject.wsu.edu/ctm.htm Accessed July 12th 2016
[19] Hake R R 1999 *American Educational Research Association's Division, Measurement Research Methodology Analyzing Change/Gain Scores* (USA: Woodland Hills)
[20] Trundle 2010 *Res. Sci. Educ.* **40** 451
[21] Brickman 2009 *Int. J. Scholarsh. Teac. Learning* **3** 2
[22] Zawadzki 2009 *Asian J. Educ. Learning* **1** 66
[23] Minderhout and Loertscher 2007 *Biochem. Mol. Biol. Educ.* **35** 172
[24] Ambarsari W, Santosa S and Maridi 2012 *J. Pendidik. Biol.* **5**
[25] Ariesta R and Supartono 2011 *J. Pendidik. Fis. Indones.* **7** 62
[26] Grant HM 2011 *The student Experience Intraditional and Inquiry Based Chemistry Labs*. Master Thesis (Bozeman: Montana State University)
[27] Khan M S, Hussain S, Ali R, Majoka M I and Ramzan M 2011 *Int. J. Academic Res.* **3** 955
[28] Slavin R E 1994 *Educational Psychology: Theory and Practice, Student Value Edition (8th Edition)* (London: Pearson Education)
[29] Santrock W J 2010 *Psikologi Pendidikan, Edisi II, Penerjemah Tri Wibowo* (Jakarta: Kencana Prenada Media Group)
[30] Kiumars A, Ebrahim M, Mostafa S, Mohammad A and Maboud O 2012 *J. Educ. Pract.* **3** 42
[31] Prince M J and Felder R M 2006 *J. Eng. Educ.* **95** 123
[32] Nicholas J S, Mark W A, Nicola J H and Paul W 2005 *Exp. Psychol.* **2** 98
[33] Prasetyowati E N and Suyatno 2016 *J. Kimia Pendidik. Kimia* **1**
[34] Johnson S and Siegel H 2010 *Teaching Thinking Skills* (New York: British Library)

IOP Conf. Series: Journal of Physics: Conf. Series **997** (2018) 012050 doi:10.1088/1742-6596/997/1/012050

Relative gravimeter prototype based on micro electro mechanical system

A S A Rozy[1], H A Nugroho[1] and M Yusuf[2]

[1]Sekolah Tinggi Meteorologi Klimatologi dan Geofisika
[2]Badan Meteorologi Klimatologi dan Geofisika

E-mail: abdussalam.alrozy@bmkg.go.id

Abstract. This research to make gravity measurement system by utilizing micro electro mechanical system based sensor in Gal order. System design consists of three parts, design of hardware, software, and interface. The design of the hardware include of designing the sensor design to measure the value of a stable gravity acceleration. The ADXL345 and ADXL335 sensors are tuned to obtain stable measurements. The design of the instrumentation system the next stage by creating a design to integrate between the sensor, microcontroller, and GPS. The design of programming algorithm is done with Arduino IDE software. The interface design uses a 20x4 LCD display to display the gravity acceleration value and store data on the storage media. The system uses a box made of iron and plate leveling to minimize measurement errors. The sensor test shows the ADXL345 sensor has a more stable value. The system is examined by comparing with gravity measurement of gravimeter A-10 results in Bandung observation post. The result of system test resulted the average of system correction value equal to 0.19 Gal. The system is expected to use for mineral exploration, water supply analyze, and earthquake precursor.

1. Introduction

Gravitational acceleration instruments are growing in their measurement methods in the 21st century. Researchers of the world produce measurement methods using laser interferometers, quartz, superconductors, and Micro-Electro-Mechanical Systems (MEMS) to measure the acceleration value of gravity with the smallest resolution possible.

Problems that occur in the field are expensive instruments that measure the rate of earth gravity acceleration provided by leading manufacturers such as L-coste and Scintrex. Another problem is the dimensions provided by the manufacturer are too large. Previous research produced several gravimeters by applying various sensors. The Massachusetts design gravimeter produces a quartz gravimeter with a resolution of 3-10 mGal[1]. The Gravimeter using MEMS designed by Mutoh yields a resolution of 0.5 mGal[2]. In this case, the design using MEMS is influenced by temperature[3]. The study was conducted with the aim of obtaining a measurement method that can produce instruments with dimensions as small as possible and low cost. Amirullah's research from Jember University entitled Design of Earth Gravity Acceleration Measuring Instrument (Gravimeter) Using Parallel Capacitors succeeded in making gravimeter by utilizing piezoelectric effect[4]. Research rocks have a resolution of up to 0.01 m/s^2 which has an accurate error value of 2.63% and a precision error of 0.045%.

This research uses ADXL sensor to measure gravity acceleration, RTC as system time sign, GPS as sign of system location, 20x4 LCD as system interface, and micro SD card as system storage media. This study aims to produce a measurement instrument of gravity acceleration value with a more efficient dimension and an affordable price. The designed system is expected to assist geophysical analysis such as mineral exploration, water supply survey, and earthquake precursor.

2. Implementation of System

This system consists of acceleration sensor ADXL 345, RTC DS1307, and GPS U-Blox Neo 6M as a component that provides input signal for microcontroller. The microcontroller processes the input signals of each component with algorithm. Microcontroller sends output signal in the form of digital data processing of input signal to LCD 20x4. The microcontroller and RTC communicate with each other at a certain baud rate to get the timing value. Microcontroller inserts the measurement data into the micro SD card. Figure 1 shows the block diagram of the system.

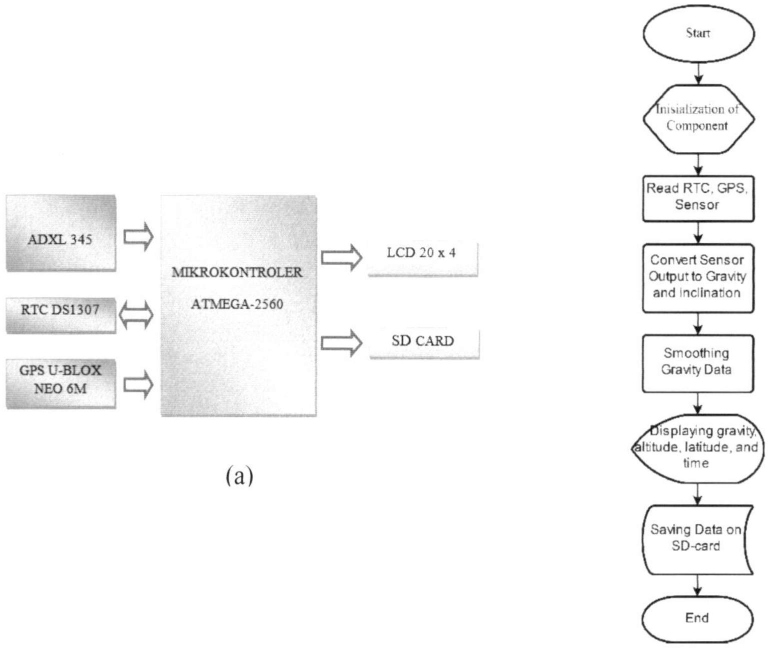

(a) (b)

Figure 1. Diagram block of system (a) and flowchart of system (b).

This system performs measurements by initializing the entire component so that the system is ready to perform data acquisition. The system performs reading of microcontroller input signals and processes algorithms to produce inclination values, gravitational acceleration, location, and time. The gravity data averaged every 40 readings for smoothing. All of data is displayed on LCD and saved on SD card. Figure 1 shows the system flowchart.

The system uses a waterproof box to protect the component so it can be carried to survey. System and has three adjustable legs for leveling. System interface using LCD located at the top of the box. The system has a small dimension to facilitate transportation to a measurement location. The system dimension is 20 cm x 15 cm x 30 cm (length x width x height). The iron box is equipped with a lock so it can be opened and closed to perform component replacements or troubleshooting the system.

3. Data Validation

The ADXL sensor test uses an inclinometer as a reference to get the value of g0. The inclinometer shows the degree of inclination of a field with the resolution on the inclinometer is 2°. Test results produce correction values which are then used in the measurement calculation. The comparison of the sensor compares the sensor output in the form of a slope angle to the Y axis (pitch) and the slope of the X ax (roll) with the plane and the inclinometer. The pitch and roll value of 0 has a straight line or perpendicular to the Z axis. Figure 2 shows the result of comparison of ADXL in degree.

The comparative results show the distribution of ADXL335 data wider, that is, has a maximum distribution value approaching 0.02° and -0.04° for the minimum distribution. ADXL345 has a relatively smaller data distribution than ADXL335 which is still around 0° (zero degree). Researchers used ADXL345 as a sensor in the measurement of gravity acceleration based on the stability of the sensor in the measurement of the test results to obtain a stable gravity acceleration value.

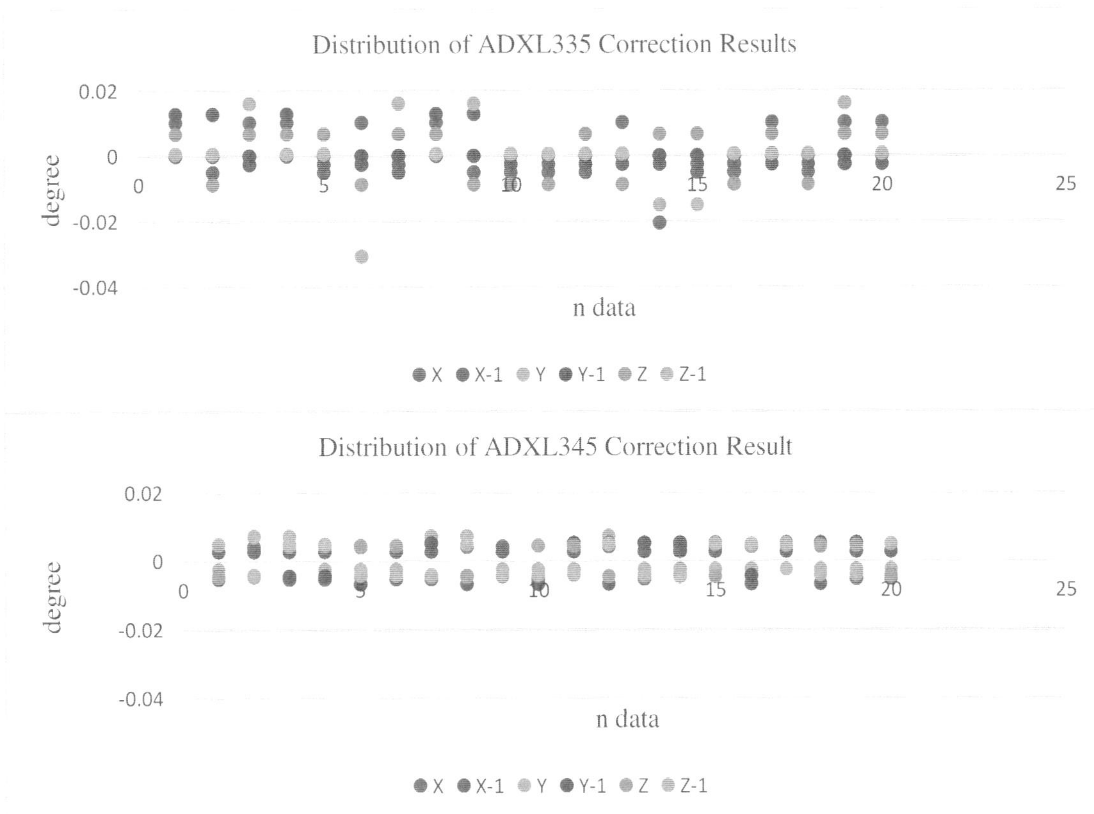

Figure 2. Comparison between ADXL345 and ADXL335.

Testing the system using the gravity observation post in Bandung as the location of testing that include Geophysical Station Bandung as a binding point measurement, post Lembang, and Post Tangkuban Perahu. Testing was conducted on August 23, 2017 from 09.00 am until 13:35 pm.

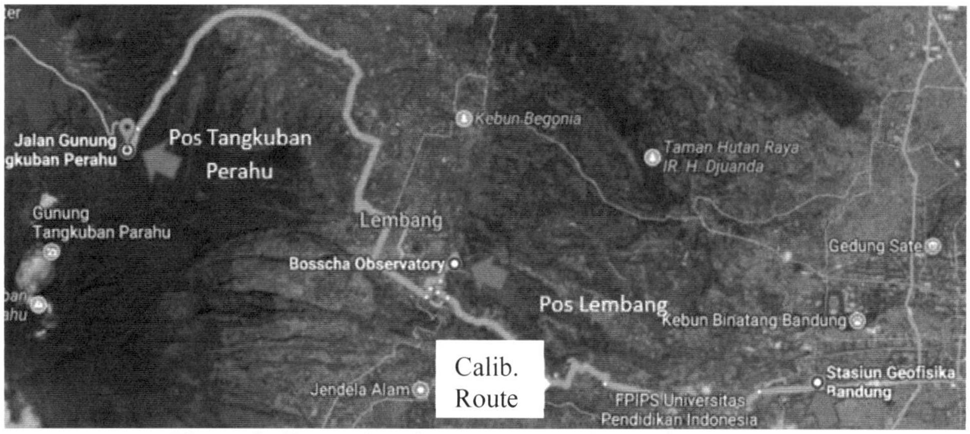

Figure 3. Comparison between ADXL345 and ADXL335.

Figure 6 shows the graph of data processing testing. The gravity value of the system shown by blue has the same trend as the A-10 measurement data, where the system follows the A-10 data pattern when there is an increase and decrease in A-10 data although there is still a correction on the system. The data pattern corresponds to the altitude of the measurement location, whereas the elevation of the location increases the acceleration value of gravity will decrease and vice versa when the location altitude decreases the acceleration value of gravity will increase.

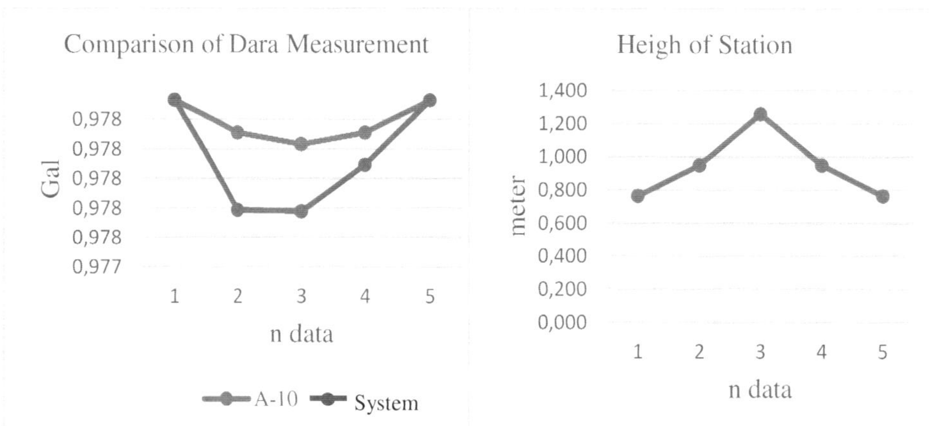

Figure 4. Comparison between ADXL345 and ADXL335.

Table 1 shows the results of calibration calculations with U95 is a 95% confidence interval and EN is the value of data received with. The calibration results show the average uncertainty value of the system of 2.8 Gal with the En value still in the range of -1 to 1. The uncertainty value indicates the range where there is a true value of a measurement as in Figure 4.12 where the first value is the standard reading value and the after value is the measurement value using the system.

Table 1. Comparison between ADXL345 and ADXL335.

Lokasi	Standart (Gal)	U95	UUT (Gal)	Drift $(^0)$	g new (Gal)	koreksi	gobs (Gal)	gobs avg (Gal)	def	U95	EN
Geof	977.965	0.00000823	837.187	0.0000	837,187	0.0000	977.965	977.967	-0.00145	2.82847	-0.00051
			837.186	-0.0009	837.187	0.0006	977.966				
			837.186	-0.0018	837.188	0.0013	977.967				
			837.186	-0.0027	837.189	0.0022	977.968				
			837.186	-0.0036	837.190	0.0031	977.969				
Lembang	977.855	0.00001059	836.688	-0.1258	836.814	-0.3730	977.592	977.594	0.26027	2.82848	0.09202
			836.688	-0.1267	836.814	-0.3721	977.593				
			836.688	-0.1276	836.815	-0.3713	977.594				
			836.688	-0.1285	836.816	-0.3704	977.595				
			836.688	-0.1294	836.817	-0.3694	977.596				
Tangkuban Perahu	977.816	0.00001071	836.606	-0.2029	836.809	-0.3779	977.588	977.589	0.22667	2.82848	0.08014
			836.606	-0.2038	836.809	-0.3772	977.588				
			836.606	-0.2047	836.810	-0.3763	977.589				
			836.606	-0.2056	836.811	-0.3754	977.590				
			836.606	-0.2065	836.812	-0.3744	977.591				
Lembang	977.855	0.00001059	836.685	-0.2814	836.966	-0.2201	977.745	977.747	0.1073	2.82848	0.03794
			836.685	-0.2823	836.967	-0.2192	977.746				
			836.685	-0.2832	836.968	-0.2182	977.747				
			836.685	-0.2841	836.969	-0.2173	977.748				
			836.685	-0.2850	836.970	-0.2164	977.749				
Geof	977.965	0.00000823	836.814	-0.3693	837.183	-0.0035	977.962	977.964	0.00178	2.82848	0.00063
			836.814	-0.3702	837.184	-0.0027	977.963				
			836.814	-0.3711	837.185	-0.0019	977.964				
			836.814	-0.3720	837.186	-0.0008	977.965				
			836.814	-0.3729	837.187	-0.0000	977.965				

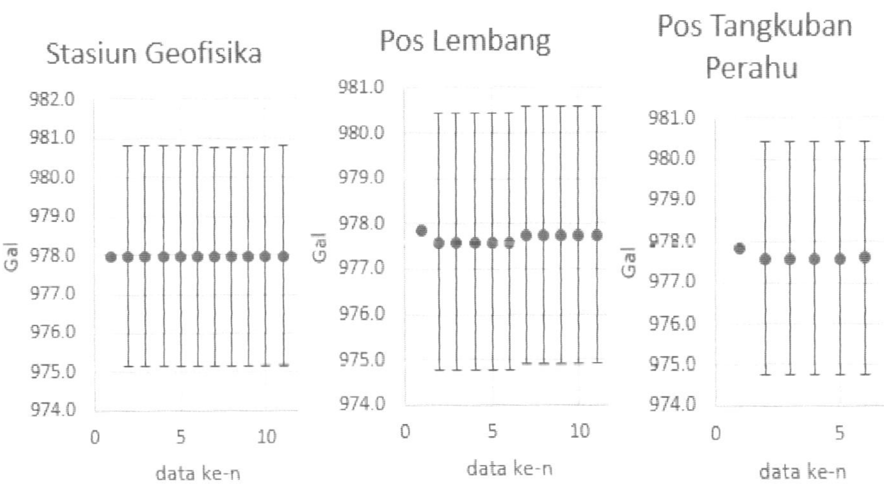

Figure 5. Comparison between ADXL345 and ADXL335.

The measurement results of gravity acceleration shows the presence of external factors and in the system that influence the measurement. The authors analyzed changes in the tool level, temperature, and position of the system in influencing the measurement of gravitational acceleration in the designed system. The value of temperature correlation to the measurement of gravity acceleration is 0.7 indicating a strong correlation. The value of the slope correlation of the system to the measurement of gravity acceleration is 0.3 indicating a low correlation. The correlation value of both factors shows the value of

0.78 which indicates that two factors interplay the measurement simultaneously. Both components of the temperature and tilt factor have a strong correlation to the drift correction value of the system.

4. Conclusion

This research shows that the ADXL345 sensor has data with a smaller deviation than ADXL335. The system which use ADXL 345 has average correction value of 0.19 Gal and 2.8 Gal of uncertainty value. The measurement result of this system has followed the pattern of the gravimeter A-10. Environmental noise i.e. the temperature and leveling system still influence in system causing noise.

Subsequent research can be developed using sensors that have better resolution and sensitivity to obtain measurements of gravity acceleration to miliGal or microGal orders. Another box material can used to minimize the environmental noise.

References

[1] Korobiichuk I, Bezvesilna O, Tkachuk A, Chilchenko T, Nowicki M and Szewczyk R 2016 *J. Autom. Mob. Robot. Intell. Syst.* **10** 43
[2] Mutoh T, Mitsuaki K, Aizawa T and Matsuoka T 2011 *Proc. Symp. Ultrason. Electron.* **32** 1
[3] Kon S, Oldham K and Horowitz R 2007 *Proc. SPIE Sens. Smart Struct. Technol. Civ., Mech. and Aerosp. Syst.* **6529**
[4] Amirullah 2007 *Rancang Bangun Alat Pengukur Percepatan Gravitasi Bumi (Gravimeter) Menggunakan Kapasitor Keping Sejajar*. Undergraduate Thesis (Jember: Universitas Jember)

IOP Conf. Series: Journal of Physics: Conf. Series **997** (2018) 012051 doi:10.1088/1742-6596/997/1/012051

The solution of 4-dimensional Schrodinger equation for polynomial inverse potential and its partner potential by using ansatz wavefunction method

A Suparmi[1], C Cari[1], Wahyulianti[1,a] and K N Wea[1]

[1]Physics Department of Post Graduate Program, Sebelas Maret University, Surakarta 57126, Indonesia

[a]E-mail: wahyuliantiyuli@yahoo.com

Abstract. The radial part of 4-dimensional Schrodinger equation for Polynomial Inverse potential and its partner potential was solved by using the Ansatz Wavefunction method. By applying of the properties and operators of the Supersymmetric Quantum Mechanic method the partner potential was constructed. The eigenvalues and eigenfunctions of the Polynomial Inverse potential and the eigenvalues and eigenfunctions of its partner potential were different. The partner potential of Polynomial Inverse potential for $n > 1$ has eigenvalue greater than Polynomial Inverse potential as its original potential.

1. Introduction

The D-dimensional Schrodinger equation is the wave equations for higher dimensions in quantum mechanics system [1-2]. The solution of D-dimensional Schrodinger equation for some potentials are exactly solved by methods such as Ansatz Wavefunction method [1,2], Nikiforov-Uvarov (NU) method [3-6], Supersymmetric Quantum Mechanic (SUSY QM) [4,8,16], Asymptotic Iteration Method (AIM) [9], Factorization methods [10] and Romanovski polynomial method [11]. The Ansatz Wavefunction method was a method for the analyze the part radial of D-dimensional Schrodinger equation [1-2]. By using and applying the operators of SUSY QM, the properties of SUSY QM and the ground state wave functions of the original potential, the partner potential was constructed [4].

This study aims is to construct partner potential and determine eigenvalue and eigenfuntion of the generalized Polynomial Inverse potential. In this paper, the solution of radial D-dimensional Schrodinger equation for Polynomial Inverse potential and its partner potential was studied using Ansatz Wavefunction method. This paper is organized as follows. Introduction is presented in Section 1. The Ansatz Wavefunction method and SUSY QM method is presented in Section 2. The results and discussion are presented in Section 3 and conclusion in Section 4.

2. Method

In this section, we review the solution of D-dimensional Schrodinger equation with Ansatz Wavefunction method and Supersymmetry Quantum Mechanics method. The Generally, Schrodinger equation in D-dimensional [1,12] given as

$$-\frac{\hbar^2}{2\mu}\nabla_D^2\psi(r,\Omega)+V(r,\Omega)\psi(r,\Omega)=E\psi(r,\Omega) \tag{1}$$

with

$$\psi(r,\Omega)=\frac{1}{r^{\frac{D-1}{2}}}U(r)Y^{l}_{l_{D-1}\cdots l_1}(\hat{x}) \qquad ;\hat{x}=\theta_1,\theta_2,\cdots\theta_{D-1} \tag{2}$$

The radial part and angular part of 4-dimensional Schrodinger equation obtained by inserted Eq. (2) into Eq. (1) and solve of using variable separation method with $Y_{\vec{l}_1,\ldots\vec{l}_3}(\theta_1,\theta_2,\theta_3)=P(\theta_1)P(\theta_2)P(\theta_3)$ is given as [13]:

$$\frac{\partial^2 U}{\partial r^2}-\left(\lambda_3+\frac{3}{4}\right)\frac{1}{r^2}U+\frac{2\mu}{\hbar^2}\left[E-V(r)\right]U=0 \tag{3}$$

$$\frac{\partial^2 P(\theta_1)}{\partial\theta_1} - V(\theta_1)P(\theta_1) + \lambda_1 P(\theta_1) = 0 \tag{4}$$

$$\frac{1}{P(\theta_2)}\left\{\frac{1}{\sin\theta_2}\left(\frac{\partial}{\partial\theta_2}\sin\theta_2\frac{\partial P(\theta_2)}{\partial\theta_2}\right)\right\} - V(\theta_2) - \frac{\lambda_1}{\sin^2\theta_2} + \lambda_2 = 0 \tag{5}$$

$$\frac{1}{P(\theta_3)}\left\{\frac{1}{\sin^2\theta_3}\left(\frac{\partial}{\partial\theta_3}\sin^2\theta_3\frac{\partial P(\theta_3)}{\partial\theta_3}\right)\right\} - V(\theta_3) + \lambda_3 - \frac{\lambda_2}{\sin^2\theta_3} = 0 \tag{6}$$

where λ_1, λ_2 and λ_3 is variable separation constant.

In this paper the potential energi of Polynomial Inverse potential generally given as [14]:

$$V(r) = \frac{a}{r^6} + \frac{b}{r^5} + \frac{c}{r^4} + \frac{d}{r^3} + \frac{e}{r^2} + \frac{f}{r} \tag{7}$$

with a, b, c, d, e and f is constant coeficient.

2.1 Review of Ansatz wavefunction method

The wavefunction that can not be completed exactly can be solved by the Ansatz wavefunction method. Ansatz wavefunction method is a method used specifically to determine the solution of the radial part [1]. Schrodinger equation The D-dimensional of the radial part of Equation (3) is reduced to two-orde differential equation (in atomic units $\hbar = 1$; $2m = 1$) [1,12] is

$$\left\{\frac{d^2}{dr^2} - \frac{l_{D-1}(l_{D-1}+D-2)+(D-1)(D-3)/4}{r^2}\right\}U(r) = \left[-E+V(r)\right]U(r) \tag{8}$$

where $\lambda_{D-1} = l_{D-1}(l_{D-1}+D-2)$, l_{D-1} is angular momentum, D is the spatial dimension and λ_{D-1} is the variable separation constant [1]. Based on Equation (8), the Schrodinger equation of the radial part for D = 4 is

$$\left\{\frac{d^2}{dr^2} - \frac{1}{r^2}\left(\lambda_3 + \frac{3}{4}\right)\right\}U(r) = \left[-E+V(r)\right]U(r) \tag{9}$$

where

$$\lambda_3 = l_3(l_3+2) \tag{10}$$

The solution of Eq. (9) as a radial wave function $U(r)$ of the Ansatz method [1,2] is expressed as:

$$U(r) = f_n(r)\exp\left[g(r)\right] \tag{11}$$

where

$$f_n(r) = \begin{cases} 1 & \text{for } n = 0, \\ \prod_{j=1}^{n}(r-\sigma_j^{(n)}) & \text{for } n = 1,2,3,\ldots \end{cases} \tag{12}$$

and $g(r)$ is the wave function in exponential form. Substitution Eq. (9) into Eq. (11), we obtained

$$-E_n + V(r) + \frac{\lambda_3 + 3/4}{r^2} = g''(r) + (g'(r))^2 + \frac{f_n''(r) + 2f_n'(r)g'(r)}{f_n(r)} \tag{13}$$

Equation (13) is a general equation that will be used in the solution of radial part the 4-dimensional Schrodinger equation to determine the energy and wave function by using the ansatz wavefunction method.

2.2 Review of SUSY QM

Supersymmetry Quantum Mechanics (SUSY QM) is the simplest model of SUSY field theory developed based on Witten's proposal [13].Witten defined two charge operators are commute with Hamiltonian (H_{ss}) of the supersymmetry quantum system [4,8,15] is geven as

$$H_{ss} = \begin{pmatrix} -\frac{d^2}{dx^2} + \frac{d\varphi(x)}{dx} + \varphi^2(x) & 0 \\ 0 & -\frac{d^2}{dx^2} - \frac{d\varphi(x)}{dx} + \varphi^2(x) \end{pmatrix} = \begin{pmatrix} H_+ & 0 \\ 0 & H_- \end{pmatrix} \tag{14}$$

with partner Hamiltonian $H_- = H_1$ and $H_+ = H_2$. By setting the new SUSY operators for raising operator $A^+ = -\frac{d}{dx} + \varphi(x)$ and lowering operator $A = \frac{d}{dx} + \varphi(x)$. The SUSY Hamiltonians of equation (14) are

$$H_-(x) = H_1 = A^+ A \text{, and } H_+(x) = H_2 = AA^+ \tag{15}$$

and

$$A\psi_0^{(-)} = A\psi_0 = 0 \tag{16}$$

The partners potential in the SUSY QM method is $V_- = V_1$ and $V_+ = V_2$ are

$$V_-(x) = V_1 = \varphi^2(x) - \varphi'(x) \text{ and } V_+(x) = V_2 = \varphi^2(x) + \varphi'(x) \tag{17}$$

where φ is superpotential. The partner potential V_1 is determined as

$$V_-(x;a_0) = V_1(x;a_0) = V_{ef}(x) - E_0 \tag{18}$$

with V_{ef} is effective potential and E_0 is the ground state energy.

By manipulating equation (17) and by applying equation (11) and equation (18), we construct the new partner potential V_2 as

$$V_2(x) = V_{ef}(x) - E_0 - 2\frac{d}{dx}\left(\frac{d\psi_0/dx}{\psi_0} \right) \tag{19}$$

where ψ_0 is the ground state wave function [4,8]. The eigenvalue and the eigenfunction of partner potential is also determined by using ansatz wavefunction method.

3. Results and Discussion

In this section, the solution of the Schrodinger equation in 4-dimensional by using ansatz wavefunction method and Supersymmetry Quantum Mechanics method. The exponential wave function $g\ (r)$ of the Polynomial Inverse potential obtained by applying the supersymmetry method is

$$g(r) = \frac{1}{2}\frac{A}{r^2} + \frac{B}{r} + Cr + D\ln r \tag{20}$$

The solution for $n = 0$ then $f_0(r) = 1$. The Equation (7) and Equation (20) substituted into Equation (13) are obtained

$$-E_0 + \frac{a}{r^6} + \frac{b}{r^5} + \frac{c}{r^4} + \frac{d}{r^3} + \frac{e}{r^2} + \frac{f}{r} = \left[\begin{aligned} &\frac{A^2}{r^6} + \frac{2AB}{r^5} + \frac{B^2 - 2AD + 3A}{r^4} + \frac{2B(1-D) - 2AC}{r^3} \\ &+ \frac{D(D-1) - 2BC - \left(\lambda_3 + \frac{3}{4}\right)}{r^2} + \frac{2CD}{r} + C^2 \end{aligned} \right] \tag{21}$$

By comparing the two sections in Eq. (21), we get

$$a=A^2,\quad b=2AB,\quad c=B^2-2AD+3A,\quad d=2B(1-D)-2AC, \tag{22}$$

$$e=D(D-1)-2BC-\left(\lambda_3+3/4\right),\quad f=2CD, \tag{23}$$

$$-E=C^2 \tag{24}$$

Furthermore, from Equation (22) we get the parameter value

$$A=\pm\sqrt{a} \tag{25}$$

The value of parameter A used in equation (25) is negative. Equation (25) is substituted into Eq. (22) the equation is obtained

$$B=-\frac{b}{2\sqrt{a}} \tag{26}$$

Then Equation (25) and Equation (26) are substituted into Eq. (22) we get

$$D=\frac{3}{2}+\frac{c}{2\sqrt{a}}-\frac{b^2}{8a\sqrt{a}} \tag{27}$$

And then Equations (25) to Eq. (27) are substituted into Equation (23)

$$\frac{b}{\sqrt{a}}C^2-\left(e+\lambda_3+\frac{9}{4}+\frac{c}{2\sqrt{a}}-\frac{b^2}{8a\sqrt{a}}\right)C+\frac{1}{2}\left(\frac{3}{2}+\frac{c}{2\sqrt{a}}-\frac{b^2}{8a\sqrt{a}}\right)f=0 \tag{28}$$

The value of C is determined from Equation (28) by using algebra and we obtained

$$C=\frac{\left(e+\lambda_3+\frac{9}{4}+\frac{c}{2\sqrt{a}}-\frac{b^2}{8a\sqrt{a}}\right)\pm\sqrt{\left(e+\lambda_3+\frac{9}{4}+\frac{c}{2\sqrt{a}}-\frac{b^2}{8a\sqrt{a}}\right)^2-4\left(\frac{3}{2}+\frac{c}{2\sqrt{a}}-\frac{b^2}{8a\sqrt{a}}\right)\frac{b}{2\sqrt{a}}f}}{\frac{2b}{\sqrt{a}}} \tag{29}$$

Furthermore, Equation (29) is substituted into Eq. (24), so that an eigenvalue we get:

$$E_0=-\frac{a}{4b^2}\left[\left(e+\lambda_3+\frac{9}{4}+\frac{c}{2\sqrt{a}}-\frac{b^2}{8a\sqrt{a}}\right)\pm\sqrt{\left(e+\lambda_3+\frac{9}{4}+\frac{c}{2\sqrt{a}}-\frac{b^2}{8a\sqrt{a}}\right)^2-4\left(\frac{3}{2}+\frac{c}{2\sqrt{a}}-\frac{b^2}{8a\sqrt{a}}\right)\frac{b}{2\sqrt{a}}f}\right]^2 \tag{30}$$

where E_0 is ground state eigenvalue. Equation (12), Eq. (20) and Eq. (25) to Eq. (27) are substituted into Eq. (11), we obtained

$$U_0(r)=N_0 r^{\left(\frac{3}{2}+\frac{c}{2\sqrt{a}}-\frac{b^2}{8a\sqrt{a}}\right)}\exp\left[-\frac{\sqrt{a}}{2}\frac{1}{r^2}-\frac{b}{2\sqrt{a}}\frac{1}{r}+Cr\right] \tag{31}$$

where $U_0(r)$ is ground state eigenfunction. Equation (30) and Equation (31) are used to determine the potential partner radial part.

The solution for $n = 1$, then $f_1(r)=r-\sigma_1^{(1)}$. The solution of this case is done with the same steps as for $n = 0$ and then we obtained

$$\begin{aligned}&-E_1+\frac{a}{r^6}+\frac{b}{r^5}+\frac{c}{r^4}+\frac{d}{r^3}+\frac{e}{r^2}+\frac{f}{r}\left(r-\sigma_1^{(1)}\right)=\\&\left(\begin{matrix}\frac{A^2}{r^6}+\frac{2AB}{r^5}+\frac{B^2-2AD+3A}{r^4}+\frac{2B(1-D)-2AC}{r^3}+\\ \frac{D(D-1)-2BC-\left(\lambda_3+3/4\right)}{r^2}+\frac{2CD}{r}+C^2\end{matrix}\right)\left(r-\sigma_1^{(1)}\right)-\frac{2A}{r^3}-\frac{2B}{r^2}+2C+\frac{2D}{r}\end{aligned} \tag{32}$$

By comparing the two sections in Eq. (32) we get

$$-a\sigma_1^{(1)} = -A^2\sigma_1^{(1)} \tag{33}$$

$$a - b\sigma_1^{(1)} = A^2 - 2AB\sigma_1^{(1)} \tag{34}$$

$$b - c\sigma_1^{(1)} = 2AB - \left[B^2 - 2AD + 3A\right]\sigma_1^{(1)} \tag{35}$$

$$c - d\sigma_1^{(1)} = B^2 - 2AD + A - \left(2B(1-\mathrm{D}) - 2AC\right)\sigma_1^{(1)} \tag{36}$$

$$d - e\sigma_1^{(1)} = -2BD - 2AC - \left(D\left(D-1\right) - 2BC - \left(\lambda_3 + \tfrac{3}{4}\right)\right)\sigma_1^{(1)} \tag{37}$$

$$e - f\sigma_1^{(1)} = D\left(D-1\right) - 2BC + 2D - \left(\lambda_3 + \tfrac{3}{4}\right) - 2CD\sigma_1^{(1)} \tag{38}$$

$$f + E\sigma_1^{(1)} = 2C(D+1) - C^2\sigma_1^{(1)} \tag{39}$$

$$-E_1 = C^2 \tag{40}$$

From Equation (34) to Eq. (39) we get parameter values

$$A = \pm\sqrt{a}, \quad B = -\frac{b}{2\sqrt{a}}, \quad D = \frac{3}{2} + \frac{c}{2\sqrt{a}} - \frac{b^2}{8a\sqrt{a}}, \tag{41}$$

$$C = \frac{\left(e + \lambda_3 + \frac{3}{4}\right) \pm \sqrt{\left(e + \lambda_3 + \frac{3}{4}\right)^2 - 4\left(\sigma_1^{(1)} + \frac{b}{2\sqrt{a}}\right)\left(\frac{3}{2} + \frac{c}{2\sqrt{a}} - \frac{b^2}{8a\sqrt{a}}\right)f}}{4\left(\sigma_1^{(1)} + \frac{b}{2\sqrt{a}}\right)} \tag{42}$$

Furthermore, Equation (42) is substituted into Eq. (40), then obtained

$$E_1 = -\left(\frac{\left(e + \lambda_3 + \frac{3}{4}\right) \pm \sqrt{\left(e + \lambda_3 + \frac{3}{4}\right)^2 - 4\left(\sigma_1^{(1)} + \frac{b}{2\sqrt{a}}\right)\left(\frac{3}{2} + \frac{c}{2\sqrt{a}} - \frac{b^2}{8a\sqrt{a}}\right)f}}{4\left(\sigma_1^{(1)} + \frac{b}{2\sqrt{a}}\right)}\right)^2 \tag{43}$$

where E_1 is the first level excited energy. Equation (12), Equation (20) and Equation (41) are substituted into Eq. (11) we get

$$U_1(r) = N_1(r - \sigma_1^{(1)})r^{\left(\frac{3}{2} + \frac{c}{2\sqrt{a}} - \frac{b^2}{8a\sqrt{a}}\right)} \exp\left[-\frac{\sqrt{a}}{2}\frac{1}{r^2} - \frac{b}{2\sqrt{a}}\frac{1}{r} + Cr\right] \tag{44}$$

where $U_1(r)$ is the eigenfunction of first level.

The solution for $n = 2$; $f_2(r) = \left(r - \sigma_2^{(2)}\right)\left(r - \sigma_1^{(2)}\right)$. The solution of this case is carried out with the same steps as for $n = 0$ and $n = 1$ and the following equation is obtained

$$a = A^2, \quad b = 2AB, \quad c = B^2 - 2AD + 3A \tag{45}$$

$$e - f\left(\sigma_2^{(2)} + \sigma_1^{(2)}\right) = D\left(D-1\right) - 2BC - \left(\lambda_3 + \tfrac{3}{4}\right) + 2 + 4D - 2C\left(D+1\right)\left(\sigma_2^{(2)} + \sigma_1^{(2)}\right) \tag{46}$$

$$f = 2C\left(D+2\right) \tag{47}$$

$$E_2 = -C^2 \tag{48}$$

By using Eq. (45) - Eq. (47) we obtain the following parameter values

$$A=\pm\sqrt{a},\quad B=-\frac{b}{2\sqrt{a}},\quad D=\frac{3}{2}+\frac{c}{2\sqrt{a}}-\frac{b^2}{8a\sqrt{a}},\tag{49}$$

$$C=\frac{\left(e+\lambda_3+\frac{3}{4}-D-2\right)\pm\sqrt{\left(e+\lambda_3+\frac{3}{4}-D-2\right)^2-4\left(\left(\sigma_2^{(2)}+\sigma_1^{(2)}\right)+\frac{b}{2\sqrt{a}}\right)Df}}{4\left(\left(\sigma_2^{(2)}+\sigma_1^{(2)}\right)+\frac{b}{2\sqrt{a}}\right)}\tag{50}$$

Furthermore, Equation (50) is substituted into Eq. (48), the eigenvalue is obtained

$$E_2=-\left(\frac{\left(e+\lambda_3+\frac{3}{4}-D-2\right)\pm\sqrt{\left(e+\lambda_3+\frac{3}{4}-D-2\right)^2-4\left(\left(\sigma_2^{(2)}+\sigma_1^{(2)}\right)+\frac{b}{2\sqrt{a}}\right)Df}}{4\left(\left(\sigma_2^{(2)}+\sigma_1^{(2)}\right)+\frac{b}{2\sqrt{a}}\right)}\right)^2\tag{51}$$

where E_2 is the second level excited energy. Equation (12), Equation (20) and Equation (49) are substituted into Eq. (11), we get

$$U_2(r)=N_2(r-\sigma_2^{(2)})(r-\sigma_1^{(2)})r^{\left(\frac{3}{2}+\frac{c}{2\sqrt{a}}-\frac{b^2}{8a\sqrt{a}}\right)}\exp\left[-\frac{\sqrt{a}}{2}\frac{1}{r^2}-\frac{b}{2\sqrt{a}}\frac{1}{r}+Cr\right]\tag{52}$$

Where $U_2(r)$ is the second level eigenfunction.

The energy equations and the wave functions in generally from Equations (30), (43), (51), (31), (44) and Eq. (52) are consecutively

$$E_n=-\left[\frac{\left(e+\lambda_3+\frac{3}{4}+(1-n)D-n(n-1)\right)\pm\sqrt{\left(e+\lambda_3+\frac{3}{4}+(1-n)D-n(n-1)\right)^2-4\left(\sum_{i=1}^{n}\sigma_i^{(n)}+\frac{b}{2\sqrt{a}}\right)Df}}{4\left(\sum_{i=1}^{n}\sigma_i^{(n)}+\frac{b}{2\sqrt{a}}\right)}\right]^2\tag{53}$$

and

$$U_n(r)=N_n\left[\prod_{i=1}^{n}(r-\sigma_i^{(n)})\right]r^D\exp\left[\frac{1}{2}\frac{A}{r^2}+\frac{B}{r}+Cr\right]\tag{54}$$

where

$$A=-\sqrt{a},\quad B=-\frac{b}{2\sqrt{a}},\quad D=\frac{3}{2}+\frac{c}{2\sqrt{a}}-\frac{b^2}{8a\sqrt{a}},$$

$$C=\frac{\left(e+\lambda_3+\frac{3}{4}+(1-n)D-n(n-1)\right)\pm\sqrt{\left(e+\lambda_3+\frac{3}{4}+(1-n)D-n(n-1)\right)^2-4\left(\sum_{i=1}^{n}\sigma_i^{(n)}+\frac{b}{2\sqrt{a}}\right)Df}}{4\left(\sum_{i=1}^{n}\sigma_i^{(n)}+\frac{b}{2\sqrt{a}}\right)},$$

E_n is eigenvalue, $U_n(r)$ adalah eigenfunction of Polynomial Inverse potential. Furthermore, from Equation (53) and Equation (54), the potential partner construction is obtained

$$V_2(r)=\frac{a}{r^6}+\frac{b}{r^5}+\frac{c}{r^4}+\frac{d+2A}{r^3}+\frac{e+\lambda_3+\frac{3}{4}+2B+2D}{r^2}+\frac{f}{r}-E_0-2C \tag{55}$$

with E_0 in Eq. (30), parameters *A*, *B*, *D* and *C* in Equations (25) to Equation (27). Equation (55) is a potential partner of the radial part of the Inverse Polynomial potential. The potential partner will be solved in the 4-dimension Schrodinger equation using the ansatz wave function method.

The exponential wave function *g* (*r*) potential partner of Polynomial Inverse potential obtained by applying the supersymmetry method is

$$g(r)=\frac{1}{2}\frac{A'}{r^2}+\frac{B'}{r}+C'r+D'\ln r \tag{56}$$

The solution for n = 0; $f_0(r)=1$. Equation (12) and Equation (56) substituted into Equation (13) we obtained

$$-E_0^{'}+\frac{a}{r^6}+\frac{b}{r^5}+\frac{c}{r^4}+\frac{d+2A}{r^3}+\frac{e+\lambda_3+\frac{3}{4}+2B+2D}{r^2}+\frac{f}{r}-E_0-2C=\frac{A'^2}{r^6}+\frac{2A'B'}{r^5}+$$
$$\frac{B^2-2A'D'+3A'}{r^4}+\frac{2B'(1-D')-2A'C'}{r^3}+\frac{D'(D'-1)-2B'C'-\left(\lambda_3^{'}+\frac{3}{4}\right)}{r^2}+\frac{2C'D'}{r}+C'^2 \tag{57}$$

By comparing the two sections in Eq. (57), we get

$$a=A'^2,\quad b=2A'B',\quad c=B'^2-2A'D'+3A',\quad d+2A=2B'(1-D')-2A'C', \tag{58}$$

$$e+\lambda_3+\frac{3}{4}+2B+2D=D'(D'-1)-2B'C'-\left(\lambda_3^{'}+\frac{3}{4}\right) \tag{59}$$

$$f=2C'D' \tag{60}$$

$$E_0^{'}=C'^2-E_0-2C \tag{61}$$

From Eq. (58) we obtain

$$A'=\pm\sqrt{a} \tag{62}$$

$$B'=-\frac{b}{2\sqrt{a}} \tag{63}$$

$$D'=\frac{3}{2}+\frac{c}{2\sqrt{a}}-\frac{b^2}{8a\sqrt{a}} \tag{64}$$

and then, from Equation (59), (60) and Eq. (62) through Eq. (64) we get

$$\frac{b}{\sqrt{a}}C'^2-\left(e+\lambda_3+\lambda_3^{'}+\frac{3}{2}-\frac{b}{\sqrt{a}}+3\left(\frac{3}{2}+\frac{c}{2\sqrt{a}}-\frac{b^2}{8a\sqrt{a}}\right)\right)C+\frac{1}{2}\left(\frac{3}{2}+\frac{c}{2\sqrt{a}}-\frac{b^2}{8a\sqrt{a}}\right)f=0 \tag{65}$$

From Eq. (65) by using algebra, we have

$$C'=\frac{\left(\lambda+3\left(\frac{3}{2}+\frac{c}{2\sqrt{a}}-\frac{b^2}{8a\sqrt{a}}\right)\right)\pm\sqrt{\left(\lambda+3\left(\frac{3}{2}+\frac{c}{2\sqrt{a}}-\frac{b^2}{8a\sqrt{a}}\right)\right)^2-4\left(\frac{3}{2}+\frac{c}{2\sqrt{a}}-\frac{b^2}{8a\sqrt{a}}\right)\frac{b}{2\sqrt{a}}f}}{\frac{2b}{\sqrt{a}}} \tag{66}$$

with

$$\lambda=e+\lambda_3+\lambda_3^{'}+\frac{3}{2}-\frac{b}{\sqrt{a}} \tag{67}$$

Equation (66) is substituted into Eq. (61) we get

$$E_0^{'}=-\frac{a}{4b^2}\left[\left(\lambda+3\left(\frac{3}{2}+\frac{c}{2\sqrt{a}}-\frac{b^2}{8a\sqrt{a}}\right)\right)\pm\sqrt{\left(\lambda+3\left(\frac{3}{2}+\frac{c}{2\sqrt{a}}-\frac{b^2}{8a\sqrt{a}}\right)\right)^2-4\left(\frac{3}{2}+\frac{c}{2\sqrt{a}}-\frac{b^2}{8a\sqrt{a}}\right)\frac{b}{2\sqrt{a}}f}\right]^2-E_0-2C \quad (68)$$

where E_0 is the ground state energy. Equation (12), Equations (56) and Equations (62) through Equation (64) are substituted to Eq. (11) we have

$$U_0^{'}(r)=N_0 r^{\left(\frac{3}{2}+\frac{c}{2\sqrt{a}}-\frac{b^2}{8a\sqrt{a}}\right)}\exp\left[-\frac{\sqrt{a}}{2}\frac{1}{r^2}-\frac{b}{2\sqrt{a}}\frac{1}{r}+C'r\right] \quad (69)$$

where $U_0(r)$ is the ground state eigenfunction.

The solution for $n = 1$; $f_1(r)=r-\sigma_1^{(1)}$. The solution of this case is done with the same steps as for $n = 0$, we obtained

$$E_1^{'}=-\left[\frac{(\lambda+2D')\pm\sqrt{(\lambda+2D')^2-4\left(\sigma_1^{(1)}+\frac{b}{2\sqrt{a}}\right)D'f}}{4\left(\sigma_1^{(1)}+\frac{b}{2\sqrt{a}}\right)}\right]^2-E_0-2C \quad (70)$$

and

$$U_1^{'}(r)=N_1(r-\sigma_1^{(1)})r^{\left(\frac{3}{2}+\frac{c}{2\sqrt{a}}-\frac{b^2}{8a\sqrt{a}}\right)}\exp\left[-\frac{\sqrt{a}}{2}\frac{1}{r^2}-\frac{b}{2\sqrt{a}}\frac{1}{r}+C'r\right] \quad (71)$$

where $E_1^{'}$ is the first level of excited energy dan $U_1^{'}(r)$ is the first level of eigenfunction.

The solution for $n = 2$; $f_2(r)=\left(r-\sigma_2^{(2)}\right)\left(r-\sigma_1^{(2)}\right)$. The solution of this case is done with the same steps as $n = 0$ and $n = 1$ we have

$$E_2^{'}=-\left(\frac{(\lambda+D'-2)\pm\sqrt{(\lambda+D'-2)^2-4\left(\left(\sigma_2^{(2)}+\sigma_1^{(2)}\right)+\frac{b}{2\sqrt{a}}\right)D'f}}{4\left(\left(\sigma_2^{(2)}+\sigma_1^{(2)}\right)+\frac{b}{2\sqrt{a}}\right)}\right)^2-E_0-2C \quad (72)$$

and

$$U_2^{'}(r)=N_2(r-\sigma_2^{(2)})(r-\sigma_1^{(2)})r^{\left(\frac{3}{2}+\frac{c}{2\sqrt{a}}-\frac{b^2}{8a\sqrt{a}}\right)}\exp\left[-\frac{\sqrt{a}}{2}\frac{1}{r^2}-\frac{b}{2\sqrt{a}}\frac{1}{r}+Cr\right] \quad (73)$$

where $E_2^{'}$ is the second level of excited energy dan $U_2^{'}(r)$ is a second level eigenfunction.

The generally, the eigenvalue and eigenfunction from Equation (68), (70), (72), (69), (71) dan Eq. (73) respectively is

$$E_n' = -\left[\frac{\left(\lambda+(3-n)D'-n(n-1)\right)\pm\sqrt{\left(\lambda+(3-n)D'-n(n-1)\right)^2-4\left(\sum_{i=1}^{n}\sigma_i^{(n)}+\frac{b}{2\sqrt{a}}\right)D'f}}{4\left(\sum_{i=1}^{n}\sigma_i^{(n)}+\frac{b}{2\sqrt{a}}\right)}\right]^2 - E_0 - 2C \quad (74)$$

and

$$U_n'(r) = N_n\left[\prod_{i=1}^{n}(r-\sigma_i^{(n)})\right]r^D \exp\left[\frac{1}{2}\frac{A'}{r^2}+\frac{B'}{r}+C'r\right] \quad (75)$$

where

$$A' = -\sqrt{a},\quad B' = -\frac{b}{2\sqrt{a}},\quad D' = \frac{3}{2}+\frac{c}{2\sqrt{a}}-\frac{b^2}{8a\sqrt{a}},$$

$$C' = \frac{\left(\lambda+(3-n)D'-n(n-1)\right)\pm\sqrt{\left(\lambda+(3-n)D'-n(n-1)\right)^2-4\left(\sum_{i=1}^{n}\sigma_i^{(n)}+\frac{b}{2\sqrt{a}}\right)D'f}}{4\left(\sum_{i=1}^{n}\sigma_i^{(n)}+\frac{b}{2\sqrt{a}}\right)},$$

E_n' is the eigenvalue and $U_n'(r)$ is the eigenfunction of the potential partner of Polynomial Inverse potential. The eigenvalue of Equation (74) has different with the eigenvalue of Equation (53) because the effect of ground state eigenvalue of Equation (30) in Equation (74). The energy spectrum of Equation (53) and Equation (74) is shown in Table 1.

Table 1. Energy spectrum for Polynomial Inverse potential and its partner potential in 4-dimensional variation radial quantum number n with $\lambda_3 = 24.4614$ and $\lambda_3' = 7.0794$.

N	a	b	c	d	e	f	$E_n\left(fm^{-1}\right)$	$E_n'\left(fm^{-1}\right)$
0	-2	1	-3	-1	2	-3	25.4706	22.9843
1	-2	1	-3	-1	2	-3	22.9957	21.6931
2	-2	1	-3	-1	2	-3	17.2794	19.2794
3	-2	1	-3	-1	2	-3	9.6340	15.0501
4	-2	1	-3	-1	2	-3	2.1214	9.6844

Table 1 shows that the increase in radial quantum numbers causes the energy spectrum to decrease. Partner potential for $n > 1$ has a larger energy spectrum than the energy spectrum of Polynomial Inverse potential as its original potential. The energy spectrum of Polynomial Inverse potential has different with the energy spectrum of its partner potential. This result is consistent with the result of research about the solution of construction of partner potential for Hylleraas potential in Ref. (4).

4. Conclusion

In this paper, we have presented the solution of 4-dimensional Schrodinger equation of radial part for Polynomial Inverse potential and its partner potential by using Ansatz wavefunction method and construction of potential by using Supersymmetric quantum mechanics method. We obtained the eigenvalue and eigenfunction from radial part solutions, which the eigenvalue and eigenfunction depend on the parameters of all components of the composed potential. Partner potential was considered to be new potential. The partner potential of Polynomial Inverse potential for n > 1 has eigenvalue greater than Polynomial Inverse potential as its original potential.

Acknowledgement
This research was partly supported by PUT-MRG of Sebelas Maret University with contract no. 632/UN27.21/LT/2017.

References

[1] Dong S H 2011 *Wave Equations in Higher Dimensions* (Springer-New York) p 97
[2] Wahyulianti, Suparmi A, Cari C and Anwar F 2017 *IOP J. Phys. Conf. Ser.* **795** 012022
[3] Yazarloo B H, Lu L, Liu G, Zarrinkamar S and Hassanabadi H 2013 *Adv. High Energ. Phys.* **2013** 5
[4] Suparmi A, Cari C, Pratiwi B N and Nugraha D A 2017 *J. Phys. Conf. Ser.* **820** 012023
[5] Ikot A N 2013 *Comm. Theo. Phys.* **59** 268
[6] Ita B I and Ikeuba A I 2013 *Appl. Math.* **4** 1
[7] Agboola D 2009 *Phys. Scr.* **80** 065304
[8] Suparmi and Cari 2014 *J. Math. Fund. Sci.* **46** 205
[9] Sari R A, Suparmi A and Cari C 2016 *Chin. Phys. B* **25** 010301
[10] Ituen B O, Ituen E E, Popoola O and Antia A D 2013 *Int. J. Recent Adv. Phys.* **2** 1
[11] Suparmi A, Cari C and Deta U A 2014 *Chin. Phys. B* **23** 090304
[12] Akpan N I, Awoga O A and Antia A D 2013 *Chin. Phys. B* **22** 020304
[13] A Suparmi, Cari C, Pratiwi B N and Nugraha D A 2017 *J. Phys. Conf. Ser.* **795** 012001
[14] Luqman H, Suparmi A, Cari C and Deta U A 2014 *J. Phys. Conf. Ser.* **539** 012019.
[15] Witten E 1981 *Nucl.Phys. B* **188** 531.
[16] Wahyulianti, Suparmi A, Cari C and Wea K N 2017 *J. Phys. Conf. Ser.* **909** 012034.

IOP Conf. Series: Journal of Physics: Conf. Series **997** (2018) 012052 doi:10.1088/1742-6596/997/1/012052

Synthesis of PANi-SiO2 Nanocomposite with In-Situ Polymerization Method: Nanoparticle Silica (NPS) Amorphous and Crystalline Phase

Munasir[1,2], N R D Luvita[1,2], D H Kusumawati[1], N P Putri[1], Triwikantoro[3] and Z A I Supardi[1,2]

[1]Department of Physics, Faculty of Mathematics and Natural Science, Universitas Negeri Surabaya, Kampus Ketintang, Jl. Ketintang, Surabaya, 60231
[2]Research Center of Advanced Materials, Universitas Negeri Surabaya, Kampus Ketintang, Jl. Ketintang, Surabaya 60231.
[3]Department of Physics, Faculty of Mathematics and Natural Science, Institut Teknologi Sepuluh Nopember Surabaya, Kampus ITS, Jl. Arif Rahman Hakim, Sukolilo, Surabaya, 60111

E-mail: munasir_physics@unesa.ac.id

Abstract: Silica which is synthesized from natural materials such as Bancar Tuban's sand composited with Polyaniline (PANi), where the silica used are silica has an amorphous phase and cristobalite phase. In this research, the composite method used is in- situ polymerization, which is silica entered during the fabrication of PANi, then automatically silica will be substitute into the chain bonding of PANi. The aim of this research is to find out the results of a composite process using in-situ methods as well as differences in the morphology of PANi/a-SiO2 and PANi/c-SiO2. For the characterization of samples tested in the form of FTIR to determine the functional groups of the composite and SEM to determine the morphology of the sample. From the test results of FTIR are known composite possibility has occurred because there are several functional groups belonging to silica also functional groups belonging polyaniline, functional group that's happened in wave numbers were almost identical between PANi/a-SiO2 and PANi/c-SiO2, but there are little differences were seen in the form of a graph generated from the peak and intensity that occurred charts for PANi/c-SiO2 has peak more pointed or sharp compared to PANi/a-SiO2 because that bond of crystal is strong, stiff and has a larger particle size than the amorphous composite. Then from the data of SEM seen clearly their morphological differences between PANi/a-SiO2 and PANi/c-SiO2 where polyaniline is composited with amorphous silica will have a fault that is not uniform or irregular different from PANi/c -SiO2 has a regular fault and this is corresponding with the nature of the typical structure of amorphous and crystalline.

1. Introduction

Nano materials have attracted much scientific interest due to the potentially new performance of the particles in nanometer (nm) scale or (<10 nm). Silica is synonymous with silicon dioxide (SiO_2). Silicon and oxygen are the two most abundant elements in the earth's crust. Silica is commonly found in nature as silica sand, rock, mudstone, mud vulcano (lusi)[1,2]. Silica exists in many different forms that can be crystalline as well as non-crystalline (amorphous). The SiO_2 nanoparticles are pore material, with a pore size that can be designed, as applicable: micropore, misopori and nanopore materials [3,4]. The SiO_2 nanoparticles have an amorphous and crystalline phase: quartz, tridymite, cristobalite, coesite, and satishovite [5]. Amorphous silicon is a kind of allotropes of silicon, silicon dioxide or silica with the chemical formula SiO_2. Crystalline silicon is usually a tetrahedron, and each silicon atom is at the top of the tetrahedron and bound covalently with four other silicon atoms. This structure can be very large stretched, thus forming a stable lattice structure. And no qualitative silicone is absent in this extended lattice structure, the lattice network between atoms is irregular. Amorphous silica is generally produced: as wet process silica, as dry process, pyrogenic, airborn fumed silica as well as, and as arc furnace silica. [6]. Silica nanoparticles are low-toxicity and biocompatible, also porous materials, so many applications, including: as drug delivery system (DDS) [7,8], as a filler on PANi composite for smart coating [9], corrosion protection [10,11], Self-healing coatings in anti-corrosion [12] , etc.

PANi has a conductive and non conductive formation. Conductive polymers have been the topic of a number of major studies over the past decade, due to their unique properties: mechanical strength, electrical conductivity, and corrosion protection. PANi organic materials can be synthesized oxidatively and electrochemically; oxidation states of polyaniline include: leucoemarldine, emeraldine-base, emarldine-salt and pernigraniline [10,13,14]. Application of conductive PANI include: solar energy conversion materials, on rechargeable batteries with SiO_2@PANi/Ag formation form [15], electrochemical sensor and biosensor[16,17], supercapasitor [18] and active corrosion protectors, smart-coating [9,19], nanowire form for chemiresistive sensing [13], support for battery electrolyte material [20], can also be used in photovoltaic applications [21] and because of its high electrical and thermal stability, this material is also the prospect for organic semiconductor LEDs [22]. PANi can be synthesized by chemical and electrochemical oxidative methods, and PANi/SiO_2 Nanoparticles can be synthesized by In-Situ polymerization methods and mixing ex-situ solution [23–25]. Many researchers interested in studying Polymer / Nano particle materials, including the most interesting are PANi/SiO_2 Nanoparticles [19], the SiO_2 Nanoparticles themselves are porous materials in which the size and shape of the porus can be modified according to the purpose of the application. Among the most popular applications currently associated with PANi-conductive/SiO_2-nanoparticle materials are as intelligent, anti-corrosion coating material [10]. Composite PANi/SiO_2 prepared by In-Situ method, obtained a good surface coating material, able to protect the steel surface against corrosive solution (such as: Strong acids, and high concentration NaCl) [10,19]. The In situ method of synthesizing nanoparticles in the polymer matrix is a simple and effective route for preparing nanocomposites materials. This method allows one step in the fabrication process of nanocomposites by in situ generating nanoparticles of the precursor accordingly. In this case, the nanoparticles can grow inside the polymer matrix. The advantage of this route is to prevent the occurrence of particle agglomeration while maintaining a good spatial distribution in the polymer matrix. The disadvantage of this method is that unreacted in-situ reaction changes can affect the nature of the end product. [25]. This study focuses only on the study of PANi/SiO_2 composites synthesized by in-situ polymerization methods, in which SiO_2 nanoparticles are prepared from silica sand by hydrothermal-co-precipitation methods. SiO_2 nanoparticles are prepared for two different phases: amorphous silica (a-SiO_2) and crystalline silica (c-SiO_2). Finally characterization of the composite material, which includes: functional group analysis with FTIR, crystal structure analysis and phase by X-ray diffraction, marine analysis with SEM.

2. Eksperimental Method

The first step of the research is to prepare the silica sand material as the base material to be processed into amorphous and crystalline phase SiO_2 Nanoparticles (NPS); and other chemicals such as aniline, APS, HCl and Acetone for polyaniline synthesis. The coprecipitation-hydrothermal method was chosen as the NPS preparation method in this study. After SiO_2 the amorphous phase, furthermore calcination will be obtained by SiO_2 crystalline phase. The second step is the formation of PANi/SiO_2

composites, where NPS and Aniline are mixed with in-situ process, polymerization process to form PANi; and the final result is obtained by composite material of PANi/SiO_2. After composite synthesis is composite characterization with FTIR and SEM. In this study, the manipulation variables used are the amorphous and crystalline silica phases and by controlling the addition of NPS compositions to the PANi/SiO_2 composites.

Table 1. The composition of Nanoparticles SiO_2 (as fillers)

Samples	SiO_2 (powders)		PANi (Wt%)
	Wt%	Mass (gram)	
1	5	0.10	95
2	10	0.20	90
3	15	0.35	85
4	20	0.50	80

3. Result and Discussion

3.1 Characterizaton with FTIR

There are 8 sample result of FTIR test, i.e. composite PANi/a-SiO2 with addition of SiO_2 by 5%, 10%, 15% and 20% then PANi/c-SiO_2 type with the same filler anyway. Following one of the FTIR results for each of the PANi/a-SiO_2 and PANi/c-SiO_2. For a moment, both of picture looks the same, even for the wavelength which is just a little bit at odds. But there is a difference in the length of the peak and peak shape in the picture, i.e. length for peak amorphous with filler tend to be shorter and not as high as peak of composite with crystal filler. This indicate that's structure of crystal filler have bigger vibration than composite with amorf filler. Then from the above data to come by tables for different types of functional groups as follows:

It can be seen in Figure 1 that the infra-red wave absorption peaks for both types of samples exhibit almost identical patterns. Composite samples of PANi/c-SiO_2 (crystal) have peaks that tend to be more pointed than the composite peak of PANi/a-SiO_2, indicating that the SiO_2 bond of the crystal is stronger than the amorphous particles, resulting in widening peaks. The most distinct infrared wave absorption pattern occurred in a sample with a 95 : 5 compositions in wt%, indicating significantly different peak width and height, for a sample of PANi/c-SiO_2 more dominantly vibrasin its functional group than for a sample of PANi/a-SiO_2. It can also be shown that the pattern of absorption of functional groups in PANi is unchanged for both types of composites, indicating that the addition of the silica particle type as a filler does not affect the PANi structure as a matrix and the characteristic properties of each base material still exist, but in general the strong vibration the functional group indicates the type of polymer bond in the sample of PANi/a-SiO_2 and PANi/c-SiO_2; the bond profile in the composite will be confirmed further with SEM data.

Infrared wave absorption pattern in the functional groups of PANi and SiO_2 can be confirmed and there is a match with the results of previous research, where the absorption pattern occurring in the C-Cl function group in the range of wave numbers 590-700 cm^{-1} [27] the results of this study samples of PANi/a-SiO_2 vibration of C-Cl functional groups occurred in wave numbers 505.37 cm^{-1}, 609.53 cm^{-1} and 690.54 cm^{-1} [Table 2]. As for the sample of PANi /c-SiO_2 the vibration of the C-Cl bonding function bands occurs at 507.95 cm^{-1} and 696.33 cm^{-1} wave numbers [Table 2]. And the vibration of CH bending occurred in the range of wave number 675-900 cm^{-1} [10, 26], the result of PANi /a-SiO_2 sample test showed that vibration of CH bending occurred at wave number 690.54 cm^{-1} and 881.5 cm^{-1} ; meanwhile for samples of PANi/c-SiO_2 the C-H bending vibrations occur at wave numbers 815.92 cm-1 and 879.57 cm-1. In the wavelength range 800-870 cm^{-1} is a group on silica [27] which appears on the composite PANi/SiO_2 (95% : 5%) both amorphous and crystalline, OH group of SiO_2 at wave number 810.13 cm^{-1} ; whereas for samples PANi /a-SiO_2 and PANi /c-SiO_2 successively occurred at wave numbers 815.92 cm^{-1} and 879.57 cm^{-1} (see Table2).

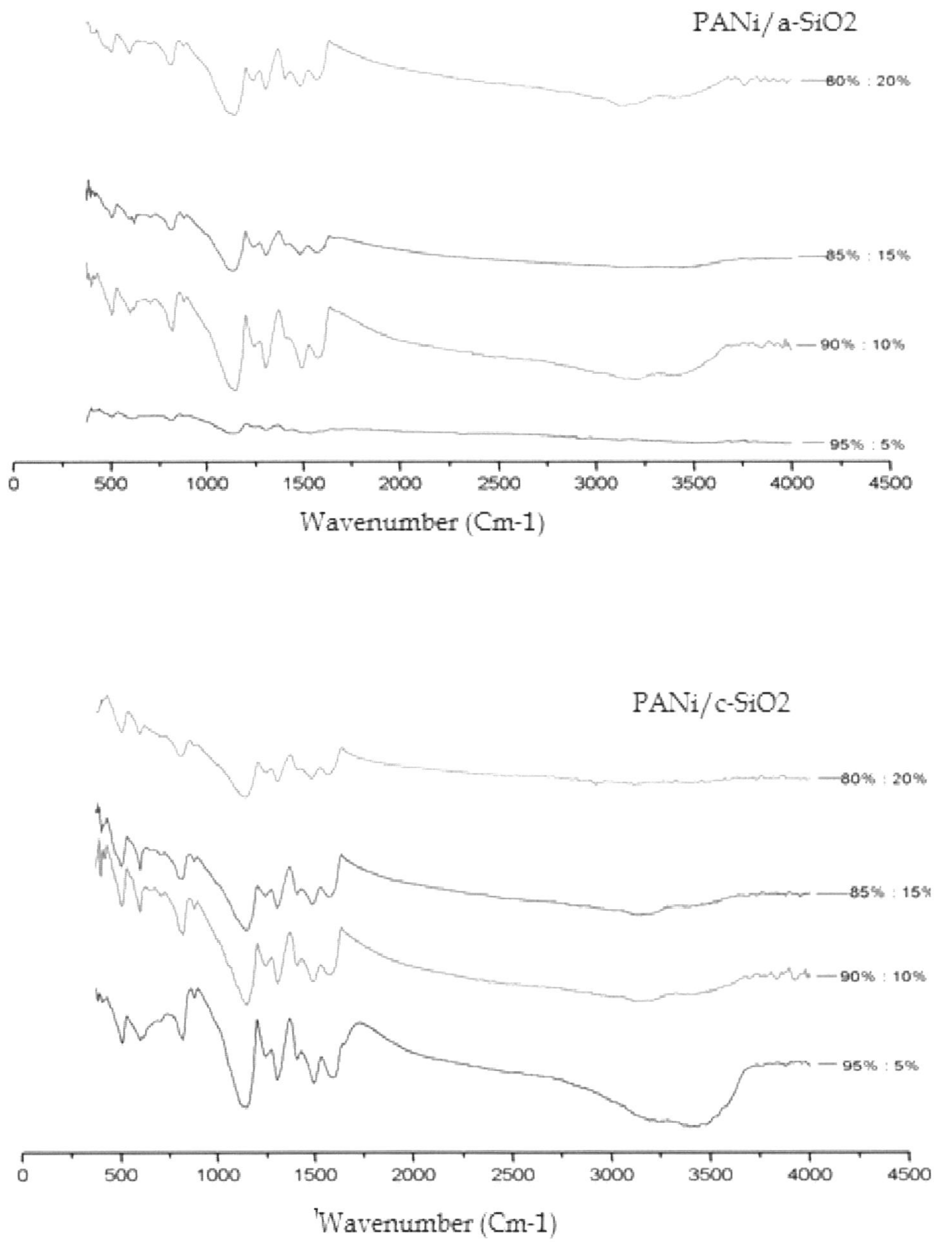

Figure 1. FTIR Pattern of samples PANi composite with reinforced: (a) amorphous silica (a-SiO_2), (b) crystalline silica (c-SiO_2) Nanoparticles

The CN stretch of benzoid ring bonding function is the absorption of the yeng wave which is the characteristic of PANi-conductive material, with formation to form emeraldine salt [10, 13, 19, 28], and this bond is in the range of 1020-1250 cm^{-1} [29, 28], in the samples prepared in this study, the CN functional groups occurred at wave numbers 1136.11 cm^{-1} and 1242.2 cm^{-1} for samples of PANi / a-SiO_2, and at wave numbers of 1147.68 cm^{-1} and 1242.2 cm^{-1} for the sample of PANi/c-SiO_2 [Table 2]. The argument is in conformity with the preceding finding, occurring in the range of 1250-1335 cm^{-1} wave numbers of C-N stretch of Q-B-Q [30, 29, 31, 32]. The functional group in the sample occurs at the same wave number of 1301.99 cm^{-1}. For the C = C quinoid ring (NBN) in the 1500-1600 cm^{-1}

wave range [30,29] it is also consistent for the sample, which occurs in the range of 1504.53 cm^{-1}, 1537.32 cm^{-1} and 1593.52 cm^{-1} for samples of PANi/a-SiO_2; and occurs at wave numbers 1591.33 cm^{-1}. And the vibration of the -OH functional group which is the prior art silica functional group occurred in the range of 3000-4000 cm^{-1} wave numbers [27] for the composite samples of PANi / a-SiO_2 ie at wave numbers 3203.87 cm^{-1} and 3392.9 cm^{-1} and for sample of PANi/c-SiO_2 at 3153.72 cm^{-1}.

Table 2. FTIR result for amorphous silica (a-SiO_2) and crystalline silica (c-SiO_2)

No	Wave Numbers (cm^{-1})				Reference
	a-SiO_2	**c-SiO_2**	**Reference**	**Functional**	
1	597.95 619.17	597.95 648.1	590 - 700	C-Cl stretching	[33, 28]
2	702.11 879.57	705.97 879.57	675 - 900	C-H bending	[33, 29]
3	815.92	817.85	800 - 870	Vibration of O-H from Si-O	[2, 27]
4	1147.68 1242.2	1147.68 1242.2	1020 - 1250	C-N stretch of benzoid ring	[33, 31, 30]
5	1300.07	1301.99	1250 - 1335	C-N stretch of Q-B-Q	[31,30,32,29]
6	1573.97	1572.04	1500 - 1600	C = C quinoid ring stretch (N-B-N)	[30, 29]
7	3169.15 3394.83	3153.72	3000 - 4000	Vibrations of O-H	[27, 34]

3.2 Result test of SEM

Shown in Figure 2, marphologi samples: (a) PANi and enlargement (a1); (b) c-SiO_2; (c) a-SiO_2; (d) PANi/c-SiO_2 composite and (e) PANi/a-SiO_2 composite. It appears that PANi marphology (Fig. 2 (a)) shows the nanometer size, and the particles form a large agglomeration and the arrangement pattern of particles that form cavities between neatly arranged particles forming a particular pattern (as a polymer material). This is different for amorphous silica and crystalline silica marphology (Fig. 2 (b) and Fig.2 (c)), thus affecting its composite microstructure; as shown in Fig. 2 (d): PANi/c-SiO_2 composite profiles and Fig. 2 (e) PANi/a-SiO_2 composite profiles. Marphologi of the PANi/c-SiO_2 appears to be more densely formed than composite particles of PANi/a-SiO_2; on the surface part of the PANi/a-SiO2 particles there are polymer fibers that exhibit irregular composite tears and ductile bonds. Instead differs in small composite PANi/c-SiO2 composite pieces which have a uniform shape indicating a strong or brittle bond due to the filler crystal phase [10,19,23].

IOP Conf. Series: Journal of Physics: Conf. Series **997** (2018) 012052 doi:10.1088/1742-6596/997/1/012052

Figure 2. Marfology of: (a) PANi, (b) a-SiO_2, (c) c-SiO_2, (d) PANi/c-SiO_2, and (e) PANi/a-SiO_2

4. Conclusion

From the results of functional group vibration analysis with FTIR it is concluded that the composite PANi/SiO_2 both with amorphous and crystalline NPS fillers prepared by in-situ polymerization method have the same group vibration pattern; there is a difference in the strength of vibration (% transmittance), which is influenced by the filler particle phase, PANi/c-SiO2 is stronger than PANi/ a-SiO_2 (95% : 5% sample). And microstructure (SEM result); there are significant differences between composite PANi/a-SiO_2 and PANi/c-SiO_2, such as size, shape and cracks profile (ductile and brittle) and particle density.

Acknowledgements

This research is supported by the Ministry of Research and Technology and Higher Education, through the Grants Competing Fund with contract number: 170/UN.38.9 /HK / LT/2015, Universitas Negeri Surabaya (Unesa); Central Laboratory, Faculty of Mathematics and Natural Sciences, Universitas Negeri Malang; Terpadu Laboratory, Faculty of Mathematics and Natural Sciences, Universitas Negeri Surabaya; and Material Laboratory, Material Engineering, ITS Surabaya.

References

[1] Munasir, Sulton A, Triwikantoro, Zainuri M and Darminto 2013 *AIP Conf. Proc.* **1555** 28
[2] Munasir, Triwikantoro, Zainuri M, Darminto 2015 *Mater. Sci. Pol.* **33** 47
[3] Santos H A, Mäkilä E, Airaksinen A J, Bimbo L M and Hirvonen J 2014 *Nanomed.* **9** 535
[4] Xu W, Thapa R, Liu D, Nissinen T, Granroth S, Närvänen A, Suvanto M, Santos H A and Lehto V P 2015 *Mol. Pharm.* **12** 4038
[5] Munasir, Dewanto S, Yulianingsih A, Saadah I K F, Supardi Z A I, Mufid A and Taufiq A 2017 *IOP Conf: Mater. Sci. Eng.* **202** 012057
[6] Kosmachev P, Vlasov V and Skripnikova N 2017 *AIP Conf. Proc.* **1800** 20016
[7] Bardi G 2011 *Drug Deliv. Lett.* **1** 9
[8] Thanh N T K and Green L A W 2010 *Nano Today* **5** 213
[9] Sambyal P, Ruhi G, Dhawan R and Dhawan S K 2016 *Surf. Coat. Technol.* **303** 362
[10] Al-Dulaimi A A and Hashim S 2012 *Int. J. Mech. Mater. Eng.* **7** 113
[11] Kumar A S, Bhandari H, Sharma C, Khatoon F and Dhawan S K 2013 *Polym. Int.* **62** 1192
[12] Stankiewicz A, Szczygieł I and Szczygieł B 2013 *J. Mater. Sci.* **48** 8041
[13] Song E and Choi J W 2013 *Nanomater.* **3** 498
[14] Kaur A and Saini D 2006 *Cell Int. J. Eng. Sci.* **1** 40
[15] Das D, Yang Y, O'Brien J S, Breznan D, Nimesh S, Bernatchez S, Hill M, Sayari A, Vincent R and Kumarathasan P 2014 *J. Nanomater.* **2014** 1
[16] Zhu Q, Gao Q, Guo Y, Yang C Q and Shen L 2011 *Ind. Eng. Chem. Res.* **50** 5881
[17] Zhu C, Yang G, Li H, Du H and Lin Y 2015 *Anal. Chem.* **87** 230
[18] Li P, Ni C, Shi G, Zhang D and Xu Y 2017 *J. Solid State Electrochem.* **22** 1
[19] Kumar A, Bhandari H, Sharma C, Khatoon H and Dhawan S K 2013 *Polym. Int.* **62** 1192
[20] Liu X X, Zhang L, Li Y B, Bian I J, Huo Y Q and Su Z 2006 *Polym. Bull.* **57** 825
[21] Babu V J, Vempati S and Ramakrishna S 2013 *Mater. Sci. Appl.* **4** 1
[22] Vivekanandan J, Ponnusamy V, Mahudeswaran A and Vijayanand P S 2011 *Arch. Appl. Sci. Res.* **3** 147
[23] Zengin H and Erkan B 2009 *Polym. Adv. Technol.* **21** 216
[24] Yilmaz H, Zengin H and Unal H I 2012 *J. Mater. Sci.* **47** 5276
[25] Guo Q, Ghadiri R, Weigel T, Aumann A, Gurevich E, Esen C, Medenbach O, Cheng W, Chichkov B and Ostendorf A 2014 *Polymers.* **6** 2037
[26] Liu F, Liu Z, Gu Y, Chen Z and Fang P 2013 *J. Appl. Polym. Sci.* **130** 2288
[27] Jal P K, Sudarshan M, Saha A, Patel S and Mishra B K 2004 *Physicochem. Eng. Asp.* **240** 173
[28] Zakaria Z, Halim N F A, Schleusingen M H V, Islam A K M S, Hashim U and Ahmad M N 2015 *J. Nanomater.* **2015** 1

[29] Abdolahi A, Hamzah E, Ibrahim Z and Hashim S 2012 *Materials.* **5** 1487
[30] Trchová M and Stejskal J 2011 *Pure Appl. Chem.* **83** 1
[31] Khan R, Khare P, Baruah B P, Hazarika A K and Dey N C 2011 *Adv. Chem. Eng. Sci.* **1** 37
[32] Gui D, Liu C, Chen F and Liu J 2014 *Appl. Surf. Sci.* **307** 172
[33] Khan A, Aldwayyan A S, Alhoshan M and Alsalhi M 2010 *Polym. Int.* **59** 1690
[34] Wang L, Wang Z, Yang H and Yang G 1999 *Mater. Chem. Phys.* **57** 260

IOP Conf. Series: Journal of Physics: Conf. Series **997** (2018) 012053 doi:10.1088/1742-6596/997/1/012053

Microstructure, thickness and sheet resistivity of Cu/Ni thin film produced by electroplating technique on the variation of electrolyte temperature

M Toifur, Y Yuningsih and A Khusnani

Magister of Physics Education Program, Ahmad Dahlan University Yogyakarta
Jl. Pramuka. No 42 Sidikan Umbulharjo Yogyakarta 55161

mtoifur@mpfis.uad.ac.id

Abstract. In this research, it has been made Cu/Ni thin film produced with electroplating technique. The deposition process was done in the plating bath using Cu and Ni as cathode and anode respectively. The electrolyte solution was made from the mixture of $HBrO_3$ (7.5g), $NiSO_4$ (100g), $NiCl_2$ (15g), and aquadest (250 ml). Electrolyte temperature was varied from 40°C up to 80°C, to make the Ni ions in the solution easy to move to Cu cathode. The deposition was done during 2 minutes on the potential of 1.5 volt. Many characterizations were done including the thickness of Ni film, microstructure, and sheet resistivity. The results showed that at all samples Ni had attacked on the Cu substrate to form Cu/Ni. The raising of electrolyte temperature affected the increasing of Ni thickness that is the Ni thickness increase with the increasing electrolyte temperature. From the EDS spectrum, it can be informed that samples already contain Ni and Cu elements and NiO and CuO compounds. Addition element and compound are found for sample Cu/Ni resulted from 70° electrolyte temperature of Ni deposition, that are Pt and PtO_2. From XRD pattern, there are several phases which have crystal structure i.e. Cu, Ni, and NiO, while CuO and PtO_2 have amorphous structure. The sheet resistivity linearly decreases with the increasing electrolyte temperature.

1. Introduction

In the husbandry industry, the existence of low-temperature thermometers is needed especially on the breeding using artificial insemination (IB). For the purposes of IB to obtain the qualified cow sperm the government is still importing. During delivery and storage, the sperm is placed in a special flask containing liquid nitrogen with the temperature of -198°C. As a part of the design of flask, there is a hole for liquid nitrogen to evaporate out to avoid the explosion. The problem is if the evaporation occurs continuously then the nitrogen content in the flask becomes less and the temperature increase. If the temperature inside the flask reaches the critical temperature i.e. -100°C then the cow sperm will die. Therefore, a low-temperature thermometer attached to the flask is needed.

Making low-temperature thermometers is not easy because it may make the equipment does not work as which should be such as freezing in the sensor so the sensor can not flow the current normally, freezing on conductor so that it can not function as a good conductor, the presence of mechanical stress, and others. These make a measurement on low-temperature need a long time compared to the measurement of medium in the room temperature [1,2]. The RTD (Resistance Temperature Detector) sensor is a sensor used to measure temperature by correlating RTD element resistance with temperature [3,4,5]. RTD elements that can be used are Platinum [6,7], Nickel [8],

Copper [9,10,11], Balco, and Tungsten. But the elements are commonly used Platinum and Copper. The advantages of RTD sensor compared to others such as thermocouple, thermistor, bimetallic include a wide temperature range from -200°C to 800°C, more accurate than those, high stability, repeatability, and resistance to electronic noise, resistance to both corrosion and oxidation. Other advantages are a simple theory not as theory on IR thermometer involving the wave length and permittivity of measured medium, simple working principles, and cheap [12,13,14,15]. This will economically influence on cost savings.

The RTD elements are generally made from either metal or alloy, in the form of coils or thin films [16,17]. A good RTD sensor is capable to perform the large resistance changes to respond the temperature changes. A Cu film sensor can already perform as a temperature sensor, but the change of its resistance to respond the change of temperature is still limited. The resistance of Cu thin film can be increased when it is combined with nickel. Making of an alloy Cu-Ni thin film or Cu coated with Ni to form Cu/Ni is an alternative material for the low-temperature sensors. Both have several conformities, i.e. both are transition metal, both materials have the same type of crystal structure fcc, both materials have almost the same atomic size, i.e. 124 pm for Cu and 120 μm for Ni. These things facilitate the occurrence of combinations, or mutual substitution mechanisms between the two materials [18]. In addition, nickel is a good material when paired with copper because of the adhesion force on the interface between Cu and Ni strong [19].

Electroplating is one method to produce Cu/Ni film. In coating Cu plate with Ni, the temperature of the solution plays a role in facilitating Ni ions to penetrate the electrolyte solution to the Cu plate functioned as cathode. The higher electrolyte temperature the more Ni ions attach to Cu cathode. It is hoped that in the presence of Ni film on Cu plate, the resistivity of Cu/Ni film greater than Cu layer resistivity alone. But with the thicker Ni, Ni film becomes more continuous and thicker.

Therefore, the addition of Ni layer thickness will produce the Ni layer more conductive, otherwise becomes less resistive. So, after the first Ni coating, the resistivity of the Cu/Ni film is greater than that for Cu film alone, but with the rising electrolyte temperature, the resistivity of Cu/Ni layer will decrease. However, this film is expected to be more accurate in responding the changes of medium temperature.

2. The Experimental details

2.1. Materials and tools

The material used in this research is copper (Cu) plate as a substrate with the size of 2.5 cm ×1.0 cm × 0.02 cm and nickel (Ni) with the size of 3 cm × 1.0 cm × 0.4 cm as a coated material. The electrolyte is made from a mixture of borax acid (H_3BO_3), nickel sulphate ($NiSO_4$), nickel chloride ($NiCl_2$), and aquadest of 7.5g, 100g, 15g, and 250ml, respectively. Alcohol is used to rinse the substrate. Cleaning powder brand Rinso is used to clean the substrate of impurities and grease sticking, polisher brand Brasso is used to polish the substrate, liquid nitrogen is used as a low temperature medium that temperature will be measured by the sensor and its performance will be studied.

The equipment used for preparation the sample consisted of 1000 ml beaker glass wich is used as plating bath, benzene burner which is used to heat the electrolyte solution, vernier thermocouple is used to control the temperature of the solution stable at 60 °C, DC voltage set on 1.5 volts as power supply, PA214-type Ohaus balance is used to measure the plate mass before and after plating, ultrasonic cleaner is used to clean the sample, and two multimeters are used to measure the voltage and current on the four point probe tool. While the equipment for characterization is SEM-EDS to know the existence of Ni and its content and four-point probe to determine the sheet resistivity of the film. Temperature sensors (vernier thermocouple) and voltage sensors equipped with logger pro software is used to determine the output voltage of the sensor at various temperatures that can be seen on the computer screen.

2.2. Experimental procedure
In preparing the sample, the copper plate with the thickness of 0.2 mm cut with the size of 10 cm $\times$1.3 cm and then sanded gradually with abrasive paper number of 1200 CW, 1500 CW, and 2000 CW. Then the sanded plates polished with autosol, cleaned with rinso solution, rinsed with aquadest, and at the final stage rinsed with alcohol in the ultrasonic cleaner for 30 minutes. Samples that have been cleaned dried with a hair dryer, after drying wrapped in tissue, and stored in plastic clips.

In the electroplating process an electrolyte with compositions as pointed above prepared. Furthermore the equipment is set by placing Ni at the anode and Cu at the cathode. The circuit connected to a 1.5 volt voltage power supply, for 1 minute and a 4 cm electrode distance. The electroplating process carried out at temperature of 40°C, 50°C, 60°C, 70°C, and 80°C.

2.3. Methods of data analysis
In this case determination of Ni film thickness formed on Cu substrate was done by indirect measurement, that is measuring the mass difference before and after coated, area and density of sample according to equation [20,21,22],

$$D = \frac{\Delta m}{\rho PL} \tag{1}$$

where D = the thickness of Ni layer,

Δm = the difference of mass between Cu/Ni and Ni films.

P = sample length

L = sample width

ρ = density of Ni

The thickness of the obtained Ni layer is confirmed by the thickness of the Ni layer from one of the samples photographed in mapping element type. The determination of the existence of Ni and Cu film both element and compound are done by EDS (*Energy-Dispersive X-Ray Spectroscopy*). The measurement of sheet resistivity was done by four point probe equipment. The current flowing in the sample is varied so a set of voltage (V_i) and current data (I_i) is obtained. Furthermore, the data of V_i and I_i are fitted in a straight line so the equation becomes:

$$V = \frac{\ln 2}{\pi} R_s I \tag{2}$$

By assuming the graphic slope equal to a that is

$$a = \frac{\ln 2}{\pi} R_s \tag{3}$$

then sheet resistivity is obtained as,

$$R_s = \frac{a\pi}{\ln 2} \tag{4}$$

Analysis on the pattern of X-ray diffraction is done by observing the crystalline phase and the direction of the hkl crystal plane. Analysis of the effect of electrolyte temperature on the sheet resistivity is done by making graph of sheet resistivity to temperature. Furthermore, from the curve determined the relationship between the two variables.

3. Results and Discussion
In Figure 1 it is shown the image of Cu substrate to be coated by Ni. The substrate is divided into 2 parts, Figure 1(a) is Cu substrat that will be used to determine the sheet resistivity and SEM-EDS and Figure 1(b) is Cu substrat that has a litography will be used to study the behaviour of material as low-temperature sensor.

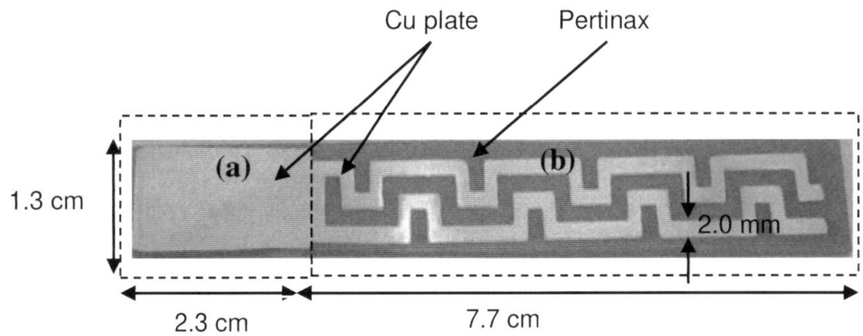

Figure 1. Lithography design of Cu substrate, (a) A part to be used as SEM-EDS analysis and determining sheet resistivity, (b) A part to be used as low-temperature sensors.

Table 1 shows the of current density in the plating process at various temperatures of the electrolyte. The current density as shown in column 2 Table 1 is obtained by measuring the electric current in the circuit using the amperemeter. The current flow occurs because the circuit is connected to a 1.5 volt voltage source. Since the current in the electrolyte solution flows between the Cu and the Ni electrodes having a cross-sectional area of 13 cm^2, the current density can be obtained from dividing the measured current by the area of the electrode cross section. The change of electrical current occurs because the temperature of the electrolyte changes.

Table 1. Current density for each rising of electrolyte temperature

Electrolyte temperature °C)	Current density (A/mm^2)
40	0.04
50	0.05
60	0.07
70	0.08
80	0.10

In Fig. 2 (a) it is shown a SEM photograph from surface of the Ni layer above the Cu substrate. This sample is resulted from Ni deposition at 60°C electrolyte temperatures, while Fig. 2 (b) is a SEM photograph of the surface of Ni layer resulting from Ni deposition at 70°C electrolyte temperatures. Photographs were taken for only 2 samples in order to know the qualitative differences in the surfaces of the two samples. From the two images, researchers have not been able to reveal the difference because there is no data obtained, except only presents the results of the photograph. Indeed, in Figure 2 (a) it has looked like islands and it are Ni materials. But in Figure 2 (b) the islands are not visible. Perhaps this is due to the poor sample surface preparation to be photographed.

In Fig. 2 (c) and Fig. 2 (d) a cross-sectional images of a Cu / Ni sample obtained from Ni deposition on a Cu substrate at electrolytic temperature of 60°C and 70°C are presented. Photos were taken with magnification of 1000x for the sample of Figure 2 (c) and 2000x for the sample of Figure 2 (d). The difference of magnification is due to technical reasons that are to obtain a clear picture, because the choice for the same magnification of one image is clearly visible, while it is for the other image looks blurred. Both photographs have not revealed any Ni layer attached to Cu substrate.

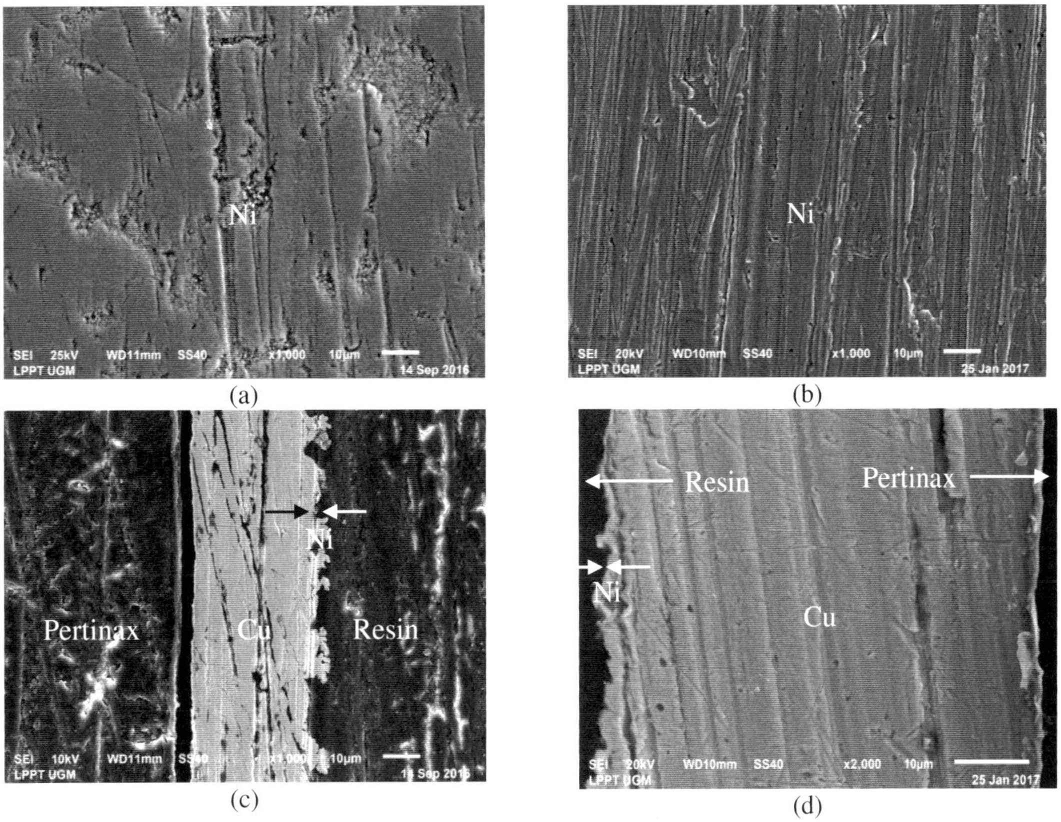

Figure 2. Image of the Ni film in the Cu/Ni sample produced from electroplating at: (a) 60°C and (b) 70 °C electrolyte temperatures.

Furthermore, the analysis was continued with EDS to determine the presence of Ni elements in the Cu/Ni sample and the content in the sample resulting from Ni deposition at 60°C and 70°C electrolyte temperatures. The results are shown in Figure 3.

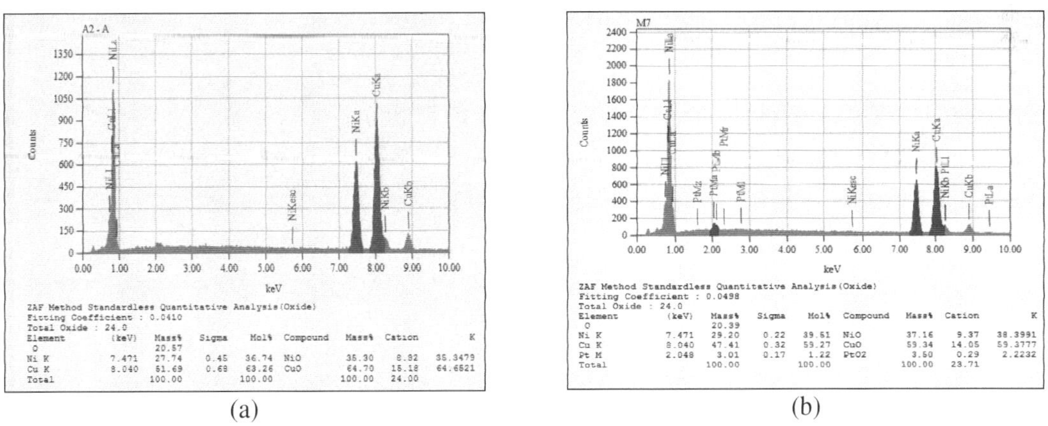

Figure 3. The EDS spectra of Cu/Ni samples resulting from Ni deposition on Cu substrate at electrolyte temperature of (a) 60°C dan (b) 70°C.

Table 2. Description of the mass composition of the elements and the mass composition of the compounds in the Cu/Ni samples

Element	Mass of Element (%)		Mass different (%)	Compound	Mass of Compound (%)		Mass different (%)
	Temp. 60 °C	Temp. 70 °C			Temp. 60 °C	Temp. 70 °C	
O	20.57	20.39	-0.18				
Ni	27.74	29.20	1.46	NiO	35.3	37.15	1.85
Cu	51.69	47.41	-4.28	CuO	64.7	59.34	-5.36
Pt	-	3.01	3.01	PtO_2	-	3.60	3.6

From the EDS spectrum it can be informed that samples already contain elements of Ni and Cu. In addition, in this sample also contains compounds NiO and CuO. Samples from Ni deposition at 70°C electrolyte temperature contain the addition element and compound that are Pt and PtO_2. From the description of the mass composition of the element and compound, it is known that the Cu/Ni sample resulting from the Ni deposition with the electrolyte temperature of 60°C has produced Ni element with the content (in percent mass) of 27.74%. This percentage rises to 29.20% or increases by 1.46% when the electrolyte temperature rises to 70°C. Although the increasing percentage is not too large that is 10°C, but if the temperature rises from 40°C to 80°C the increase of mass percentage of Ni is expected to be larger. This indicates that the electrolyte temperature has an effect on the speed of Ni layer formation on the Cu surface. The higher the electrolyte temperature, the easier the Ni ions penetrate the electrolyte so that the current density becomes greater. Due to the increasing of Ni content, the Cu content decrease from 51.69% to 47.41% or decreased by 4.28%.

The presence of NiO and CuO compounds is the result of a reaction between nickel and oxygen from the electrolyte. This oxygen comes from the solvent element of borax acid (H_3BO_3) and nickel sulphate ($NiSO_4$). Both of these materials are needed in the electroplating process. Borax acid is needed to control the electrolyte so that the electrolyte remains in base state, while nickel sulphate is a contributor of Ni ions to produce Ni layers on Cu substrate. In the electroplating process, this temperature variation also affects the content of NiO and CuO compounds. The content of NiO compound is increased from 35.3% (percent mass) to 37.15% or increased by 1.85% while the content of CuO compounds decreased from 64.7% to 59.34% or decreased by 5.36%. At Ni deposition at electrolyte temperature of 70°C is obtained another element that is Pt. The present of Pt is thought to be an element originated from Ni plates on the anode. To process Pt into Pt^{2+} it takes a power of 2.28 eV (corresponding to its ionization energy), while the magnitude of the voltage of the power supply in the electroplating process is 1.5 volts less than ionization energy. Therefore, the electrolyte temperature of 70°C contributes to the ionization energy of Pt through thermal energy, which then the Pt ion moves toward the Cu substrate together with the Ni ions to form the layer on the Cu surface. From the EDS spectrum it can also be known that at a temperature below 70°C Pt ions have not appeared.

Furthermore, in Table 3 it is shown the thickness of Ni layer for each electrolyte temperature. The thickness of Ni film is determined using equation (1) with a cross-sectional area of A = 7.61 cm^2 and density of Ni = 8.9 g/cm^3.

Table 3. Measurements of mass and determination of thickness of Ni film

No	Electrolyte temperature, T (°C)	Mass of Cu substrat, m_1(g)	Mass of Cu/Ni film, m_2(g)	The thickness of Ni film, D ($\times 10^{-6}$ mm)
1	40	3.1347 ± 0.00003	3.1369 ± 0.00003	32.5 ± 4.80
2	50	3.1023 ± 0.00003	3.1055 ± 0.00003	47.2 ± 6.96
3	60	3.7090 ± 0.00001	3.7125 ± 0.00003	51.7 ± 7.58
4	70	3.1454 ± 0.00001	3.1498 ± 0.00001	65.0 ± 9.40
5	80	3.2028 ± 0.00001	3.2076 ± 0.00001	70.9 ± 5.30

From Table 2 column 4, it can be seen that along with the increase of the electrolyte temperature, the Ni film formed thicker. To see the relationship between the thicknesses of Ni film to electrolytic temperature, in Figure 3 is showed data (D_{id}, T_i). The trend of the data follows a straight line, so that the data is fitted according to the straight-line equation $y = ax + b$ where $y = D$ and $x = T$.

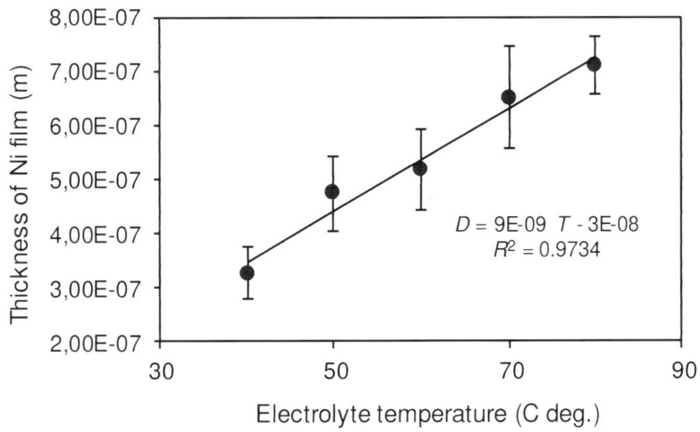

Figure 4. The curve of the relationship between the thickness of the Ni layer and the electrolyte temperature

The relationship between the electrolyte temperature (T) and the thickness of the Ni film (D) is,

$$D = 0.946\times10^{-6}T - 3.3\times10^{-5} \text{ (mm)} \qquad (5)$$

This equation has a determination index $R^2 = 0.97$ so that the thickness of Ni film is strongly proportional to the electrolyte temperature. The higher the electrolyte temperature the thicker the Ni film.

To see the crystal structure formed on the sample, Figure 5 shows the X-ray diffraction pattern for the Cu/Ni sample resulting from deposition at various electrolyte temperatures. As a result, in Cu / Ni samples there are several phases which have crystal structure i.e. Cu, Ni, and NiO. The dominant peak of Cu occurs at a diffraction angle 2θ of about 43.3° with the orientation of the hkl plane (111) while the dominant peak of Ni occurs at a diffraction angle of about 39.04 ° with the orientation of the hkl plane (111). No Ni phase shift is caused by the variation of electrolyte temperature. The NiO phase appears at an angle of 41.5° with the orientation of the hkl plane (111), while the CuO and Pt, and PtO phases are not found. If this is confirmed by the element and compound resulted on the EDS spectrum analysis where on the EDS spectrum it is found Pt element, and CuO and PtO_2 compounds, we conclude that the Pt, CuO and PtO_2 are exist in not crystalline structure but amorphous.

The ordering of the crystal structure is related to the characteristics of the material, whether physical or electrical. In films that have a regular crystal arrangement will provide an easy flow of electric current to flow so that the resistivity is relatively smaller than the film that has the less regular or irregular crystal structure.

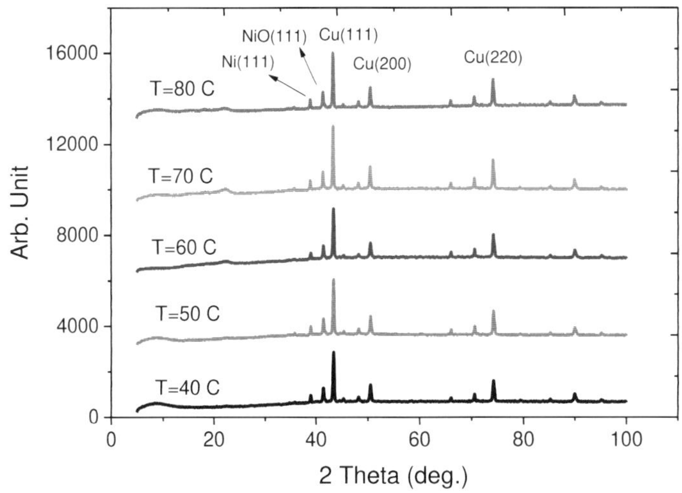

Figure 5. Pattern of X-ray diffraction for Cu/Ni samples resulted from Ni deposition at various electrolyte temperature.

Figure 7 shows the sheet resistivity of the Cu/Ni sample. Before coated with Ni, the resistivity of substrate Cu = $(1{,}19 \pm 0{,}08) \times 10^{-3}$ Ω/sq and after coated with Ni the sheet resistivity of all Cu/Ni samples become increase greater than sheet resistivity of Cu substrate. This is in accordance with Afsarimanesh's expression that one of methods to increase Cu resistance is combining it with other materials to form alloys [9]. According to the Hume-Rothery formula if two materials have the atoms with nearly the same size, having a relative atomic size difference that less than 15%, having the same crystal structure, and the having an almost identical electronegativities, the two materials are easier to form the alloy because its are possible to diffuse substitutionally. In this process, the formation of this alloy is theoretically possible because the size of atoms of Ni and Cu is almost equal that are 0.1246Å for Ni and 0.1278Å for Cu, relative difference of the radius of the two atoms is 2.57%, while the crystal structure is the same i.e. fcc (*face centered cubic*), and its electronegativity is almost the same, that is 1.91 for Ni and 1.9 for Cu. So that there will be the diffusion process.

Table 4. The change of sheet resistivity before and after electroplating process.

No	Electrolyte temperature (°C)	$(Rs \pm S_{Rs}) \times 10^{-3}$ Ω/sq	
		Cu film	Cu/Ni film
1	40	1.150 ± 0.3	1.585 ± 0.09
2	50	1.234 ± 0.4	1.520 ± 0.08
3	60	1.197 ± 0.7	1.482 ± 0.08
4	70	1.204 ± 0.8	1.400 ± 0.08
5	80	1.145 ± 0.5	1.334 ± 0.07

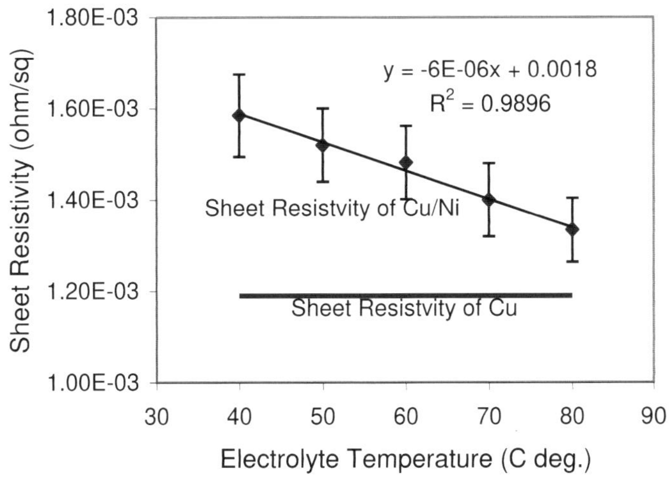

Figure 6. Sheet resistivity of Cu/Ni thin film on various electrolyte temperatures

In Figure 6 it appears that the sheet resistivity is inversely proportional to the electrolyte temperature. The relationship the two quantity follows a straight-line equation

$$R_s = -\,0.00006T + 0.0018 \qquad (6)$$

Where T is electrolyte temperature and R_s is sheet resistivity. The value R^2 = 0.9896 indicates the temperature of the electrolyte has a strong effect on the sheet resistivity of the Cu/Ni film.

By knowing the profile of sheet resistivity on the electrolyte temperature variation, the Ni plating on Cu can increase the original Cu resistivity $(1.19 \pm 0.08) \times 10^{-3}$ Ω/sq to become larger, but with the increasing the Ni thickness due to increasing of the electrolyte temperature then the Cu/Ni resistivity becomes reduced. However, the Cu/Ni resistivity is still greater than Cu resistivity at all electrolyte temperature. According to the purpose of plating Ni to Cu is to increase the resistivity of the material so in this study the goal has been achieved. It is expected that with this achievement the research can be continued on the performance of Cu / Ni film as the low-temperature sensor.

4. Conclusion

In this research, we have succeeded in producing Cu/Ni films by electroplating method on the electrolyte temperature variation from 40°C up to 80°C. On the formation of Ni layer, the temperature variation affects the thickness of the formed Ni film. The thickness of Ni film corresponds linearly with the electrolyte temperature. The higher the electrolyte temperature the thicker the Ni layer. On the study on the the relationship between the electrolyte temperature to the sheet resistivity of Cu/Ni film, it is obtained the reverse linear relationship, i.e. the higher the temperature of the electrolyte the smaller the sheet resistivity. However, all of the resulted Cu/Ni films have a greater sheet resistivity than that for Cu film alone at all the variation of electrolyte tempera`tures. With this result, the study is expected to be continued on the sensor performance testing in measuring the low temperature medium.

References

[1] Yeager C J and Courts S S 2001 *IEEE Sens. J.* **1** 352
[2] Fisher B H and Malocha D C 2007 *2007 IEEE Int. Freq. Control Symp. Joint 21st Eur. Freq. Time Forum*·(Geneva: IEEE International) pp 505-510
[3] Chowdhury, Tasnuva and Bulbul H 2010 *Int. J. Eng. Sci. Technol.* **2** 7896

[4] Toifur M, Haryadi B and Riswanto. *Proc. 2014 Int. Conf. Phys.Appl.* (Paris: Atlantis Press) pp 135-137
[5] Toifur M, Haryadi B, Rahmadani W 2015 *Contemp. Eng. Sci.* **8** 1351
[6] Maher A, Velusamy V, Riordan D and Walsh J. 2014. *Proc. 8th Int. Conf. Sens. Technol.* (Liverpool: Massey University) pp 19-22
[7] Lebioda M and Rymaszewski J. 2015 *Prz. Elektrotechniczny* **91** 225
[8] Nusse D, Noetzel R and Vdi 2008 *Sensoren Und Messsysteme* **2008** 891
[9] Afsarimanesh N and Ahmed P Z 2011 *Int. J. Smart Sens. Intell. Syst.* **4** 726
[10] Santos E J P, Vasconcelos I B 2008 *Proc. 26th Int. Conf. Microelectron.* (Serbia: IEEE International) pp 333-336
[11] Post J W and Bhattacharyya A 2012 *Microelectron. Reliab.* **52** 1389
[12] Chen S, Brown L, Levendorf M, Cai W, Ju S, Edgeworth J, Li X, Magnuson C, Velamakanni A, Piner R R, Park J and Ruoff S F 2011 *ACS Nano* **5** 1321
[13] Rahimi F, Rahmati R and Mardani S 2014 *Soft Nanosci. Lett.* **4** 1
[14] Nenkov M R and Pencheva TG 2008 *Central Eur. J. Phys.* **6** 332
[15] Sultan M Z and Sultana N 2015 *Electr. Electron. Syst.* **4** 160
[16] Yan W, Li H, Kuang K, Du L, Guo J 2008 *J. Alloy. Compound.* **449** 210
[17] Fraden J 2003 *Handbook of Modern Sensors: Physics, Designs, and Applications*. (New York: Springer)
[18] Schleich W 2005 *Application Of Copper-Nickel Alloy Uns C70600 For Seawater Service* (online) Available on https://www.copper.org/applications/marine/cuni/alloys/uns_c70600.html accessed March 13th 2016
[19] Okamoto N, Wang F and Watanabe T 2004 *Mater. Transactions* **45** 3330
[20] Piegari A, Masetti E 1985 *Thin Solid Films* **124** 249
[21] Røli R and Flood P R 1978 *J. Microsc.* **112** 359
[22] Phillip S R 2012 *Measuring the Thickness of Thin Metal Films* (Idaho: Department of Physics and Astronomy Brigham Young University)